Advances in
Growth Hormone and
Growth Factor Research

Advances in Growth Hormone and Growth Factor Research

EDITED BY

EUGENIO E. MÜLLER
DANIELA COCCHI
VITTORIO LOCATELLI
Department of Pharmacology, Chemotherapy
and Toxicology, University of Milan School of
Medicine, 20129 Milan, Italy

Pythagora Press
Roma Milano

Springer-Verlag Berlin
Heidelberg GmbH

Eugenio E. Müller
Daniela Cocchi
Vittorio Locatelli

Department of Pharmacology,
Chemotherapy and Toxicology,
University of Milan School of Medicine,
Via Vanvitelli, 32
20129 Milan, Italy

Distribution rights world-wide:
Springer-Verlag Berlin Heidelberg New York London Paris Tokyo

ISBN 978-3-662-11056-0   ISBN 978-3-662-11054-6 (eBook)
DOI 10.1007/978-3-662-11054-6

© Pythagora Press Rome Milan and
  Springer-Verlag Berlin Heidelberg 1989

Originally published by Pythagora Press Rome Milan and Springer-Verlag Berlin Heidelberg in 1989

FOREWORD

In all experimental sciences, but perhaps in none to the extent as in the biological sciences, is it true that fields dealing with different phenomena and problems suddenly merge into a unique area of endeavour. A typical instance of such a unified approach is the study of hormones, somatomedins and specific growth factors in the past decade. Although in recent years the common features of polypeptides listed in one of the categories of biologically active macromolecules have been stressed at an ever-increasing number of symposia, a major merit of this volume is to have brought to the fore the similarity in the control mechanisms in cell proliferation, differentiation, and function elicited by all of the peptides. Another equally important contribution is that such a remarkably extensive number of essays, produced by highly competent investigators in each one of these areas of reseach, have been collected.

The first section of the volume deals with recent developments in the study of the growth hormone which, ever since its discovery, has been the model of choice for the study of hormone production, mechanisms of action, and spectrum of activity. The exploitation of the powerful recombinant DNA and monoclonal antibody techniques has unveiled additional most important aspects of its evolutionary origin, regulatory mechanisms at the level of its coding gene, and modulatory effect on the proliferative activity of different cells.

The second and third sections dealing respectively with recent advances in the exploration of the somatomedins and of the ever-lengthening list of specific growth factors, shed light on the manifold properties of these substances as revealed in normal and experimentally manipulated animal models or exemplified by rigorous measurements of blood serum of humans suffering from different endocrine dysfunctions.

The fourth section is almost entirely devoted to a thorough analysis of the neural regulation of growth hormone secretion, a topic of utmost interest. This raises the question of to what extent the opposite is also true: Is the developing and fully differentiated nervous system in turn dependent on growth hormone and other factors belonging to the second and third categories listed above? It is all too well known that steroid hormones play a most important role in modulating the formation of some brain centers and of their functional properties. To what extent, if any, does growth hormone modulate the formation and function of the same or of other brain centers and of their connecting circuits? The findings (not reported in this volume) that NGF is endowed with a most important role in the regulation and function of brain cholinergic neurons[1] and that its depletion through the injection of NGF-specific antibodies during fetal life impairs the normal differentiation of the neuroendocrine axis[2] suggest extending to other hormones, somatomedins, and growth factors the study of their possible role in the formation of that immensely complex network which forms the central nervous system.

The fifth and sixth sections bring to light a new and most needed change in the criteria guiding the selection of speakers in biological symposia and the inclusion of their reports in the proceedings. In this volume they supplement studies reported in the previous sections with others dealing with the aberrant functions in the synthesis and secretion of the most extensively investigated hormone: the growth hormone. Thus, the time-honored concept that basic and clinical (or applied) science belong to two distinct categories is timely defeated. There is only one sound criterium which ought to direct the selection of speakers by conference organizers and the acceptance of their reports in the proceedings: that is, that there exist only two main classes of scientific contribution: one consists of excellently acquired and meaningful data, the other fails to meet these prerequisites. Obviously, only the former should be included in the proceedings. The present volume honors this unwritten, and not always followed, guiding rule.

1 – Thoenen H, Bandtlow C, Heumann R (1987) The physiological function of nerve growth factor in the central nervous system: comparison with the periphery. *Rev Physiol Biochem Pharmacol* **109**: 145-178
2 – Aloe L, Cozzari C, Calissano P, Levi-Montalcini R (1981) Somatic and behavioural postnatal effects of fetal injections of nerve growth factor antibodies in the rat. *Nature* **291**: 413-415

Rome, August 1988   Rita Levi-Montalcini

PREFACE

A series of international symposia held in Milan in the last 20 years under the chairmanship of Dr. C.H. Li on growth, growth hormones, and related fields, testify to major advances made in this important area of biology and medicine. In the first meeting "Growth Hormone" (1967), the major events were the reports of the final determination of the chemical structure of human growth hormone (hGH) and the official recognition and introduction into clinical practice of GH radioimmunoassay. The second meeting, "Growth and Growth Hormone" in 1971, focused on the newly discovered human chorionic mammosomatotropin (hCS) and heralded the recognition of prolactin as a chemical and biologic entity separate from GH. Four years later, the third symposium, "Growth Hormone and Related Peptides," was the appropriate forum to discuss the manifold properties of the newcomer somatostatin and to report progress on the chemical and biological study of somatomedins. In 1979, the last symposium, "Growth Hormone and Other Biologically Active Peptides" dealt mainly with somatomedins and a new family of pituitary hormones, the endorphins.

Since then, after an initial slow down, the last few years have witnessed a renaissance of interest and accomplishments in this field.

Among them are the availability of synthetic GH obtained by recombinant DNA technology, allowing more extensive clinical use in GH-deficiency states and extrasomatotropic pathologies; the isolation and characterization of GHRH, coupled with a better understanding of the neurohumoral control of GH secretion; new knowledge on the chemical and biological nature and potential clinical applications of somatomedins; and the recognition of their belonging to a large family of hormone-like molecules, the growth factors. Finally, the clinical consequences of these discoveries which will allow a more precise diagnosis and discrete approach to therapy of growth disorders and other disease states.

This volume is devoted to the proceedings of a congress held in Milan during the fall of 1987, "Advances in Growth Hormone and Growth Factor Research," where outstanding speakers discussed the aforementioned problems. In the first section on Native and Biosynthetic Growth Homones, Parks and coworkers discuss the mechanisms responsible for the extent and tissue specificity of hGH and hCS gene expression, Tripputi et al. deal particularly with the regulation of GH gene expression, Bennett et al. recapitulate the steps involved in the biosynthesis and the properties of the recombinant hGH expressed in E. coli, *Kawauchi and Yasuda characterize the structural similarities between GH and prolactin at many phylogenetic levels, Baumann reports on the properties and possible function of the newly discovered binding protein for hGH, and Schwartz and colleagues discuss direct and independent metabolic effects of GH and insulin-like growth factors (IGFs) on adipocytes.*

In the section Somatomedins, Hynes et al. apply hybridization analysis to study biosynthesis of IGF-I and IGF-II and their tissue regulation, Pòvoa et al. focus on the somatomedin binding protein throughout life, pregnancy and in some disease states, Rosenfeld describes the structure and physiologic roles of IGF-I and IGF-II receptors in mediating a wide variety of anabolic and mitogenic actions, Zapf et al. describe the metabolic and growth-promoting effects of recombinant human IGF-I in rats and humans.

The important topic of Growth Factors is addressed by Donaldson et al., who summarize recent developments of epidermal growth factor and its receptor related research, while Betsholtz et al. discuss the structure, biological functions, and relatedness to oncogene products of human platelet-derived growth factors, Moses et al. extensively review the biological effects of transforming growth factors and their putative role in several disease states involving abnormal proliferation.

The contribution of maternal and neonatal compartments to the regulation of growth in the fetus and neonate is dealt with by Wehrenberg and Gaillard, in the section Neural Regulation of Growth Hormone Secretion. Here Mayo reports on studies of cDNA GHRH cloning, isolation, and characterization of the rat GHRH gene and its expression in transgenic mice, Berthelier et al. review the neurotransmitter and neuropeptide regulation of GH secretion and postreceptor coupling mechanisms, Frohman and coworkers examine the complex neuroendocrine regulation of GH secretion and its derangement under pathologic conditions, and Camanni and colleagues focus on noradrenergic and cholinergic control in animals and humans and the implications of these findings for the diagnosis and therapy of GH deficiency.

In a major clinical section Blizzard reviews the etiology and characteristics of GH deficiency and GH-like deficiency states, Bierich and colleagues analyze the contribution of GH provocative tests or spontaneous GH secretion to the diagnosis of GH deficiency, Rappaport et al. report on the GH secretory dysfunction in children after hypothalamic and pituitary irradiation. After a critical state of the art review by Milner, Takano, Ross, Pintor and their colleagues describe advantages and drawbacks of the diversified approaches to therapy of GH deficiency, by way of recombinant hGH, GHRH, and CNS-acting compounds. The discussion generated from these contributions adds a great deal to the appeal of the book.

Finally, in the section Growth Hormone Hypersecretory States, Shibasaki, Spada, Chiodini and their colleagues deal with aspects of the neural regulation, morphology, and function of the hypersecreting somatotroph and the medical therapy of acromegaly, del Pozo et al. describe the effect of a long-acting somatostatin analog on glucose homeostasis in normal subjects, and Reichlin reviews a number of pathological states (diabetes, anorexia nervosa, hepatic cirrhosis, etc.) in which excess secretion of GH occurs.

A subtle analysis of past research, present knowledge, and future perspectives in this exciting field is finally presented by Friesen.

In all, the book provides an invaluable forum where important aspects of the biological nature of growth and development, and the pathophysiology, diagnosis, and therapy of growth hormone deficiency and excess states are addressed and critically evaluated. It should therefore be of particular interest not only to basic and clinical endocrinologists, neuroendocrinologists, and pediatric endocrinologists, but also to the increasing number of students of cell biology.

We wish to thank Kabi Vitrum, Pierrel, and Sandoz for generous financial support, and the Italian Council of Research (CNR, Rome).

Milan, January 1989   The Editors

CHOH HAO LI

1913 - 1987

Choh Hao Li – CH – is no longer amongst us. He was the last of the large group of biochemists who between 1940 - 1970 put pituitary hormones into the limelight in endocrine research. He was in my opinion one of the best. The door to his laboratory has been closed and the name plate taken down.

As a scientist, CH belongs to an exclusive group of great men in his area; PHILIP E. SMITH, who introduced hypophysectomizing rats as a working method (1926-30) and was the first to demonstrate the effects of the operation; HERBERT M. EVANS, who produced the first purified hormone extract from the pituitary; BERNARDO A. HOUSSAY, who shed light on the pituitary's importance for experimental diabetes; ROSALYN S. YALOW and SOLOMON A. BERSON who introduced the RIA method in order to measure peptide hormones in the blood.

Within the space of a short obituary it is difficult to give an account of a research achievement which is distinguished by more than 1,000 scientific publications. Neither is it made any easier because CH worked within an area which during his life time underwent an explosive development and which was highly competitive. I shall therefore restrict myself to those achievements that I myself regard as milestones in CH's scientific career.

His earlier achievements, together with Evans, in the purification of ACTH and GH were in themselves milestones in research on peptide hormones even if the hormones produced later on – when new techniques became available – showed to be far from "pure".

CH was one of the few – may be the only one – who consistently studied pituitary peptides from the first stage, that is the purification, to structural analysis and synthesis. This concerned ACTH and MSA. He found himself at the forefront of every stage, from the beginning to the end, even if other researchers with the unlimited resources of drug companies could contend his success. He remained the central figure within the area.

CH was the first one to discover, isolate, purify and structurally analyze a pituitary hormone, LPH, an achievement which can be described as innovative.

CH's early discoveries concerning the pituitary hormone's structure and analysis of the biological effect of certain amino acid sequences opened a new field within the peptide hormone area at that time. This made it possible to explain overlapping effects of different peptides isolated from the pituitary. The latest example of many other discoveries was that sequences within the β-LPH molecule were identical with endorphins.

One can summarize his achievements chronologically in this way: 1940 purification of LH; 1944 isolation of bovine GH; 1949 isolation of FSH; 1956 isolation of MSA and also isolation of monkey and human GH; 1964 isolation of lipotropin; 1976 isolation of human β-lipotropin, LPH; in later years total synthesis of IGF-I and II.

Something which has always fascinated me was CH's obstinacy in his work – he "pursued" his peptides year after year, even when the odds were not especially propitious. This is illustrated with LPH. He isolated it together with Birk in 1964, really as a byproduct during the purification of ACTH. They found that the peptide consisted of 59 amino acid residues and had evident but not especially high lipolytic activity. Their continued work resulted in the isolation of β-LPH in 1965 with al least 90 amino acid residues (later on 91) and 1966 in γ-LPH (58 residues) where the sequencing of the 18 C-terminal amino acids was identical with the sequence in β-MSH. That opened up the possibility that β-LPH could be a large molecule, a "prohormone", which by cleaving could be changed to γ-LPH and β-MSH: β-LPH (91 amino acid residues) → γ-LPH (58) → β-melanotropin (39)

This was the first time one could see β-LPH as a precursor for other peptides.

CH continued to work his favourite object in recent years, that is LPH. The area exploded again in 1975, when Hughes and other scientists showed that enkephalin, the ligand for the recently discovered opiate receptors in the brain, consisted of two pentapeptides, of which one has a sequence which corresponds with the 61-65 sequence of the β-LPH which CH had isolated. This discovery instigated a stream of studies in different parts of the world. CH participated in such studies and found, together with A. Goldstein and B.M. Cox, that the β-LPH sequence 61-91 had opiate receptor-binding capacities, and with Guillemin et al. that the sequences 61-77 (γ-endorphin) and 61-76 (α-endorphin) had similar capacities. The field now lay wide open, which the literature bears witness to.

It is quite clear today that if CH had not discovered β-LPH, determined its sequences and attached weight to these different fragments after cleaving, it would have been a long time before the endorphins would had been discovered.

CH was born in Canton in China 1913. He came to the University of California in Berkeley 1935, and was awarded a Ph.D. in physico-organic chemistry 1938. After this he began his career with the legendary Herbert M. Evans and stayed there until 1950, when he then organized his own domain in Berkeley, **The Hormone Research Laboratory**. This was moved in 1967 to the San Francisco campus of the University of California. He remained there till his death, as active, innovative and stimulating as ever before. Our thoughts go to his wife Annie, her support and stimulation during their long life together, and his children.

The step from the material world to the world of memory is short, infinitely short. CH was in our material world a rich source of common research and friendship. In our memories, he will be as rich a source for inspiration and happiness.

Stockholm, August 1988   Rolf Luft

Contents

Native and Biosynthetic Growth Hormones

Somatomedins

Growth Factors

Neural Regulation of Growth Hormone Secretion

Growth Hormone Deficiency States

Round Table-Approach to Therapy

Growth Hormone Hypersecretory States

Contributors

Althoff P.H.
University Medical Clinic
Frankfurt am Main
Federal Republic of Germany

Arosio M.
Department of Endocrinology
School of Medicine
University of Milan
Via F. Sforza, 35
20122 Milan, Italy

Asakawa K.
Department of Medicine
Institute of Clinical Endocrinology
Tokyo Women's Medical College
10 Kawada-Cho
Shinkjuku-ku, Tokyo, Japan

Bascom C.C.
Department of Cell Biology
Vanderbilt University
School of Medicine
Nashville, TN 37232, U.S.A.

Bassetti M.
CNR Center of Cytopharmacology
Department of Pharmacology
University of Milan
Via Vanvitelli, 32
20129 Milan, Italy

Baumann G.
Center for Endocrinology
Metabolism and Nutrition
Department of Medicine
Northwestern University
303 East Chicago Avenue
Chicago, ILL 60611, U.S.A.

Bennett B.
Genentech Inc.,
460 Point San Bruno Boulevard South
San Francisco CA 94080, U.S.A.

Berthelier C.
Unitè de Neuroendocrinologie
Institut National de la Santè
et de la Recherche Medicale
INSERM U 159
Centre Paul Broca
2 ter, rue d'Alésia
75014 Paris, France

Berthold H.
Experimental Therapeutics Department
Clinical Research
Sandoz Ltd.
CH-4002 Basel, Switzerland

Bertrand P.
Unitè de Neuroendocrinologie
Institut National de la Santè
et de la Recherche Medicale
INSERM U 159
Centre Paul Broca
2 ter, rue d'Alésia
75014 Paris, France

Besser G.M.
Department of Endocrinology
St. Bartholomew's Hospital
West Smithfield
London EC1A 7BE
United Kingdom

Betsholtz C.
Department of Pathology
University Hospital
S– 751 85 Uppsala, Sweden

Bierich J.R.
Department of Pediatrics
University of Tübingen
D-74 Tübingen
Federal Republic of Germany

Blizzard R.M.
Department of Pediatrics
University of Virginia Medical School
Box 386
Charlottesville, VA 22901, U.S.A.

Bluet-Pajot M.T.
Unitè de Neuroendocrinologie
Institut National de la Santè
et de la Recherche Medicale
INSERM U 159
Centre Paul Broca
2 ter, rue d'Alésia
75014 Paris, France

Boccardi E.
Servizio di Neuroradiologia
Ospedale Niguarda
Piazza Ospedale Maggiore, 3
20162 Milan, Italy

Brauner R.
Hôpital des Enfants Malades
149 Rue de Sèvres
F-75015 Paris, France

Brooks P.J.
Department of Physiology
and Curriculum on Neurobiology
University of North Carolina at
Chapel Hill,
Chapel Hill, NC 27514. U.S.A.

Brügmann G.
Department of Pediatrics
University of Tübingen
D-74 Tübingen,
Federal Republic of Germany

Bywater M.
Department of Pathology
University Hospital
S-751 85 Uppsala, Sweden

Camanni F.
Dipartimento di Biochimica
Endocrino-Metabolica e
Gastroenterologica,
Divisione di Endocrinologia
Università di Torino
Corso Polonia, 14
10126 Turin, Italy

Canova-Davis E.
Genentech Inc.,

460 Point San Bruno
Boulevard
South San Francisco CA
94080, U.S.A.

Carpenter G.
Department of Biochemistry
Vanderbilt University
School of Medicine
Nashville, TN 37232, U.S.A.

Carter-Su C.
Department of Physiology,
Pediatrics and Biochemistry
University of Michigan
Medical School
Ann Arbor, MI 48109, U.S.A.

Cella S.G.
Department of Pharmacology
University of Milan
Via Vanvitelli, 32
20129 Milan, Italy

Chakel J.
Genentech Inc., 460 Point San
Bruno Boulevard
South San Francisco CA
94080, U.S.A.

Chiodini P.G.
Divisione di Endocrinologia
Ospedale Niguarda
Piazza Ospedale Maggiore, 3
20162 Milan, Italy

Chloubek R.
Genentech Inc., 460 Point San
Bruno Boulevard
South San Francisco CA
94080, U.S.A.

Chomczynski P.
Division of Endocrinology and Metabolism
University of Cincinnati
231 Bethesda Avenue
Cincinnati, OH, 45267, U.S.A.

Clauser H.
Unitè de Neuroendocrinologie
Institut National de la Santè
et de la Recherche Medicale
INSERM U 159
Centre Paul Broca
2 ter, rue d'Alésia
75014 Paris, France

Cocchi D.
Department of Pharmacology
University of Milan
Via Vanvitelli, 32
20129 Milan, Italy

Coffey Jr. R.J.
Department of Biochemistry
Vanderbilt University
School of Medicine
Nashville, TN 37232, U.S.A.

Collins V.P.
Ludwig Institute for Cancer Research,
Stockholm Branch
Stockholm, Sweden

Cozzi R.
Divisione di Endocrinologia
Ospedale Niguarda
Piazza Ospedale Maggiore, 3
20162 Milan, Italy

Dallabonzana D.
Divisione di Endocrinologia
Ospedale Niguarda
Piazza Ospedale Maggiore, 3
20162 Milan, Italy

De Gennaro Colonna V.
Department of Pharmacology
University of Milan
Via Vanvitelli, 32
20129 Milan, Italy

del Pozo E.
Sandoz Research Institute
Monbijoustrasse 115
3001 Bern, Switzerland

Donaldson R.W.
Department of Biochemistry
Vanderbilt University
School of Medicine
Nashville, TN 37232, U.S.A.

Downs T.R.
Division of Endocrinology and Metabolism
University of Cincinnati
Department of Internal Medicine
231 Bethesda Avenue
Cincinnati, Ohio 45267, U.S.A.

Durand D.
Unitè de Neuroendocrinologie
Institut National de la Santè
et de la Recherche Medicale
INSERM U 159
Centre Paul Broca
2 ter, rue d'Alésia
75014 Paris, France

Elahi F.R.
CNR Center of Cytopharmacology
Department of Pharmacology
University of Milan
Via Vanvitelli, 32
20129 Milan, Italy

English J.
Department of Physiology
and Curriculum in Neurobology
University of Nprth Carolina
at Chapel Hill
Chapel Hill
27514, NC, U.S.A.

Enjalbert A.
Unitè de Neuroendocrinologie
Institut National de la Santè
et de la Recherche Medicale
INSERM U 159
Centre Paul Broca
2 ter, rue d'Alésia
75014 Paris, France

Epelbaum J.
Unitè de Neuroendocrinologie
Institut National de la Santè
et de la Recherche Medicale
INSERM U 159
Centre Paul Broca
2 ter, rùe d'Alésia
75014 Paris, France

Fontoura M.
Hôpital des Enfants Malades
149 Rue de Sèvres
F-75015 Paris, France

Foster C.M.
Department of Physiology,
Pediatrics and Biochemistry
University of Michigan
Medical School
Ann Arbor, MI 48109, U.S.A.

Friesen H.G.
Department of Physiology
University of Manitoba
Winnipeg, MB R3E OW3, Canada

Froesch E.R.
Metabolic Unit
Department of Medicine
University Hospital
Rämistrasse 100
CH-8091 Zürich, Switzerland

Frohman L.A.
Division of Endocrinology and Metabolism
Department of Internal Medicine
University of Cincinnati
231 Bethesda Avenue
Cincinnati, OH 45267, U.S.A.

Gaillard R.C.
Clinique Medicale et Division
d'Endocrinologie
Hôpital Cantonal Universitaire
de Geneve
Geneva, Switzerland

Gellefors P.
Kabi Vitrum Peptide Hormones AB
S-112 87 Stockholm, Sweden

Ghigo E.
Dipartimento di Biochimica
Endocrino-Metabolica e
Gastroenterologica,
Divisione di Endocrinologia
Università di Torino
Corso Polonia, 14
10126 Turin, Italy

Giannattasio G.
CNR Center of Cytopharmacology
Department of Pharmacology
University of Milan
Via Vanvitelli, 32
20129 Milan, Italy

Goffi S.
Dipartimento di Biochimica
Endocrino-Metabolica e
Gastroenterologica,
Divisione di Endocrinologia
Università di Torino
Corso Polonia, 14
10126 Turin, Italy

Guerin S.
Department of Molecular Biology
Massachusetts General Hospital
Boston, MA 02114, U.S.A.

Guler H.P.
Metabolic Unit
Department of Medicine
University Hospital
Rämistrasse 100
CH-8091 Zürich, Switzerland

Hall K.
Department of Endocrinology
Karolinska Institute
Karolinska Hospital
S-104 01 Stockholm, Sweden

Hancock W.S.
Genentech Inc.,
460 Point San Bruno
Boulevard South
San Francisco CA 94080, U.S.A.

Harris R.
Genentech Inc.,
460 Point San Bruno
Boulevard South
San Francisco CA 94080, U.S.A.

Heldin C.-H.
Ludwig Institute for Cancer Research
Box 595, Biomedical Center
S-753 23 Uppsala, Sweden

Hizuka N.
Department of Medicine
Institute of Clinical Endocrinology
Tokyo Women's Medical College
10 Kawada-Cho
Shinkjuku-ku, Tokyo 162, Japan

Hotta M.
Department of Medicine
Institute of Clinical Endocrinology
Tokyo Women's Medical College
10 Kawada-Cho
Shinjuku-ku, Tokyo 162, Japan

Horikawa R.
Department of Medicine
Institute of Clinical Endocrinology
Tokyo Women's Medical College
10 Kawada-Cho
Shinkjuku-ku, Tokyo 162, Japan

Hynes M.A.
Howard Hughes Medical Institute
Research Laboratory
Neurobiology and Behavior Center
Columba University
New York, NY U.S.A.

Imaki T.
Department of Medicine
Institute of Clinical Endocrinology
Tokyo Women's Medical College
10 Kawada-Cho
Shinjuku-ku, Tokyo 162, Japan

Imperiale E.
Dipartimento di Biochimica
Endocrino-Metabolica e

Gastroenterologica,
Divisione di Endocrinologia
Università di Torino
Corso Polonia, 14
10126 Turin, Italy

Jansson J.-O.
Division of Endocrinology and Metabolism
Department of Internal Medicine
University of Cincinnati
231 Bethesda Avenue,
Cincinnati, OH 45267, U.S.A.

Johnson C.
Department of Pediatrics
Emory University
School of Medicine
2040 Ridgewood Dr.
Atlanta, GA 30323, U.S.A.

Kassels M.
Department of Pediatrics
Emory University
School of Medicine
2040 Ridgewood Dr.
Atlanta, GA 30323, U.S.A.

Katakami H.
Division of Endocrinology and Metabolism
Department of Internal Medicine
University of Cincinnati
231 Bethesda Avenue
Cincinnati, OH 45267, U.S.A.

Kawauchi H.
Laboratory of Molecular Endocrinology
School of Fisheries Sciences
Kitasato University
Sanriku, Iwate 02201, Japan

Keck R.
Genentech Inc.,
460 Point San Bruno
Boulevard South
San Francisco CA 94080 U.S.A.

Keski-Oja J.
Department of Virology
University of Helsinki,
Helsinki 29, Finland

Kiessling E.
Department of Pediatrics
University of Tübingen
D-74 Tübingen
Federal Republic of Germany

Kordon C.
Unitè de Neuroendocrinologie
Institut National de la Santè
et de la Recherche Medicale
INSERM U. 159
Centre Paul Broca
2 ter, rue d'Alésia
75014 Paris, France

Kurtz A.
Institute of Physiology
University of Zürich
CH-8057 Zürich, Switzerland

Lampis A.
Department of Pediatrics,
Chair of Pediatric Endocrinology
University of Cagliari
Via Jenner
09100 Cagliari, Italy

Lancranjan I.
Department of Neuroendocrinology
Sandoz Ltd.
CH-4002 Basel, Switzerland

Levi Montalcini R.
Istituto di Biologia Cellulare
Via G. Romagnosi, 18/A
00196 Rome, Italy

Liuzzi A.
Divisione di Endocrinologia
Ospedale Niguarda
Piazza Ospedale Maggiore, 3
20162 Milan, Italy

Locatelli V.
Department of Pharmacology
University of Milan
Via Vanvitelli, 32
20129 Milan, Italy

Loche S.
Department of Pediatrics,

Chair of Pediatric Endocrinology
University of Cagliari
Via Jenner
09100 Cagliari, Italy

Lund P.K.
Department of Physiology
University of North Carolina
School of Medicine
Medical Research Wing 206H
Chapel Hill, NC 27514, U.S.A.

Lyons R.M.
Department of Cell Biology
School of Medicine
Vanderbilt University
Nashville, TN 37232, U.S.A.

Martina V.
Dipartimento di Biochimica
Endocrino-Metabolica e
Gastroenterologica,
Divisione di Endocrinologia
Università di Torino
Corso Polonia, 14
10126 Turin, Italy

Massara F.
Dipartimento di Biochimica
Endocrino-Metabolica e
Gastroenterologica,
Divisione di Endocrinologia
Università di Torino
Corso Polonia, 14
10126 Turin, Italy

Masuda A.
Department of Medicine
Institute of Clinical Endocrinology
Tokyo Women's Medical College
10-Kawada-Cho
Shinljuku-Ku, Tokyo 162, Japan

Mayo K.E.
Department of Biochemistry,
Molecular Biology and Cell Biology
Northwestern University
2153 Sheridan Road
Evanston, ILL 60201, U.S.A.

Mazza E.
Dipartimento di Biochimica
Endocrino-Metabolica e
Gastroenterologica,
Divisione di Endocrinologia
Università di Torino
Corso Polonia, 14
10126 Turin, Italy

McKean M.C.
Department of Pediatrics
Emory University
School of Medicine
2040 Ridgewood Dr.
Atlanta, GA 30323, U.S.A.

Meacham L.
Department of Pediatrics
Emory University
School of Medicine
2040 Ridgewood Dr.
Atlanta, GA 30323, U.S.A.

Milner R.D.G.
Department of Paedriatrics
University of Sheffield
Children's Hospital
Sheffield S10 2TH
United Kingdom

Moore D.D.
Department of Molecular Biology
Massachusetts General Hospital
Boston, MA 02114, U.S.A.

Moses H.L.
Department of Cell Biology
Vanderbilt University
School of Medicine
Nashville, TN 37232, U.S.A.

Müller E.E.
Department of Pharmacology
University of Milan
Via Vanvitelli, 32
20129 Milan, Italy

Neufeld M.
Experimental Therapeutics Department

Clinical Research
Sandoz Ltd.
CH-4002 Basel, Switzerland

Nishibe S.
Department of Biochemistry
Vanderbilt University
School of Medicine
Nashville, TN 37232, U.S.A.

Oppizzi G.
Divisione di Endocrinologia
Ospedale Niguarda
Piazza Ospedale Maggiore, 3
20162 Milan, Italy

Parks J.S.
Department of Pediatrics
Emory University
School of Medicine
2040 Ridgewood Dr.
Atlanta, GA 30323, U.S.A.

Parks J.T.
Department of Pediatrics
Emory University
School of Medicine
2040 Ridgewood Dr.
Atlanta, GA 30323, U.S.A.

Pavlu B.
Kabi Vitrum Peptide
Hormones AB
S-112 87 Stockholm, Sweden

Petroncini M.M.
Divisione di Endocrinologia
Ospedale Niguarda
Piazza Ospedale Maggiore, 3
20162 Milan, Italy

Pintor C.
Department of Pediatrics,
Chair of Pediatric Endocrinology
University of Cagliari
Via Jenner
09100 Cagliari, Italy

Pòvoa G.
Centro Biomèdico – UFES
Caixa Postal 780
29000 – Vitòria – ES Brazil

Preece M.A.
Department of Growth and Development,
Institute of Child Health
London, United Kingdom

Puggioni R.
Department of Pediatrics,
Chair of Pediatric Endocrinology
University of Cagliari
Via Jenner
09100 Cagliari, Italy

Rappaport R.
Hôpital des Enfants Malades
149 Rue de Sèvres
F-75015 Paris, France

Rerat E.
Unitè de Neuroendocrinologie
Institut National de la Santè
et de la Recherche Medicale
INSERM U 159
Centre Paul Broca
2 ter, rue d'Alésia
75014 Paris, France

Reichlin S.
Endocrine Division
Department of Medicine
New England Medical Center
Tufts University School of Medicine
750 Washington Street
Boston, MA 02111, U.S.A.

Rorsman R.
Department of Pathology
University Hospital
S-751 85 Uppsala, Sweden

Rosenfeld R.G.
Department of Pediatrics
Stanford University Medical Center
Stanford, CA 94305, U.S.A.

Ross R.J.M.
Department of Endocrinology
St. Bartholomew's Hospital
West Smithfield
London EC1A 7BE
United Kingdom

Savage M.O.
Department of Endocrinology
St. Bartholomew's Hospital
West Smithfield
London EC1A 7BE
United Kingdom

Schmid C.
Metabolic Unit
Department of Medicine
University Hospital
Rämistrasse 100
CH-8091 Zürich, Switzerland

Schwartz J.
Department of Physiology,
Pediatrics and Biochemistry
University of Michigan
Medical School
Ann Arbor, MI 48109, U.S.A.

Shafer J.A.
Department of Physiology,
Pediatrics and Biochemistry
University of Michigan
Medical School
Ann Arbor, MI 48109, U.S.A.

Shibasaki T.
Department of Medicine
Institute of Clinical Endocrinology
Tokyo Women's Medical College
10 Kawada-Cho
Shinjuku-ku, Tokyo 162, Japan

Shizume K.
Department of Medicine
Institute of Clinical Endocrinology
Tokyo Women's Medical College
10 Kawada-Cho
Shinkjuku-ku, Tokyo 162, Japan

Sieber C.
Experimental Therapeutics Department
Clinical Research
Sandoz Ltd.
CH-4002 Basel, Switzerland

Sipes N.J.
Department of Cell Biology
School of Medicine
Vanderbilt University
Nashville, TN 37232, U.S.A.

Spada A.
Department of Endocrinology
School of Medicine
University of Milan
Via F. Sforza, 35
20122 Milan, Italy

Sukegawa I.
Department of Medicine
Institute of Clinical Endocrinology
Tokyo Women's Medical College
10 Kawada-Cho
Shinkjuku-ku, Tokyo 162, Japan

Takano K.
Department of Medicine
Institute of Clinical Endocrinology
Tokyo Women's Medical College
10 Kawada-Cho
Shinkjuku-ku, Tokyo 162, Japan

Tripputi P.
Department of Molecular Biology
Massachusetts General Hospital
Boston, MA 02114, U.S.A.

Vallar L.
CNR Center of Cytopharmacology
Department of Pharmacology
University of Milan
Via Vanvitelli, 32
20129 Milan, Italy

Van Wyk J.J.
Department of Pediatrics
University of North Carolina at Chapel Hill,
Chapel Hill,
NC 27514, U.S.A.

Verde G.
Divisione di Endocrinologia
Ospedale Niguarda
Piazza Ospedale Maggiore, 3
20162 Milan, Italy

Wehrenberg W.B.
Department of Health Sciences
University of Wisconsin-Milwaukee
Milwaukee, WI 53201, U.S.A.

Westermark B.
Department of Pathology
University Hospital
University of Uppsala
S-751 85 Uppsala, Sweden

Yasuda A.
Laboratory of Molecular Endocrinology
School of Fisheries Sciences
Kitasato University
Sanriku, Iwate 02201, Japan

Zapf J.
Metabolic Unit
Department of Medicine
University Hospital
Rämistrasse 100
CH-8091 Zürich, Switzerland

List of Abbreviations

(Other abbreviations used are defined in the text.)

A

AC	adenylate cyclase
ACh	acetycholine
AITT	arginine-insulin tolerance test
ALL	acute lymphoblastic leukemia

B

BP	binding protein
Br	bromocriptine

C

CDGA	constitutional delayed growth and adolescence
CI	continuous infusion
CLO	clonidine
CNS	central nervous system
CS	somatomammotropin
CRF	corticotropin-releasing factor
CSF	cerebrospinal fluid
CT	computed tomography

D

DA	dopamine

E

ECP	Escherichia Coli protein
EGF	epidermal growth factor
EIA	enzyme immunoassay

G

GAL	galanin
GH	growth hormone
GHD	growth hormone deficiency
GHRH	growth hormone-releasing hormone
GnRH	gonadotropin releasing hormone
GRF	growth hormone-releasing hormone or factor
GRH	growth hormone releasing hormone

H

HPLC	high performance liquid chromatography
HV	height velocity

I

IDDM	insulin-dependent diabetes
IH	insulin hypoglycemia
IGF	insulin-like growth factor
IEF	isoelectric focusing
ISHH	in situ hybridization histochemistry

M

mf-hGH	methionine free hGH
m-hGH	methyonil hGH
MI	multiple injections
MPO	medial preoptic area

MSA   multiplication stimulating activity

R

r-hGH   recombinant hGH

N

NE   norepinephrine
NRS   normal rabbit serum
NSC   normal short children
NSGHD   neurosecretory growth hormone deficiency
NVSS   normal variant short stature

S

SDS-PAGE   sodium dodecyl sulphate-polyacrylamide gel electrophoresis
Sm-C   somatomedin-C
SMS   SMS 201-995
SRIF   growth hormone-inhibiting factor
SS   somatostatin
α-SUB   α-subunit
SSV   simian sarcoma virus

O

OPI   opioid neurons

T

TBI   total body irradiation
TFA   trifluoroacetic acid
TGF   transforming growth factor
TPA   12-o-tetradecanoyl-phorbol-13-acetate
TRH   thyrotropin releasing hormone

P

P III P   serum procollagen III propeptide
Parlodel LAR   long-acting repeateable bromocriptine
Parlodel MR   modified release bromocriptine
PD   pyridostigmine
PDGF   platelet-derived growth factor
PL   placental lactogen
POMC   proopiomelanocortin
p-hGH   pituitary extracted hGH
PRL   prolactin
PSS   psychosocial short stature

U

UEP   unit evolutionary period

V

VIP   vasoactive intestinal polypeptide

NATIVE AND BIOSYNTHETIC GROWTH HORMONES

Advances in Growth Hormone and Growth Factor Research,
edited by E.E. Müller, D. Cocchi and V. Locatelli
Pythagora Press, Roma-Milano and Springer Verlag, Berlin-Heidelberg © 1989

Evolution and structure of the growth hormone gene cluster*

J. S. Parks, M. Kassels, M. C. McKean, J. T. Parks, C. Johnson and L. Meacham

Division of Pediatric Endocrinology, Department of Pediatrics, Emory University School of Medicine, Atlanta, Georgia, U.S.A.

INTRODUCTION

Within the past 10 years, molecular genetic techniques have yielded a wealth of new information about growth hormones (GHs), prolactins, and placental lactogens (Pls). Cloned cDNA copies of messenger RNAs permit sequence comparisons of related peptide hormones within and between species. Genomic clones reveal details of gene structure and, in some cases, disclose clusters of duplicated genes. The purpose of this review is to summarize recent findings about this large family of hormone genes, with special emphasis upon evolution of the human GH and chorionic somatomammotropin (CS) gene cluster.

ORGANIZATION OF GH AND PROLACTIN GENES

All GH, prolactin and PL genes follow the pattern of organization shown for hGH in Fig. 1 (DeNoto et al. 1981). Each gene encodes a mature protein of approximately 200 amino acids, preceded by a signal peptide of about 25 amino acids. The 5 exons or coding sequences, designated I-V, are interrupted by 4 introns designated A-D. Exon/intron boundaries occur at equivalent sites with respect to amino acid codons. In hGH, exon I contains 60 nucleotides of 5' untranslated sequences, codons -26 to -24 of the signal peptide, and the first nucleotide of codon -23. The second exon codes for the rest of the signal peptide and residues $1 - 31$. Exons III and IV encode amino acids $32 - 71$ and $72 - 126$. Exon V specifies amino

* The work described in this report was supported in party by March of Dimes Birth Defects Foundation Clinical Research Grant 6-474 and by an Emory University MSRTP Grant to M.K.

acids 127 – 191 and extends to a polyadenylation site approximately 100 bp 3' to the translational stop signal TAG. The entire sequence is transcribed as a pre-messenger RNA and the intron sequences, beginning with GT and ending with AG, are removed by splicing to yield a mature messenger RNA. This mRNA is translated to produce a prehormone. Post-translational processing removes the signal peptide and yields the mature hormone. Human pituitary GH is not glyco-sylated, but there is evidence for glycosylation of human prolactin and placental GH.

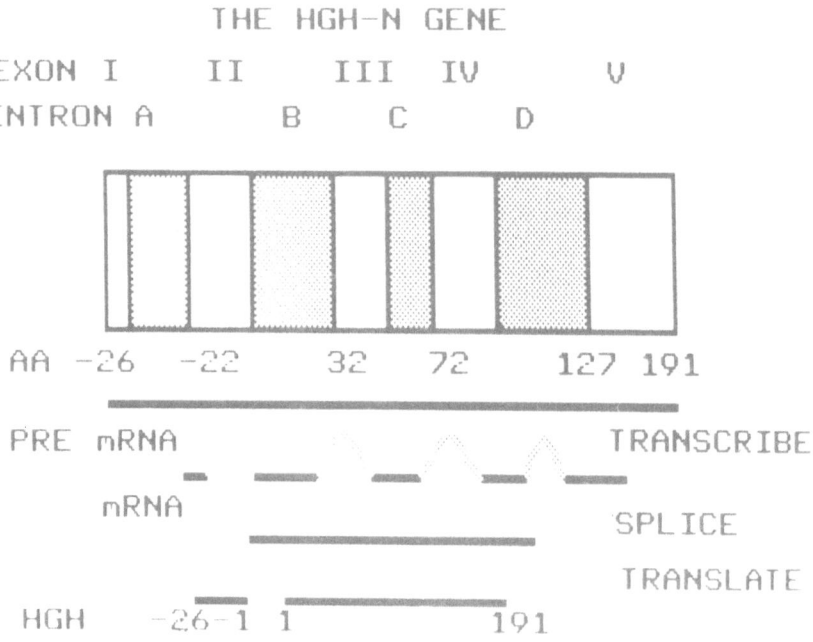

Fig. 1. The hGH-N gene. The gene is divided into exons (I – V, *open boxes*) and introns (A – D, *hatched boxes*). The first complete codon in each intron is indicated below the diagram of the gene. The gene is transcribed into a pre mRNA (*solid line*) and the mRNA sequences corresponding to introns (*hatched lines*) are removed by splicing. Following translation, the signal peptide (-26 – -1) are removed from the mature hGH peptide (1 – 191).

Niall et al. (1971) suggested, on the basis of amino acid homologies, that the ancestor of GH and prolactin genes might have evolved through duplication of a gene encoding a much smaller peptide domain. Miller and Eberhardt (1983) and Selby et al. (1984) have extended this argument through analysis of homologies between exons II, IV and V as well as short repeating sequences flanking these exons. A peptide domain corresponding to the current exon II is duplicated and reduplicated to form a precursor gene with 4 similar exons. An intron separating two 3' domains is removed to form the ancestor of exon V and reduce the gene to 3 exons. A different domain, corresponding to the current exon III, is then inserted. Finally, different domains containing regulatory elements and very short coding sequences are inserted to yield different exons I for GH and prolactin.

DIVERGENCE OF GH AND PROLACTIN GENES

Despite similarities in gene organization, GH and prolactin sequences show considerable divergence. The amino acid homologies between GH and prolactin in human (Cooke et al. 1981), rat (Cooke et al. 1980), mouse (Linzer and Talamantes 1985), and bovine (Miller et al. 1981; Miller 1982) species range from 17% to 24%. Interspecies homologies for GH are much higher, from 64% between rat and human GH (Seeburg et al. 1977), to 91% between bovine and porcine GH (Seeburg et al. 1983) and 95% between rat and mouse GH (Linzer and Talamantes 1985). Similarly, comparisons of amino acid sequences for prolactin in distantly related species show between 55% and 82% homology (Miller and Eberhardt 1983; Linzer and Talamantes 1985).

The distant relationship between GH and prolactin genes is reflected in different chromosomal locations and a seven-fold difference in gene sizes. In man, the GH gene is located on chromosome 17 (Owerbach et al. 1980; George et al. 1981; Harper et al. 1982) and the prolactin gene is located on chromosome 6 (Owerbach et al. 1981). Growth hormone genes are generally compact, with all the necessary information being contained in a distance of less than 2 kb (Woychik et al. 1982). The introns of prolactin genes are much larger and account for gene sizes of about 10 kb in man (Truong et al. 1984) and the rat (Chien and Thompson 1980; Cooke and Baxter 1982).

There are several theoretical approaches to estimating the time since divergence of related proteins or genes (Fig. 2). The simplest is the unit evolutionary period (UEP), which Wilson et al. (1977) define as the time in millions of years required to generate a 1% amino acid sequence divergence in two related peptides. Use of a UEP of 4.5 for GH and prolactin (Cooke et al. 1981; Miller et al. 1981) and multplying by 76 (for the percent divergence between human GH and prolactin) suggests divergence 342 million years ago. Calculations of evolutionary distance from nucleotide sequence comparisons (Perler et al. 1980; Kimura 1980; Kimura, 1981) give similar results (Miller et al. 1981). The evolutionary distance between GH and prolactin is consistent with the fact that both hormones are present in fish as well as birds and mammals.

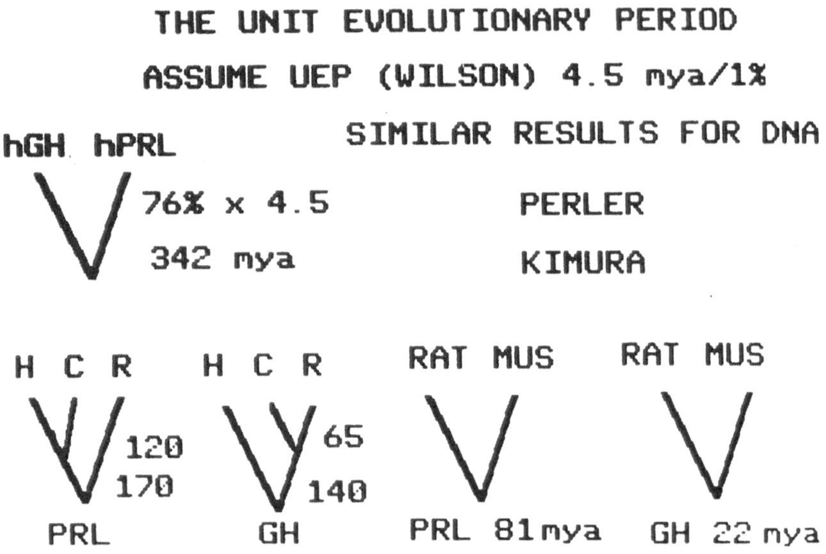

THE UNIT EVOLUTIONARY PERIOD

ASSUME UEP (WILSON) 4.5 mya/1%

Fig. 2. The unit evolutionary period (UEP). Diagrams show the estimated times since divergence of peptide precursors in human (H), bovine (C), rat (R) and mouse (Mus) species, based upon a figure of 4.5 million years for each divergence of 1% in amino acid sequence.

Estimates of the evolutionary distance between species based on comparison of GH and prolactin sequences give conflicting results. For example, mouse and rat GH have 95% amino acid sequence homology but the corresponding prolactin sequences have only 82% homology. The first would suggest a 22- million year and the second an 81- million year interval since a common ancestor. Prolactin comparisons suggest that man is 120 million years from the cow and 170 from the rat, while GH comparisons suggest only a 65 million years between the cow and the rat and 140 million years between their common ancestor and man (Miller and Eberhardt 1981). Fossil data and thermal dissociation of DNA suggest a branch point about 85 million years ago, at the time of the mammmalian radiation. The evolutionary clock cannot be read accurately from the narrow vantage point of the GH and prolactin genes.

EVOLUTION OF PLACENTAL LACTOGENS

Placental lactogens are GH- and prolactin-related hormones produced by fetal placental tissue. Their presence appears to be universal among mammals, but their evolutionary histories differ (Fig. 3). Among primates, PL, or CS, is closely related to GH.With few exceptions, other mammals have only a single GH-related gene. Southern blotting provides evidence for a second GH-related gene in the cow

AMINO ACID HOMOLOGIES

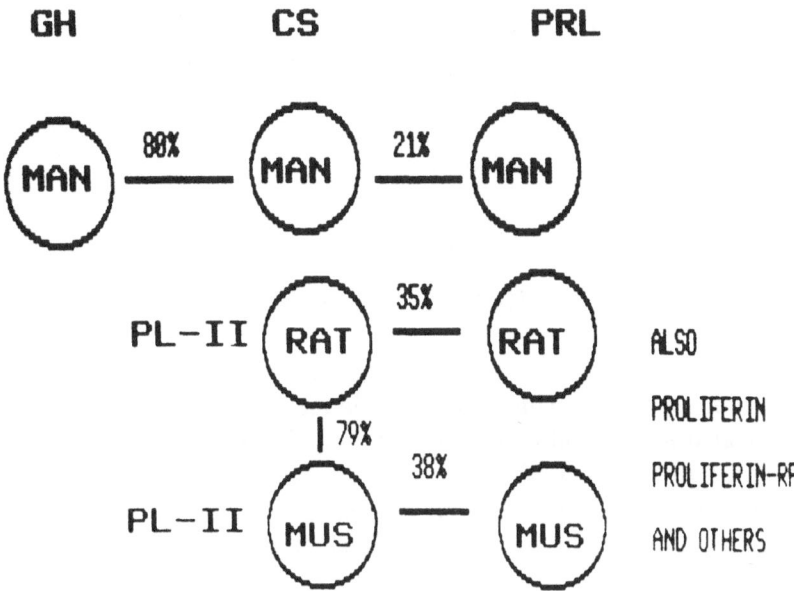

Fig. 3. Amino acid homologies for growth hormones, chorionic somatomammotropins, placental lactogens and prolactins. Species are indicated within the *circles* and the per cent amino acid sequence divergence is indicated above the connecting lines.

(Gordon et al. 1983), but none in dogs (J. T. Parks, unpublished observations), rats or mice (Barta et al. 1981; Parks et al. 1982). Among non-primates, placental lactogens are more closely related to prolactin.

Both mice and rats express at least 2 PLs. The mid-pregnancy lactogens, termed PL-I, appear transiently, are highly glycosylated, and have apparent molecular weights of approximately 40,000. The mouse and rat PL-Is have not been cloned. Late pregnancy PLs, termed PL-II, increase until birth, are not glycosylated, and have molecular weights of about 20,000. Their cDNAs have been cloned. Mouse PL-II has 38% amino acid sequence homology with mouse prolactin (mPRL), but only 20% homology with mGH (Jackson et al. 1986). Rat PL-II has 35% homology with rGH, but only 19% with rGH (Duckworth et al. 1986). There is 79% homology between mPL-II and rPL-II. Duckworth et al. (1986a) have cloned four other prolactin-related rat placental cDNAs and sequenced one which they term "rat prolactin (rPRL) -like protein A". It has about 30% sequence homology with rPL-II, rPRL, mouse proliferin, and mouse proliferin-related protein. Consideration of substitutions of equivalent amino acids raises the degrees of homology to about 45%.

The plethora of placental prolactin-like peptides is intriguing. It is not known whether their genes are clustered or dispersed among several chromosomes. Overall, the 65% sequence divergence implies that their precursor and that of prolactin diverged nearly 300 million years ago. The region corresponding to the middle of exon III is most subject to variation in length and sequence and may confer different biological activities. Apparent redundancy may, in fact, reflect a need for different hormones to support maternal carbohydrate and fat utilization, preparation for lactation, and fetal growth. .

STRUCTURE AND FUNCTION OF THE HGH AND HCS GENE CLUSTER

It is not known whether human beings and other primates have placental lactogens corresponding to those described in rats. They do have a family of genes closely related to GH and clustered over a small portion of the genome.

The modern hGH and hCS gene is shown in Fig. 4. This cluster contains five genes aligned in the same transcriptional orientation over a distance of about 50 kb on the long arm of chromosome 17 (Barsh et al. 1983). Reading 5' to 3', they

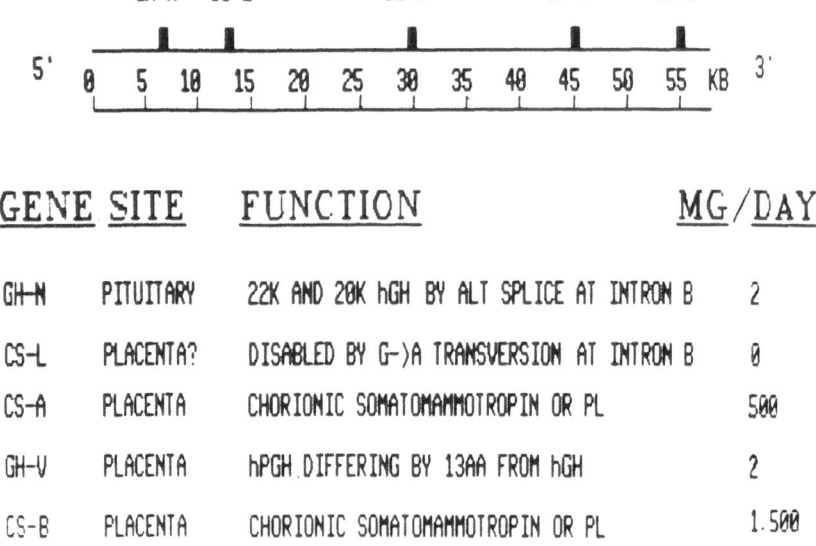

Fig. 4. Structure and function of the human GH and CS gene cluster. The 5 related genes and their locations are indicated by *black boxes*, situated on a line above a scale indicating distance in kilobase pairs (*KB*). Gene designations, sites of expression, encoded peptides, and estimates of daily peptide production are indicated below the regional map of the gene cluster.

are hGH-N, hCS-L, hCS-A, hGH-V and hCS-B. Synonyms are hGH-1, hCS-5, hCS-1, hGH-2 and hCS-2 (Hirt et al. 1987). The hGH-N gene encodes 22K pituitary hGH and, through alternative mRNA splicing of intron B, 20K hGH (DeNoto et al. 1981). A single G to A transversion of the donor splice site at the beginning of the second intron has disabled the hCS-L gene (Hirt et al. 1987). The hCS-L message has not been found in pituitary or placental mRNA, but the gene is transcribed efficiently in a cell-free system (Selvanayagam et al. 1984). There is a potential alternative splice site 45 bases downstream which could give rise to a protein 15 amino acids longer than the hCS proteins. Both hCS-A and hCS-B specify the same mature hCS peptide and share in its production (Barrera-Saldena et al. 1983). Concentrations of hCS in the maternal circulation reach 6000 ng/ml near term, a level 1000 times that of mean hGH concentrations. The hGH-V gene encodes a variant, more basic, placental GH differing at 13 of 191 amino acids from pituitary GH (Seeburg, 1982). Dot blotting of selective oligodeoxynucleotide probes to mRNAindicates that hGH-V is expressed in fetal placental tissue (Frankenne et al. 1987). The level of expression is less than 1/1,000 that of hCS-A or hCS-B. However, the placenta is roughly 1,000 times as large as the anterior pituitary and the hGH-V peptide appears to replace pituitary hGH in the maternal circulation in the second half of pregnancy (Hennen et al. 1985). A transcriptional enhancer located 2 kb 3' to hCS-B is involved in the tissue-specific expression of the hCS genes (Rogers et al. 1986) and possibly to the hGH-V gene. Pituitary expression of hGH-N depends upon control elements 180 – 280 bp 5' to the gene (Cattini et al. 1986; Bodner and Karin, 1987).

MOLECULAR EVOLUTION OF THE HGH AND HCS GENE CLUSTER

Possible steps in the evolution of the hGH and hCS gene cluster were proposed by Barsh et al. (1983) and delineated in greater detail by Hirt et al. (1987). These steps are depicted in Fig. 5. The argument is based upon the locations and lengths of middle repetitive *Alu* sequences as well as upon differences in non-repetitive flanking sequences within the cluster. hGH-N and hGH-V have *Alu* sequences of approximately 270 nucleotides flanking their 3' ends, while similar sequences flanking hCS-L, hCS-A, and hCS-B end after about 25 bp. In step 1, an hGH/hCS precursor gene residing on a 2.6 kb *Eco* RI fragment and with a complete *Alu* sequence in the 3' flanking region combines with an identical gene through unequal crossing-over with an *Alu* sequence 5' to the second copy of the gene. The 3' *Alu* sequence of the right hand gene is then deleted and the gene pair is recognizable as a 2.6 kb *Eco* RI fragment containing a pre-GH gene and a 2.9 kb *Eco* RI fragment containing a pre-CS gene. Step 2 involves a second duplication event with unequal crossing- over between an *Alu* sequence several kb 3' to the pre-CS gene and an

Alu sequence 5' to the pre-GH gene. The result is a four- member gene cluster containing precursors of GH-N, CS-L, GH-V and CS-B. A fifth gene is added in step 3 through crossing over between *Alu* sequences 3' to one CS-L allele and 5' to the other. A later gene conversion event causes CS-A to assume a closer resemblance to CS-B than to CS-L in the 3' direction from a point within intron D. Loss of two *Eco* RI sites 3' to CS-L and other minor changes lead to the current hGH and hCS locus.

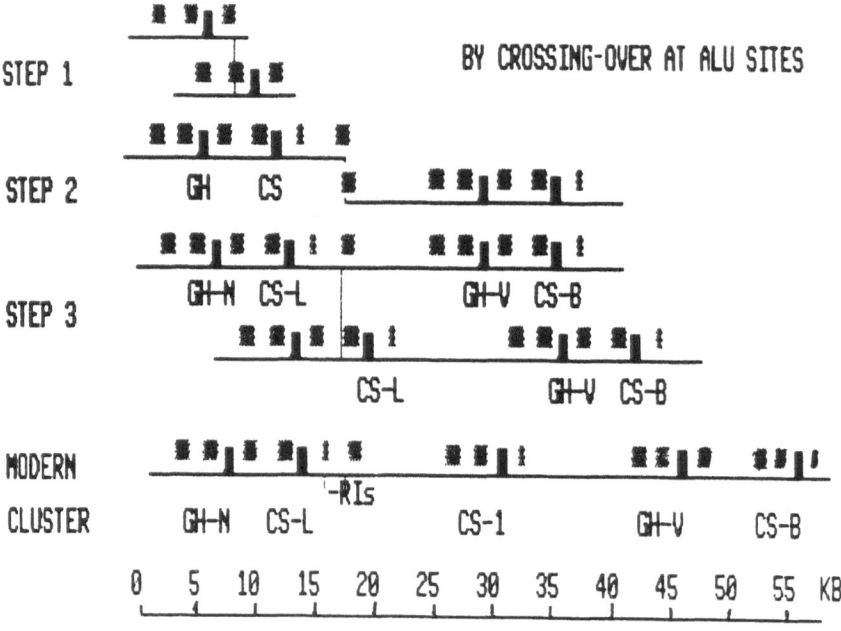

Fig. 5. Evolution of the human GH and CS gene cluster. GH and CS clusters (*filled boxes*) and middle repetitive *Alu* I sequences (*hatched boxes*) are indicated for successive evolutionary steps. The vertical lines indicate postulated sites of unequal crossing-over at meiosis.

THE ARGUMENT FOR RECENT GENE DUPLICATIONS

Two sets of observations have been used to argue for duplication of hGH and hCS genes within recent times. The first is based upon linkage disequilibrium for restriction fragment length polymorphisms (RFLPs) within the cluster. Chakravarti et al. (1984) found that the polymorphic *Msp* I sites located 3' to the hCS-L and hCS-A genes were either both present or both absent on a single chromosome. Less than 5% of chromosomes could be identified as having one site present and the other absent. This, together with an excess of site presence in U.S. blacks and

additional linkage to polymorphic *Hin* cII and *bgl* II sites, led to the suggestion
that the duplications to form five- member hGH and hCS gene clusters occurred
independently within the past 115,000 years and after the divergence of the major
races of man. Chakravarti et al. (1984) also suggested that instances of absence
of genes from the right hand side of the cluster might reflect persistence of ancestral
forms of the gene cluster rather than deletions.

Several types of abnormal hGH and hCS gene clusters have been described
in association with hormone deficiency. There have been at least seven reports of

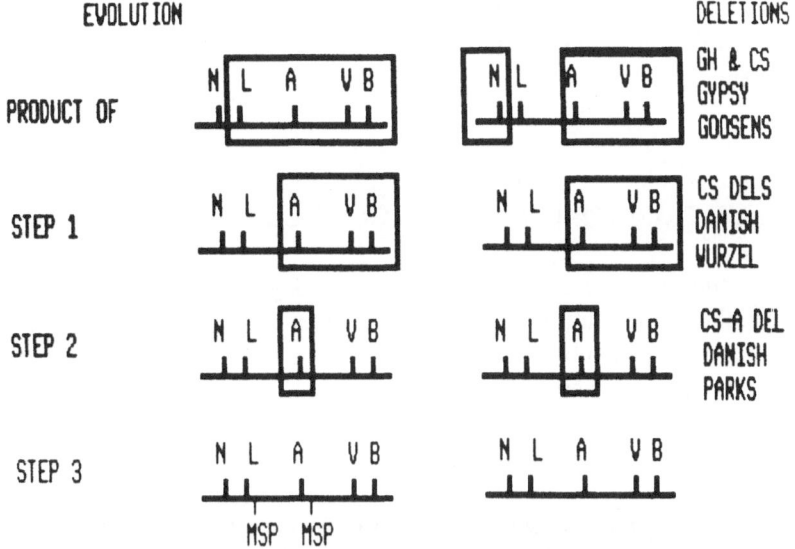

Fig. 6. Comparison of evolutionary steps and deletions within the human GH and CS gene
cluster. The genes present at each stage are located in *open areas*, while missing genes are
enclosed within *boxes*.

homozygosity for absence of the hGH-N gene, with (Phillips et al. 1981; Nishi
et al. 1984; Rivarola et al. 1984; Frisch et al. 1986; Goosens et al. 1986) and
without(Laron et al. 1985; Braga et al. 1986) development of interfering antibodies
during hGH treatment. There are at least two general types of human gene clusters
lacking hCS genes. In one, the hCS-A, hGH-V and hCS-B genes are absent, leaving
an abnormal marker *Eco* rI fragment of either 10.0 kb (Wurzel et al. 1982; Simon
et al. 1986; Goossens et al. 1986) or 7.0 kb (Parks et al. 1983; Wohlk et al. 1984)
containing the hCS-L gene. In the other type of cluster only the hCS-A gene is

missing. When paired with the first type of cluster, this results in either a 75% reduction (Parks et al. 1985) or an absence (Simon et al. 1986) of hCS immunoreactivity in the maternal circulation. As shown in Fig. 6, these clusters bear a strong resemblance to postulated stages in the evolution of the hGH and hCS gene cluster. In addition, the placenta from a child with homozygosity for the first type of cluster contains small quantities of immunoreactive hGH-N peptide and a peptide with hCS immunoreactivity, but anomalous behavior on ion exchange chromatography (Frankenne et al. 1986). These findings suggest retention of an earlier version of the gene cluster in which both hGH-N and a still functional hCS-L gene are expressed in placenta. It is intriguing to suppose that the two member cluster is a "living fossil" representative of an early step in evolution of the hGH and hCS gene cluster.

STUDIES OF PRIMATE GH AND CS GENE CLUSTERS

We reasoned that examination of GH and CS genes in other primates could provide insight into the timing of gene duplications in man. If the third duplication event leading to a 5 member hGH and hCS gene cluster occurred within the past 115,000 years, then restriction fragments representing the hCS-A gene should not have close counterparts in other primate species.

White blood cell or placental DNA was prepared from the six primate species rhesus (*Macacus rhesus*), orangutan (*Pongo pygmaeus*), chimpanzee (*Pan troglodytes*), pygmy chimpanzee (*Pan paniscus*), gorilla (*Gorilla gorilla*) and man (*Homo sapiens*) and digested individually with the six restriction enzymes *Bam* HI, *Bg*lII, *Eco* RI, *Hin* cII, *Sst* I, and *Xba* I. Homologous fragments were detected with a 550 bp hCS cDNA probe (Shine et al. 1977) that recognizes both GH and CS genes.

Restriction patterns were scored for number and similarities in fragment size. The rhesus fragment pattern was most different from that of man. There were more fragments, suggesting a more complex GH and CS cluster and considerable divergence in gene duplication events. The proportion of similar size fragments was intermediate in comparisons between man and chimpanzee or man and gorilla, and greatest in comparisons between chimpanzee and pygmy chimpanzee.

The results of *Eco* RI and *Bam* HI digestion of normal and hCS deletion human DNA are compared with orangutan and chimpanzee digests in Fig. 7. Chimpanzee DNA digests suggest a five-, or possibly six-, member gene cluster with genes corresponding to each of the five human genes. In contrast, the orangutan lacks the 9.5 kb *Eco* RI and the 8.3 kb *Bam* HI equivalents of the hCS-L gene. Digestions with other enzymes support a model in which the orangutan has a smaller gene cluster with analogs of the hGH-N, hCS-A, hGH-V, and hCS-B, but not the hCS-L gene. The distinctive fragment pattern of hCS-L is postulated (Hirt et al. 1987)

GH AND CS RESTRICTION PATTERNS
FOR MAN AND OTHER PRIMATES

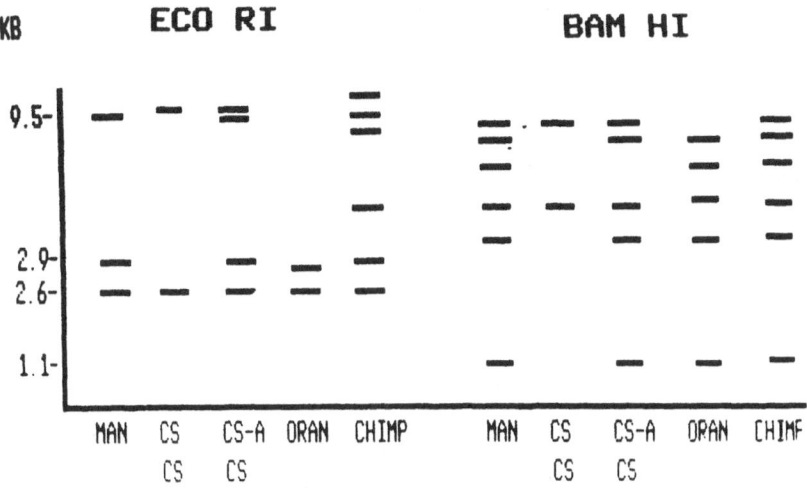

Fig. 7. GH and CS restriction patterns for man and other primates. Fragment size in kilobases is indicated in a log scale on the x-axis. Sources of DNA for *Eco* RI and *Bam* HI digestion are indicated on the y-axis.

PRIMATE EVOLUTIONARY TREE
BASED ON DNA ANNEALING

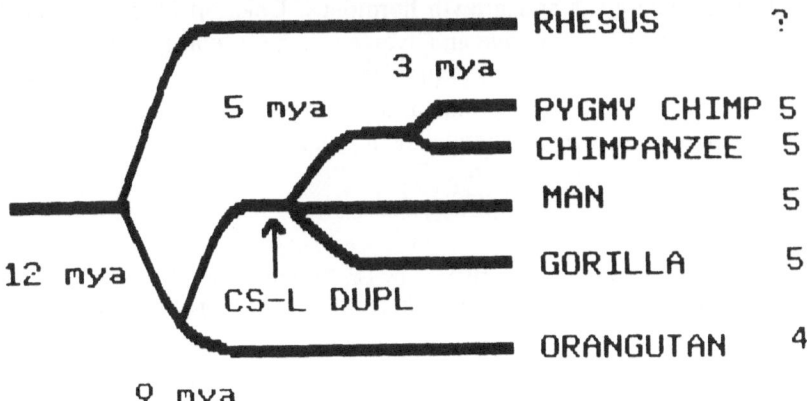

Fig. 8. Primate evolutionary tree for GH and CS. Branch points indicate the time of divergence of common ancestors, as based on DNA annealing and protein polymorphisms. The apparent numbers of GH and CS genes for each primate species are indicated at the right. *Mya*, million years ago.

to reflect loss of *Eco* RI and other restriction sites after a third duplication event. Common ancestry with the orangutan may antedate either the 5 member cluster or else the subsequent modification of these sites.

There are many possible alternatives in the interpretation of restriction patterns and certainty about evolutionary relationships would require cloning and analysis of large genomic fragments from the different primate species. Genes present in equivalent positions may have different functions in the primate species. It is not known, for instance, whether the second gene of another great ape cluster is a disabled CS and the fourth a placental growth hormone. However, the restriction patterns of these genes are consistent with the cladogram for primate evolution (Zihlman et al. 1978) shown in Fig. 8 and suggest that the third duplication event in the cluster occurred 5 to 9 million years ago, after divergence of common ancestors of orangutans and great apes, but before the divergence of ancestors of the chimpanzee, the gorilla, and man. This evidence runs counter to the hypothesis of independent duplications following divergence of the major races of man.

SUMMARY AND CONCLUSION

Current growth hormone and prolactin genes are descended from a common ancestor. Their divergence occurred more than 300 million years ago and they have retained their distinctive characters in many species. These genes have been modified to encode placental lactogens. The chorionic somatomammotropins of man and other primates are derived from growth hormone genes and do not seem to have counterparts in other mammals. There is evidence for great redundancy among placental lactogens and growth hormones. Deletion of the hCS-A, hGH-V, and hCS-B genes does not have an adverse effect on fetal growth. It is entirely possible that future investigations will disclose a family of human genes and proteins more closely related to prolactin and proliferin that play important roles in the regulation of growth and fetal development.

Acknowledgment

We gratefully acknowledge the help afforded by Dr. Kenneth Gould, Miss Carol Alle and their associates at the Yerkes Primate Research Center in providing primate specimens for DNA analysis.

REFERENCES

Barrera-Saldena HA, Seeburg PH, Saunders GF (1983) Two structurally different genes produce the same human placental lactogen hormone *J Biol Chem* **258**: 3787-3793
Barsh GS, Seeburg PH, Gelinas RE (1983) The human growth hormone gene family: structure and evolution of the chromosomal locus. *Nucleic Acids Res* **11**: 3939-3958

Barta A, Richards RI, Baxter JD, Shine J (1981) Primary structure and evolution of the rat growth hormone gene. *Proc Natl Acad Sci USA* **78**:4867-4871

Bodner M, Karin M (1987) A pituitary-specific tra*ns*-acting factor can stimulate transcription from the growth hormone promoter in extracts of nonexpressing cells. *Cell* **50**: 267-275

Braga S, Phillips JA III, Joss E, Schwarz H, Zuppinger K (1986) Familial growth hormone deficiency resulting from a 7.6 kb deletion within the growth hormone gene cluster. *Am J Med Genet* **25**: 443-452

Cattini PA, Peritz LN, Anderson TR, Baxter JD, Eberhardt NL (1986) The 5' flanking sequences of the human growth hormone gene contain a cell-specific control element. *DNA* **5**: 503-509

Chakravarti A, Phillips JA III, Mellits KH, Buetow KH, Seeburg PH (1984) Patterns of polymorphism and linkage disequilibrium suggest independent origins of the human growth hormone gene cluster. *Proc Natl Acad Sci USA,* **81**: 6085-6089

Chien Y, Thompson EB (1980) Genomic organization of rat prolactin and growth hormone genes. *Proc Natl Acad Sci USA* **77**: 4583-4587

Cooke NE, Coit D, Weiner R.I, Baxter J.D, Martial J.A. (1980) Structure of cloned DNA complementary to rat prolactin messenger RNA. *J Biol Chem* **255**: 6502-6510

Cooke NE, Coit D, Shine J, Baxter JD, Martial JA (1981) Human prolactin: cDNA structural analysis and evolutionary comparisons. *J Biol Chem* **256**: 4006-4016

Cooke NE, Baxter JD (1982) Structural analysis of the prolactin gene suggests a separate origin for its 5' end. *Nature* **297**: 603-606

DeNoto FM, Moore DD, Goodman HM (1981) Human growth hormone DNA sequence and mRNA structure: possible alternative splicing. *Nucleic Acids Res* **9**: 3719-3730

Duckworth ML, Kirk KL, Friesen HG (1986a) Isolation and identification of a cDNA clone of rat placental lactogen II. *J Biol Chem* **261**: 10871-10878

Duckworth ML, Peden LM, Friesen HG (1986b) Isolation of a novel prolactin-like cDNA from developing rat placenta. *J Biol Chem* **261**: 10879-10884

Frankenne F, Hennen G, Parks JS, Nielsen PV (1986) A gene deletion in the hGH/hCS gene cluster could be responsible for the placental expression of hGH and/or hCS like molecules absent in normal subjects. Program and Abstracts, 68th Annual Meeting, Endocrine Society. Abstract 388

Frankenne F, Rentier-Delrue F, Scippo M-L, Martial J, Hennen G (1987) Expression of the human growth hormone variant gene in human placenta. *J Clin Endocrinol Metab* **65**: 635-637

Frisch H, Phillips JA III (1986) Growth hormone deficiency due to GH-N gene deletion in an Austrian family. *Acta Endocrinol* [Suppl] (Copenh) **279**: 107-112

George DL, Phillips JA III, Francke U, Seeburg PH (1981) The genes for human growth hormone and chorionic somatomammotropin are on the long arm of human chromosome 17 in region q21 to ter. *Hum Genet* **57**: 138-141

Goossens M, Brauner R, Czernichow P, Duquesnoy P, Rappaport R (1986) Isolated growth hormone (GH) deficiency type 1A associated with a double deletion in the human GH gene cluster. *J Clin Endocrinol Metab* **62**: 712-716

Gordon DF, Quick DP, Erwin CR, Donelson JE, Maurer RA (1983) Nucleotide sequence of the bovine growth hormone chromosomal gene. *Mol Cell Endocrinol* **33**: 81-95

Harper ME, Barrera-Saldena HA, Saunders GF (1982) Chromosomal localization of the human placental lactogen-growth hormone gene cluster to 17q22-24. *Am J Hum Genet* **34**: 227-234

Hennen G, Frankenne F, Closet J, Gomez F, Pirens G, El Khayat N (1985) A human placental

GH: increasing levels during second half of pregnancy with pituitary GH suppression as revealed by monoclonal antibody assays. *Int J Fertil* **30**: 27-33

Hirt H, Kimelman J, Birnbaum MJ, Chen EY, Seeburg PH, Eberhardt NL, Barta A (1987) The human growth hormone gene locus: structure, evolution and allelic variation. *DNA* **6**: 59-70

Jackson LL, Colosi P, Talamantes F, Linzer DI (1986) Molecular cloning of mouse placental lactogen cDNA. *Proc Natl Acad Sci USA* **83**: 8496-8500

Kimura M (1980) A simple method for estimating evolutionary rates of base substitution through comparative studies of nucleotide sequences. *J Mol Evol* **16**: 111-120

Kimura M (1981) Estimation of evolutionary distances between homologous nucleotide sequences. *Proc Natl Acad Sci USA* **78**: 454-458

Laron Z, Kelijman M, Pertzelan A, Keret R, Shoffner JM, Parks JS (1985) Human growth hormone gene deletion without antibody formation or growth arrest during treatment - a new disease entity? *Isr J Med Sci* **21**: 999-1006

Linzer DI, Talamantes F (1985) Nucleotide sequence of mouse prolactin and growth hormone mRNAs and expression of the mRNAs during pregnancy. *J Biol Chem* **260**: 9574-9579

Miller WL (1982) Bovine prolactin: corrected cDNA sequence and genetic polymorphism. *DNA*, **2**: 313-314

Miller WL, Eberhardt NL (1983) Structure and evolution of the growth hormone gene family. *Endocr Rev* **4**: 97-130

Miller WL, Coit D, Baxter JD, Martial JA (1981) Cloning of bovine growth hormone cDNA and evolutionary implications of its sequence. *DNA* **1**: 37-50

Nishi Y, Aihara K, Usui T, Phillips III JA, Mallonee RL, Migeon CJ (1984) Isolated growth hormone deficiency type 1A in a Japanese family. *J Pediatr* **104**: 885-889

Owerbach D, Rutter WJ, Martial JA, Baxter JD, Shows TB (1980) Genes for growth hormone, chorionic somatomammotropin, and growth hormone-like gene on chromosome 17 in humans. *Science* **209**: 289-292

Owerbach D, Rutter WJ, Cooke NE, Martial JA, Shows TB (1981) The prolactin gene is located on chromosome 6 in humans. *Science* **212**: 815-816

Parks JS, Herd JE, Wurzel JM, Martial JA (1982) Structural analysis of rodent growth hormone genes: application to genetic forms of hypopituitarism. *Endocrinology* **110**: 1672-1675

Parks JS, Herd JE, Nielsen PV (1983) A new deletion of the hGH-Variant and hCS genes *DNA* **2**: 78 (Abstract)

Parks JS, Nielsen PV, Sexton LA, Jorgensen EH (1985) An effect of gene dosage on production of human chorionic somatomammotropin. *J Clin Endocrinol Metab* **60**: 994-997

Perler D, Efstratiadia A, Lomedico P, Gilbert W, Kolodner R, Dodgson J (1980) The evolution of genes: the chicken proinsulin gene. *Cell* **20**: 555-566

Phillips III JA, Hjelle BL, Seeburg PH, Zachmann M (1981) Molecular basis for familial isolated growth hormone deficiency. *Proc Natl Acad Sci USA* **78**: 6372-6376

Rivarola MA, Phillips III JA, Migeon CJ, Heinrich JJ, Hjelle B (1984) Phenotypic heterogeneity in familial isolated growth hormone deficiency type I-A. *J Clin Endocrinol Metab* **59**: 34-40

Rogers BL, Sobnosky MG, Saunders GF (1986) Transcriptional enhancer within the human placental lactogen and growth hormone multigene cluster. *Nucleic Acids Res* **14**: 7467-7659

Seeburg PH, Shine J, Martial JA, Baxter JD, Goodman HM (1977) Nucleotide sequence and amplification in bacteria of the structural gene for rat growth hormone. *Nature* **270**: 486-494

Seeburg PH (1982) The human growth hormone gene family: nucleotide sequences show recent divergence and predict a new polypeptide hormone. *DNA* **1**: 239-249

Seeburg PH, Sias S, Adelman J, de Boer HA, Hayflick J, Jhurani P, Goeddel DV, Heynecker HL (1983) Efficient bacterial expression of bovine and porcine growth hormones. *DNA* **2**: 37-45

Selby MJ, Barta A, Baxter JD, Bell GI, Eberhardt NL (1984) Analysis of a major human chorionic somatomammotropin gene. Evidence for two functional promoter elements. *J Biol Chem* **259**: 13131-13138

Selvanayagam CS, Tsai SY, Tsai M-J, Selvanayagam P, Saunders GF (1984) Multiple origins of transcription for the human placental lactogen genes. *J Biol Chem* **259**: 14642-14646

Shine J, Seeburg PH, Martial JA, Baxter JD, Goodman HM (1977) Construction and analysis of recombinant DNA for human chorionic somatomammotropin. *Nature* **270**: 494-499

Simon P, Decoster C, Brocas H, Schwers and Vassart G (1986) Absence of human chorionic somatomammmotropin during pregnancy associated with two types of gene deletion. *Hum Genet* **74**: 235-238

Truong AT, Duez C, Belayew A, Renard A, Picter R, Bell GI, Martial JA (1984) Isolation and characterization of the human prolactin gene. *EMBO J* **3**:429-437

Wilson AC, Carlson SS, White TJ (1977) Biochemical evolution. *Ann Rev Biochem* **46**: 573-639

Wohlk P, Nexo E, Jorgensen EH, Chemnitz J, Nielsen PV, Parks JS (1984) Lavt eller manglende S-placentalaktogent hormon i to normalt forlobende graviditeter. *Ugeskrift for Laeger* **146**: 727-730

Woychik RP, Camper SA, Lyons RH, Horowitz S, Goodwin EC, Rottman FM (1982) Cloning and nucleotide sequencing of the bovine growth hormone gene. *Nucleic Acids Res* **10**: 7197-7210

Wurzel JM, Parks JS, Herd JE, Nielsen PV (1982) A gene deletion is responsible for absence of human chorionic somatomammotropin. *DNA* **1**: 251-257

Zihlman AL, Cronin JE, Cramer DL, Sarich VM (1978) Pygmy chimpanzee as a possible prototype for the common ancestor of humans, chimpanzees and gorillas. *Nature* **275**: 744-746

Advances in Growth Hormone and Growth Factor Research,
edited by E.E. Müller, D. Cocchi and V. Locatelli
Pythagora Press, Roma-Milano and Springer Verlag, Berlin-Heidelberg © 1989

Cell-type-specific regulation of growth hormone gene expression

P. TRIPPUTI, S. GUERIN and D. MOORE

Department of Molecular Biology, Massachussets General Hospital and Department of Genetics Harvard Medical School, Boston, Massachusetts, U.S.A.

INTRODUCTION

Growth hormone (GH) is produced specifically in the anterior lobe of the pituitary gland–no other tissue produces a significant amount of GH. This cell-type specificity of gene expression is faithfully reproduced by rat pituitary cells in culture, which can produce more than 10^8 times higher amounts of GH than similarly cultured rat liver cells (Ivarie et al. 1983).

Growth hormone is involved in the regulation of growth and metabolism during development (Daughaday 1985), and GH levels have been shown to be regulated by a variety of effectors, including the hypothalamic factors GH-releasing hormone (GHRH) and somatostatin (SRIF), which exert stimulatory and inhibitory influences respectively; thyroid hormone (triiodothyronine, T3), which increases GH transcription and mRNA stability (Diamond and Goodman 1986; Flug et al. 1987); and glucocorticoids, which increase GH expression by stabilizing mRNA and, at least weakly, increasing transcription (Diamond and Goodman 1985).

Transfection of a variety of chimeric gene constructs into differentiated tissue culture cells originally led to the discovery of various *cis*-acting genetic elements involved in the establishment of cell-type and tissue-specific transcription (Banerji et al. 1983; Gillies et al. 1983; Queen and Stafford 1984; Edlund et al. 1985). Both positively and negatively acting elements have been described for cell-type regulation of rat or human GH expression (Larsen et al. 1986; Bodner and Karin 1987; Lufkin and Bancroft, 1987). At least in the case of rat GH, the cell-type-specific negative elements seem to lie further away from the promoter than the positively acting elements.

Expression of GH in Transfected Cells

To map the sequences required for regulation of cell-type- specific expression of rat growth hormone (rGH), a series of plasmids was constructed containing variable amounts of the rGH promoter and 5' flanking sequences fused to a CAT expression vector (Larsen et al. 1986). These plasmids were transfected into a variety of cells lines including the rGH- secreting pituitary cells lines GC or GH_4C1, the mouse fibroblast cell line LTK⁻, the rat fibroblast cell line XC, and the monkey kidney cell line CV-1. Four days after transfection, CAT activity was measured. To avoid variability in CAT expression due to variations in transfection efficiency, the cells were also cotransfected with a control plasmid which constitutively expressed human growth hormone (hGH), and CAT expression was normalized to amount of hGH expression.

In these experiments, an approximately 1.8-kilobase (kb) fragment containing the rGH promoter and upstream sequences showed strongly pituitary specific expression. The plasmid containing this large rGH fragment, pRGH1753, directed at least 280 times more CAT activity in GC cells then in LTK⁻ fibroblasts, when normalized to the level of hGH directed by equivalent amounts of the internal control plasmid pXGH5 (Larsen et al. 1986).

In contrast, three plasmids containing much shorter rGH promoter fragments, pRGH309, pRGH237, and pRGH183 (containing rGHfragments of 309, 237 and 183 base pairs (bp) respectively), directed significant levels of CAT expression in fibroblast and kidney cells as well as in pituitary cells. The lost of specificity is most striking for pRGH237, which directs over 100 times more CAT expression than pRGH1753 in fibroblasts, and over 700 times in CV-1 cells. The large relative increase in CAT expression directed by the shortest rGH fragments in the LTK⁻ cells suggest that the additional upstream sequences present on the larger fragments act to drastically repress expression from an otherwise functional rGH promoter.

Since the 237-bp fragment in pRGH237 generates the highest levels of CAT in both pituitary and nonpituitary cells, such repressive sequences must lie upstream of the 237. Initial localization of such sequences was provided by analysis of two plasmids containing large deletions, one extending from -1236 to -554, which had virtually no effect on expression in LTK⁻ or CV-1 cells, and another extending from -1236 to -210, which significantly increased CAT expression in the nonpituitary cells. This indicates that at least some repressive elements lie in the interval between -554 and -237, while sequences further upstream (which contain a variety of repeats of simple, short sequences that are present in large numbers elsewhere in the genome) are not involved.

To confirm this localization, three plasmids were constructed in which an rGH *PstI-Bgl*II fragment, extending from - 526 to -237 was inserted upstream of the minimal rGH promoter in pRGH237. pRGH330A⁺ and pRGH 330 A⁺ A⁺ contain either one or two copies of the fragment in normal orientation respectively, and pRGH330A⁻ contains, a single copy in inverted orientation. As indicated in Figure 1, insertion of the single copy in pRGHA⁺ decreased CAT expression 20- to 50-fold compared with pRGH237 in transient transfections of mouse fibroblast L cells

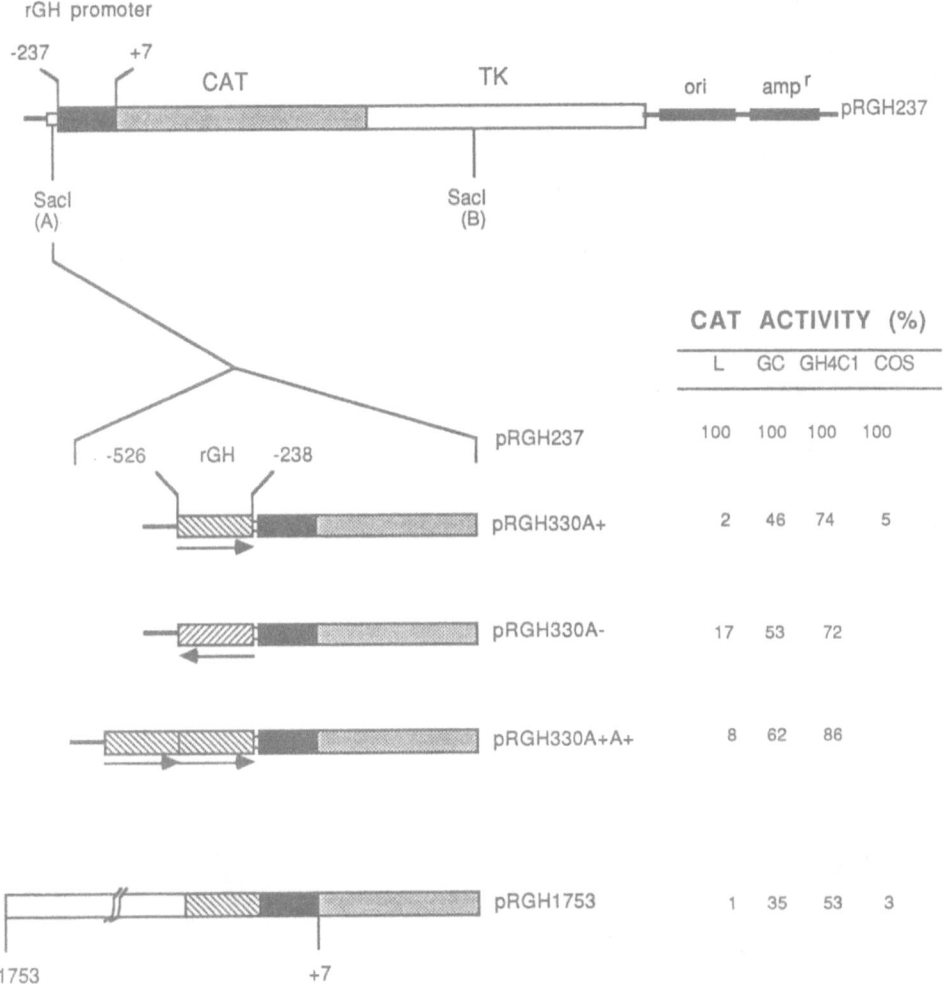

Fig. 1. Orientation analysis of the rGH negative sequence. The rGH PstI/BglII restriction fragment spanning the region from - 238 to -526 and represented by the *hatched boxes* in the figure was inserted into a *Sac*I site immediately upstream from the previously cloned rGH minimal promoter region from the CAT expression vector pRGH237 (Larsen et al. 1986). A single copy of this fragment in each orientation and two copies, both in normal orientation, were inserted into this plasmid. Following transient transfection into pituitary (GC and GH₄) and nonpituitary cell lines (LTK and COS cells) the levels of CAT expression determined by these plasmids were measured and expressed as percent of CAT activity compared with the level obtained from pRGH237, which constitutively expressed CAT activity. pRGH1753, which contains the rGH upstream region from +7 to -1753, was also used as a control. *Arrows* indicate the orientation of the cloned rGH fragment, while the *black boxes* represent the minimal rGH promoter region covering the rGH DNA sequence from -237 to +7.

or monkey kidney-derived cos cells, but only two fold in GC or GH_4 C1 rat pituitary cells. Addition of another copy of the fragment in pRGH 330A$^+$ A$^+$ did not decrease the level of expression further in L cells. Significant but somewhat lower cell type specific repression was also seen with pRGH330A$^-$, showing that the negative regulatory element can function in an orientation- independent fashion.

To define more precisely the minimal sequences sufficient for the silencer effect, a series of plasmid-containing 5' deletions in the -526 to -237 fragment was created using Bal31. As diagrammed in Figure 2, deletions extending to -370 had

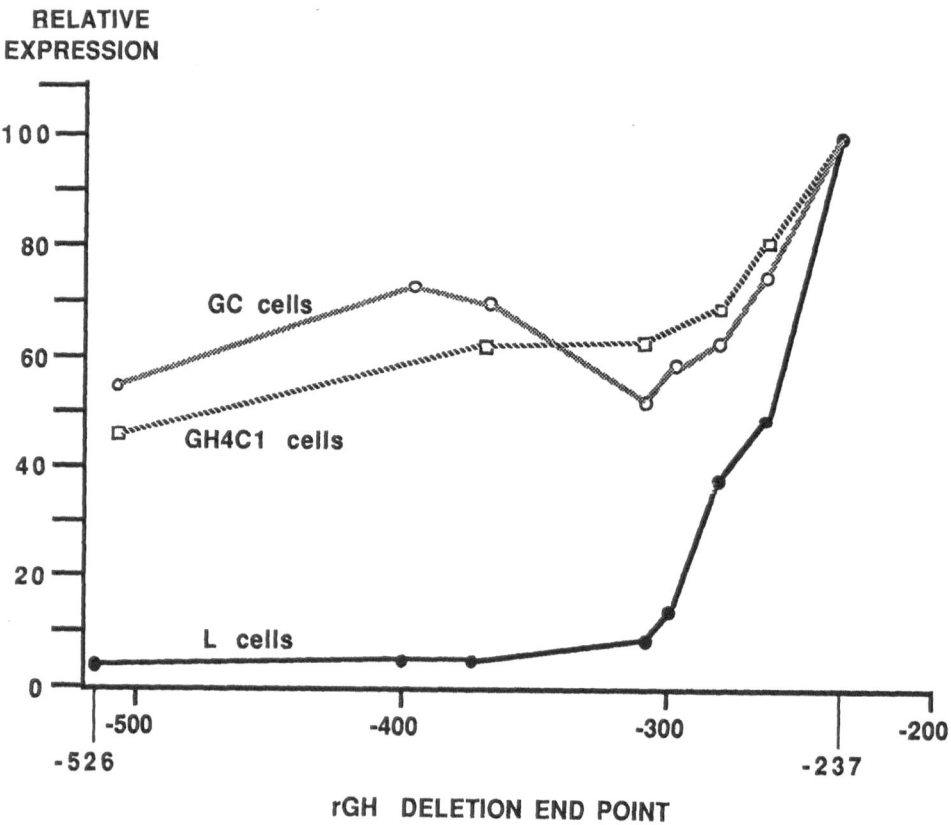

Fig. 2. Bal31 deletion mapping of the rGH silencer. The rGH negative regulatory sequence covering the region from -237 to - 526 was progressively digested from its 5' end using Bal31 exonuclease. Resulting fragments were subcloned into the SacI site (site A, Fig. 3) immediately upstream of the rGH promoter in pRGH237. Plasmids containing insert sizes ranging from -500 to -265 were selected and transiently transfected into pituitary (GC and GH$_4$) and nonpituitary cells (LTK$^-$). Levels of CAT expression were determined, and the average value obtained from each plasmid used was plotted against the position of the corresponding deletion end point on the rGH upstream regulatory region.

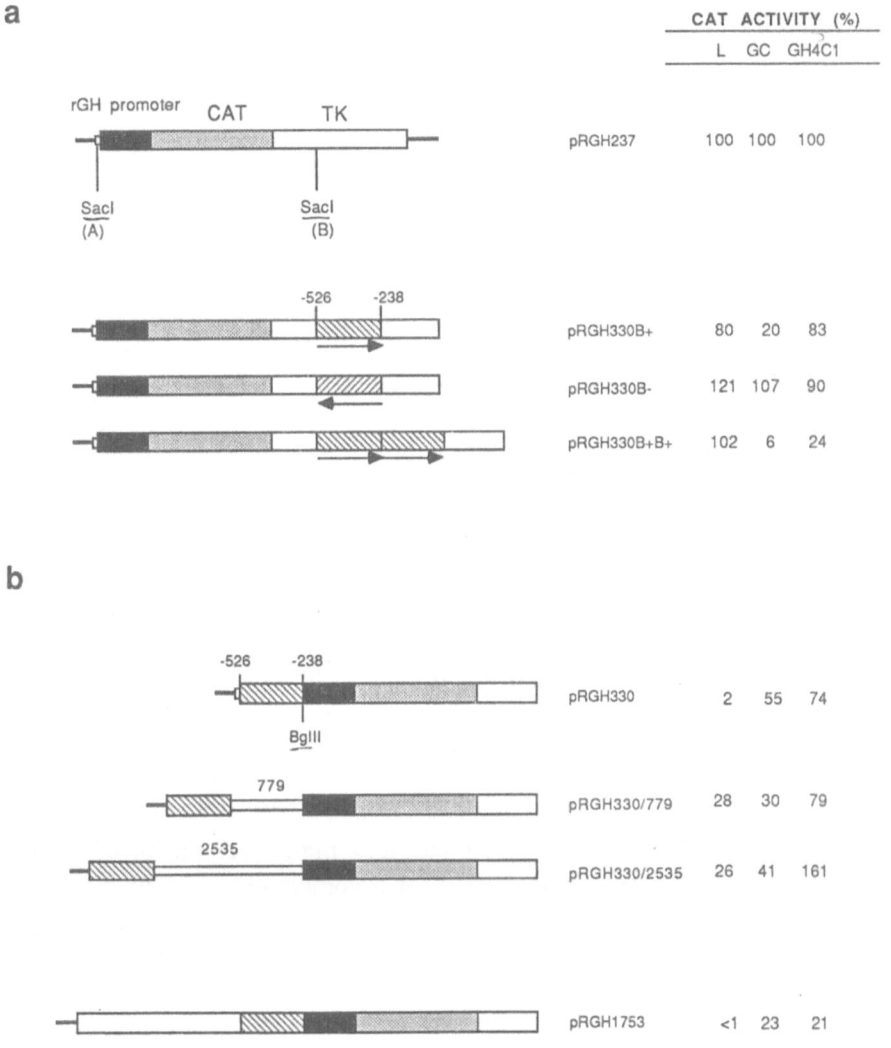

Fig. 3a, b. Position analysis of the rGH negative regulatory element. **a** The rGH PstI/BglII fragment from position -238 to -526 (*hatched box*) was inserted 1.6 kb downstream from the rGH promoter region (*black box*) into the second SacI site (B) of pRGH237. One or two copies of this fragment were inserted in normal or inverted orientation, as shown by the *arrows*. After transient transfection into pituitary and nonpituitary cells, CAT activity was determined and expressed as percent of CAT activity generated by pRH237 (considered to be 100%). **b** XhoII restriction fragments of 779 and 2535 bp were inserted between the rGH promoter and the complete or partially deleted silencer element by ligation into the unique BglII site from either pRGH330 or pRGH75. Following transient transfection into the same cell lines as in **a**, levels of CAT expression were determined and compared with the level from pRGH237. pRGH1753 was also used as control.

little or no effect on cell-type-specific negative regulation. However, deletion to -295 slightly increased CAT expression, and further deletions significantly increased CAT expression in L cells. None of the deletions had a significant effect on expression in pituitary cell lines. These results show that the 5' boundary of a silencer element lies near -295 and map the element to the segment between -320 and -237.

To determine whether the silencer could act in a position independent manner, one or two copies of the -526 to -237 fragment were inserted in either orientation into a *Sac* I site in the 3' untranslated segment of the predicted CAT mRNA, 1.6 kb downstream of the rGH promoter (site B, Fig. 3). No silencer activity was observed in the plasmids containing such downstream insertion (Fig. 3a). The silencer fragment was also separated from the promoter by insertion of 779- or 2535-bp stuffer fragments into the *Bg*II site at -237. Plasmids containing these fragments showed a moderate decrease in CAT expression relative to controls in both pituitary and fibroblast cells but essentially no specific silencer function (Fig. 3b). Thus, the rGH silencer does not function when located downstream or far upstream of the promoter.

This silencer is apparently not the only element involved in cell-type-specific regulation of rGH expression. Other studies have shown that sequences in the first 237 bp of the promoter region are specifically involved in positive regulation of rGH expression in pituitary cells (Nelson et al. 1986; West et al. 1987). In addition, a specific interaction of proteins presumed to be positive transcriptional regulatory factors with such proximal sites has been detected (West et al. 1987; Ye and Samuels, 1987). It seems likely that the very high level of cell-type-specific regulation of rGH expression is a result of a combination of both positive and negative mechanisms, mediated by sites which are close to each other as shown for the human B-interferon (Goodbourn et al. 1986) and rat insulin 1 (Nir et al. 1986) genes.

Expression of GH in Somatic Cell Hybrids

Fusion between cells of different types can lead to the formation of either true hybrids (in which the nuclei of the parental cells fuse to form a single nucleus) or heterokaryons (in which the nuclei of the parental cells exist as separate entities in a common cytoplasm). In true hybrids between different cell types, it has frequently been observed that the expression of genes unique to one particular parental cell type is shut off. This phenomenon is termed "extinction".

A detailed analysis of the factors underlying the phenotypic differences between the hybrids and the parental cell lines can be carried out by mapping the sequences required for extinction to specific chromosomes of the nonexpressing parent. Experiments of this type have suggested that the shutoff of differentiated products is a consequence of expression of specific negative regulatory genes, termed "tissue-specific extinguishers" (tse), by the fibroblast parents (Killary and Fournier 1984).

To determine the relationship, if any, between the phenomenon of extinction and the negative regulation of rGH expression described above, we performed somatic cell hybridization experiments between pituitary and fibroblast cell lines. As a first step along this line, the rat pituitary cell line, GH_4, was exposed to the drugs 8-azaguanine (AG) or 5-bromodeoxyuridine (BUDR) to generate hypoxanthine-guanine- phosphoribosyltransferase-deficient (HGPRT-) or thymidine-kinase- deficient (TK-) cell lines respectively. Cells lacking HGPRT- or TK- can be isolated by their ability to grow in the presence of AG or BUDR, respectively, since the wild-type cells incorporate these nucleotide analogues into the DNA via the salvage pathway and are killed (Littlefield 1964).

Using these resistant pituitary cell lines we made two types of hybrids, one between HGPRT-GH_4 cells and LMTK- mouse fibroblasts and the other between HGPRT-GH_4 cells and FF7 human fibroblasts. Briefly, 10^7 GH_4 cells and 10^7

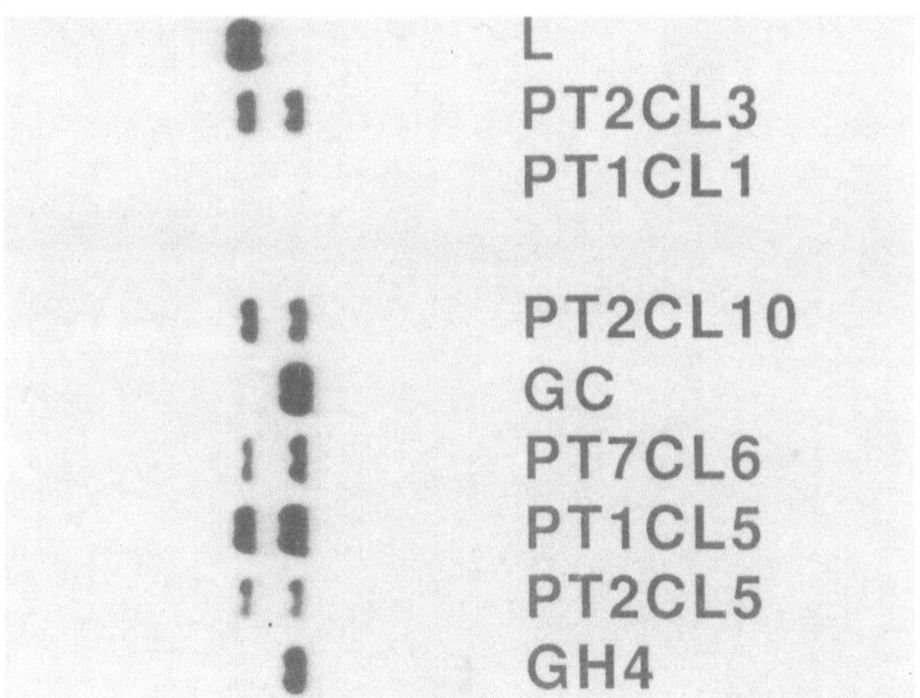

L
PT2CL3
PT1CL1

PT2CL10
GC
PT7CL6
PT1CL5
PT2CL5
GH4

Fig. 4. Southern blot analysis of DNA from GH_4, L, and GH_4 x L hybrid cell lines (PTxCLx, as indicated) probed with rGH cDNA.

fibroblasts were mixed in conical centrifuge tubes and resuspended in serum-free DMEM medium. One milliliter of polyethylene glycol (PEG, MW150) was added to the cell pellet for 1 min. After centrifugation, the PEG was removed and the cells were plated in selective medium containing HAT (hypoxanthine, aminopterin, and thymidine) and, for the FF7 fusion, ouabain. The aminopterin blocks the *de novo* DNA synthesis pathway, killing cells deficient in one of the salvage pathway enzymes since they cannot utilize the exogenous nucleotides. The normal human fibroblasts are 10^4 times more sensitive to the drug ouabain, which inhibits the Na^+-/K^+-activated membrane ATPase, than are rodent cells, so these, too, are killed. Thus, this medium selects directly for hybrids, since only cells which are fused and which complement each other's deficiency can survive.

The first hybrids were detected microscopically 15 - 20 days after fusion. Rat-mouse hybrids were examined for the presence of GH genes from both species and for rGH expression. As shown in Figure 4, a Southern blot analysis of *Hind*III-digested DNA from GH_4, L cells, and GH_4x L hybrids probed with a rat GH cDNA showed hybridizing bands of 5.9 kb and 4.5 kb in all somatic cell hybrids. This

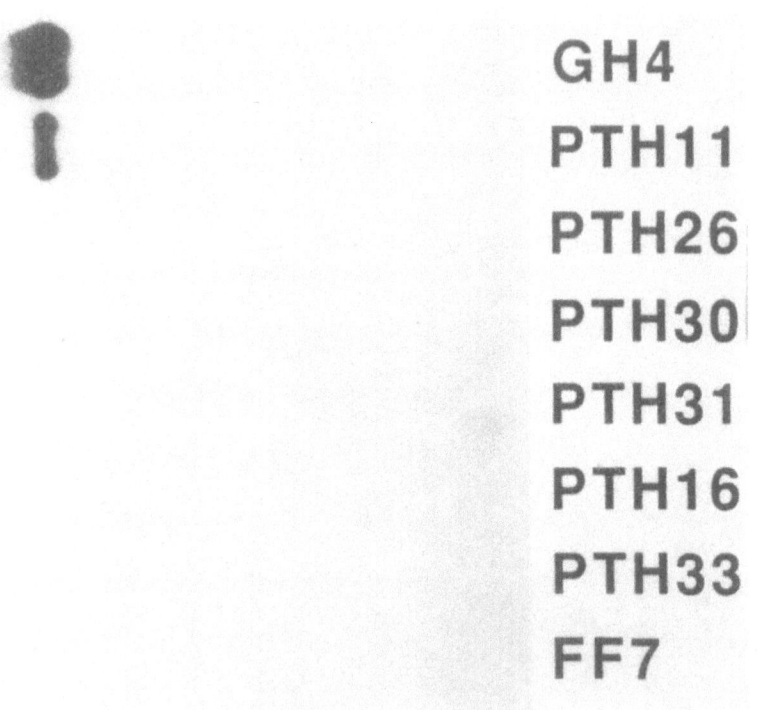

GH4

PTH11

PTH26

PTH30

PTH31

PTH16

PTH33

FF7

Fig. 5. Northern blot analysis of total RNA from GH_4, FF7, and GH_4 x FF7 hybrid cell lines (PTHx, as indicated) probed with rGHcDNA.

shows that both rat and mouse GH genes are present in all of the hybrids. Direct analysis of the mRNA expressed by the hybrids by Northern blotting showed that GH production is completely repressed in all of the hybrids between GH_4 and L cells (not shown). These results confirm and extend previous studies of similar L x GH hybrids.

Hybrids between human FF7 fibroblasts and rat GH_4 cells were also isolated. The human chromosomal content of the hybrids was determined using both Southern blot analysis, with human probes derived from a number of different human chromosomes, and karyotypic analysis, using a combination of trypsin Giemsa and G11 banding techniques (Tripputi et al. 1985).

Extinction of rGH expression is observed in the hybrids with a large human chromosome content. However, as shown in Figure 5, re- expression of rGH is observed in hybrids which have lost a number of human chromosomés. This strongly suggests that one or more of the human chromosomes encodes a negative regulatory factor, or tse, responsible for repressing rGH expression. Studies to date indicate that such a hypothetical GH tissue- specific extinguisher gene or genes must reside on a small subset of human chromosomes. Future studies will map this locus more precisely and determine its mode of action at the molecular level.

REFERENCES

Banerji J, Olson L, Schaffner W (1983) A lymphocyte- specific cellular enhancer is located downstream of the joining region in immunoglobulin heavy-chain genes. *Cell* 33: 729-740

Bodner M, Karin M (1987) A pituitary-specific trans-acting factor can stimulate transcription from the growth hormone promoter in extracts of nonexpressing cells. *Cell* 50: 267-275

Daughaday WH (1985) The anterior pituitary. In: Wilson JD, Foster DW (eds) *Textbook of Endocrinology*. Saunders, Philadelphia, pp 568-613

Diamond DJ, Goodman HM (1985) Regulation of growth hormone messenger RNA synthesis by dexamethasone and triiodothyronine: transcriptional rate and mRNA stability changes in pituitary tumor cells. *J Mol Biol* 181: 41-62

Edlund T, Walker MD, Barr PJ, Rutter WJ (1985) Cell-specific expression of the rat insulin gene: evidence for role of two distinct 5' flanking elements. *Cell* 230: 912-916

Flug F, Copp RP, Casandova J, Horowitz ZD, Janocko L, Plotnik M, Samuels HH (1987) *Cis*-acting elements of the rat growth hormone gene which mediate basal and regulated expression by thyroid hormone. *J Biol Chem* 262: 6373-6382

Gillies SD, Massison SL, Oi VT, Tonegawa S (1983) A tissue- specific transcription enhancer element is located in the major intron of a rearranged immunoglobulin heavy-chain gene. *Cell* 33: 717-728

Goodbourn S, Burnstein H, Maniatis T (1986) The human β–interferon gene enhancer is under negative control. *Cell* 345: 601-610

Ivarie RD, Schachter BS, O'Farrel P (1983) The level of expression of the rat growth hormone gene in liver tumor cells is at least eight orders of magnitude less than that in anterior pituitary cells. *Mol Cell Biol* **3**: 1460-1467

Killary AM, Fornier REK (1984) A genetic analysis of extinction: transdominant loci regulate expression of liver- specific traits in hepatoma hybrid cells. *Cell* **38**: 523-534

Larsen PR, Harney JW, Moore DD (1986) Sequences required for cell-type specific thyroid hormone regulation of rat growth hormone promoter activity. *J Biol Chem* **261**: 14373-14376

Littlefield J (1964) Selection of hybrids from mating of fibroblasts *in vitro* and their presumed recombinants *Science* **145**: 709

Lufkin T, Bancroft C (1987) Identification by cell fusion of gene sequences that interact with positive trans-acting factors. *Science* **237**: 283-286

Nelson C, Crenshaw III EB, Franco R, Lira SA, Albert VR, Evans RM, Rosenfeld MG (1986) Discrete *cis*-active genomic sequences dictate the pituitary cell-type-specific expression of rat prolactin and growth hormone genes. *Nature* **322**: 557-562

Nir Um Walker MD, Rutter WJ (1986) Regulation of rat insulin 1 gene expression: evidence for negative regulation in nonpancreatic cells. *Proc Natl Acad Sci USA* **83**: 3180-3184

Queen C, Stafford J (1984) Fine mapping of an immunoglobulin gene activator. *Mol Cell Biol* **4**: 1042-1049

Tripputi P, Blasi F, Verde P, Cannizzaro LA, Emanuel BS, Croce CH (1985) Human urokinase gene is located on the long arm of chromosome 10. *Proc Natl Acad Sci USA* **82**: 4448-4452

West BL, Catanzaro DF, Mellon SH, Cattini PA, Baxter JD, Reudelhuber TL (1987) Interaction of a tissue-specific factor with an essential rat growth hormone gene promoter element. *Mol Cell Biol* **7**: 1193-1201

Ye ZS, Samuels HH (1987) Cell-and sequence-specific binding of nuclear proteins to 5'-flanking DNA of the rat growth hormone gene. *J Biol Chem* **262**: 6313-6317

Advances in Growth Hormone and Growth Factor Research,
edited by E.E. Müller, D. Cocchi and V. Locatelli
Pythagora Press, Roma-Milano and Springer Verlag, Berlin-Heidelberg © 1989

Characterization of natural–sequence recombinant human growth hormone

W.F. Bennett[1], R. Chloupek[1], R. Harris[1], E. Canova-Davis[1], R. Keck[1], J. Chakel[1], W.S. Hancock[1], P. Gellefors[2], and B. Pavlu[2]

[1] *Genentech, Inc., South San Francisco, California, USA;* and [2] *KabiVitrum AB, Stockholm, Sweden*

INTRODUCTION

Improved understanding of bacterial physiology in recent years has permitted novel approaches to the biosynthesis of recombinant proteins in general, and of human growth hormone (hGH) in particular. Originally, plasmid constructions for hGH synthesis coded for a 192- residue protein consisting of the natural 191 amino acid sequence preceded by an N-terminal methionine residue. This additional amino acid was a necessity of direct protein expression in *Escherichia coli* (Goeddel et al. 1979). The protein thus expressed and purified has been shown to be structurally and functionally very similar to the pituitary- derived hormone, and considerable clinical experience has shown methionyl -hGH (mhGH) to be a safe and effective therapeutic agent (see Fryklund et al. 1986 for a review). However, understanding of the mechanism by which bacteria secrete proteins has now made it feasible to produce a secreted recombinant hGH having the 191-residue "natural" sequence (Gray et al. 1984).

This was accomplished, in this case, by the addition of a "secretion signal" at the N-terminus of the protein. As the protein is synthesized, the signal directs it to the cell membrane, whereupon the protein is secreted into the periplasmic space, and the secretion signal is enzymatically removed (Chang et al. 1986). The mature, 191-residue protein can then be extracted from the periplasm and purified. In this communication, recombinant hGH (rhGH) expressed in this manner has been characterized and compared with pituitary-derived and recombinant methionyl-hGHs. The covalent properties of the proteins were analyzed with respect to protein sequence, tryptic peptide map, and amino acid composition. Conformation-dependent attributes were analyzed with respect to chromatographic behavior, isoelectric point, and ultraviolet absorption.

In the course of developing this secreted product, we discovered a variant form of rhGH in which a proteolytic clip (between amino acids Thr 142 and Tyr 143) had converted the single polypeptide chain into a two-chain molecule. The disulfide bond between cysteines 53 and 165 prevents the chains from separating, and as long as this bond is intact, the protein behaves similarly to hGH. Since this clip is at an unusual proteolytic site, but is in a region of the molecule where other clips have been demonstrated to occur (see Lewis 1984 for review), we have compared some of its properties with those of single-chain rhGH.

MATERIAL AND METHODS

Materials

Recombinant hGH, two-chain rhGH and methionyl hGH (mhGH) were produced by Genentech, Inc. or KabiVitrum AB. Pituitary-derived hGH (pit-hGH) was produced by KabiVitrum AB (Crescormon) and Serono (Asellacrin). Amino acid standards, constant boiling HCl, and trifluoroacetic acid (TFA) came from the Pierce Chemical Co. Other organic solvents were from Burdick and Jackson.

Equipment

Buffer exchange was carried out using PD-10 columns (Pharmacia). A Savant Speed-Vac was used to evaporate samples. High Performance Liquid Chromatography (HPLC) analysis was done either with Waters 510 pumps and a 440 absorbance monitor with an extended wavelength module, or with a Hewlett-Packard 1090 liquid chromatograph with a diode array detector.

Amino Acid Analysis

Samples of hGH were exchanged into 5% acetic acid, and protein content was estimated by absorbance at 280 nm. Aliquots (0.13 nmol) were dried, and 150 μl 6M HCl was added to the samples. Evacuated vials were incubated at 110°C for 24, 48, or 72 h. HCl was removed under vacuum, and the hydrolysates were reconstituted with 0.2 N sodium citrate buffer, pH 2.2. The samples were analyzed on a Beckman 6300, using post-column Ninhydrin detection. External standard mixtures containing 2 nmol of each amino acid were used as the basis for quantitative analysis of the hGH samples.

Slight modifications of the above procedures were employed for analysis of cysteine and tryptophan. Cysteines were oxidized to cysteic acid using performic acid as previously described (Glazer et al. 1975) in some samples. Quantitation of tryptophan was achieved by addition of thioglycolic acid to the samples during 24-h acid hydrolysis (Matsubara and Sasaki 1969).

Carboxy-Terminal Sequencing

Recombinant hGH was exchanged into 0.2 M N-ethylmorpholine acetate, pH 8.5, containing 0.1% sodium dodecyl sulfate (SDS). Carboxypeptidase A was added such that the final concentrations of hGH and carboxypeptidase A were 0.5 mg/ml and 0.05 mg/ml respectively. Digestion was carried out at ambient temperature and was halted by the addition of four volumes of 2 M acetic acid at 100°C, with incubation at that temperature for 10 min. After centrifugation, samples were evaporated to dryness and prepared for amino acid analysis by reconstitution with 0.2 M sodium citrate, pH 2.2

Amino-Terminal Sequencing

About 20 nmol rhGH was equilibrated with 0.1 M ammonium bicarbonate, pH 8.3, and applied to the Beckman 890 C spinning cup sequencer. Forty cycles of Edman degradation (0.1 M Quadrol program) were run and analyzed as previously described (Rodriguez et al. 1984).

Tryptic Digestion

Samples of hGH were exchanged into 100 mM sodium acetate, 10 mM Tris and 1 mM CaCl$_2$, pH 8.3, Trypsin was added to give 100:1 ratio by weight of substrate to trypsin, and the samples were incubated at 37°C. After 2 h, a second aliquot of trypsin (equal to the first) was added. Digestion was stopped 2 h later by the addition of phosphoric acid (to 10% by volume), and the samples were stored at 2°-8°C until analyzed by hPLC or mass spectrometry.

In the cases where reduction and carboxymethylation were carried out prior to trypsin digestion, aliquots of rhGH were treated with 10 mM dithiothreitol in 0.1 M ammonium bicarbonate for 4 h, followed by alkylation with 25 mM iodoacetic acid for 30 min. The samples were then exchanged into digestion buffer and treated with trypsin as described above.

Tryptic Mapping

All samples were analyzed on a 4.46 mm x 15 cm Nucleosil C-18, 5-μm particle size column packed by Alltech Associates. Two procedures were employed, depending on the aqueous mobile phase. In the TFA procedure, the aqueous mobile phase was 0.1% TFA, the column temperature was 40°C, and the flow rate was 1 ml/min. Elution was performed with a linear gradient between the aqueous phase and acetonitrile containing 0.08% TFA, at a rate of 0.5% acetonitrile per min for 120 min, followed by an isocratic hold for 10 min at 60% acetonitrile.

In the sodium phosphate system, the aqueous mobile phase consisted of 50 mM sodium phosphate, pH 2.85. The flow rate and column temperature were the same as in the TFA procedure. Elution was carried out by a linear gradient of acetonitrile of 0.3% per min for 120 min.

Mass Spectrometry

Tryptic peptide peaks were collected manually using the TFA mobile phase procedure. Volatile salt was removed by repeated lyophilization, and the samples were reconstituted in a minimum volume of 0.1% acetic acid. This material was added to glycerol in the probe of a JEOL HX11OHF tandem mass spectrometer with a DA 5000 data system. Data were acquired over a mass range of 100 – 4300 daltons.

Isoelectric Focusing

Isoelectric focusing (IEF) analysis was carried out using preformed Ampholine PAG plates (pH 4-6.5, LKB) prefocused at 25 W until a reading of 850 V was attained. Samples of hGH (20 μg) were applied on sample wicks approximately 1 cm from the anode. Focusing was carried out at 1200 V (max 25 W) and was stopped after 180 min. Gels were fixed in 11.5% trichloroacetic acid/3.5% sul-fosalicylic acid for 1 h at 4°C and then washed in destain solution (10% methanol, 10% HAc) for 5 min at room temperature. Coomassie staining was carried out in an aqueous solution of Coomassie blue R250 (0.04%), cupric sulfate (0.5%), ethanol (27%), and acetic acid (10%). For silver staining, the procedure of Morrissey (1981) was used.

SDS Polyacrylamide Gel Electrophoresis

SDS polyacrylamide gel electrophoresis (SDS-PAGE) analysis was carried out essentially as described by Laemmli (1970). Staining was performed either with Coomassie Brilliant Blue or with silver staining reagents (Oakley et al. 1980; Morrissey 1981).

Hydropobic Interaction Chromatography

The column used was a 0.75 x 7.5 cm TSK-phenyl 5 PW (10-μm particles, 1000-Å pores) with a flow rate of 0.5 ml/min at 30°C. Initial equilibration of the column was with buffer A: 500 mM sodium sulfate, 30mM Tris-HCL, 2% (v/v) acetonitrile, pH8.0. hGH was loaded as a 0.5-mg/ml solution in buffer A; the injection volume was 0.15 ml. Elution with solution B [30 mM Tris-HCl, 5% (v/v) acetonitrile, 0.075% (v/v) Brij-35, pH 8.0] was carried out by means of the following gradient: a linear increase of 1.8% B/min for 10 min, followed by an isocratic hold for 4 min at 18% B; a linear increase of 5.125% B/min for 16 min, followed by an isocratic hold at 100% B for 5 min.

Ultraviolet Absorption Spectrophotometry

Samples of hGH were exchanged into 0.2 M Tris-HCL, pH 8.1, and were adjusted to an A_{278} of approximately 1.0. A Perkin-Elmer model 552 spectro-photometer, with the temperature controlled at 25°C, was used. Absorbance measurements were recorded in 0.1-nm increments using a 1-nm spectral band-width. Three individual spectra for each sample were averaged; the average spectra were corrected for light scattering using a least square extrapolation of a log/log plot of absorbance vs wavelength. The 400 points between 360 and 320 nm were used for this correction. The corrected spectra were smoothed and the second derivative spectra were calculated using a nine-point cubic fit algorithm following Steiner et al.'s (1972) modification of the Savitsky and Golay (1964) procedure.

RESULTS AND DISCUSSION

Amino Acid Analysis

Recombinant hGH was analyzed for amino acid composition after acid hydrolysis. The data from time-course analysis, combined with separate determi-nations for cysteine and tryptophan (as described in Materials and Methods) are shown in Table 1. The residues were calculated based on complete recovery of alanine and leucine. These results are in excellent agreement with published compositions for pit-hGH (Li 1975; Fryklund et al. 1986), and with the theoretical composition predicted from the cDNA sequence (Goeddel et al. 1979). While the amino acid composition is of relatively minor utility in a structural analysis of a 20000 dalton protein, it does provide a straightforward, albeit somewhat insensitive method to detect pre- or posttranslational modifications in the protein. According to the data in Table 1, no such modifications are indicated for *E. coli*-derived rhGH.

Table 1. Composition of rhGH: residues per mole

Amino acid	24 h	48 h	72 h	Oxidized[a]	Trp[b]	Observed	Theoretical value[c]
Asx	19.9	–	–	–	–	19.9	20
Thr	9.97	9.83	9.45	–	–	10.3[d]	10
Ser	15.88	15.12	13.85	–	–	17.0[d]	18
Glx	27.1	–	–	–	–	27.1	27
Pro	–	–	–	7.8	–	7.8	8
Gly	8.0	–	–	–	8.0	8.0	8
Ala	7.0	7.0	7.0	7.1	7.0	7.0[e]	7
1/2-Cys	–	–	–	4.0	–	4.0	4
Val	6.8	7.0	7.0	–	–	7.0	7
Met	2.7	–	–	–	2.8	2.7	3
Ile	7.4	7.5	7.5	–	7.3	7.5	8
Leu	26.1	26.1	26.0	25.7	26.1	26.1[e]	26
Tyr	7.7	–	–	–	8.0	7.7	8
Phe	13.0	–	–	–	12.8	13.0	13
His	3.1	–	–	–	3.0	3.1	3
Lys	9.0	–	–	–	9.1	9.0	9
Trp	–	–	–	–	0.8	0.8	1
Arg	11.8	–	–	–	–	11.8	11

[a] Performic acid oxidation.
[b] Thioglycolate addition.
[c] Based on cDNA-derived composition.
[d] Determined by extrapolation to zero time of hydrolysis.
[e] Residues were calculated based on the complete recovery of alanine and leucine. The average of these two residues was used to calculate the recoveries of all the amino acids.

Carboxy-Terminal Sequencing

The predicted C-terminus of rhGH is Cys-Gly-Phe (Bewley and Li 1975). The predominant amino acid released by carboxypeptidase A digestion of rhGH was phenylalanine (Fig. 1). This is consistent with phenylalanine being the C-terminal residue (position 191). A slow rise in the level of glycine is consistent with the known substrate specificity of the enzyme (Ambler 1967) and with the presence of a glycine residue in position 190. The detection of the cysteine residue in position 189 would not be expected, even if it were cleaved, because a free amino acid would not be released due to the presence of the disulfide linkage. No other residues were detected in this experiment, suggesting that secondary "clips" (which would be expected to generate C- termini) are not present in detectable quantities.

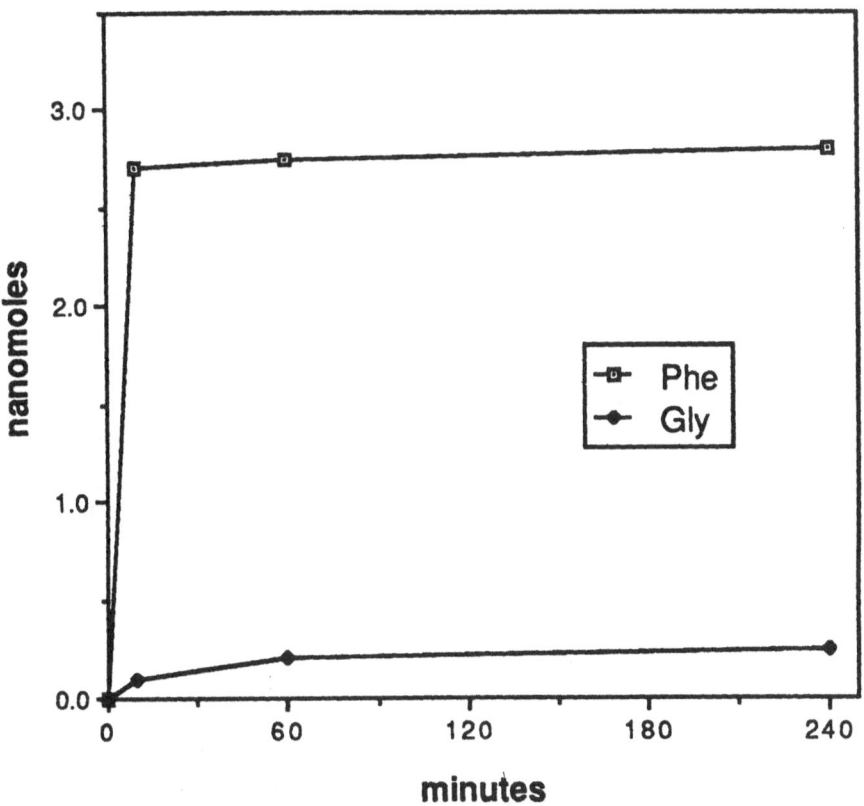

Fig. 1. Carboxypeptidase digestion of rhGH. Digestion of 3 nmol rhGH and analysis of the products were carried out as described in Materials and Methods.

Amino-Terminal Sequencing

Edman degradation was used to confirm the identity and integrity of the N-terminal sequence of rhGH. A number of lots of rhGH were subjected to ten cycles of automated Edman degradation to verify the N-terminal sequence (F-P-T-I-P-L-S-R-L-F), which corresponds to the pit-hGH sequence (Li 1975). Forty cycles of Edman degradation were also performed with 18 nmol (0.4 mg) of protein. The sequence data confirmed the predicted cDNA sequence (F-P-T-I-P-L-S-R-L-F-D-N-A-M-L-R-A-H-R-L-H-Q-L-A-F-D-T-Y-Q-E-F-E-E-A-Y-I-P-K-E-Q). These data indicate no significant additional sequences present due to proteolytic processing at the N- terminus, at internal cleavage sites, or due to contaminating proteins.

Tryptic Mapping

Table 2 indicates that trypsin should cleave at about 20 sites on hGH. Figure 2 show the elution profile for the tryptic digestion products of 200μg rhGH, using a 0.1% TFA mobile phase. About 20 major peaks are resolved. The identities of the peaks were determined by compositional analysis (Table 3), and in cases where mixtures produced ambiguities the identities were confirmed by sequencing and fast atom bombardment mass spectrometry (Table 4). The peptides were assigned T numbers based on the positions of the theoretical trypsin cleavage sites (Table 2), beginning with T1 at the N-terminus, and ending with T21. All the predicted trypsin cleavage products were found with the following exceptions: T5, a hydrophilic tripeptide EQK, was not found, It is probably not sufficiently hydrophobic to be retained by a C18 reversed-phase column. Likewise, T17, a single lysine residue, would not be expected to be retarded under these conditions. The position of T17 in the molecule, however, is confirmed as a result of one of several incomplete trypsin cleavages. A peptide encompassing T17-T19, a result of two

Table 2. rhGH tryptic peptides

	From	To	Sequence
T1	1	8	FPTIPLSR
T2	9	16	LFDNAMLR
T3	17	19	AHR
T4	20	38	LHQLAFDTYQEFEEAYIPK
T5	39	41	EQK
T6	42	64	YSFLQNPQTSLCFSESIPTPSNR
T7	65	70	EETQQK
T8	71	77	SNLELLR
T9	78	94	ISLLLIQSWLEPVQFLR
T10	95	115	SVFANSLVYGASDSNVYDLLK
T11	116	127	DLEEGIQTLMGR
T12	128	134	LEDGSPR
T13	135	140	TGQIFK
T14	141	145	QTYSK
T15	146	158	FDTNSHNDDALLK
T16	159	167	NYGLLYCFR
T17	168	168	K
T18	169	172	DMDK
T19	173	178	VETFLR
T20	179	183	IVQCR
T21	184	190	SVEGSCGF

Table 3. Amino acid analysis of rHGH tryptic peptides

	T1	T1c	T2[a]	T3	T4
CYA					
ASX	0.1		1.9(2)		1.0(1)
THR	1.0(1)	1.0(1)			1.0(1)
SER	0.9(1)				
GLX		0.1		0.1	5.0(5)
PRO	2.0(2)	2.0(2)			1.0(1)
GLY		0.1			
ALA			1.0(1)	1.0(1)	2.0(2)
CYS					
VAL					
MET			0.9(1)		
ILE	1.0(1)	1.0(1)			1.0(1)
LEU	1.1(1)	1.0(1)	2.0(2)		2.1(2)
TYR					1.8(2)
PHE	1.0(1)	1.0(1)	0.9(1)		2.0(2)
HIS				1.0(1)	1.0(1)
LYS				0.1	1.0(1)
ARG	1.1(1)		1.1(1)	1.0(1)	
Recovery	6 nmol	2 nmol	8 nmol	3 nmol	8 nmol

	T6-T16	T6-T16c	T7	T8	T9	T10
CYA		0.7	0.1			
ASX	3.0(3)	2.3(2)		1.0(1)	0.1	4.1(4)
THR	2.1(2)	2.6(2)	1.1(1)			
SER	4.4(5)	4.9(5)		0.8(1)	1.8(2)	3.8(4)
GLX	3.1(3)	3.1(3)	4.0(4)	1.0(1)	3.0(3)	
PRO	3.1(3)	3.2(3)			1.0(1)	
GLY	1.0(1)	1.4(1)	0.1			1.2(1)
ALA		0.1	0.1			2.0(2)
CYS	1.8(2)	CYA(2)				
VAL					1.0(1)	2.9(3)
MET						
ILE	1.0(1)	1.0(1)			1.9(2)	
LEU	4.1(4)	4.2(4)	0.1	3.0(3)	4.9(5)	3.1(3)
TYR	2.8(3)	2.2(3)				1.8(2)
PHE	3.0(3)	3.4(3)			1.0(1)	1.0(1)
HIS			0.1			
LYS			1.0(1)			1.0(1)
ARG	2.1(2)	1.3(1)		1.0(1)	1.0(1)	
Recovery	4 nmol	0.1 nmol	8 nmol	8 nmol	4 nmol	4 nmol

Table3 (continued)

	T10c1	T10c2	T11	T12	T13	T14
CYA						
ASX	1.0(1)	2.8(3)	1.0(1)	1.0(1)		
THR			1.1(1)		1.1(1)	1.0(1)
SER	0.9(1)	2.8(3)		0.9(1)		0.9(1)
GLX			3.2(3)	1.0(1)	1.0(1)	1.0(1)
PRO				1.0(1)		
GLY		0.8(1)	2.0(2)	3.4(1)	1.0(1)	
ALA	1.0(1)	1.0(1)				
CYS						
VAL	1.1(1)					
MET			0.9(1)			
ILE			1.0(1)		0.9(1)	
LEU		2.9(3)	2.0(2)	1.0(1)		
TYR		1.8(2)				1.0(1)
PHE	1.0(1)				1.0(1)	
HIS						
LYS		1.0(1)			1.0(1)	1.0(1)
ARG			1.1(1)	1.1(1)		
Recovery	3 nmol	3 nmol	8 nmol	9 nmol	9 nmol	6 nmol

	T14a	T14c	T15	T17-T18-T19	T18-T19*	T19	T20-T21
CYA	0.2					0.1	
ASX	0.3		4.9(5)	1.9(2)	2.0(2)	0.2	
THR	1.1(1)	1.0(1)	1.0(1)	1.0(1)	1.0(1)	1.0(1)	
SER	0.9(1)	0.1	0.9(1)	0.1		0.1	1.9(2)
GLX	1.0(1)	1.1(1)	0.1	1.0(1)	1.0(1)	1.0(1)	2.2(2)
PRO			0.1				
GLY				0.1		0.1	2.2(2)
ALA			1.0(1)	0.1		0.2	
CYS			0.1				2.0(2)
VAL			0.1	1.0(1)	1.0(1)	1.0(1)	1.6(2)
MET				0.8(1)	1.0(1)		
ILE							0.5(1)
LEU			2.0(2)	1.0(1)	1.0(1)	1.0(1)	
TYR	1.0(1)	0.9(1)					
PHE	0.2		1.0(1)	0.9(1)	1.0(1)	1.0(1)	1.1(1)
HIS			1.0(1)				
LYS	0.9(1)		1.0(1)	1.9(2)	1.0(1)	0.1	
ARG				1.0(1)	1.0(1)	1.0(1)	1.1(1)
Recovery	1 nmol	2 nmol	8 nmol	2 nmol	6 nmol	1 nmol	6 nmol

10 nmol load; tryptics identified with a are pyroglutamic-acid-containing; those with c are nontryptic cleavage.
* Coelutions reported separately.

Table 4. Mass spectral analysis of tryptic peptides of rhGH

rhGh peptides	R.T. mins[a]	Theoretical mass[b]	Observed
T1	57.8	930.55	930
T2	55.8	979.50	979
T3	Void	383.21	Not analyzed
T4	70.0	2342.14	2343
T5	Void	404.22	Not analyzed
T6-T16	74.6	3761.77	3761
T7	13.5	762.36	762
T8	51.6	844.49	844
T9	93.8	2055.20	2056
T10	72.6	2262.13	2262
T11	65.9	1361.67	1361
T12	29.2	773.38	773
T13	40.5	693.39	693
T14	24.9	626.32	626
T15	46.7	1489.69	1489
T17	Void	147.11	Not analyzed
T18-T19	55.8	1253.62	1253
T20-T21	46.0	1400.65	1400

[a] See Fig. 2
[b] Monoisotopic

incomplete cleavages, is shown in Fig. 2 at about 55 min. Sequencing and mass spectral analysis shows this peptide to be KDMDKVETFLR (Table 5).

Other peptides detectable under these conditions can be shown to result from atypical cleavages by trypsin. For example, the peptide T6-T16c is formed as a result of the removal of the two C-terminal amino acids from peptide T6, wile retaining the disulfide bond with an intact peptide T16. This structure has been confirmed by compositional analysis. The site of this cleavage, between Ser_{62} and Asn_{63}, is not typical for trypsin; however, our analysis of this cleavage using highly purified trypsin has led us to conclude that it is indeed trypsin, and not a contaminating protease, that produces this unusual clip (C. du Mee, K. Mulhollad, W. Bennett, unpublished results). Other atypical cleavages on peptides T1, T10, and T14 are shown in Table 5. Each of these, interestingly, involves a serine residue at one of the positions around the cleavage. The peptide T14a is apparently the result of cyclization of the N-terminal glutamine of peptide T14 to pyroglutamic acid. A blocked N-terminus was observed in sequence analysis, and the mass (Table 5) is consistent with the loss of 17 mass units that one would expect from such a cyclic product.

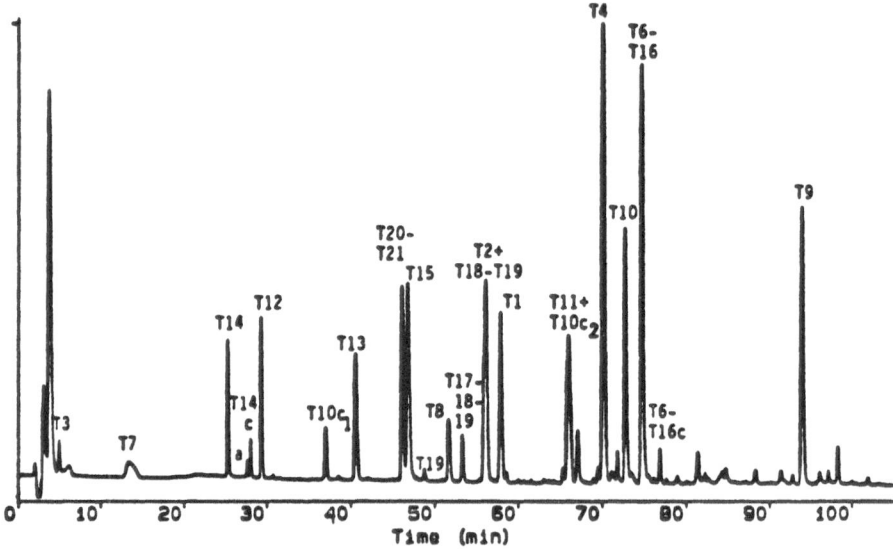

Fig 2. Tryptic mapping of rhGH. Ten nmol rhGH were digested and analyzed using the TFA mobile phase. Peaks were collected manually and identified using a combination of amino acid analysis, N-terminal sequencing, and mass spectral analysis. This figure shows the 220-nm trace of the reversed-phase elution.

Table 5. Mass spectral analysis of peptides produced from rhGH by nontryptic cleavages

rhGH peptides[a]	From	To	Sequence	Theoretical masses[b]	Observed
T1c	1	6	FPTIPL	687.42	687
T10c$_1$	95	99	SVFAN	537.27	537
T10c$_2$	100	115	SLVYGASDSNVDLLK	1743.90	1743
T14a	141	145	(pQ)TYSK	609.29	609
T14c	141	143	QTY	411.19	411
T17-18-19	168	178	KDMDKVETFLR	1381.71	1381
T6-T1bc	42	62	YSFLQNPQTSLCFSESIPTPS	3491.63	3491
	and				
	159	167	NYGLLYCFR		

(pQ), pyroglutamic acid.
[a] The following notation was used: a for pyroglutamic-acid-containing and c for nontryptic-like cleavage.
[b] Monoisotopic.

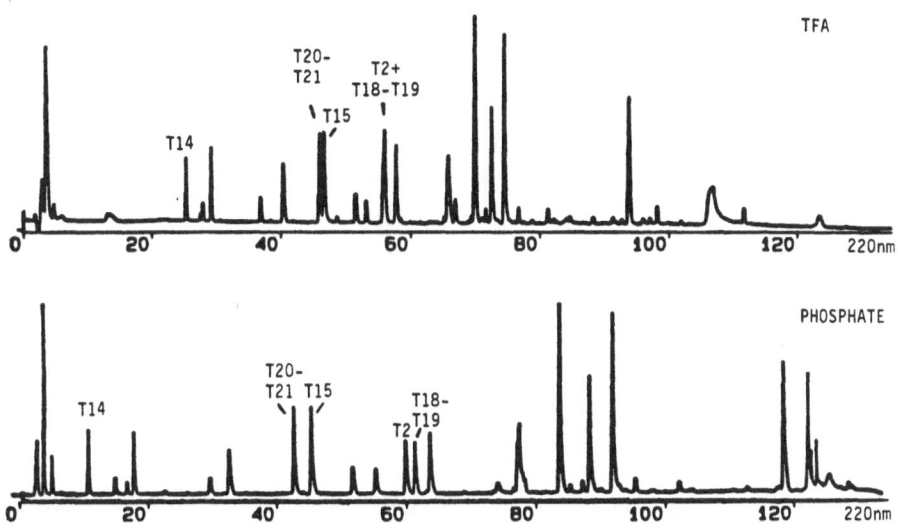

Fig. 3. Comparison of the TFA- and phosphate-containing mobile phases for resolution of tryptic fragments of rhGH. Absorbance traces at 220 nm are shown.

The TFA mobile phase, while its advantages for preparative work are substantial, does not provide optimal resolution of hGH tryptic peptides. For this reason, a sodium phosphate mobile phase was developed. Figure 3 shows a comparison between the TFA and sodium phosphate mobile phases. Major improvements in resolution are apparent, especially with respect to peptides T20- T21 and T15 and peptides T2 and T18-T19. The sodium phosphate mobile phase "spreads out" the peptides and improves the overall resolution. This mobile phase was used for the analysis of disulfide bonds.

Cysteine-containing peptides were characterized by comparing reversed-phase behavior in the sodium phosphate system, under either reduced or nonreduced conditions. Figure 4 shows the peptide profiles for dithiothreitol-reduced, reduced and carboxymethylated, and nonreduced rhGH peptides. The major differences among the profiles are in the peptides of the large disulfide loop, T6-T16, and of the small, carboxymethylated disulfide loop, T20-T21. Reduced, carboxymethylated peptides of T6-T16 appear in unoccupied regions of the map, and a comparison of the nonreduced with the reduced samples suggests that very little free sulfhydryl is present in the nonreduced protein.

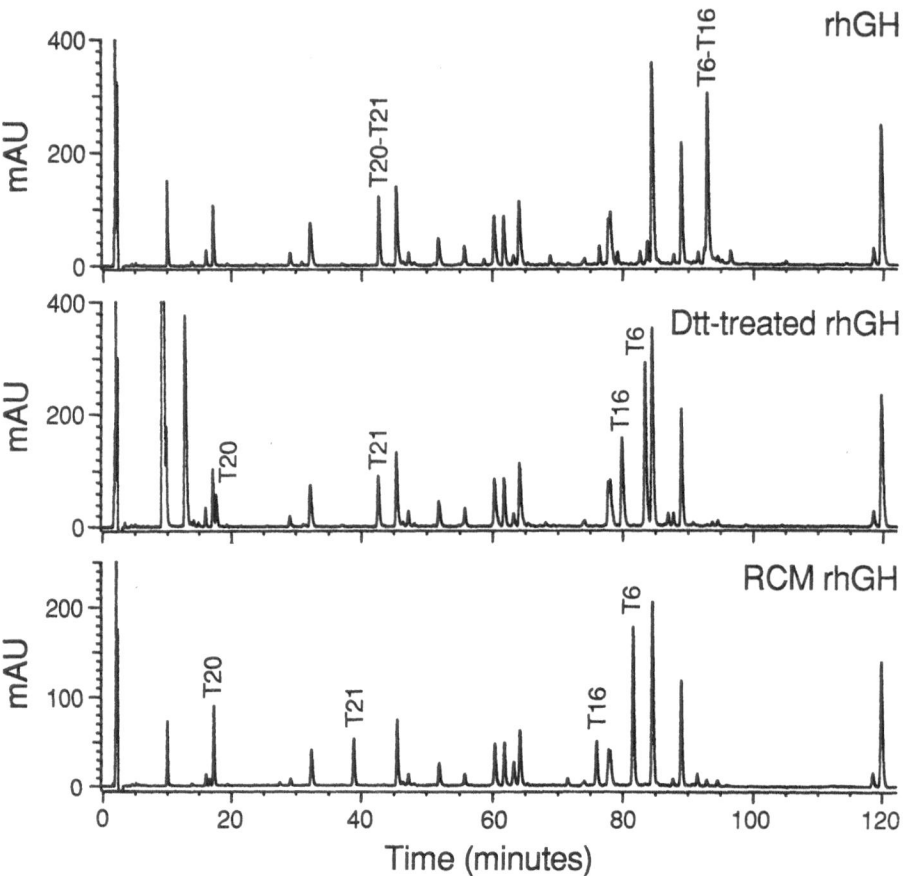

Fig. 4. Tryptic mapping of nonreduced, reduced (Dtt-treated), and reduced carboxymeth-ylated (RCM) rhGH. Reduction with carboxymethylation was carried out on the intact protein before digestion. Dithiothreitol (Dtt) treatment was carried out on trypsin-digested, nonreduced protein 1 h prior to HPLC analysis, using the sodium phosphate mobile phase. The positions of the cysteine-containing peptides are shown in the three profiles. Absorbance traces at 220 nm are shown.

SDS-Polyacrylamide Gel Electrophoresis

Polyacrylamide gel electrophoresis in SDS is a powerful tool for estimating the molecular mass and relative purity of a protein, but it is not usually thought of as being a means of conformational analysis. In the case of growth hormone, it is useful for all these functions, since the mobility on SDS gels is known to be sensitive not only to molecular mass, but to the redox state of the disulfide bonds (the reduced molecule behaves as if its mass were about 2000 daltons greater than that of the nonreduced form), and the presence or absence of clips in the 135-145

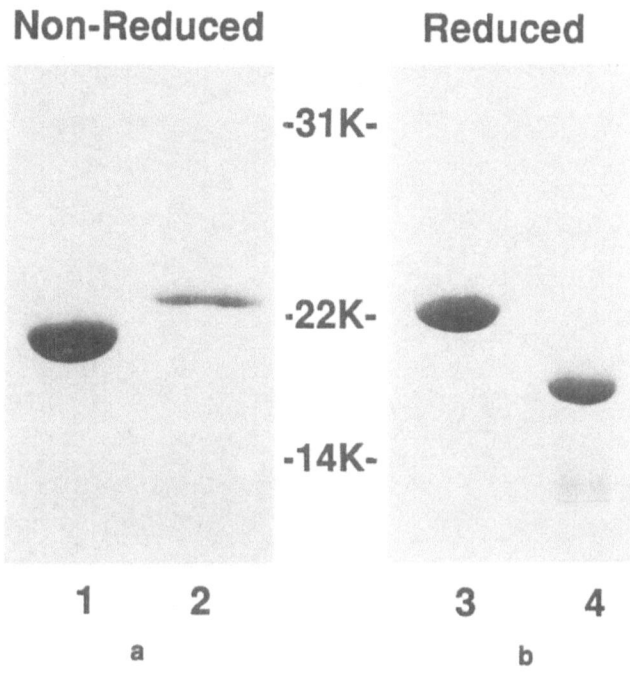

Fig. 5. a, b. SDS-PAGE of single-chain and two-chain forms of rhGH. Reduction was carried out by addition of Dtt to a final concentration of 10 nM. Samples were heated in 1% SDS, with or without Dtt, at 75°C for 10 min before being loaded onto the gel. *Lanes* 1, 3, rhGH; *lanes* 2, 4, two-chain rhGH.

region of the protein (clipped forms generally migrating as if their mass were about 2000 daltons greater). Figure 5a shows a nonreduced gel that illustrates this effect.

Comparison with Fig. 5b (reducing conditions) shows that the reduced one-chain protein has an apparent mass of about 20000 daltons, compared with an apparent mass of about 18000 daltons for the nonreduced protein. The portion of the protein from positions 1 to 142 can be seen in the 14000 dalton region of the reduced gel. Under these conditions, the remaining portion is at the dye front.

Analysis using SDS-PAGE of four preparations of hGH (Fig. 6: rhGH, mhGH, two preparations of pit-hGH) shows identical behavior among these samples. This indicates that the protein is properly folded and that the disulfide bonds are correctly formed in the majority of the molecules. Silver staining of the gels (not shown) indicates slightly different impurity patterns among the preparations, but supports the interpretation that these preparations are rather homogeneous with respect to covalent structure.

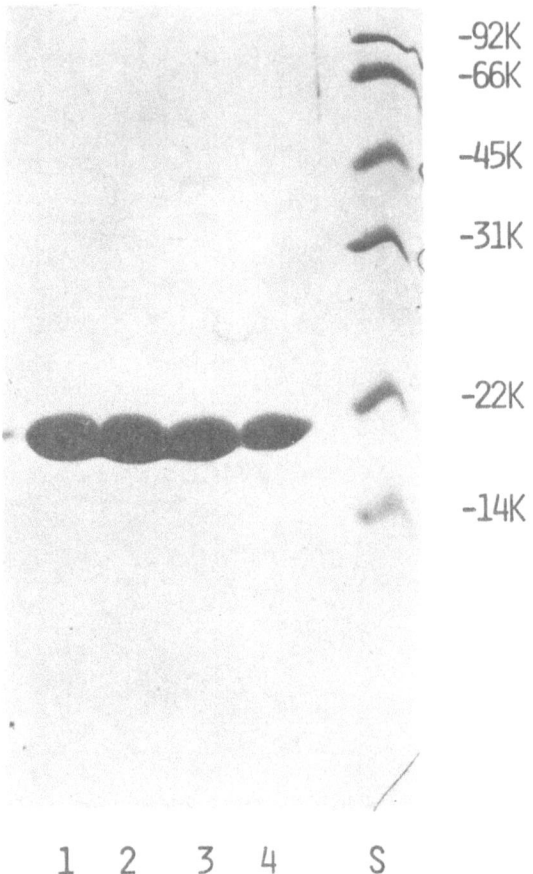

Fig. 6. Nonreduced SDS-PAGE of four different preparation types of hGH. Lyophilized samples were reconstituted with water to a protein concentration of 1 mg/ml, and SDS sample buffer was added. Each lane represents 20 µg protein. *Lane* 1, rhGH; *lane* 2, mhGH (Protropin); *lane* 3, pit-hGH (Crescormon).; *lane* 4, pit-hGH (Asellacrin).

Isoelectric Focusing

The isoelectric point of a protein is not necessarily a conformation-dependent property, but in the case of hGH, there is a strong suggestion of conformational influences on the apparent isoelectric point. Figure 7 is an IEF gel showing the pit-hGH, rhGH, and two-chain rhGH. All the full-length molecules have the same apparent isoelectric point (approximately 5.0). The two- chain molecule, in contrast, exhibits an apparent isoelectric point about 0.2 pH units higher. There is no evidence that any charge modification has occurred, and it is therefore likely that the conformational change brought about by the one-chain to two-chain conversion influences the apparent isoelectric point of the protein. Comparison of the other lanes reveals that rhGH, pit-hGH, and mhGH have identical isoelectric points. This suggests that the overall conformations of the proteins are very similar. The fact mhGH and the natural sequence hormones have identical isoelectric points suggests that the N-terminus of the protein is not a major participant in either the charge properties or the conformation of the protein; if it were, as it apparently is with interleukin-2 (Kato et al. 1985), a discernible difference in the isoelectric point would be observed for mhGH.

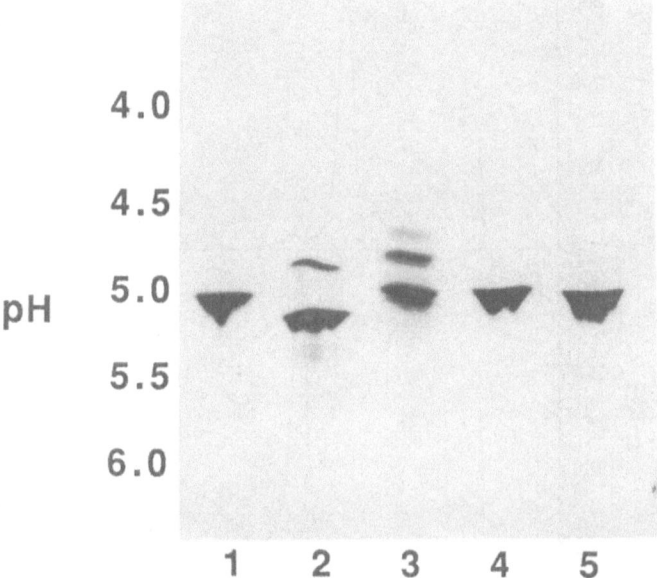

Fig. 7. IEF analysis of hGH. Lyophilized samples were reconstituted to 1 mg/ml with water, and after prefocusing, 20 µg protein was loaded onto the gel. Staining was with Coomassie brilliant blue. *Lane* 1, rhGH; *Lane* 2, two- chain rhGH; *lane* 3, pit-hGH (Crescormon).; *lane* 4, rhGH; *lane* 5, mhGH (Protropin).

Hydrophobic Interaction Chromatography

The nonpolar properties of proteins can be examined using hydrophobic interaction chromatography, under conditions unlikely to cause denaturation of the protein commonly seen in reversed- phase chromatography. We have compared the elution of rhGH, mhGH, and pit-hGH from a TSK-phenyl 5 PW column, using a reverse (high- to-low) sodium sulfate gradient.

Gradients of acetonitrile (from 2% to 5%) and Brij-35 (0- 0.75%) are used to sharpen the separations.

Figure 8 shows that rhGH and pit-hGH have identical retention times (about 13 min) for the main protein species. Pituitary hGH contains several minor peaks that have also been observed using other analytical methods. However, mixing experiments have demonstrated that the main protein peaks are indistinguishable

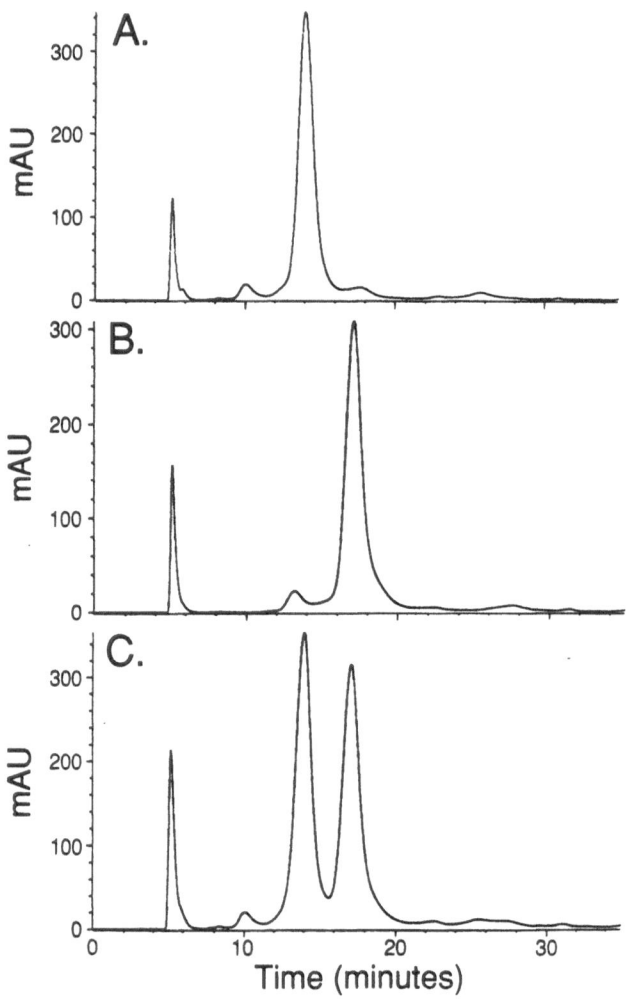

Fig. 8 a-c. Hydrophobic chromatography of hGH. The 280-nm elution traces are shown. *A*, pit- hGH (Crescormon); *B*, mhGH (Somatonorm); *C*, a mixture of rhGH and mhGH.

between the two preparations. In contrast, recombinant mhGH is readily distinguished by its retention time of 17 min. Mixing experiments show resolution of two distinct peaks. This demonstrates that, under these conditions, mhGH behaves as if it were more hydrophobic than either the pituitary-derived or recombinant natural-sequence material.

Ultraviolet Absorbance Spectrophotometry

A protein's ultraviolet absorbance spectrum represents a summation of the electronic transitions of the individual aromatic residues. These individual spectra can differ considerably amongst themselves because of their singular environments within the protein. Human growth hormone is a good subject for ultraviolet spectral analysis because the environment of the single tryptophan can be observed relatively unimpeded by other tryptophan residues (Bewley 1979). The eight

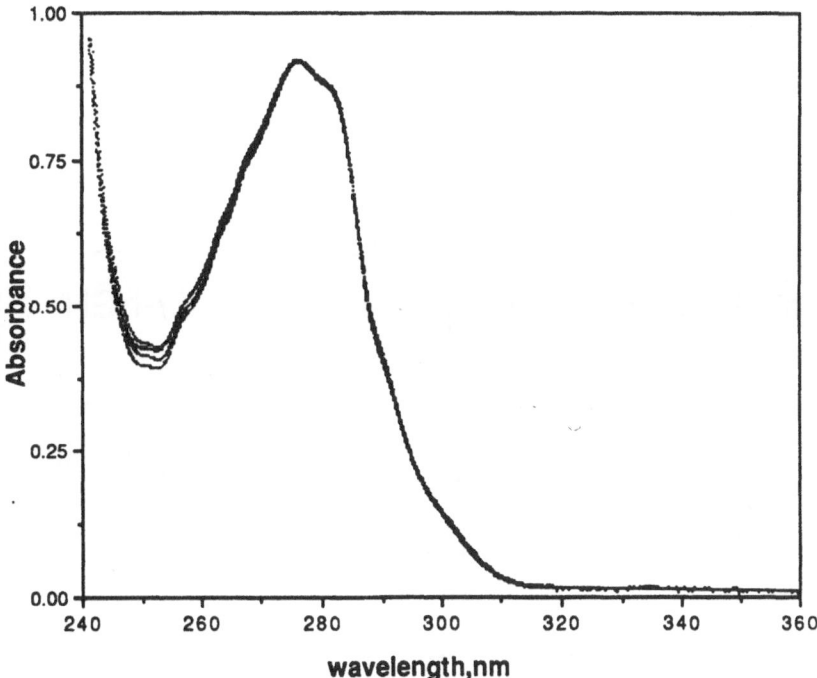

Fig. 9. Ultraviolet absorbance spectral analysis of of hGH. Four spectra are superimposed: rhGH, mhGH (Protropin), pit-hGH (Crescormon), and two-chain rhGH. Each was reconstituted from a lyophilized cake at about 2 mg/ml with water, exchanged into 0.1 M Tris-HCl, pH 8.1, and diluted to about 1 AU/ml with the same buffer. Data were collected as described in Materials and Methods. The individual traces are not identified in the figure.

tyrosines, on the other hand, must be analyzed as a group. Figure 9 shows the light-scattering-corrected absorbance spectra for rhGH, pit-hGH, and mhGH. No significant differences can be observed among the three spectra, suggesting that the tryptophan and tyrosine environments are very similar in these proteins.

Second derivative plotting of the UV absorbance data yields the curves shown in Fig. 10. The peaks on such a plot are representative of the individual transitions from phenylalanine, tyrosine, and tryptophan. The second derivative spectrum allows a more detailed analysis of the wavelenght of the individual transitions (and therefore the wavelenght shifts produced by different environments). No such shifts are detectable in the second derivative plots for rhGH, mhGH, or pit-hGH (mhGH spectrum omitted for clarity). It is therefore likely that the structures of these proteins are very similar, possibly identical in the regions surrounding the aromatic residues. Corroboration of these conclusion has been obtained by circular dichroism analysis (not shown), which likewise yields completely superimposable spectra for rhGH, mhGH and pit-hGH.

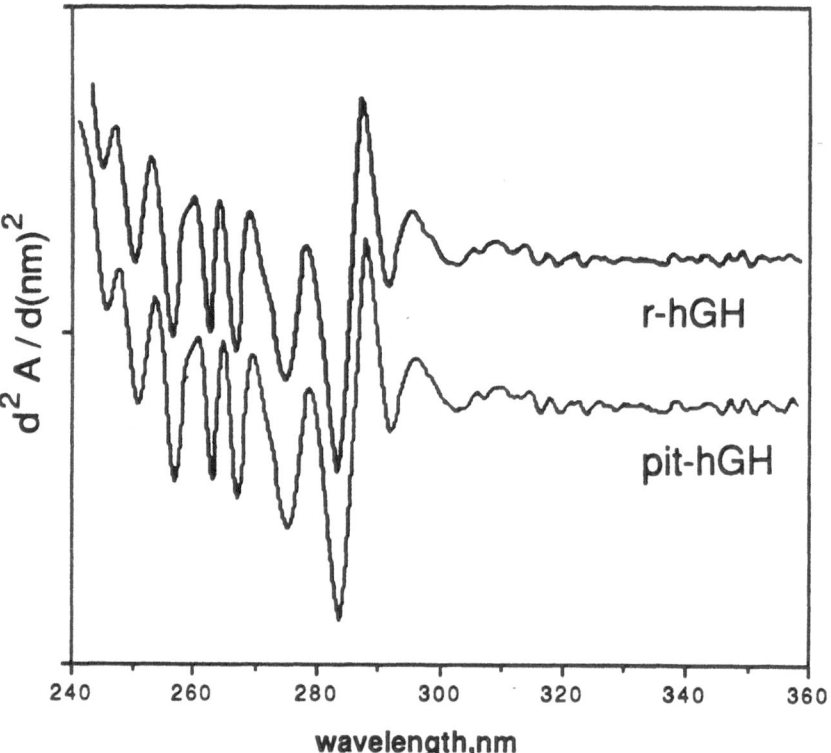

Fig. 10. Ultraviolet absorbance, second derivative spectral analysis. The data were derived as described in the text. The spectra have been vertically offset for clarity. The range for both spectra was -0.005 to 0.005 Å/nm².

CONCLUSION

The recombinant forms of hGH behave in most biochemical analyses exactly as do the pituitary-derived materials. Methionyl hGH can be shown to behave as a slightly more hydrophobic protein than the natural-sequence species, probably as a result of an interaction of the N-terminus with a hydrophobic surface. It is evident from the data presented that the primary sequence, the disulfide bond pattern, and, to the extent that they are determinable, the overall folded structure of the recombinant and natural materials are identical.

Acknowledgments

The authors would like to acknowledge contributions to this work by Tom Bewley, Louisette Basa, Tom Doherty, Cordelia Leonard, and Rong-Chag Pai. The support of Dr. Linda Fryklund is also gratefully acknowledged.

REFERENCES

Ambler RP (1967) Carboxypeptidases A and B. *Meth Enzymol* **11**: 436-445

Bewley TA (1979) Circular dichroism of pituitary hormones. *Rec Prog Horm Res* **35**: 155-213

Bewley TA, Li CH (1975) The chemistry of human pituitary growth hormone. *Adv Enzymol* **42**: 73-83

Chang JYH, Pai RC, Bennett WF, Keck RG, Bochner BR (1986) In: Leive L (ed) *Microbiology 1986*. American Society for Microbiology, Washington, DC, pp 324-329

Fryklund LM, Bierich JR, Ranke MB (1986) Recombinant human growth hormone. *Clin Endocrinol Metabol* **15**: 511-535

Glazer AN, Delange RJ, Sigman DS (1975) In: Work TS, Work E (eds) *Chemical Modification of Proteins*. American Elsevier, New York, pp 13-37

Goeddel DV, Heyneker HL, Hozumi T, Arentzen R, Itakura K, Yansura DG, Ross MJ, Miozzari G, Crea R, Seeburg PH (1979) Direct expression in *Escherichia coli* of a DNA sequence coding for human growth hormone. *Nature* **281**: 544-548

Gray GL, McKeown KA, Jones AJS, Seeburg PH, Heyneker HL (1984) *Biotechnology* **2**: 161-165

Kato K, Yamada T, Kawahara K, Onda H, Asano T, Sugino H, Kakinuma A (1985) Purification and characterization of recombinant human interleukin-2 produced in *Escherichia coli*. *Biochem Biophys Res Commun* **130**: 692-699

Laemmli UK (1970) Cleavage of structural proteins during the assembly of the head of bacteriophage T4. *Nature* **227**: 680-685

Lewis UJ (1984) Variants of growth hormone and prolactin and their posttranslational modifications. *Annu Rev Physiol* **46**: 33-42

Li CH (1975) The chemistry of human pituitary growth hormone: 1967-1973. In: Li CH (ed) *Hormonal Proteins and Peptides,* vol. 3. Academic, New York, pp 1-40

Matsubara H, Sasaki RM (1969) High recovery of tryptophan from acid hydrolysates of proteins. *Biochem Biophys Res Commun* **13**: 175-181

Morrissey JH (1981) Silver stain for proteins in polyacrylamide gels: a modified procedure with enhanced uniform sensitivity. *Anal Biochem* **117**: 307-310

Oakley BR, Kirsh DR, Morris NR (1980) A simplified ultrasensitive silver stain for detecting proteins in polyacrylamide gels. *Anal Biochem* **105**: 361-363

Rodriguez H, Kohr WJ, Harkins RN (1984) Design and operation of a completely automated Beckman microsequencer. *Anal Biochem* **140**: 538-547

Savitsky A, Golay MJE (1964) Smoothing and differentiation of data by simplified least squares procedures. *Anal Chem* **36**: 1627-1639

Steiner J, Termonia Y, Deltour J (1972) Comments on smoothing and differentiation of data by simplified least squares procedure. *Anal Chem* **44**: 1906-1909

Advances in Growth Hormone and Growth Factor Research,
edited by E.E. Müller, D. Cocchi and V. Locatelli
Pythagora Press, Roma-Milano and Springer Verlag, Berlin-Heidelberg © 1989

Evolutionary aspects of growth hormones from nonmammalian species

H. KAWAUCHI and A. YASUDA

Laboratory of Molecular Endocrinology, School of Fisheries Sciences, Kitasato University, Sanriku, Iwate, Japan

INTRODUCTION

Growth hormone (GH) and prolactin are a family of peptide hormones that are secreted from the pituitary glands of all vertebrate animals and share a number of common structural and biological characteristics. GH is involved in the regulation of postnatal somatic growth and the maintenance of nitrogen, lipid, carbohydrate, and mineral metabolism in most vertebrates. Prolactin has diverse functions, ranging from the initiation and maintenance of lactation in mammals to the maintenance of intercellular osmolality in modern bony fish (teleosts) (Bern, 1983). Human GH is known to exhibit both somatic and lactogenic actions. In addition, it has been demonstrated that administration of bovine GH stimulated lactation in dairy cows (Peel et al. 1981). On the basis of the amino acid sequences (Bewley and Li 1971; Niall et al. 1971) and gene structures (Miller and Eberhardt 1983) of the mammalian hormones, it has been postulated that they have evolved from a common ancestral gene by duplication followed by evolutionary divergence. Accordingly, characterization of the homologous hormones at many phylogenetic levels may allow us to gain better insight into the diversity of functions. As well, structure-function relationships can be analayzed with a view toward understanding the consequences of molecular evolution.

Until recently, however, chemical knowledge of GH and prolactin has been mainly confined to those of mammals, although there have been several attempts to isolate nonmammalian GHs and prolactins (birds: Farmer et al. 1974; Scanes et al. 1975; Harvey and Scanes 1977; reptiles: Papkoff and Hayashida 1972; Farmer et al. 1976a; Chang and Papkoff 1985; amphibians: Farmer et al. 1977a; Yamamoto and Kikuyama 1981; primitive bony fish: Lews et al. 1972; Farmer et al. 1981; modern bony fish (teleosts): Farmer et al. 1976a, 1977a,b; Cook et al. 1983; Wagner et al. 1985; Specker et al. 1984, 1985; Kawauchi et al. 1983a, 1986; Prunet and

Houdebine 1984; Kawazoe et al. 1988; Rand-Weaver et al. 1988). We recently determined the complete amino acid sequences of these teleost GHs and prolactins (Yasuda et al. 1986, 1987; Yamaguchi et al. 1987; Noso et al. 1988) as well as the GH of two phylogenetically important species, the sea turtle (Yasuda et al. 1988) and the blue shark (Yamaguchi et al. 1988b).

Moreover, the recently developed recombinant-DNA technique has now been applied to the lower vertebrate GHs and prolactins. GHcDNAs of chicken (Souza et al. 1984), chum salmon (Sekine et al. 1985), coho salmon (Nicoll et al. 1988), rainbow trout (Agellon and Chen 1986), eel (Saito et al. 1985), yellow tail (Watabiki et al. 1988), and tuna (Sato et al. 1988), and prolactin cDNA of salmon (Kuwana et al. 1986; Song et al. 1988) were cloned. The paper reviews the present knowledge concerning the chemical, physiological, and immunological characteristics of GHs and prolactins of lower vertebrates, bony fish, and deals with the molecular evolution of the hormones on the basis of the complete amino acid sequences.

Isolation of Teleost Growth Hormone and Prolactin

One of the major limitations to comparative research has been the lack of purified hormones from nonmammalian vertebrates. The development of purification procedures should expedite hormone isolation from many species. The availability of highly purified hormones will significantly aid investigation into their functions in homologous system and their molecular evolution. There are now theee methods for preparing the hormones. The classical method involves extraction of the hormones from pituitary glands. The second method involves recovery of the hormones released from the organ-cultured pituitary glands. The third method is gene technology, involving cloning and expression of the genes in bacteria.

Glandular Hormones

An alkaline extraction procedure described for mammalian GH by Papkoff and Li (1958) has been employed exclusively for the isolation of fish GHs such as those of the shark (Lewis et al. 1972), the tilapia (Farmer et al. 1976a), the sturgeon (Farmer et al. 1981), the salmon (Wagner et al. 1985; Kawauchi et al. 1986), the carp (Cook et al. 1983), and of advanced marine fish (Kawazoe et al. 1988; Rand-Weaver et al. 1988; Noso et al. 1988).

Since the molecular weight of GH is approximately 21-22 K, an appropriate gel filtration step is effective for partial purification. Major contaminants in the

GH fraction are the glycoprotein hormones. If pituitary gonadotropin content is high, as in glands from mature fish, the contaminants are so significant that ion-exchange chromatography should be employed. Knowledge of the isoelectric points of GH is essential to choosing the type of ion-exchange chromatography, and these can be determined using the protein purified from the initial gel filtration by analytical high-performance liquid chromatography (HPLC) in an early stage of the isolation procedure. Bonito GH (Noso et al. 1988), yellow tail GH (Kawazoe et al. 1988), and cod GH (Rand-Weaver et al. 1988) have an isolelectric point (pI) around 7; they were therefore purified first by DEAE-cellulose and finally with preparative HPLC.

Alkaline extraction of chum salmon GH from pituitary glands resulted in only limited success. This may be due to the highly hydrophobic nature of this molecule. To reduce the chromatographic procedure, a two-step extraction procedure was employed, namely prolactin (Kawauchi et al. 1983a), POMC-related hormones (Kawauchi 1983), melanin-concentrating hormone (Kawauchi et al. 1983b), isotocin, and Arg-vasotocin (Kawauchi et al. 1984) were extracted with acid-acetone (Kawauchi et al. 1986). An isoelectric fractionation procedure was employed to remove the glycoproteins. The two components, salmon GH I and II, could then be separated by HPLC on a reverse-phase column. The isoelectric points of chum salmon GH I and II were estimated to be 5.6 and 6.0 respectively. They exhibited identical molecular weights of 22 K.

Teleost prolactins thus far characterized are notably basic proteins. The isoelectric points of chum salmon prolactin (10.3), chinook salmon prolactin (8.5), carp prolactin (7.5), tilapia 20-K prolactin (6.7), and tilapia 24-K prolactin (8.5) are higher than those of mammalian prolactins; the pI of ovine prolactin is 5.6. Indeed, teleost prolactins could be effectively extracted from the pituitary gland with acid-acetone, which was originally employed for the isolation of prolactin (Cole and Li 1955). Chum salmon and carp prolactins were purified from the acidic extract by gel filtration on Sephadex G-25 and ion-exchange chromatography on CM-Sephadex C-25 or DEAE-cellulose. Two variants of chum salmon prolactin could be separated simply by reverse-phase HPLC, although they are almost indetical in amino acid composition, molecular weight (23 K), and isoelectric point (10.3).

Secreted Hormones

It has been well established that the secretion of prolactin from mammalian pituitary glands is primarily under the inhibitory control of the hypothalamus. Thus, prolactins could be isolated from cultured pituitary glands of the mouse (Kohmoto, 1975; Shoer et al. 1978) and the hamster (Colosi et al. 1981). Similarly, Specker et al. (1984, 1985) isolated two variants of tilapia prolactin and a single tilapia GH from the medium of cultured rostral pars distalis (prolactin cell region) and proximal pars distalis (GH and gonadotropin cell regions) of the pituitary glands respectively. The two prolactin variants differ in molecular weight at 20 K and 24 K, and have been found to originate in separate genes. The 20-K variant seems to be identical to the tilapia prolactin extracted from the pituitary gland (Farmer et al. 1977b). Interestingly, Kishida et al. (1987) obtained two molecular forms of eel GH by the gel filtration and chromatofocusing technique, but no eel prolactin has been isolated from the cultured pituitary gland. The two eel GHs are almost identical in amino acid composition and molecular weight (23 K), but differ in pI (6.3 and 6.7). Amino acid sequence analysis revealed that they differ in the processing of the terminal portion; eel GH I has a three-residue extension at the N-terminal of eel GH II. Nicoll et al. (1988) have also prepared coho salmon GH using a similar technique with the GH cell region of coho salmon pituitary glands. The GH was purified by reverse- phase HPLC.

Cloned Hormones

Human and bovine GHs can now be produced by genetically engineered microorganisms. Teleost GH has also been synthesized by means of such techniques. Oligo-DNA complementary to chum salmon GHmRNA was used as a probe to identify cDNA clones in a salmon pituitary gland cDNA library (Sekine et al. 1985). This cloned gene was expressed in *E. coli*. The GH synthesized by *E. coli* containing a plasmid for sGH was estimated to be about 15% of the total cellular protein present, and the protein formed inclusion bodies in cells. The particles were easily recovered and purified from sonicated cells by centrifugation. The particles collected by centrifugation were solubilized in 8M urea solution and renatured by dilution and dialysis, as described by Marston et al. (1984). The synthesized GH is identical to natural chum salmon GH I in physicochemical and biological properties.

BIOLOGICAL ACTIVITIES

Growth

Growth hormone has been considered to be concerned almost exclusively with the stimulation of animal growth in the vertebrates. The rat tibia test has been employed as a standard assay of mammalian GHs (Greenspan et al. 1949). Purified GH from representative species of tetrapods (Farmer et al. 1974, 1976a, 1977a), as well as from primitive bony fish such as elasmobranchs (Hayashida and Lewis

1978), and sturgeon (Farmer et al. 1981), is able to stimulate cartilaginous growth in the assay. Purified modern bony fish GH is also capable of showing significant stimulation in the rat tibia assay, but only when high doses, on the order of 20 times higher than a standard bovine GH, were employed (Farmer et al. 1976a). These results suggest that GHs are classified into two groups in terms of rat-tibia stimulating activity and the GHs of primitive fish appear to be less divergent from the main line of evolution leading to tetrapod GHs than are GHs from modern bony fish (Hayashida 1970).

Growth-promoting activities of purified teleost GHs have been examined by administering intraperitoneal injections to juvenile rainbow trout at doses of 0.01 − 10μg/g body wt. at 1− week intervals. Teleost GH elicited profound stimulation of both weight gain and length increase in a dose-related manner. The effects of chum salmon GH were more remarkable on weight gain than on length increase. A similar effect of GH has been shown in the carp (Cook et al. 1983; Kawauchi et al. 1986). The recombinant chum salmon GH is equipotent to natural chum salmon GH in promoting increases in the weight and length of rainbow trout.

Prolactin is known to exhibit growth-stimulating activity in some species of vertebrates. The growth-promoting activities of chum salmon GH and prolactin were examined using juvenile rainbow trout. Chum salmon prolactin was found to be an active stimulator of weight gain but not of length increase. Of the tilapia hormones, both the 24-K prolactin and GH stimulated weight gain, but only the larger prolactin stimulated an increase of length in intact juvenile tilapia (Specker et al. 1984).

Osmoregulation

It is well established that prolactin plays an important role in maintaining the hydromineral balance of euryhaline teleost fish in fresh-water, since the original finding by Pickford and Phillips (1959) that ovine prolactin enables the survival of hypophysectomized killifish in fresh water. The specific actions of prolactin in maintaining plasma sodium levels in freshwater fish and in increasing plasma sodium levels in seawater-adapted fish have been shown in several teleost species. However, most of the sodium-retaining actions of prolactin have been based on studies with mammalian prolactins, usually ovine and bovine. Recently, the sodium-retaining activity of purified teleost prolactins (tilapia and salmon) was examined in euryhaline teleosts (Hirano 1986). Hasegawa et al. (1986) showed the sodium- retaining activities of chum salmon prolactin in hypophysectomized *Fundulus*. After hormone or saline injection, fish were maintained in 50% seawater for 1 h, transferred to fresh water, and held for 24 h until blood sampling. Chum salmon prolactin was 100 times more potent than ovine prolactin in maintaining plasma sodium levels. The effects of prolactins were parabolic, high doses of the hormones being less effective than low doses. Similar results were obtained for coho salmon prolactin in the killifish assay (Grau et al. 1984) and for tilapia prolactin in the tilapia sodium-retaining assay (Farmer et al. 1977b).

In contrast, GH has been implicated in the control of seawater adaptation, since it enhances survival and reduces plasma Na^+ concentration of various fresh-water-adapted salmonids transferred to seawater (Komourdjian et al. 1976). However, it is not clear whether the effect is a direct or an indirect result of growth, as seawater survival is known to be a function of size (Mahnken et al. 1982). Bolton et al. (1987) have shown that the seawater adapting actions of GH are independent of growth, using fish with a narrow range of body weights and injected with hormones over a short time period. Plasma Na^+ levels 24 h after transfer to 80% seawater were reduced significantly by chum salmon GH and ovine GH, whereas the same dose of chum salmon prolactin had no significant effect. Both ovine and salmon GHs significantly reduced plasma Mg^{2+} and Ca^{2+} levels. These effects of GH on plasma electrolyte levels are shown to be dose dependent. Chum salmon prolactin also had no effect on plasma Ca^{2+} or Mg^{2+} levels.

IMMUNONOLOGICAL PROPERTIES

The immunochemical relatedness of GH from species representing the major vertebrate classes was extensively investigated by Hayashida with the use of both radioimmunoassay and immunodiffusion techniques, employing monkey antiserum to rat GH (Hayashida and Lagios 1969; Hayashida 1969, 1970, 1971, 1973, 1977; Hayashida et al. 1975). In general, decreasing immunochemical relatedness was observed with increasing phylogenetic distance from the mammal. Interestingly, the GHs of primitive fish, such as sharks (Hayashida and Lewis, 1978), have all shown significant immunochemical relatedness to mammalian GHs, whereas in the same studies, GHs of several species of teleost did not show any significant relatedness to those of mammals. Recently developed radioimmunoassays of teleost GH and prolactin confirmed that the teleost hormones show neither cross-reaction nor parallel displacement curves with the mammalian hormones or even with other teleost species (Bolton et al. 1986; Hirano et al. 1985; Wagner and McKeown 1986).

In a noncompetitive binding technique, such as immunocytochemistry, hormone specificity was evident while no species specificity was found. The immunocytochemical staining of teleost pituitary glands specifically localized GH immunoreactivity to the putative GH cells in the proximal pars distalis for antisera against salmon GH (Kawauchi et al. 1986) and eel GH (Kishida et al. 1987). Immunoreactivities of antisera against tilapia prolactin (Nagahama et al. 1981), salmon prolactin (Naito et al. 1983), and carp prolactin (Miyajima et al. 1988) were localized to the follicular cells in the rostral pars distalis. No cross-reactivity has been seen between these antisera. Naito et al. (1983) showed that the anti-chum salmon prolactin cross-reacted immunocytochemically with the prolactin of nonsalmonid teleost species. These results are evidence that teleost GH and prolactin acquired the structural characteristics of each hormone.

PRIMARY STRUCTURES OF TELEOST HORMONES

The complete amino acid sequences of eel GH (Yamaguchi et al. 1987) and bonito GH (Noso et al. 1988), and the base sequences of cDNAs of chum salmon GH I (Sekine et al. 1985), rainbow trout GH (Agellon and Chen 1986), coho salmon GH (Nicoll et al. 1987), tuna GH (Sato et al. 1988) and yellow tail GH (Watabiki et al. 1988) have been determined. Fish GHs display general sequence similarities to mammalian GH. This finding suggests that the structure of GH was conserved during evolution. The complete amino acid sequence of chum salmon prolactin (Yasuda et al. 1986), carp prolactin (Yasuda et al. 1987), and tilapia prolactins (Yamaguchi et al. 1988a), and the cDNA sequences of chum salmon prolactin II (Sekine et al. 1985) and chinook salmon prolactin (Song et al. 1988) have been elucidated. These teleost prolactins differ slightly from mammalian prolactins in that one disulfide loop is absent from the N-terminal region. Thus, an overall molecular feature of fish prolactins is more similar to vertebrate GHs than it is to other prolactins and there appears to have been considerably more structural divergence of prolactin than of GH during the evolution of vertebrates. This suggests that the prolactin gene may have diverged from an ancestral GH-like gene. The comparison of base sequences of the chum salmon GHcDNA and prolactin cDNA clearly supports this speculation (Fig. 1). Each cDNA code consisted of a polypeptide of 210 amino acids whose sequence identity was 24% between two pre-hormones. A putative signal peptide of 22 amino acids is cleaved from a pre-GH of 210 residues to produce the 188 amino acids of GH I, and a signal peptide of 23 amino acids is cleaved from a pre-prolactin of 210 residues to produce the 187 amino acids of prolactin II. This comparison provides the first tandem-matched structural information for this family of hormones in lower vertebrates.

In teleosts, although structural information of the GH prolactin family is limited to only a few of over 20,000 present- day species, it is unique that these teleosts were found to have two molecular forms of GH and/or prolactin: chum salmon GHs (Kawauchi et al. 1986), salmon prolactins (Yasuda et al. 1986), tilapia prolactins (Specker et al. 1985; Yamaguchi et al. 1988a), and carp prolactins (Yasuda et al. 1987). These differ from the N-terminal heterogeniety observed in preparations of eel and mammalian GHs. It has been noted that the chum salmon pituitary glands also secrete two distinct molecules, but with great homology, of

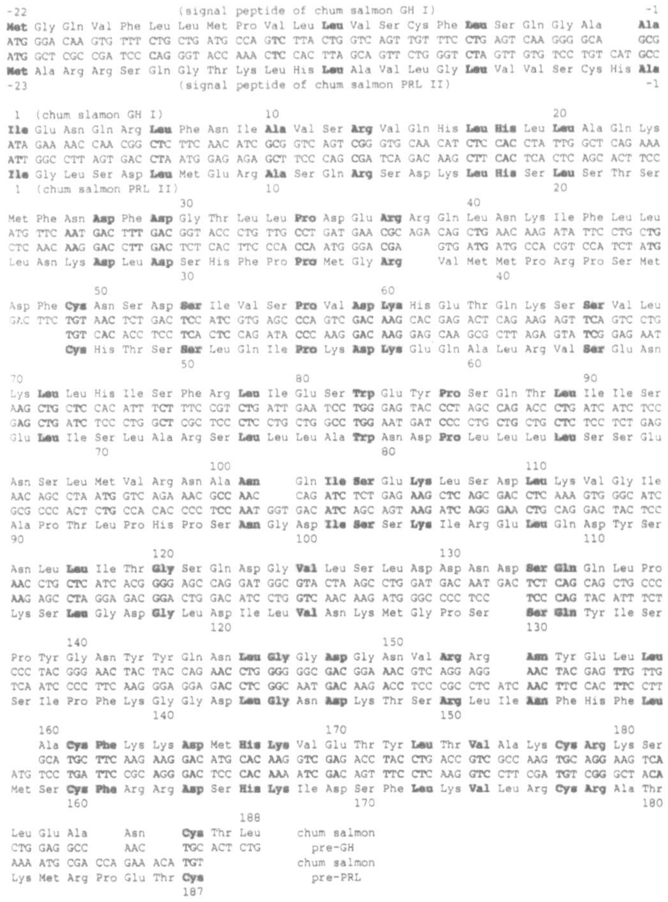

```
  -22                        (signal peptide of chum salmon GH I)                          -1
  Met Gly Gln Val Phe Leu Leu Met Pro Val Leu Leu Val Ser Cys Phe Leu Ser Gln Gly Ala    Ala
  ATG GGA CAA GTG TTT CTG CTC ATG CCA GTC TTA CTG GTC AGT TGT TTC CTG AGT CAA GGG GCA    GCG
  ATG GCT CGC CGA TCC CAG GGT ACC AAA CTC CAC TTA GCA GTT CTG GGT CTA GTT GTG TCC TGT CAT GCC
  Met Ala Arg Arg Ser Gln Gly Thr Lys Leu His Leu Ala Val Leu Gly Leu Val Val Ser Cys His Ala
  -23                        (signal peptide of chum salmon PRL II)                        -1

  1  (chum slamon GH I)              10                                       20
  Ile Glu Asn Gln Arg Leu Phe Asn Ile Ala Val Ser Arg Val Gln His Leu His Leu Ala Gln Lys
  ATA GAA AAC CAA CGG CTC TTC AAC ATC GCG GTC AGT CGG GTG CAA CAT CTC CAC CTA TTG GCT CAG AAA
  ATT GGC CTT AGT GAC CTA ATG GAG AGA GCT TCC CAG CGA TCA GAC AAG CTT CAC TCA CTC AGC ACT TCC
  Ile Gly Leu Ser Asp Leu Met Glu Arg Ala Ser Gln Arg Ser Asp Lys Leu His Ser Leu Ser Thr Ser
  1  (chum salmon PRL II)          10                                       20
  
  Met Phe Asn Asp Phe Asp Gly Thr Leu Leu Pro Asp Glu Arg Arg Gln Leu Asn Lys Ile Phe Leu Leu
  ATG TTC AAT GAC TTT GAC GGT ACC CTG TTG CCT GAT GAA CGC AGA CAG CTG AAC AAG ATA TTC CTG CTG
  CTC AAC AAG GAC CTT GAC TCT CAC TTC CCA CCA ATG GGA CGA     GTG ATG ATG CCA CGT CCA TCT ATG
  Leu Asn Lys Asp Leu Asp Ser His Phe Pro Pro Met Gly Arg     Val Met Met Pro Arg Ser Met
                        30                                   40

  Asp Phe Cys Asn Ser Asp Ser Ile Val Ser Pro Val Asp Lys His Glu Thr Gln Lys Ser Ser Val Leu
  GAC TTC TGT AAC TCT GAC TCC ATC GTG AGC CCA GTG GAC AAG CAC ACT CAG AAG AGT TCA GTC CTG
  TGT CAC ACC TCC TCA CTC CAG ATA CCC AAG GAC AAG GAG CAA GCG CTT AGA GTA TCG GAG AAT
  Cys His Thr Ser Ser Leu Gln Ile Pro Lys Asp Lys Glu Gln Ala Leu Arg Val Ser Glu Asn
                        50                                   60

  70                                   80                                   90
  Lys Leu Leu His Ile Ser Phe Arg Leu Ile Glu Ser Trp Glu Tyr Pro Ser Gln Thr Leu Ile Ile Ser
  AAG CTG CTC CAC ATT TCT TTC CGT CTG ATT GAA TCC TGG GAG TAC CCT AGC CAG ACC CTG ATC ATC TCC
  GAG CTG ATC TCC CTG GCT CGC TCC CTC CTG CTG GCC TGG AAT GAT CCC CTG CTG CTG CTC TCC TCT GAG
  Glu Leu Ile Ser Leu Ala Arg Ser Leu Leu Leu Ala Trp Asn Asp Pro Leu Leu Leu Leu Ser Ser Glu
                        70                                   80

  Asn Ser Leu Met Val Arg Asn Ala Asn     Gln Ile Ser Glu Lys Leu Ser Asp Leu Lys Val Gly Ile
  AAC AGC CTA ATG GTC AGA AAC GCC AAC     CAG ATC TCT GAG AAG CTC AGC GAC CTC AAA GTG GGC ATC
  GCG CCC ACT CTG CCA CAC CCC TCC AAT GGT GAC ATC AGC AGT AAG ATC AGG GAA CTG CAG GAC TAC TCC
  Ala Pro Thr Leu Pro His Pro Ser Asn Gly Asp Ile Ser Ser Lys Ile Arg Glu Leu Gln Asp Tyr Ser
  90                                   100                                  110

  120                                  130
  Asn Leu Leu Ile Thr Gly Ser Gln Asp Gly Val Leu Ser Leu Asp Asp Asn Asp Ser Gln Gln Leu Pro
  AAC CTG CTC ATC ACG GGG AGC CAG GAC GGT GTA CTA AGC CTG GAT GAC AAT GAC TCT CAG CTG CTG CCC
  AAG AGC CTA GGA GAC GGA CTG GAC ATC CTG GTC AAC AAG ATG GGC CCC TCC     TCC CAG TAC ATT TCT
  Lys Ser Leu Gly Asp Gly Leu Asp Ile Leu Val Asn Lys Met Gly Pro Ser     Ser Gln Tyr Ile Ser
                        120                                  130

  140                                  150
  Pro Tyr Gly Asn Tyr Tyr Gln Asn Leu Gly Gly Asp Gly Asn Val Arg Arg     Asn Tyr Glu Leu Leu
  CCC TAC GGG AAC TAC TAC CAG AAC CTG GGG GGC GAC GGA AAC GTC AGG AGG     AAC TAC GAG TTG TTG
  TCA ATC CCC TTC AAG GGC GGA GAC CTC GGC AAT AAG ACC TCC CGC CTC ATC AAC TTC CAC TTC CTT
  Ser Ile Pro Phe Lys Gly Gly Asp Leu Gly Asn Asp Lys Thr Ser Arg Leu Ile Asn Phe His Phe Leu
  140                                  150

  160                                  170                                  180
  Ala Cys Phe Lys Lys Asp Met His Lys Val Glu Thr Tyr Leu Thr Val Ala Lys Cys Arg Lys Ser
  GCA TGC TTC AAG AAG GAC ATG CAC AAG GTC GAG ACC TAC CTG ACC GTC GCC AAG TGC AGG AAG TCA
  ATG TCC TGA TTC CGC AGG GAC TCC CAC AAA ATC GAC AGT TTC CTC AAG GTC CTT CGA TGT CGG GCT ACA
  Met Ser Cys Phe Arg Arg Asp Ser His Lys Ile Asp Ser Phe Leu Lys Val Leu Arg Cys Arg Ala Thr
  160                                  170                                  180

                              188
  Leu Glu Ala     Asn     Cys Thr Leu        chum salmon
  CTG GAG GCC     AAC     TGC ACT CTG        pre-GH
  AAA ATG CGA CCA GAA ACA TGT               chum salmon
  Lys Met Arg Pro Glu Thr Cys               pre-PRL
                      187
```

Fig. 1. Complete cDNA sequences of chum salmon pre-GH I and chum salmon pre-PRL II. Identical amino acid residues are shown in gray.

proopiomelanocortin-related hormones (Kawauchi 1983), which are coded on two separate genes. Such unusual multiplicity of molecular forms has been accounted for by the tetraploid hypothesis of salmonids.

It is well known that the mammalian placenta produces placental lactogen (PL), a member of the GH/ prolactin family. Human PL has been found to be more similar to human GH than to human prolactin, having 85% identity with human GH and only 35% with human prolactin (Bewley et al. 1972). Two genes for human GH and three for human PL are found clustered at the same chromosomal location at band q22–24 on chromosome 17 (George et al. 1981), whereas human prolactin genes are located on chromosome 6 (Cooke et al. 1981). This suggests that PL arose as a consequence of another, more recent duplication of the GH gene (Miller and Eberhardt 1983). Lewis et al. (1980) have demonstrated the existence of a variant of human GH (20-K variant) in which residues 32–46 are missing from 22- K GH. Seeburg (1982) demonstrated that the two different mRNAs for 20-K and 22-K human GH are derived from the same gene, by utilizing a consensus slice site in the precursor RNA.

Recently, another new member of this family, namely proliferin, has been found in cultures of mouse fibroblasts in which proliferation was stimulated by serum growth factors (Linzer and Nathans 1984). The sequence of proliferin is more similar to that of prolactin than to that of GH, but the homology is less than that between the various mammalian prolactins, suggesting divergence from the prolactin gene (Linzer et al. 1985). Such multiplicity of genes related to GH and prolactin suggests that gene duplications have taken place more often within this family than was originally envisaged. It would be expected that one of the duplicate genes could diverge more significantly or rapidly, as long as the other gene continued to perform the normal function. Consequently, the divergent gene could become established as a new functional gene, as in the cases of GH/prolactin, GH/ PL, and prolactin/proliferin, or could sometimes become inactive as a pseudogene. The two variants of salmon prolactin and carp prolactin differ from each other by only a limited number of replacements, whereas the 20-K and 24-K variants of tilapia prolactin differ significantly in amino acid sequence as well as molecular weight (Yamaguchi et al. 1988a). Since divergence rate in a homologous protein may be constant, the small extent of mutation between two variants of carp and chum salmon GH/prolactin families probably reflects a shorter evolutionary time after gene duplication than in the case of tilapia prolactin.

PHYLOGENETIC TREE OF GROWTH HORMONE

A phylogenetic tree of GH can ben constructed on the basis of sequence comparison. Several procedures have been devised for comparing amino acid homology (Dayhoff et al. 1972; Wilson et al. 1977). The most widely used system is the "Unit evolutionary period" (UEP), which directly incorporates the concept of the evolutionary clock. It is simply the length of time (in millions of years) required for 1% of the amino acids to change in two related proteins. Miller ad Eberhardt (1983) estimated a UEP of 4.5 for GH and prolactin on the basis of comparisons of the amino acid sequences of bovine, rat, and human hormones. This suggests that the divergence of GH and prolactin took place about 350 million years ago which is probably too recent, since primitive bony fish have both GH and prolactin in their pituitary glands. To minimize this kind of error, the sequence comparison should be made between distant species in the phylogeny. Comparing the primary sequences of mammalian and teleost hormones, the UEP was estimated to be 6.3 in the case of both GH and prolactin. The divergence of prolactin and

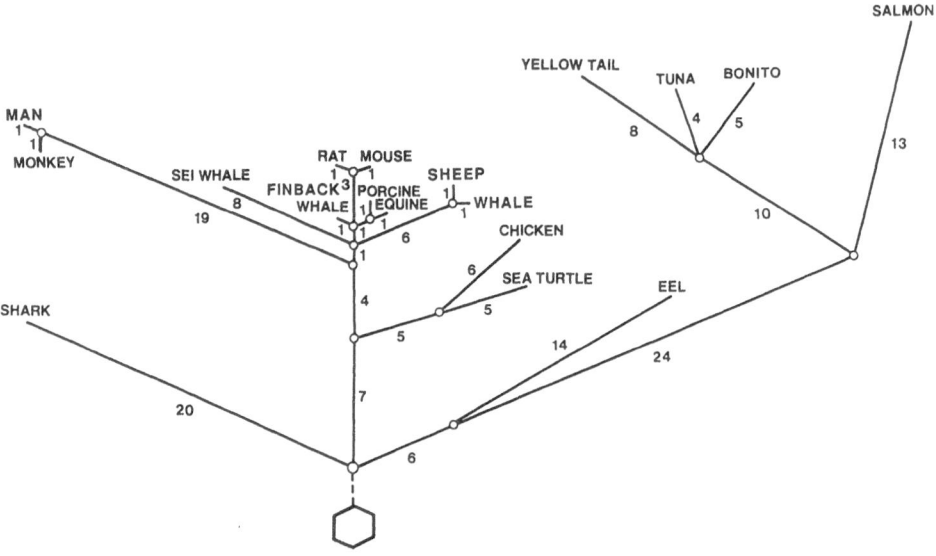

Fig. 2. Molecular phylogenetic tree of the GH on the basis of sequence comparison. The UEP for the tetrapod and the teleost trunk are estimated to be 9.6 and 6.3 respectively. The complete amino acid sequences of GHs are taken from man (Martial et al. 1979), monkey (Li et al. 1986), cattle (Graf and Li 1974, Miller et al. 1980), sheep (Li et al. 1973; Seeburg et al. 1983), pig (Seeburg et al. 1983), horse (Zakin et al. 1976), finback whale (Tsubokawa and Kawauchi 1985), sei whale (Pankov et al. 1982), rat (Seeburg et al. 1977), mouse (Linzer and Talamantes 1986), chicken (Souza et al. 1984), sea turtle (Yasuda et al. 1988), tuna (Sato et al. 1988), yellow tail (Watabiki et al. 1988), eel (Yamaguchi et al. 1987), and blue shark (Yamaguchi et al. 1988b).

GH, based on this value, was estimated to have occurred 460-480 million years ago. This estimation is consistent with the observation that all vertebrates have both GH and prolactin in their pituitary glands, except cyclostomes.

Very recently, we determined the complete amino acid sequences of the blue shark (Yamaguchi et al. 1988b), and the sea turtle GH (Yasuda et al. 1988) in addition to the teleosts mentioned above. Now, we have substantial information from representative species through phylogeny. Figure 2 shows the phylogenetic tree of GH on the basis of amino acid sequences of GH from 18 species. Figure 3 illustrates such comparison of the GHs from seven nonmmalian and three mammalian species, and of the prolactins from two teleost and three mammalian species.

Thus far, 204 alignment positions have been identified. GHs could be clearly classified into two families on the basis of sequence identity, the predicted secondary structure by the method of Chou and Fasman (1978), and the hydrophilicity profile by the method of Hopp and Woods (1981): one is the teleost family

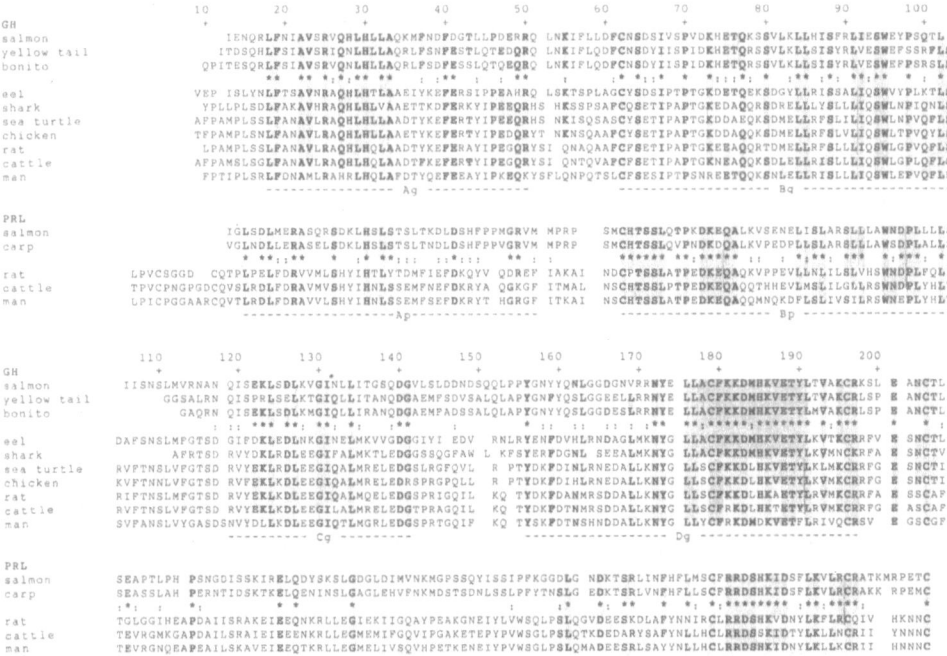

Fig. 3. The complete amino acid sequences of the GHs and the prolactins (PRLs) of various vertebrate species. GHs in the phylogenetic tree have been classified into two groups, the teleost trunk and the tetrapod, in terms of sequence identity, the predicted secondary structure, and the hydrophilicity profiles. Identical and highly conserved residues among all the GHs and prolactins are marked with asterisks and colons respectively. Four highly conserved domains for GHs – Ag, Bg, Cg, and Dg – are identified. Three highly conserved domains for prolactins – Ap, Bp, and Dp – are also identified. The complete amino acid sequences of prolactins are taken from chum salmon (Yasuda et al. 1986), carp (Yasuda et al. 1987), rat (Cooke et al. 1980), cattle (Wallis, 1974), and man (Cooke et al. 1981).

and the other is the tetrapod family. Interestingly, the eel GH is more similar to the shark and tetrapod GHs, with 54% – 60% identity, than to the teleost GHs, with 43% – 48% identity. Moreover, shark GH shows about 40% sequence identity with teleost and 65% identity with tetrapod GHs. Thus, the location of the shark and eel GHs are the key constructing a molecular phylogenetic tree of GH. Assuming that teleost and tetrapod GHs all evolved with a similar UEP, the shark GH branched between the eel GH and the sea turtle GH. Therefore, we should hypothesize that the phylogenetic tree of GH consists of several trunks which have their own UEPs. The UEP of the teleost trunk (6.3) is lower than that of the tetrapod trunk (9.6). This observation is in good accordance with the results of rat tibial activities and immunochemical relatedness of GH from various vertebrate species. Hayashida and Lewis (1978) suggested that the blue shark GH appears to be less divergent from the main line leading to tetrapod GHs than are GHs from teleosts.

The sequence comparison indicated that the eel belongs to a primitive family of teleost fish.

CONSERVED DOMAINS

Structure-function relationships of both GH and prolactin have been investigated for the mammalian hormones, either by chemical modification of amino acid residues or by chemical or enzymatic fragmentation. However, once the structures of the related hormones from the other verbebrates have been elucidated, knowledge of the comparative hormone structure may provide cues for much more precise strategies to predict possible relationships to biological activity. Nicoll et al. (1986) have discussed the structure and activity relationships of GH and prolactin molecules on the basis of amino acid sequences of the mammalian hormones. We have proposed the conserved domains for GH and prolactin molecules (Yasuda et al. 1987; Yamaguchi et al. 1987) throughout vertebrate evolution (Fig. 3). GHs have four highly conserved domains which are located in aligned positions: 18 – 50 (A g), 61 – 102 (B g), 119 – 141 (C g) and 156 – 197 (D g). The hydrophilicity profiles have also been conserved in these domains. Except for that of domain A, the predicted secondary structure is identical to all GHs. It is striking that domain Cg has common secondary structure through the teleost and tetrapod trunk, no matter how many deletions of amino acid residues there are in the N-terminal side of the domain. Similarly, three highly conserved domains were identified for prolactins located in the aligned positions 15 – 45 (A p), 62 – 102 (B p), and 161 – 196 (D p). The locations of three domains, A, B, and D, are common to both GH and prolactin, while domain C g is specific to GHs. Yamazaki et al. (1975) demonstrated that a tryptic peptide of bovine GH, composed of residues 96 – 134, retains about 10% of the bioactivity of the intact hormone. The fragment corresponds to domain C g of this analysis.

It should be noted that domain B and D are bound together by the disulfide linkage between aligned positions 62 and 179 and are therefore closely apposed in the tertiary structure. Plasmin digestion of human GH resulted in the removal of a hexapeptide, residues 134 – 140, divided by Arg and/or Lys residues (Singh et al. 1974; Li and Graf 1974). The remaining molecule, composed of two chains linked by one disulfide bridge, retains full biological activity, while the N-terminal fragment (1 – 134) exhibits weak biological activity. These data suggest that each of the conserved domains may play an essential role in maintaining biological activity. The conservation of disulfide bridges suggests the critical importance of the tertiary structure for both GHs and prolactins.

CONCLUSION

Growth hormones and prolactins have been purified from several teleost fish and subjected to biochemical, physiological, and immunological characterization. Structural determination of the proteins and the cDNAs of teleost and tetrapod GHs provides a broader evolutionary picture of the GH-prolactin family. It has been speculated that prolactin genes diverged from an ancestral GH-like gene 460 – 480 million years ago.

It seems significant that gene duplication can play such a central role in evolution as suggested by Ohno (1970): following gene duplication events, one of the duplicate genes is less restrained and can diverge more significantly or rapidly, as long as the other gene continues to perform the normal functions. Consequently, the divergent gene can become established as a new functional gene, as in the cases of GH/prolactin and GH/PL, or it can sometimes become inactive as in the case of a pseudogene.

On the basis of the amino acid sequence of GHs, a molecular phylogenetic tree has been constructed. There are two trunks in this tree; one includes teleosts and the other includes tetrapods and primitive bony fish. This structural information is in good accordance with the established rat tibial activities and immunochemical relatedness of GH between two trunks.

Tilapia prolactin and some reptilian and amphibian prolactins have shown growth promoting activities. Teleost GH stimulates seawater adaptation of teleosts by lowering Na^+ levels in serum, whereas teleost prolactins are far more potent in teleost Na^+ retaining activities. GHs and prolactins have four and three conserved domains respectively; the locations of three domains are common between GHs and prolactins, while one domain is specific to GHs. To obtain a precise picture of the structure-function relationships, further studies on the comparative physiology and chemistry of GHs and prolactins are obviously called for.

REFERENCES

Agellon LB, Chen TT (1986) Rainbow trout growth hormone: molecular cloning of cDNA and expression in *Escherichia coli. DNA* **5**: 463-471

Bern HA (1983) Functional evolution of prolactin and growth hormone. *Am Zool* **23**: 663-671

Bewley TA, Li CH (1971) Sequence comparison of human pituitary growth hormone, human chorionic somatomammotropin and ovine pituitary lactogenic hormone. *Experientia* **27**: 1368-1371

Bewley TA, Dixon JS, Li CH (1972) Sequence comparison of human pituitary growth hormone, human chorionic somatomammotropin, and ovine pituitary growth and lactogenic hormone. *Int J Pept Protein Res* **4**: 281-287

Bolton JP, Takahashi A, Kawauchi H, Kubota J, Hirano T (1986) Development and validation of a salmon growth hormone radioimmunoassay. *Gen Comp Endocrinol* **62**: 230-238

Bolton JP, Collie NL, Kawauchi H, Hirano T (1987) Osmoregulatory actions of growth hormone in rainbow trout (*Salmo gairdneri.*). *J Endocr* **112**: 63-68

Chang YS, Papkoff H (1985) Isolation and properties of sea turtle (*Chelonia mydas*) pituitary prolactin. *Gen Comp Endocrinol* **60**: 372-379

Chou PY, Fasman GD (1978) Empirical predictions of protein conformation. *Annu Rev Biophys* **47**: 251-276

Cook AF, Wilson SW, Peter RE (1983) Development and validation of a carp growth hormone radioimmunoassay. *Gen Comp Endocrinol* **50**: 335-347

Cooke NE, Coit D, Weiner RI, Baxter JP, Martial JA (1980) Structure of cloned DNA complementary to rat prolactin messenger. *J Biol Chem* **225**: 6502-6510

Cooke NE, Coit D, Shine J, Baxter JD, Martial JA (1981) Human prolactin: cDNA structural analysis and evolutionary comparisons. *J Biol Chem* **256**: 4007-4016

Cole RD, Li CH (1955) Pituitary lactogenic hormone. XIV. A simplified procedure of isolation. *J Biol Chem* **213**: 197-201

Colosi P, Markhoff E, Levy A, Ogren L, Shine N, Talamantes F (1981) Isolation and partial characterization of secreted hamster pituitary prolactin. *Endocrinology* **108**: 850-854

Dayhoff MO, Eck RV, Park CM (1972) A model of evolutionary change in proteins. In: Dayhoff MD (ed) *Atlas of Protein Sequence and Structure,* vol. 5. National Biomedical Research Foundation, Washington DC, pp 89-99

Farmer SW, Papkoff H, Hayashida T (1974) Purification and properties of avian growth hormones. *Endocrinology* **95**: 1560-1565

Farmer AW, Papkoff H, Hayashida T (1976a) Purification and properties of reptilian and amphibian growth hormones. *Endocrinology* **99**: 692-700

Farmer SW, Papkoff H, Hayashida T, Bewley TA, Nishioka RS, Bern HA, Li CH (1976b) Isolation and properties of teleost growth hormone. *Gen Comp Endocrinol* **30**: 91-100

Farmer SW, Licht P, Papkoff H (1977a) Biological activity of bullfrog growth hormone in the rat and the bullfrog (*Rana catesbeiana*). *Endocrinology* **101**: 1145-1150

Farmer SW, Papkoff H, Bewley TA, Hayashida T, Nishioka RS, Bern HA, Li CH (1977b) Isolation and properties of teleost prolactin. *Gen Comp Endocrinol* **31**: 60-71

Farmer SW, Hayashida T, Papkoff H, Polenov AL (1981) Characteristics of growth hormone isolated from sturgeon (*Acipenser guldenstadti*) pituitaries. *Endocrinology* **108**: 377-381

George DL, Phillips JA, Francke U, Seeburg PH (1981) The genes for growth hormone and chorionic somatomammotropin are on the long arm of human chromosome 17 in region q21-qter. *Hum Genet* **57**: 138-141

Graf L, Li CH (1974) On the primary structure of pituitary bovine growth hormone. *Biochem Biophys Res Commun* **56**: 168-176

Grau EG, Prunet P, Gross T, Nishioka RS, Bern HA (1984) Bioassay for salmon prolactin using hypophysectomized *Fundulus heteroclitus. Gen Comp Endocrinol* **32**: 427-431

Greenspan FS, Li CH, Simpson ME, Evans HM (1949) Bioassay of hypophyseal growth hormone: the tibia test. *Endocrinology* **45**: 455-463

Harvey S, Scanes CG (1977) Purification and radioimmunoassay of chicken growth hormone. *J Endrocr* **73**: 321-329

Hasegawa S, Hirano T, Kawauchi H (1986) Sodium-retaining activity of chum salmon prolactin in some euryhaline teleosts. *Gen Comp Endocrinol* **63**: 309-317

Hayashida T (1969) Relatedness of pituitary growth hormone from various vertebrate species. *Nature* **222**: 254-255

Hayashida T (1970) Immunological studies with rat pituitary growth hormone (RGH). II. Comparative immunochemical investigation of GH from representatives of various vertebrate classes with monkey antiserum to RGH. *Gen Comp Endrocrinol* **15**: 432-452

Hayashida T (1971) Biological and immunochemical studies with growth hormone in pituitary extracts of holostean and chondrostean fishes. *Gen Comp Endocrinol* **17**: 275-280

Hayashida T (1973) Biological and immunochemical studies with growth hormone in pituitary extracts of elasmobranchs. *Gen Comp Endocrinol* **20**: 377-385

Hayashida T (1977) Immunochemical and biological studies with growth hormone in a pituitary extract of the coelacanth, *Latimeria chalumnae smith*. *Gen Comp Endocrinol* **32**: 221-229

Hayashida T, Lagios MD (1969) Fish growth hormone: a biological, immunochemical and ultrastructural study of sturgeon and paddlefish pituitaries. *Gen Comp Endocrinol* **13**: 403-411

Hayashida T, Lewis UI (1978) Immunochemical and biological studies with antiserum to shark growth hormone. *Gen Comp Endocrinol* **36**: 530-542

Hayashida T, Farmer SW, Papkoff H (1975) Pituitary growth hormone: further evidence for evolutionary conservatism based on immunochemical studies. *Proc Natl Acad Sci USA* **72**: 4322-4326

Hirano T (1986) The spectrum of prolactin action in teleosts. In: Ralph CL (ed.) *Comparative Endocrinology: Developments and Directions*. Liss, New York, pp. 53-74

Hirano T, Prunet P, Kawauchi H, Takahashi A, Ogasawara T, Kubota J, Nishioka RS, Bern HA, Takada K, Ishii S (1985) Development and validation of salmon prolactin radioimmunoassay. *Gen Comp Endocrinol* **59**: 266-276

Hopp TP, Woods KR (1981) Prediction of protein antigenic determinations from amino acid sequences. *Proc Natl Acad Sci USA* **78**: 3824-3828

Kawauchi H (1983) Chemistry of proopiocortin-related peptides in the chum salmon pituitary. *Arch Biochem Biophys* **227**: 343-350

Kawauchi H, Abe K, Takahashi A, Hirano T, Hasegawa S, Naito N, Nakai Y (1983a) Isolation and properties of chum salmon prolactin. *Gen Comp Endocrinol* **49**: 446-458

Kawauchi H, Kawazoe I, Tsubokawa M, Kishida M, Baker BI (1983b) Caracterization of melanin-concentrating hormone from chum salmon pituitaries. *Nature* **305**: 321-323

Kawauchi H, Kawazoe I, Adachi Y, Buckley DI, Ramachandran J (1984) Chemical and biological characterization of salmon melanocyte-stimulating hormones. *Gen Comp Endocrinol* **53**: 37-48

Kawauchi H, Moriyama S, Yasuda A, Yamaguchi K, Sirahata K, Kubota J, Hirano T (1986) Isolation and characterization of chum salmon growth hormone. *Arch Biochem Biophys* **244**: 542-552

Kawazoe I, Noso T, Kuriyama S, Akasaka A, Kawauchi H (1988) Growth hormone from yellow tail (*Seriola quinegeradiata*): isolation and characterization. *Nippon Suisan Gakkaishi* **54**: 393-399

Kishida M, Hirano T, Kubota J, Hasegawa S, Kawauchi H, Yamaguchi K, Shirahata K (1987) Isolation of two forms of growth hormone secreted from eel pituitaries *in vivo*. *Gen Comp Endocrinol* **65**: 478-488

Kohmoto K (1975) Mouse prolactin obtained by pituitary organ culture in a serum-free medium. *Endocrinol Jpn* **22**: 465-469

Komourdjian MP, Saunders RL, Henwich C (1976) The effect of porcine somatotropin on growth and survival in seawater of Atlantic salmon (*Salmo salar*) parr. *Can J Zool* **54**: 531-535

Kuwana Y, Kuga T, Sekine S, Sato M, Itoh S, Kawauchi H (1986) Cloning and expression of chum salmon PRLcDNA. Program of 8th Annual Meeting of the Agricultural Chemical Society of Japan. Kyoto, Abstract xxxc

Lewis UJ, Singh RNP, Seavey BK, Pickford GE (1972) Growth hormone and prolactin-like proteins of the blue shark (*Prionace glauca*). *Fish Bull* **70**: 933-939

Lewis UJ, Singh RNP, Tutwiler GF, Sigle MB, VanderLaan EF, VanderLaan WP (1980) Human growth hormone: a complex of proteins. *Recent Prog Horm Res* **36**: 477-508

Li CH, Graf L (1974) Human pituitary growth hormone: isolation and properties of two biologically active fragments from plasmin digests. *Proc Natl Acad Sci USA* **71**: 1197-1201

Li CH, Gordon D, Knorr J (1973) The primary structure of sheep pituitary growth hormone. *Arch Biochem Biophys* **156**: 493-508

Li CH, Chung D, Lahm HW, Stein S (1986) The primary structure of monkey pituitary growth hormone. *Arch Biochem Biophys* **245**: 287-291

Linzer DIH, Nathans D (1984) Nucleotide sequence of a growth-related mRNA encoding a member of the prolactin-growth hormone family. *Proc Natl Acad Sci USA* **81**: 4255-4259

Linzer DIH, Talamantes F (1986) Nucleotide sequence of mouse prolactin and growth hormone mRNAs, and expression of those mRNAs during pregnancy. *J Biol Chem* **260**: 9574-9579

Linzer DIH, Lee S, Ogren L, Talamantes F, Nathans D (1985) Identification of proliferin mRNA and protein in mouse placenta. *Proc Natl Acad Sci USA* **82**: 4356-4359

Mahnken C, Prentice E, Waknitz W, Monan G, Sims C, Williams J (1982) The application of recent smoltification research to public hatchery release: an assessment of size/time requirements for Columbia river hatchery coho salmon (*Oncorhynchus kisutch*). *Aquaculture* **28**: 251-268

Marston FAO, Lower PA, Doel MT, Schoemaker JM, White S, Angal S (1984) Purification of cals prochymosin (prorennin) synthesized in *Escherichia coli*. *Bio/Technology* **2**: 800-804

Martial JA, Hallewell RA, Baxter JD, Goodman HM (1979) Human growth hormone: complementary DNA cloning and expression in bacteria. *Science* **205**: 602-607

Miller WL, Eberhardt NL (1983) Structure and evolution of the growth hormone gene family. *Endocr Rev* **4**: 97-130

Miller WL, Martial JA, Baxter JD (1980) Molecular cloning of DNA complementary to bovine growth hormone mRNA. *J Biol Chem* **255**: 7521-7524

Miyajima K, Yasuda A, Sweanson P, Kawauchi H, Cook H, Kaneko T, Peter RE, Suzuki R, Hasegawa S, Hirano T (1988) Isolation and characterization of carp prolactin. *Gen Comp Endocrinol* **70**: 407-417

Nagahama Y, Olivereau M, Farmer SW, Nishioka RS, Bern HA (1981) Immunocytochemical identification of the prolactin- and growth hormone-secreting cells in the teleost pituitary with antisera to tilapia prolactin and growth hormone. *Gen Comp Endocrinol* **44**: 389-395

Naito N, Takashashi A, Nakai Y, Kawauchi H, Hirano T (1983) Immunocytochemical identification of the prolactin-secreting cells in the teleost pituitary with an antiserum to chum salmon prolactin. *Gen Comp Endocrinol* **50**: 282-291

Niall HD, Hogan M, Sauer R, Rosenblum IY, Greenwood FC (1971) Sequences of pituitary and placental lactogenic and growth hormones: evolution from a primordial peptide by gene duplication. *Proc Natl Acad Sci USA* **68**: 866-869

Nicoll CS, Mayer GL, Russel SM (1986) Structural feature of prolactins and growth hormones that can be related to their biological properties. *Endocr Rev* **7**: 169-203

Nicoll CS, Steiny SS, King DS, Nishioka RS, Mayer GL, Eberhardt NL, Baxter JD, Yamanaka MK, Miller JA, Sweilhamer JJ, Schilling JW, Johnson LK (1988) The primary structure of coho salmon growth hormone and its cDNA. *Gen Comp Endocrinol* **68**: 387-399

Noso T, Yasuda A, Kavazoe I, Takehara H, Sakai K, Kawauchi H (1988) Isolation and characterization of growth hormone from a marine fish, bonito (*Ktsuwonus pelamis*). *Int J Pept Protein Res* (in press)

Ohno S (1970) Evolution by gene duplication: Springer, Berlin Heidelberg New York

Pankov YA, Bulatov AA, Osipova TA (1982) Primary structure of sei whale pituitary somatotropin. *Int J Pept Protein Res* **20**: 396-399

Papkoff H, Li CH (1958) Isolation and characterization of growth hormone from anterior lobes of whale pituitaries. *J Biol Chem* **231**: 367-377

Papkoff H, Hayashida T (1972) Pituitary growth hormone from the turtle and duck: purification and immunochemical studies. *Proc Soc Exp Biol Med* **140**: 251-255

Peel CJ, Bauman DE, Gorewith RC, Sniffen CJ (1981) Effects of exogenous growth hormone on lactational performace in high-yielding dairy cows. *J Nutr* **111**: 1662-1671

Pickford G, Phillips JG (1959) Prolactin, a factor in promoting survival of hypophysectomized killifish in fresh water. *Science* **130**: 454-455

Prunet P, Houdebine LM (1984) Purification and biological characterization of chinook salmon prolactin. *Gen Comp Endocrinol* **53**: 49-57

Rand-Weaver M, Walther BT, Kawauchi H (1988) Isolation and characterization of growth hormone from Atlantic cod (*Gadus morhua*). *Gen Comp Endocrinol* (in press)

Saito A, Sekine S, Okada Y, Sato M, Ito S, Hirano T (1985) Cloning and expression of eel GHcDNA. Program of Annual Meeting of the Molecular Biology Society of Japan. Tokyo, Abstract 3E-17

Sato N, Watanabe K, Murata K, Sakaguchi M, Kariya U, Kimura S, Nonoka M, Kimura A (1988) Molecular cloning and nucleotide sequence of tuna growth hormone cDNA *Biochim Biophys Acta* **949**: 35-42

Scanes CG, Bolton NT, Chadowick A (1975) Purification and properties of an avian prolactin. *Gen Comp Endocrinol* **27**: 371-379

Seeburg PH (1982) The human growth hormone gene family: nucleotide sequences show recent divergence and predict a new polypeptide hormone. *DNA* **1**: 239-249

Seeburg PH, Shine J, Martial JA, Baxter JP, Goodman HM (1977) Nucleotide sequnce and amplification in bacteria of the structural gene for rat growth hormone. *Nature* **270**: 486-494

Seeburg PH, Sias S, Adelman J, Deboer HA, Hyflick J, Jhurai P, Goedel DV, Heyneker HL (1983) Efficient bacterial expression of ovine and porcine growth hormones. *DNA* **2**: 37-45

Sekine S, Mizukami T, Nishi T, Kuwana Y, Saito A, Sato M, Itoh S, Kawauchi H (1985) Cloning and expression of cDNA for salmon growth hormone in *E coli*. *Proc Natl Acad Sci USA* **82**: 4306-4310

Shore LF, Shine NR, Talamantes F (1978) Isolation and partial characterization of secreted mouse pituitary prolactin. *Biochem Biophys Acta* **537**: 336-347

Singh RNP, Seavey BK, Rice VP, Lindsey TT, Lewis UJ (1974) Modified forms of human growth hormone with increased biological activities. *Endocrinology* **94**: 883-891

Song S, Trinh K-Y, Hwang S-J, Belkhole S, Idler DR (1988) Molecular cloning and expression of salmon prolactin cDNA. *Eur J Biochem* **172**: 279-285

Souza LM, Boone TC, Murkock D, Langley K, Wypych J, Fenton D, Jonson S, Lai PH, Everet R, Hsu RY, Bosselman R (1984) Application of recombinant DNA technologies to studies on chicken growth hormone. *Exp Zool* **232**: 465-473

Specker JL, King DS, Rivas RJ, Young BK (1984) Partial characterization of two prolactins from a cichlid fish. In: MacLeod RM, Scapagnini U, Thorner MO (eds) *Prolactin.* Springer, Berlin Heidelberg New York, pp. 427-435

Specker JL, King DS, Nishioka RS, Shirahata K, Yamaguchi K, Bern HA (1985) Isolation and partial characterization of a pair of prolactins released in vitro by the pituitary of a cichlid fish, *Oreochromis mossambicus. Proc Natl Acad Sci USA* **82**: 7490-7494

Tsubokawa M, Kawauchi H (1985) Complete amino acid sequence of fin whale growth hormone. *Int J Pept Protein Res* **25**: 297-304

Wagner GF, McKeown BA (1986) Development of a salmon growth hormone radioimmunoassay. *Gen Comp Endocrinol* **62**: 452-458

Wagner GF, Fargher RC, Brown JC, McKeown BA (1985) Further characterization of growth hormone from chum salmon (*Oncorhynchus keta*).*Gen Comp Endocrinol* **60**: 27-34

Watabiki M, Tanaka M, Masuda, Yamakawa M, Yoneda H, Kakashima K (1988) cDNA cloning and primary structure of yellow tail (*Seriola quinqueradiata*) pregrowth hormone. *Gen Comp Endocrinol* **70**: 401-406

Wilson AC, Carlson SS, White JT (1977) Biochemical evolution. *Annu Rev Biochem* **46**: 537-639

Yamaguchi K, Yasuda A, Kishida M, Hirano T, Sano H, Kawauchi H (1987) Primary structure of eel (*Anguilla japonica*) growth hormone. *Gen Comp Endocrinol* **66**: 447-453

Yamaguchi K, Specker J, King DS, Yokoo Y, Nishioka R, Hirano T, Bern HA (1988a) Complete amino acid sequences of a pair of fish (tilapia) prolactins, tPRL177 and tPRL188. *J Biol Chem* **263**: 9113-9121

Yamaguchi K, Yasuda A, Lewis UJ, Yokoo Y, Kawauchi H (1988b) The complete amino acid sequence of growth hormone from an elasmobranch, the blue shark (*Prionace glauca*). *Gen Comp Endocrinol* (in press)

Yamamoto K, Kikuyama S (1981) Purification and properties of bullfrog prolactin. *Endocrinol Jpn* **28**: 59-64

Yamazaki N, Kangawa S, Kobayashi S, Kikutani M, Sonenberg M (1972) Amino acid sequence of biologically active fragment of bovine growth hormone. *J Biol Chem* **240**: 2874-2880

Yasuda A, Itoh H, Kawauchi H (1986) Primary structure of chum salmon prolactins: occurrence of highly conserved regions. *Arch Biochem Biophys* **244**: 528-541

Yasuda A, Miyajima K, Kawauchi H, Peter RE, Lin HR, Yamaguchi K, Sano H (1987) Primary structure of common carp prolactins. *Gen Comp Endocrinol* **66**: 280-290

Yasuda A, Yamaguchi K, Papkoff H, Yokoo Y, Kawauchi H (1988) The complete amino acid sequence of growth hormone from the sea turtle *Chelonis mydas. Gen Comp Endocrinol* (in press)

Zakin MM, Pokus E, Langton AA, Ferrara P, Santome JA, Dellacha JM, Paladini AC (1976) Primary structure of equine growth hormone. *Int J Pept Protein Res* **8**: 435-444

Advances in Growth Hormone and Growth Factor Research,
edited by E.E. Müller, D. Cocchi and V. Locatelli
Pythagora Press, Roma-Milano and Springer Verlag, Berlin-Heidelberg © 1989

*Circulating binding proteins for human growth hormone**

G. BAUMANN

Center for Endocrinology, Metabolism and Nutrition, Department of Medicine, Northwestern University Medical School, Chicago, Illinois, USA

Growth hormone (GH) promotes a complex array of seemingly unrelated biological effects ultimately leading to somatic growth. Its mechanism(s) of action and biological fate are at best poorly understood. The recent discovery of binding proteins for GH in human plasma has added a new degree of complexity to the physiology of GH. This field is only about two years old, with the first report appearing in 1985 (Baumann et al. 1985a). The information available about GH binding proteins is therefore still limited. It is useful to briefly contemplate the history of their discovery and the reasons for the delay in recognition of GH binding proteins.

HISTORICAL PERSPECTIVE

The possibility of carrier proteins for polypeptide hormones, including GH and insulin, was considered at the time when radiolabeled hormones became available in the late 1950's and 1960's. Several groups presented evidence for binding of GH to plasma proteins (Irie and Barrett 1962; Touber and Maingay 1963; Hadden and Prout, 1964, 1965; Collipp et al. 1964; Collipp and Kaplan, 1966; MacMillan et al. 1967), and Hadden and Prout suggested the existence of a specific GH-binding α_2-globulin. However, these findings were attacked as artifacts, due either to structurally altered iodinated GH, to denatured GH, or to contaminating serum proteins in GH preparations. The ensuing controversy was resolved in favor of the view that polypeptide hormones, including GH, in blood generally existed in the

* This work was supported in part by Grants DK1W0699, DK38128, and RR0537 from the National Institutes of Health, and by a McGaw Medical Center Interinstitutional Grant.

free state (Berson and Yalow 1966, 1968). This dictum has remained dogma for two decades. Periodically, reports appeared in support of GH binding to plasma proteins (Antoniades 1975; Beitins et al. 1977; Bieler et al. 1977), but these observations were largely relegated to the artifactual. Peeters and Friesen (1977) described a low affinity binding protein for GH in the pregnant mouse, but this was believed to be unique to murine pregnancy. The discovery of binding proteins for somatomedins (insulin-like growth factors) in plasma (Zapf et al. 1975; Megyesi et al. 1975; Hintz and Liu 1977) was the first example of a protein-bound peptide hormone, in apparent conflict with prevailing dogma. This was followed by the description of binding proteins for platelet-derived growth factor (Huang et al. 1984, Raines et al. 1984). These cases were given special dispensation because the peptides involved were not real hormones, but rather "growth factors".

The size heterogeneity of polypeptide hormones in plasma, which was first recognized in the late 1960s and early 1970s, again raised the possibility of carrier proteins. However, such size heterogeneity was generally attributed to hormone precursors, such as in the case of insulin, or to polymers, as in the case of GH (Yalow 1974). In favor of this interpretation was the fact that similar size isomers could be demonstrated in (serum-free) hormone preparations extracted from the endocrine gland. The first reports on "big" GH (Goodman et al. 1972; Gorden et al. 1973a) showed a greater proportion of "big" GH in plasma than in pituitary, but this was not further pursued. The existence of "big-big" GH (>60 K) was not fully recognized until later. Subsequent studies dealing with large molecular-weight (mol. wt.) GH forms were primarily concerned with whether they represented precursors or aggregates (Frohman et al. 1972; Stachura and Frohman 1973; Wright et al. 1974; Yalow 1974; Benveniste et al. 1975) and with their biological properties (Gorden 1973b; Guyda 1975; Soman and Goodman 1977). The general conclusion was that the "big" GH forms represented GH aggregates, although this was based solely on analysis of pituitary GH, with plasma "big" and "big-big" GH remaining largely uncharacterized. Even among the pituitary "big" forms, only one such variant (a disulfide dimer) has been characterized in detail (Lewis et al. 1977).

Studies in my laboratory have been concerned with GH heterogeneity and its biological significance for several years. As part of these studies we investigated the composition of circulating GH, focusing first on the monomeric forms (Baumann et al. 1983, 1985b), and later on the "big" forms (Stolar et al. 1984a, 1984b). This effort led to a detailed characterization of the "big" and "big-big" forms as an oligomeric series up to pentameric GH, with previously recognized monomeric variants as building blocks (Stolar et al. 1984a). Although we considered the possibility of complex formation between GH and plasma proteins and recognized the fact that our data did not categorically exclude such an interaction, we were unable to demonstrate conversion of native (unlabeled) GH to high molecular weight immunoreactivity upon incubation with plasma. This was in agreement with earlier work by others (Berson and Yalow 1968; Yalow 1974) and, reinforced in our interpretation by immunochemical evidence of similarity between circulating and pituitary GH oligomers (Stolar and Baumann 1986), we concluded that there was no evidence for binding of GH to plasma components (Stolar et al. 1984a). However, in the course of these experiments, we repeatedly observed that mixtures of radioiodinated GH with plasma yielded radioactive peaks eluting near the void volume of the column (i.e. in the range of "big-big" GH) during gel filtration in Sephadex G-100. As had others before us, we attributed that to the known propensity of radioiodinated peptides to aggregate, which might somehow be enhanced in plasma. Furthermore, void volume peaks are always suspicious as being artifacts. Nevertheless, we were intrigued by the following three observations: a) the void volume peak appeared proportionally larger in short columns than in long columns; b) frequently a double peak was present; and c) "aggregation" of [^{125}I]GH occurred many times faster in the presence of plasma than in its absence (Baumann and Amburn 1986a). We finally decided to investigate these phenomena under rigorous conditions. The ensuing experiments, much to our surprise, led to the discovery of the GH binding proteins (Baumann et al. 1985a, 1986).

Herington and coworkers in Australia had been working on cellular GH receptors for a number of years. In the course of their studies with rabbit liver membranes, they found that not only membrane-bound GH receptors were relatively soluble in aqueous media (Herington et al. 1981), but that a very similar GH-binding protein was present in the cytosolic fraction of rabbit liver and other tissues (Ymer et al. 1984; Herington et al. 1986a). They then demonstrated that rabbit serum also contained the same or a similar GH receptor-like protein (Ymer and Herington 1985). Extension of their investigations to human serum then revealed circulating binding proteins for GH in that species. Thus, via a completely different approach, Herington et al. independently demonstrated the existence of GH binding proteins in human plasma, their report appearing only a few months after ours (Herington et al. 1986b).

Additional strong evidence for the GH binding proteins was reported at about the same time by Nixon and Jordan (Nixon and Jordan 1986), who compared gel filtration pattern of GH contained in the cerebrospinal fluid (CSF) of a patient with a large suprasellar GH-secreting pituitary tumor with that of GH in a simultaneously obtained blood specimen. The CSF pattern corresponded to monomeric GH, while the plasma contained high molecular weight forms. Mixing CSF with GH-poor plasma resulted in partial conversion of monomeric GH to "big" and "big-big" forms. While these authors did not directly prove the existence of the binding proteins, their observations are in complete agreement with ours and with those of Herington et al. It is of interest that after two decades of general conviction that GH existed in free form in plasma, three groups almost simultaneously performed the critical experiments which documented the existence of GH binding proteins in human blood.

REASONS FOR THE DELAYED RECOGNITION
OF GH BINDING PROTEINS

The question must be asked why GH- binding proteins were not recognized for so many years. The evidence for the binding phenomenon is quite clear (Fig. 1), and the experimental conditions to show binding are simple enough to be reproduced in any laboratory. As already alluded to above, the evidence for the binding phenomenon has been published before, but was never pursued in detail.

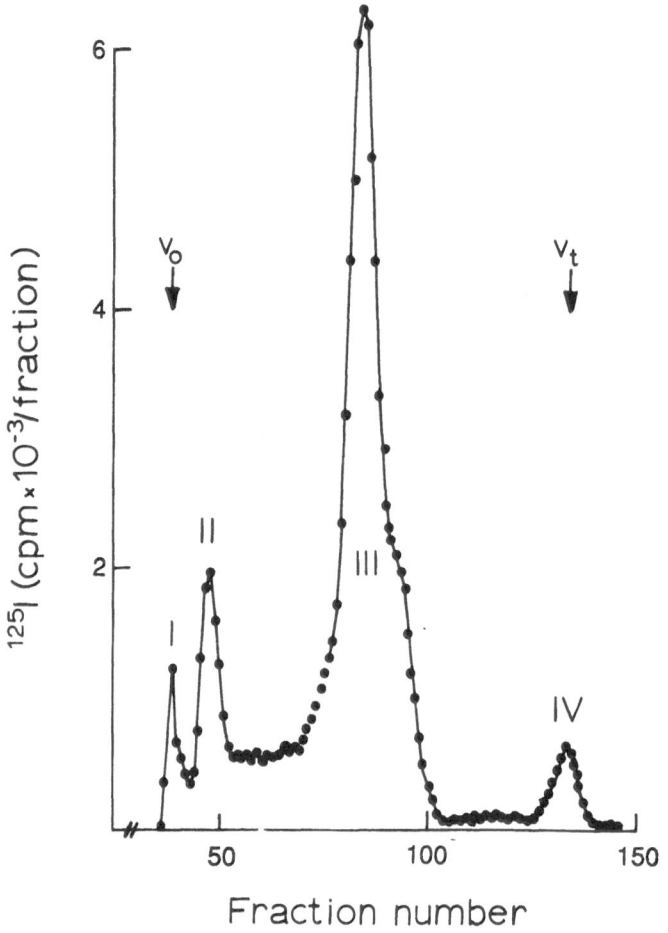

Fig. 1. Typical gel filtration profile of a plasma-[^{125}I]hGH mixture on Sephadex G-100. Peaks I and II represent the two complexes of GH with the binding proteins, peak III free [^{125}I]hGH, and peak IV iodide. The plateau area between peaks II and III represents [^{125}I]GH dissociating from the complexes during gel filtration. V_o denotes the void volume, V_t the total volume of the column.

Several factors came together to prevent detection of the binding proteins. Probably, first and foremost among these was the prevailing dogma about the state of peptide hormones in blood. Because of the intensity of the argument regarding this issue in the 1960s (see above), any deviation from accepted teachings was considered either artifact or heresy. This made it difficult and unattractive for investigators to pursue such deviating observations.

A second reason lies in the widespread lack of full appreciation of the reversibility of binding interactions, and of how this affects the behavior of complexes during analysis. In the case of complexes linked by forces of moderately high affinity, such as, for example, GH and its binding proteins, dissociation during column chromatography is substantial (Baumann et al. 1986). Consequently, the amount of residual binding after column separation of complexed from free hormone varies depending on time, temperature, dilution factor, and column size. Taken to its extreme conclusion, the absence of a peak corresponding to the complex in no way indicates that no complex existed in the unfractionated sample. (While I use a gel filtration column as an example here, similar considerations apply to other analytical techniques that, by their very nature, disrupt the binding equilibrium). Dissociation is obviously much less of a problem with strongly interacting systems, such as antigen-antibody complexes or the complexes between thyroid hormones and their specific binding proteins. In the case of GH in plasma, most of the techniques used tended to minimize the residual bound fraction which was then easily overlooked. Indeed, efforts to use high resolution systems (such as longer and bigger columns) to better separate the various size isomers of GH had the opposite effect in terms of recognizing complexed GH because of greater complex dissociation. This was certainly one reason for our own failure to detect the complex in a study designed to analyze the nature of "big" plasma GH (Stolar et al. 1984a). It is evident that the considerations discussed here are relevant to other potential hormone-binding protein complexes.

A third reason for the difficulty in identifying GH-binding proteins is the similarity between their molecular size and that of GH oligomers. Thus, complexed GH coelutes with "big-big" GH upon gel filtration, which makes it difficult to differentiate GH-binding protein-complexes from GH oligomers. The proportion of "big-big" GH is highly variable depending on the age, treatment, and history of the plasma sample (see Stolar et al. 1984a for review). Therefore, it was difficult to recognize the contribution of complexed GH to the proportion of "big-big" GH against this variable background. A further factor contributing to variable immunoreactivity in the "big-big" fraction is nonspecific interference of plasma proteins in the radioimmunoassay for GH. This is particularly prevalent in plasma fractions eluting near the void volume of a column which contain most of the plasma proteins, i.e., again in the "big-big" fraction. For all these reasons, it was easy to overlook the contribution of complexed GH to the total immunoreactivity in the "big-big" fraction. The use of radiolabeled GH as a probe would be free of the problems just discussed but has its own set of problems, as indicated below.

Experiments using radioiodinated GH were generally considered to be not representative for native GH. Whereas a number of investigators showed apparent binding of iodo-GH to plasma proteins (cited above), this was deemed an artifact due to either aggregation or anomalous binding of a chemically altered GH molecule. With the advent of newer mild iodination procedures, molecular integrity of the labeled product is high (Rogol and Chrambach 1975) and aggregation is a very slow process (Baumann and Amburn 1986a). Demonstration of binding of biosynthetically labeled [³H]GH and unlabeled GH was required to correct misconceptions about [^{125}I]GH that were derived from early preparations of lesser quality.

Finally, the decreased immunoreactivity of GH-binding protein complexes should be mentioned as a potential reason for missing them. This would be a problem only when unlabeled GH is used as the probe. Indeed, one of the GH-binding protein complexes (peak I, see below) is poorly immunoreactive. However, the principal complex (peak II) is fully recognized by anti-GH antibodies (Baumann et al. 1986). Therefore, altered immunoreactivity of complexed GH is only a minor factor contributing to the delay in uncovering the GH- binding proteins.

CHEMICAL AND FUNCTIONAL CHARACTERISTICS OF THE GH- BINDING PROTEINS

When human plasma is incubated with monomeric [^{125}I]hGH and then fractionated on a Sephadex G-100 column, a characteristic and highly reproducible pattern is seen (Fig. 1). In addition to free [^{125}I]GH (peak III) and iodide (peak IV), two additional radioactive components are present (peak I and peak II). Peak II represents GH complexed with the main binding protein in plasma, peak I is a secondary complex. The main binding protein has high specificity for hGH, high affinity and limited binding capacity. Its characteristics are summarized in Table I. Chemical cross linking studies between either whole plasma or purified binding proteins and [^{125}I]GH have identified the main (peak II) binding protein as a single-chain protein (Fig. 2) (Herington et al. 1986c; Baumann and Shaw 1986). These studies have also established that cross-linked complexes underestimate the molecular weight of the binding protein and its complex with GH due to molecular constriction induced by cross-linking (Baumann and Shaw 1986). On the other hand, cross-linking does not appear to substantially affect molecular charge, so that a pI of 5 derived from the cross-linked complex is fairly representative of the native complex (Baumann et al., submitted for publication). As already noted in our initial description (Baumann et al. 1986), the complex is fully immunoreactive with anti-hGH antisera, which indicates either that the complex dissociates upon exposure to antibody, or that the critical epitope(s) on the GH molecule remain(s) exposed when complexed with the binding protein. We recently showed that even a covalently cross-linked complex exhibits full immunoreactivity (Baumann et al. 1987b), suggesting that the latter interpretation is correct.

The peak-I complex represents a minority of the complexed GH in plasma. The binding protein responsible for peak-I is less well characterized than the peak-II binding protein. It copurifies with peak-II binding protein in GH-affinity

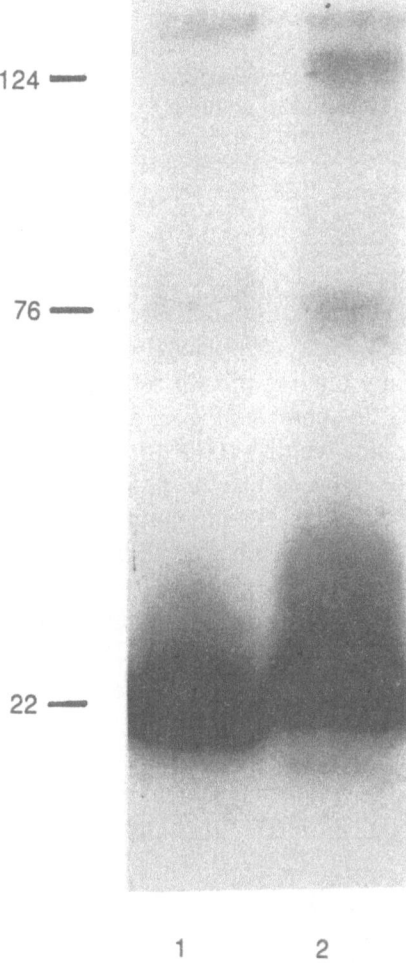

Fig. 2. Autoradiograph of an SDS-PAGE gel containing [^{125}I]GH crosslinked to a purified binding protein preparation. The numbers on the left denote the mol. wt. in kilodaltons. *Lane* 1 contains a nonreduced sample, *lane* 2 a sample reduced with dithiothreitol. The three radioactive components are free GH (22 K), the peak-II complex (76 K), and the peak-I complex (124 K).

Table 1. Characteristics of the main GH-binding protein in human plasma

---

1. Specificity: human GH_{22K} and hGH_{20K}
    does not bind human placental lactogen or prolactin
    does not bind animal GHs
    does not bind a variety of other polypeptide hormones
2. Affinity: $K_a = 3 \times 10^8/M$ for hGH_{22K}
    $K_a = 1.2 \times 10^7/M$ for hGH_{20K}
3. Binding capacity: 20 ng/ml plasma
4. Association rate: 75%-80% of maximal binding reached in 5 min at 37°C
5. Dissociation rate: 50% dissociated within 40 min at 21 C
6. Binding stoichiometry: 1 mole GH per mole binding protein
7. Molecular weight: 61 K, complex 80-85 K
8. Isoelectric point: ~ 5 (cross-linked complex)
9. Calculated plasma concentration: 1 nM

Data derived from Baumann et al. 1986; Herington et al. 1986b; Baumann and Shaw 1986.

chromatography (Baumann et al. 1986; Baumann and Shaw 1986), but was initially thought to be a nonspecific, nonsaturable binding component (Baumann et al. 1986). However, recent studies in my laboratory indicate that peak-I is a specific but relatively low affinity binding protein for GH. This binding protein has also been crosslinked to GH; the complex has an approximate mol. wt. of 124 K (Fig. 2), thereby indicating a mol. wt. of ~ 100 K for the binding protein itself (Baumann and Shaw, 1986). Its tentative isoelectric point is about 7 (Baumann and Shaw, 1986). GH complexed with peak-I binding protein is only about 30% immunoreactive (Baumann et al. 1986). Little other detailed information is presently available about this secondary binding protein.

It is not known whether GH oligomers bind to either of the binding proteins. Preliminary experiments to answer this question have been performed in our laboratory but have not yet yielded conclusive results.

EFFECT OF THE BINDING PROTEINS ON THE STATE OF GH IN BLOOD

Part of the circulating GH associates with the binding proteins to form a complex. The precise distribution between the three moieties, free GH, peak-II complex and peak-I complex is of physiological interest. Unfortunately, conventional techniques do not permit accurate estimation of free and bound GH fractions in undiluted plasma because of the above-mentioned dissociation of complexes during analysis. We have used methodology that minimizes disruption of binding equilibria, such as frontal analysis, or that allows for a correction to the undissociated state, such as gel filtration in prelabeled columns, to estimate the bound and free fractions at 37°C. We found that under basal conditions (GH level <5 ng/ml) 45%-50% of $hGH2_{2K}$ and 25%-30% of hGH_{20K} is bound to the binding proteins (Baumann and Amburn 1986b). The majority of the bound fraction (80%-85%) is complexed with peak-II binding protein. At high GH levels, the bound proportion declines due to saturation of the (primarily peak-II) binding proteins, but this effect does not become marked until levels above 20 ng/ml are reached. This increased proportion of free hormone at high GH levels is probably responsible for the previously reported (Gorden et al. 1976) higher proportions of "little" GH in the plasma of acromegalic patients compared with that in normal subjects. We have found a high degree of correlation between calculated free GH and measured "little" GH values in acromegaly (Baumann, 1987).

Since such a substantial fraction of GH in plasma is bound, the question arises to what degree the presence of the binding proteins affects the measurement of GH levels by radioimmunoassay. As already indicated above, the peak-II complex is fully immunoreactive and thus fully measured. On the other hand, peak-I complex has diminished immunoreactivity. However, since peak-I complex represents only a small fraction of total plasma GH, the underestimation of GH concentration by radioimmunoassay is minor and, in my estimation, does not exceed 10%.

PHYSIOLOGICAL SIGNIFICANCE OF THE GH- BINDING PROTEINS

The biological function of the GH- binding protein is largely unknown at present. One such function appears to relate to the in vivo turnover of GH. We have shown in a rat model that GH complexed with purified binding proteins is cleared from blood at a substantially slower rate than free GH, has a much smaller distribution space, and is degraded at a slower rate than free GH (Baumann et al. 1987a). We recently further confirmed this finding with a covalent GH-binding protein-complex of proper 1:1 stoichiometry (Baumann 1987b). Thus, it appears that the binding proteins prolong the in vivo survival of GH, restrict its dilution and distribution in the body, and modulate its fluctuations in serum by acting as a circulating buffer and hormone reservoir.

Another biological effect of the binding proteins thus far demonstrated is their inhibitory influence on GH binding to tissue receptors. This was shown with whole plasma by Herington et al. (Herington et al. 1986b), and with purified binding protein in our laboratory (G. Baumann, unpublished observations). This effect of the binding protein is not unexpected since the affinity of the main (peak-II) binding protein for GH is similar to that of GH receptors. Thus, the binding proteins compete with the receptor for GH binding. This property is another example of how the binding proteins can modulate GH action. The sum of the two effects described here on overall GH action and economy is complex and difficult to predict. The discovery of further biological effects of the binding proteins is likely to increase the complexity of our knowledge about GH action.

It is unclear whether the binding proteins have any relationship to the puzzling disparity between GH-like bioactivity and immunoreactive GH in plasma (Ellis et al. 1978), but this remains an intriguing, albeit speculative possibility.

SOURCE AND REGULATION OF THE GH- BINDING PROTEINS

The GH- binding proteins are present at fairly constant levels, at least as assessed by a GH-binding assay, in all normal plasma samples examined to date. There appear to be no major differences between men and women, between pregnant and nonpregnant women, or between children and adults (Shaw et al. 1987). An exception are newborns who have low GH binding activity in their serum (Shaw et al. 1987). A variety of disease states, such as infection, uremia, acromegaly and hypopituitarism, are also attended by largely normal binding protein activity in plasma (Shaw et al. 1987). Patients with liver cirrhosis have variable but generally decreased binding protein activity, suggesting the liver as a possible organ of origin for the binding proteins (Shaw et al. 1987). Based on these preliminary data, and in consideration of their low plasma concentration, it appears that the GH- binding proteins are tightly regulated to yield a constant plasma level.

The main binding protein (peak-II) is completely absent - or non-functional - in Laron dwarfism (Baumann et al. 1987c; Daughaday and Trivedi, 1987), which is the only condition identified to date with such an abnormality. The minor (peak-I) binding protein appears to be normal in Laron dwarfism (Baumann et al. 1987c). This finding has interesting implications with regard to the nature, source, and function of the lacking binding protein. Since Laron dwarfism is believed to be caused by a deficiency in GH receptors, the parallel deficiency of the plasma binding protein in that condition suggests a relationship between the receptor and the binding protein. The coexistence of severe growth failure and deficient binding protein could also indicate that the binding protein is important for normal growth to occur.

The source of the two circulating binding proteins is obscure at this time. It has been suggested that they may be derived from tissue GH receptors by shedding of receptor subunits or fragments into the extracellular space (Baumann et al. 1986; Herington et al. 1986b). There is evidence both for and against this possibility. The strongest argument for a derivation from receptors comes from the observations in Laron dwarfs just described. Absence of the receptor in this disorder would also result in absence of the binding protein if the former was the source of the latter. However, this would apply only to the peak-II binding protein. Another point in favor of a receptor connection is the similar hormone specificity of the main (peak-II) binding protein and the GH receptor, as well as their similar affinity for GH. In the rabbit, immunochemical similarity between the GH receptor and the plasma binding protein has been demonstrated (Barnard and Waters, 1986). The molecular size of the rabbit plasma binding protein corresponds to the receptor subunit and the cytosolic GH binding protein in one study (Barnard and Waters 1986) but not in another (Ymer and Herington 1985). It is not clear to what extent the rabbit is representative for man. In man, the molecular size of the peak-II binding protein does not correspond to the human receptor or its subunit (Asakawa et al. 1985; Baumann and Shaw 1986; Herington et al. 1986c; Hughes et al. 1983). This does not necessarily argue against a receptor source of the binding protein; it is possible that a receptor fragment containing the GH binding domain is released from cell membranes. The molecular size of the peak-I binding protein does correspond to that of the human GH receptor subunit (Baumann and Shaw 1986), but, ironically, this is the binding protein that is present in Laron dwarfism. Peak-I binding protein also has much lower affinity for GH than the receptor. Perhaps the most compelling argument against the receptor origin of the binding proteins is the low abundance of GH receptors in human tissues. Indeed, it is very difficult to demonstrate any binding of GH to normal human cells or tissues which are clearly responsive to GH, with the exception of, liver the where binding is very low and variable (Carr and Friesen 1976; Baumann and Abramson, unpublished data). In view of this dearth of demonstrable GH receptors, it is hard to visualize how so much binding activity in plasma can be derived from so little binding activity in tissues. In my opinion, the connection between GH receptors and the binding proteins, while being an attractive concept, cannot presently be considered firmly established.

SUMMARY AND CONCLUSIONS

Within the last two years, binding proteins for hGH have been identified in human plasma. The principal binding protein has high specificity, high affinity and limited binding capacity for hGH; it has a mol. wt. of about 60 K. A second, larger (100 K) binding protein has lower affinity for GH. Almost half of circulating GH is bound under physiological conditions, principally to the higher affinity binding protein. The GH-binding protein -complex constitutes part of "big-big" GH and, to a lesser extent, "big" GH in plasma, with other constituents of those fractions being GH oligomers. The binding proteins strongly affect the in vivo kinetics of GH, prolonging its half-life, restricting its distribution, and slowing its degradation. The binding proteins inhibit the interaction of GH with its tissue receptors. Circulating binding protein levels appear well- preserved in a variety of physiological and pathological conditions. Levels are low in neonates and variably decreased in liver disease. The principal binding protein is completely absent in Laron dwarfism. The tissue source of the binding proteins has not been determined; they may be derived from the GH receptor. The physiological significance of the GH- binding proteins remains to be elucidated.

Note added in proof

Since the submission of this article, the relationship between the main (peak II) GH-BP and the GH receptor has been clarified. Both structural (Leung et al. 1987) and immunochemical (Baumann and Shaw 1988) evidence indicate that the two are related. The BP appears to correspond to the extracellular portion of the GH receptor (Leung et al. 1987).

REFERENCES

Antoniades H (1975) Conversion of [^{125}I] growth hormone into high molecular weight forms *in vivo*. *Endocrinology* **96**: 799-802

Asakawa K, Grunberger G, McElduff A, Gorden P (1985) Polypeptide hormone receptor phosphorylation. Is there a role in receptor-mediated endocytosis of human growth hormone? *Endocrinology* **117**: 631-637

Barnard R, Waters MJ (1986) Serum and liver cytosolic growth-hormone-binding proteins are antigenically identical with liver membrane 'receptor' types 1 and 2. *Biochem J* **237**: 885-892

Baumann G (1987) Molecular heterogeneity of circulating growth hormone in acromegaly. In: Robbins RJ, Melmed S (eds) *Acromegaly: A Century of Scientific and Clinical Progress*. Plenum, New York, pp 35-43

Baumann G, Amburn K (1986a) The autodecomposition of radiolabeled growth hormone. *J Immunoassay* **7**: 139-149

Baumann G, Amburn K, Shaw MA (1988) The circulating growth hormone (GH)-binding protein complex: a major constituent of plasma GH in man. *Endocrinology* **122**: 976-989

Baumann G, Shaw MA (1986) The circulating growth hormone binding proteins: partial purification and structural characterization by affinity crosslinking. *Clin Res* **34**: 949A

Baumann G, Shaw MA (1988) Immunochemical similarity of the human plasma growth hormone-binding protein and the rabbit liver growth hormone receptor. *Biochem Biophys Res Commun* **152**: 573-578

Baumann G, MacCart JG, Amburn K (1983) The molecular nature of circulating growth hormone in normal and acromegalic man: evidence for a principal and minor monomeric forms. *J Clin Endocrinol Metab* **56**: 946-952

Baumann G, Amburn K, Stolar MW (1985a) A growth hormone binding protein in human plasma. *Clin Res* **33**: 567 (Abstract)

Baumann G, Stolar MW, Amburn K (1985b) Molecular forms of circulating growth hormone during spontaneous secretory episodes and in the basal state. *J Clin Endocrinol Metab* **60**: 1216-1220

Baumann G, Stolar MW, Amburn K, Barsano CP, DeVries BC (1986) A specific growth hormone-binding protein in human plasma: initial characterization. *J Clin Endocrinol Metab* **62**: 134-141

Baumann G, Amburn KD, Buchanan TA (1987a) The effect of circulating growth hormone-binding protein on metabolic clearance, distribution, and degradation of human growth hormone. *J Clin Endocrinol Metab* **64**: 657-660

Baumann G, Shaw MA, Buchanan TA (1987b) Growth hormone binding proteins in human plasma: metabolic clearance of a covalently linked growth hormone-binding protein complex. *Clin Res* **35**: 884A

Baumann G, Shaw MA, Winter RJ (1987c) Absence of the plasma growth hormone-binding protein in Laron-type dwarfism. *J Clin Endocrinol Metab* **65**: 814-816

Beitins IZ, Rattazzi MC, MacGillivray MH (1977) Conversion of radiolabeled human growth hormone into higher molecular weight moieties in human plasma *in vivo* and *in vitro. Endocrinology* **101**: 350-359

Benveniste R, Stachura ME, Szabo M, Frohman LA (1975) Big growth hormone (GH) conversion to small GH without peptide bond cleavage. *J Clin Endocrinol Metab* **41**: 422-425

Berson SA, Yalow RS (1966) State of human growth hormone in plasma and changes in stored solutions of pituitary growth hormone. *J Biol Chem* **241**: 5745-5749

Berson SA, Yalow RS (1968) Peptide hormones in plasma. *Harvey Lect* **62**: 107-163

Bieler EU, Pitout MJ, Stroud SW, VanRooyen RJ (1977) Conversion of monomeric human growth hormone and big growth hormone into different molecular weight forms *in vitro* and after injection into humans. *Hormone Res* **8**: 29-36

Carr D, Friesen HG (1976) Growth hormone and insulin binding to human liver. *J Clin Endocrinol Metab* **42**: 484-493

Collipp PJ, Kaplan SA, Boyle DC, Shimizu CSN (1964) Protein-bound human growth hormone. *Metabolism* **13**: 532-538

Collipp PJ, Kaplan SA (1966) Interaction of ^{14}C-labeled human and bovine growth hormone with serum proteins. *Biochem Biophys Acta* **117**: 416-423

Daughaday WH, Trivedi B (1987) Absence of serum growth hormone binding protein in patients with growth hormone receptor deficiency (Laron dwarfism) *Proc Natl Acad Sci USA* **84**: 4636-4640

Ellis S, Vodian MA, Grindeland RE (1978) Studies on the bioassayable growth hormone-like activity of plasma. *Recent Prog Horm Res* **34**: 213-238

Frohman LA, Burek L, Stachura ME (1972) Characterization of growth hormone of different molecular weights in rat, dog, and human pituitaries. *Endocrinology* **91**: 262-269

Gorden P, Hendricks CM, Roth J (1973a) Evidence for "big" and "little" components of human plasma and pituitary growth hormone. *J Clin Endocrinol Metab* **36**: 178-184

Gorden P, Lesniak MA, Hendricks CM, Roth J (1973b) "Big" growth hormone components from human plasma: decreased reactivity demonstrated by radioreceptor assay. *Science* **182**: 829-831

Gorden P, Lesniak MA, Eastman R, Hendricks CM, Roth J (1976) Evidence for higher proportion of "little" growth hormone with increased radioreceptor activity in acromegalic plasma. *J Clin Endocrinol Metab* **43**: 364-373

Goodman AD, Tanenbaum R, Rabinowitz D (1972) Existence of two forms of immunoreactive growth hormone in human plasma. *J Clin Endocrinol Metab* **35**: 868-878

Guyda HJ (1975) Heterogeneity of human growth hormone and prolactin secreted in vitro: immunoassay and radioreceptor assay correlations. *J Clin Endocrinol Metab* **41**: 953-967

Hadden DR, Prout TE (1964) A growth hormone binding protein in normal human serum. *Nature* **202**: 1342-1343

Hadden DR, Prout TE (1965) Studies on human growth hormone II. The effect of human serum on growth hormone labeled with radioactive iodine. *Bull J Hopkins Hosp* **116**: 122-131

Herington AC, Elson D, Ymer S (1981) Water soluble receptors for human growth hormone from rabbit liver. *J Recept Res* **2**: 203-220

Herington AC, Ymer S, Roupas P, Stevenson J (1986a) Growth hormone-binding proteins in high-speed cytosols of multiple tissues of the rabbit. *Biochem Biophys Acta* **881**: 236-240

Herington AC, Ymer S, Stevenson J (1986b) Identification and characterization of specific binding proteins for growth hormone in normal human sera. *J Clin Invest* **77**: 1817-1823

Herington AC, Ymer SI, Stevenson JL (1986c) Affinity purification and structural characterization of a specific binding protein for human growth hormone in human serum. *Biochem Biophys Res Commun* **139**: 150-155

Hintz RL, Liu F (1977) Demonstration of specific plasma protein binding sites for somatomedin. *J Clin Endocrinol Metab* **45**: 988-995

Huang JS, Huang SS, Deuel TF (1984) Specific covalent binding of platelet-derived growth factor to human plasma α_2-macroglobulin. *Proc Natl Acad Sci USA* **81**: 342-346

Hughes JP, Simpson JSA, Friesen HG (1983) Analysis of growth hormone and lactogenic binding sites cross-linked to iodinated human growth hormone. *Endocrinology* **112**: 1980-1985

Irie M, Barrett RJ (1962) Immunologic studies of human growth hormone. *Endocrinology* **71**: 277-287

Leung DW, Spencer SA, Cachianes G, Hammonds G, Collins C, Henzel WJ, Barnard R, Waters, MJ, Wood WI (1987) Growth hormone receptor and serum binding protein: purification, cloning and expression. *Nature* **330**: 537-543

Lewis UJ, Peterson SM, Bonewald LF, Seavey BK, VanderLaan WP (1977) An interchain disulfide dimer of human growth hormone. *J Biol Chem* **252**: 3697-3702

MacMillan DR, Schmid JM, Eash SA, Read CH (1967) Studies on the heterogeneity and serum binding of human growth hormone. *J Clin Endocrinol Metab* **27**: 1090-1094

Megyesi K, Kahn CR, Roth J, Gorden P (1975) Circulating NSILA-s in man: preliminary studies of stimuli in vivo and of binding to plasma components. *J Clin Endocrinol Metab* **41**: 475-484

Nixon DA, Jordan DM (1986) Conversion of CSF monomeric growth hormone to large growth hormone with exposure to serum. *Acta Endocrinol* (Copenh) **111**: 289-295

Peeters S, Friesen HG (1977) A growth hormone binding factor in the serum of pregnant mice. *Endocrinology* **101**: 1164-1183

Raines EW, Bowen-Pope DF, Ross R (1984) Plasma binding proteins for platelet-derived growth factor that inhibit its binding to cell surface receptors. *Proc Natl Acad Sci USA* **81**: 3424-3428

Rogol AD, Chrambach A (1975) Radioiodinated human pituitary and amniotic fluid prolactins with preserved molecular integrity. *Endocrinology* **97**: 406-417

Shaw MA, Amburn K, Baumann G (1987) Plasma growth hormone binding proteins in various physiological and pathological states. *Clin Res* **35**: 845A

Soman V, Goodman AD (1977) Studies of the composition and radioreceptor activity of "big" and "little" human growth hormone. *J Clin Endocrinol Metab* **44**: 569-581

Stachura ME, Frohman LA (1973) Large growth hormone: evidence for the association of growth hormone with another protein moiety in the rat pituitary. *Endocrinology* **92**: 1708-1713

Stolar MW, Amburn K, Baumann G (1984a) Plasma "big" and "big-big" growth hormone (GH) in man: an oligomeric series composed of structurally diverse GH monomers. *J Clin Endocrinol Metab* **59**: 212-218

Stolar MW, Baumann G, Vance ML, Thorner MO (1984b) Circulating growth hormone forms after stimulation of pituitary secretion with growth hormone releasing factor in man. *J Clin Endocrinol Metab* **59**: 235-239

Stolar MW, Baumann G (1986) Big growth hormone forms in human plasma: immuno-chemical evidence for their pituitary origin. *Metabolism* **35**: 75-77

Touber JL, Maingay D (1963) Heterogeneity of human growth hormone. Its influence on a radio-immunoassay of the hormone in serum. *Lancet* **i**: 403-405

Wright DR, Goodman AD, Trimble KD (1974) Studies on "big" growth hormone from human plasma and pituitary. *J Clin Invest* **54**: 1064-1073

Yalow RS (1974) Heterogeneity of peptide hormones. *Recent Prog Horm Res* **30**: 597-633

Ymer SI, Stevenson JL, Herington AC (1984) Identification of a rabbit liver cytosolic binding protein for human growth hormone. *Biochem J* **221**: 617-622

Ymer SI, Herington AC (1985) Evidence for the specific binding of growth hormone to a receptor-like protein in rabbit serum. *Mol Cell Endocrinol* **41**: 153-161

Zapf J, Waldvogel M, Froesch ER (1975) Binding of nonsuppressible insulin-like activity to human serum. *Arch Biochem Biophys* **168**: 638-645

Advances in Growth Hormone and Growth Factor Research,
edited by E.E. Müller, D. Cocchi and V. Locatelli
Pythagora Press, Roma-Milano and Springer Verlag, Berlin-Heidelberg © 1989

Direct actions of growth hormone and insulin-like growth factor in cultured adipocytes*

J. Schwartz, C. Carter-Su, C.M. Foster and J. A. Shafer

Departments of Physiology, Pediatrics and Biochemistry, University of Michigan, Medical School, Ann Arbor, Michigan, U.S.A.

The recent availability of growth hormone (GH) produced by recombinant DNA techniques has greatly facilitated our ability to study the mechanism of action of GH. Nevertheless, we have made only limited progress in this regard, despite the fact that GH was first identified over 60 years ago. Progress has been difficult in part because GH has diverse actions, including growth promotion, induction of diabetes and stimulation of lipolysis (for recent reviews see Davidson 1987; Green et al. 1985; Isaksson et al. 1985). In addition, suitable cellular systems for analyzing GH action have been lacking. While rat adipose tissue and cells have been among the most sensitive *in vitro* targets for GH, these preparations have limited viability *in vitro* (hours), while the physiological effects of GH are typically delayed in onset (days). Accordingly, several laboratories have used cell culture techniques to gain insight into delayed actions of growth hormone in adipocytes (Nyberg and Smith 1977; Maloff et al. 1980; Schwartz 1984; Walton et al. 1986; Doglio et al. 1986).

In our laboratory, we have been using the 3T3-F442A cultured cell line and have found these cells excellent for analyzing rapid as well as delayed actions of GH. The 3T3-F442A cells, which are of embryonic mouse origin, grow as preadipocyte fibroblasts and, when confluent, undergo an adipose conversion, expressing the differentiated adipocyte phenotype (Green and Kehinde 1974). These cells are particularly interesting from our standpoint, since GH is one of the major serum factors required for their adipose conversion (Morikawa et al.

* These studies were supported by grants form the National Institutes of Health (AM 34171 and AM 35249, from the National Science Foundation) (DCB 8609773), from the American Diabetes Association, Michigan Affiliate, and from the Rackham School of Graduate Studies of the University of Michigan. Dr. Carter-Su is a recipient of a Career Development Award from the Juvenile Diabetes Foundation. Dr. Foster is a recipient of a postdoctoral fellowship (AM 07245) and a Clinical Associate Physician Award (5M01RR3) from the National Institutes of Health.

1982; Nixon and Green 1984). Our aim was to use the differentiated adipocyte form of the cells as a target for GH, and to examine whether GH altered glucose and lipid metabolism in the cultured adipocytes as it did in rat and human adipocytes. This would provide a point of departure for detailed analysis of cellular mechanisms of GH action.

GROWTH HORMONE ALTERS CARBOHYDRATE METABOLISM IN CULTURED ADIPOCYTES

To obtain differentiated adipocytes in culture, the 3T3-F442A fibroblasts were grown in Dulbecco's modified Eagle's Medium (DMEM) in the presence of calf serum until confluent, and were then exposed to DMEM containing fetal calf serum, as well as insulin, dexamethasone, and methylisobutylxanthine to hasten adipose conversion. After 48 h, cells were maintained in medium containing only fetal calf serum until they were used in experiments. For 18-24 h prior to incubations with hormones, cells were deprived of serum by incubation in DMEM containing bovine serum albumin.

To determine whether GH altered metabolism in the 3T3-F442A adipocytes, we first measured the conversion of [^{14}C] D-glucose to carbon dioxide and lipid, since these parameters were sensitive to GH in rat adipocyte preparations. When glucose oxidation was measured cumulatively over 4h, GH produced a dose-related increase in glucose metabolism (Fig. 1). This corresponds to the stimulation of glucose oxidation by GH observed in adipose tissue and muscle from GH-deficient preparations, an insulin-like effect of GH (Goodman 1968; Davidson 1987). The stimulation was observed in the 3T3 adipocytes only after the cells had been deprived of serum for the preceding 18-24h. As observed in other systems, the magnitude of this insulin-like response to GH was typically 50-100% above control values.

Since the insulin-like responses to GH in other systems are transient, and subside despite the continued presence of GH, we incubated cells with GH for longer periods of time to determine whether the stimulation of glucose oxidation in the cultured adipocytes was also transient. The cultured cells are particularly amenable to long-term incubation with hormone. Figure 2 shows that after 48 h total incubation, GH actually *inhibited* glucose oxidation. In these experiments, the cell monolayers were incubated with or without GH for 44 h. The cells were then suspended and used for measurements of glucose oxidation during the final 4 h, in the presence or absence of GH as before. The inhibition of glucose oxidation was dose dependent, and the maximum inhibition was typically to 50% of control values. Similar results were obtained when the conversion of glucose to lipid was measured cumulatively in the cultured adipocytes. Inhibition was significant at

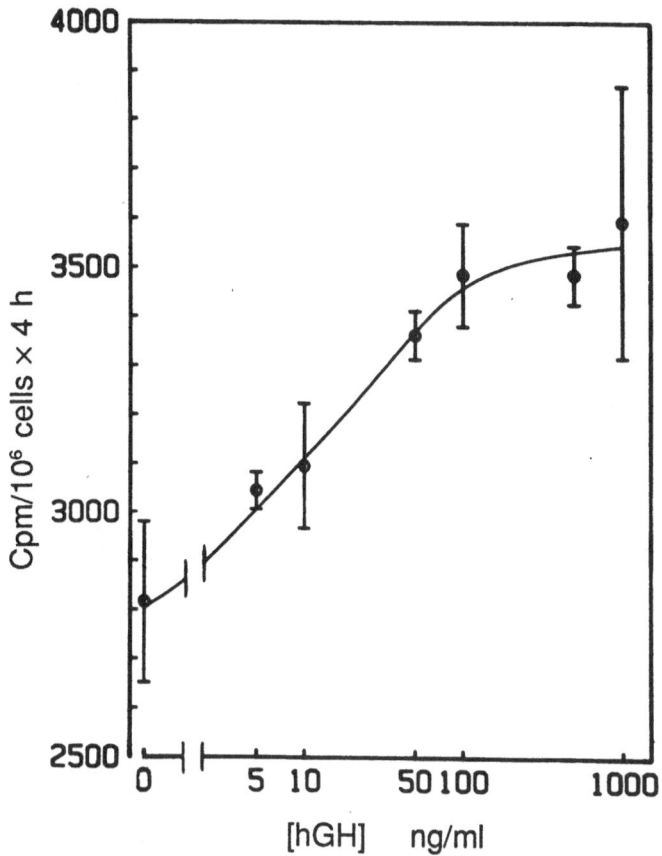

Fig. 1. Stimulation of glucose oxidation by GH in 3T3-F442A adipocytes. Cells were incubated for a total of 4 h. Each point shows the mean ± SEM of four observations. This experiment was repeated four times. (Reproduced from Schwartz et al. [1985]).

concentrations of GH as low as 0.5 ng/ml. The inhibition of the conversion of glucose to lipid or carbon dioxide was sustained for at least 7 days in cells maintained in the presence of GH (Schwartz 1984).

These results indicate, first of all, that the increase in glucose oxidation due to GH is transient and subsides within 48 h. In requiring prior deprivation of fetal calf serum, which contains GH, it appears to be similar to other insulin-like responses to GH which are evident only with prior GH deficiency. Second, these results indicate that the 3T3-F442A adipocytes are highly sensitive to a sustained inhibitory effect of GH on glucose metabolism. Such inhibition, considered an anti-insulin effect of GH, was predicted from earlier studies on diabetogenic actions of GH (Altszuler 1974), but had been difficult to demonstrate *in vitro*. Similar inhibition has been reported in human, rat, and porcine adipose tissue, after several

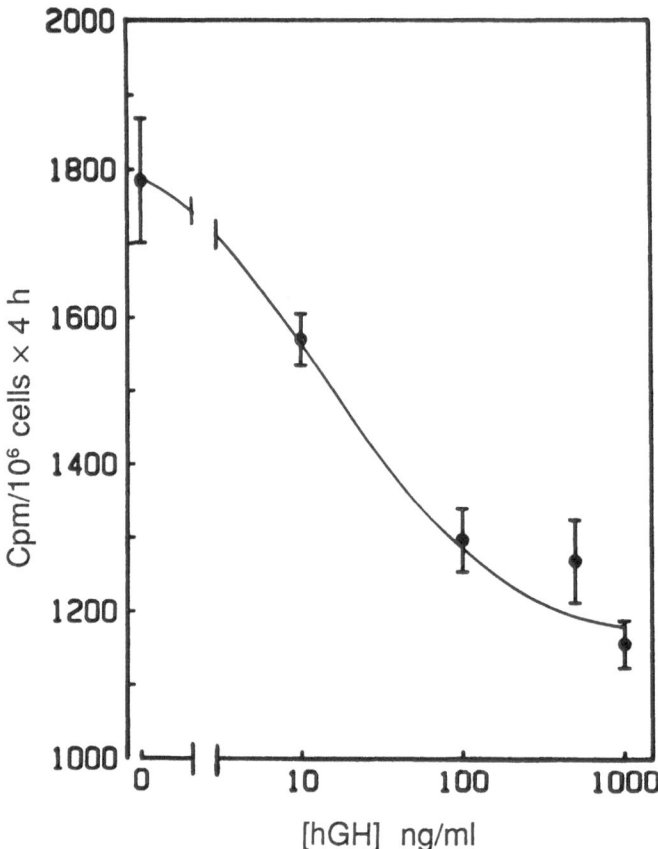

Fig. 2. Inhibition of glucose oxidation by GH in 3T3-F442A adipocytes after 48 h total incubation. Glucose oxidation was measured during the final 4 h. (Reproduced from Schwartz et al. [1985]).

days of culture with GH (Nyberg and Smith 1977; Maloff et al. 1980; Walton et al. 1986). The 3T3-F442A adipocytes thus provide a reproducible cellular model demonstrating both the characteristic insulin-like and anti-insulin responses to GH.

Pituitary human GH (hGH) was used in all of the experiments described above. Recombinant-DNA-derived methionyl-hGH (met-hGH) produced comparable insulin-like and anti-insulin responses to GH in the 3T3 adipocytes (Schwartz and Foster 1986). Figure 3 shows the comparison of the ability of pituitary and met-hGH to inhibit the conversion of glucose to lipid after 48-h incubation. The inhibition by pituitary hGH was evident at 5 ng/ml in this experiment. The met-hGH was compared with a control containing mannitol, since mannitol was present in the met-hGH preparations. The 22,000 mol. wt. form of met-hGH inhibited lipid accumulation as effectively as native (22,000) pituitary hGH. In addition, the

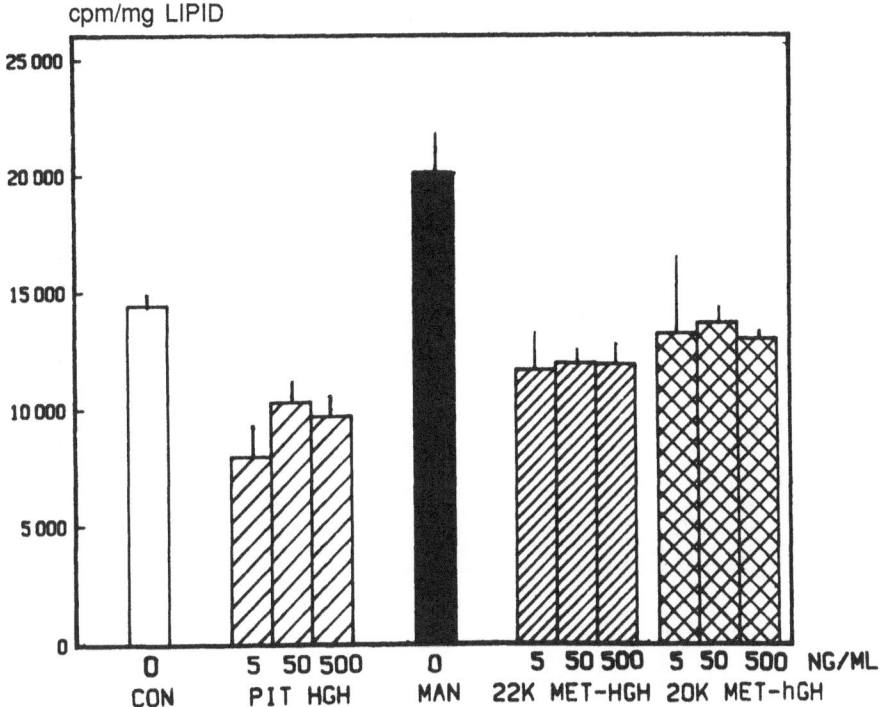

Fig. 3. Comparison of the inhibition of lipid accumulation in 3T3-F442A adipocytes by pituitary and recombinant-DNA-derived human GH preparations. Each bar shows the mean + SEM of triplicate observations. Similar results were obtained in three other experiments. (Reproduced from Schwartz and Foster [1986] by permission).

20,000-dalton form of met-hGH had inhibitory activity comparable to that of the other two hGH preparations. These data in the 3T3-F442A adipocytes agree with *in vivo* comparisons of the diabetogenic activity of pituitary and recombinant-DNA-derived hGH preparations (Kostyo et al. 1985, Ader et al. 1986). In the studies described below, the 22,000 mol. wt. form of met-hGH, which was dialyzed to remove mannitol, was used, unless indicated otherwise.

GROWTH HORMONE RAPIDLY ALTERS GLUCOSE UPTAKE

In order to determine the mechanism of GH action with more precision, we felt it necessary to obtain greater temporal refinement in measuring responses to GH in the 3T3 adipocytes. Our previous assays generally required a minimum of 4 h. Measuring the uptake of glucose in the 3T3 adipocytes served two purposes. It not only provided an indicator of GH action with resolution of 5 min or less

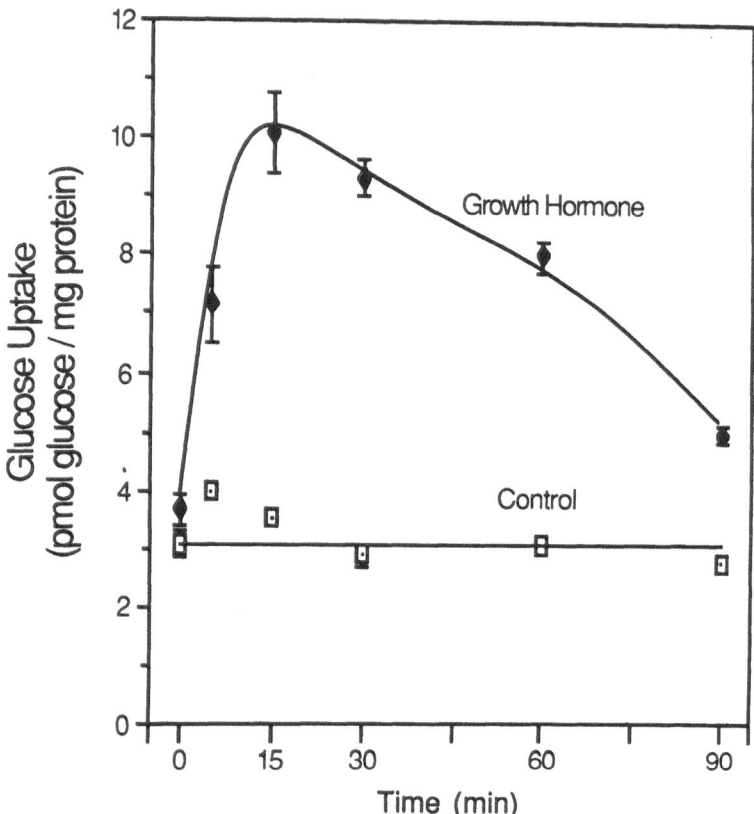

Fig. 4. Stimulation of glucose uptake by met-hGH in 3T3-F442A adipocytes. Cells incubated with met-hGH (500 ng/ml) for the times indicated. The uptake of [^{14}C] D-glucose (558 nM) was measured during the subsequent 5 min. Each point represents the mean ± SEM for triplicate observations. Similar results were obtained in more than ten experiments. (Reproduced from Schwartz and Carter-Su [1988] by permission).

(duration of the uptake assay); it also provided mechanistic information on the contribution of glucose uptake to the GH-induced changes in glucose metabolism measured previously. Figure 4 indicates that as little as 5 min incubation with GH stimulates the uptake of [^{14}C] D-glucose (Kashiwagi et al. 1983; Gliemann et al.1984), measured during the subsequent 5 min. The stimulation after 5 min incubation with GH was approximately two fold and continued to increase to a peak three fold increase after 15 min incubation with GH. Then, as is characteristic of insulin-like responses to GH, glucose uptake subsided during the subsequent hour, despite the continued presence of GH. Both the rapidity and the magnitude of the stimulation of glucose uptake by GH have been highly reproducible.

However, after 24-h incubation with GH, glucose uptake was inhibited in the 3T3-F442A adipocytes. The inhibition was evident even when fresh GH was added

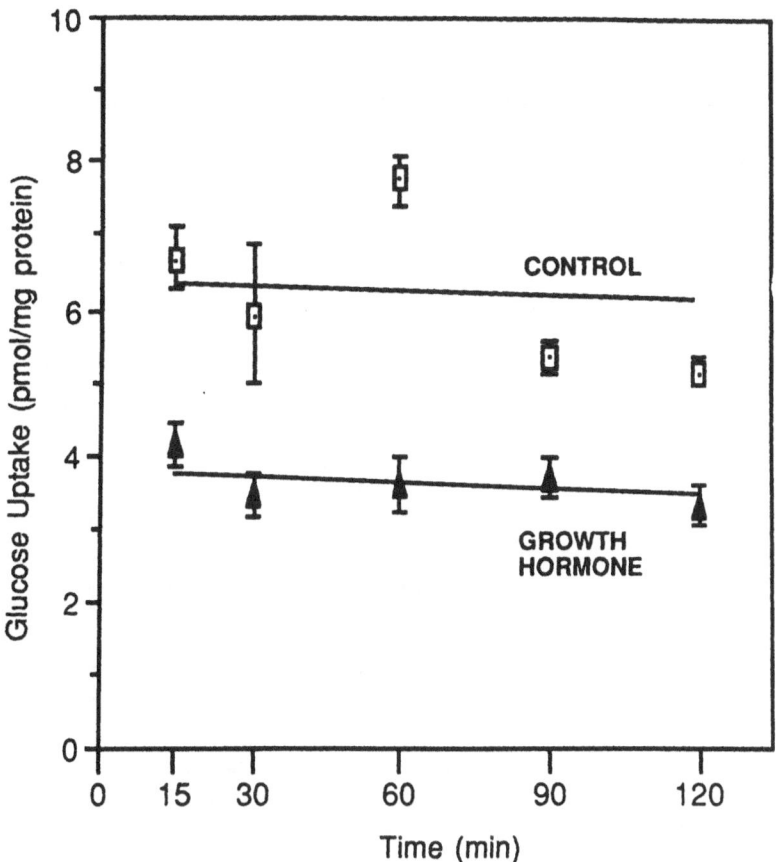

Fig. 5. Inhibition of glucose uptake by GH in 3T3-F442A adipocytes after long-term incubation. Cells incubated with or without met-hGH (500 ng/ml) for 24 h. Fresh medium of the same composition was added for the indicated amounts of time prior to assay for glucose uptake. Same format as for Fig. 4. (Reproduced from Schwartz and Carter-Su [1988] by permission).

for various times after the initial 24-h, as shown in Fig. 5. These findings indicate that changes in glucose uptake contribute to the changes in glucose metabolism in the 3T3 adipocytes. Furthermore, since GH stimulates glucose uptake significantly within 5 min, our findings make it most likely that the initial effect of GH on glucose uptake is a direct effect of the hormone.

GROWTH HORMONE AND INSULIN-LIKE GROWTH FACTOR I

Since insulin-like growth factor I (IGF-I) is thought to mediate some of the actions of GH (Van Wyk and Underwood 1978; Daughaday 1980), the abilities

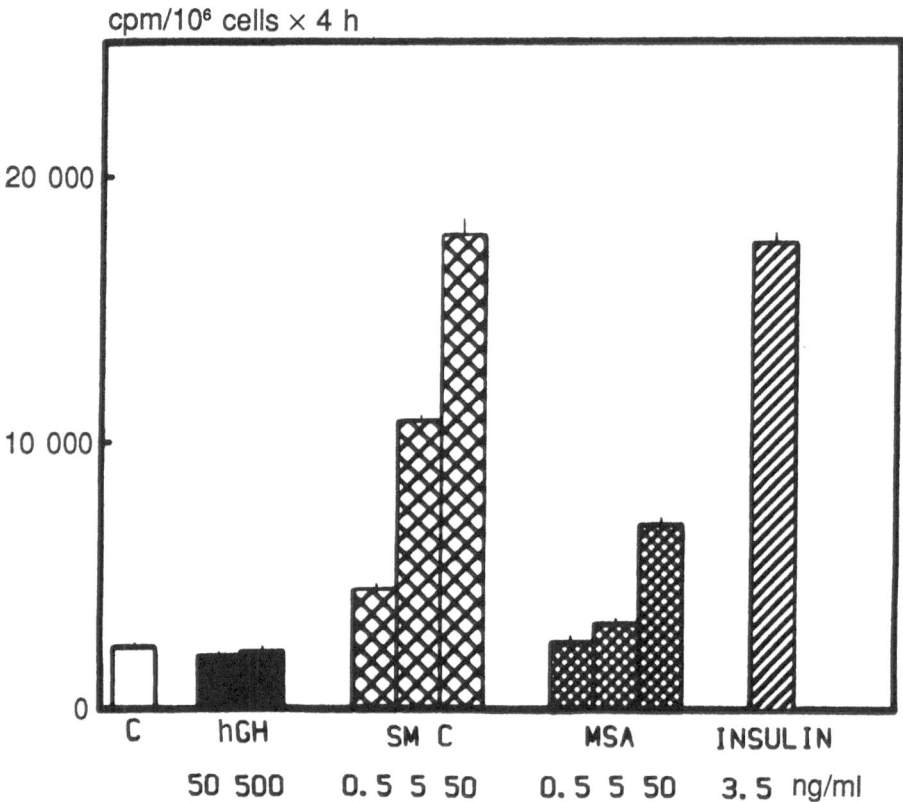

Fig. 6. Comparison of responses of 3T3-F442A adipocytes to hGH, IGF-I (SM C), rIGF-II (MSA) and insulin after 48 h total incubation. Glucose oxidation was measured during the final 4 h. Each *bar* represents the mean + SEM of four observations. Similar results were obtained in four other experiments. The inhibition by pituitary hGH (500 ng/ml) and the stimulation by hIGF-I, rIGF-II and insulin were significant except for IGF-II at 0.5 ng/ml. (Reproduced from Schwartz et al. [1985]).

of GH and IGF-I were compared in the 3T3 adipocytes. Figure 6 shows that after 48-h incubation, when GH is inhibitory, IGF-I stimulates glucose oxidation. Similar results were obtained when glucose uptake and conversion to lipid were measured. Further, when GH and IGF-I were tested in combination after 48 h, GH inhibited IGF-I-stimulated glucose oxidation and lipid accumulation (Schwartz et al. 1985). These findings are consistent with GH and IGF-I exerting their effects in 3T3-F442A adipocytes by different mechanisms after 24-48 h of incubation.

In contrast, after short-term incubations, both GH and IGF-I (SM C), as well as IGF-II (MSA) and insulin, stimulated glucose oxidation in 4 h in the 3T3 adipocytes (Fig. 7). The magnitude of the stimulation by GH was less than with

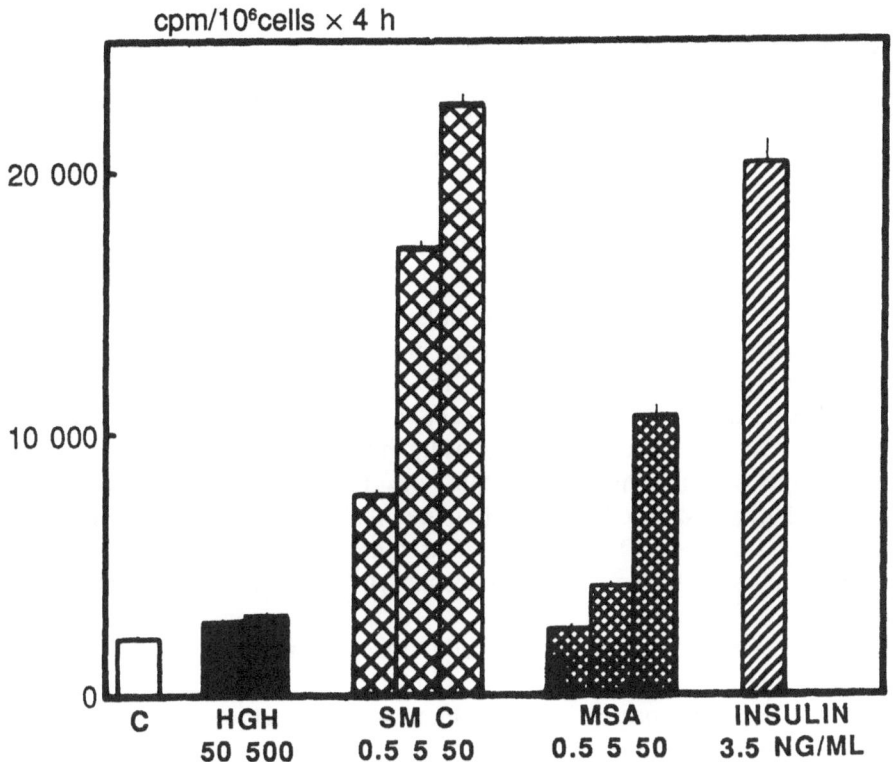

Fig. 7. Comparison of the stimulation of glucose oxidation by hGH, IGF-I (SM C), rIGF-II (MSA) and insulin after 4 h total incubation. Same experiment as in Fig. 6. The stimulation was significant for all conditions except rIGF-II at 0.5 ng/ml. (Reproduced from Schwartz et al. [1985]).

any of the other hormones. Similar results were obtained after 15 min when the effects of GH and IGF-I on glucose uptake were compared. Then, the stimulation of glucose uptake by GH subsided, typical of the transient nature of its insulin-like effects, while the stimulation by IGF-I was sustained throughout incubation with the growth factor. When maximally effective concentrations of IGF-I and GH were tested in combination after 4 h to evaluate whether their ability to stimulate glucose metabolism exhibited additivity, the stimulation of the conversion of glucose to lipid was no greater with GH and IGF-I than with IGF-I alone. The non-additivity of the stimulation by GH and IGF-I suggests that, unless the cells were incapable of being stimulated beyond the level produced by IGF-I alone, the two hormones share a common mechanism in their ability to stimulate glucose metabolism. However, the stimulation of lipid accumulation or glucose uptake by GH was not altered by addition of a monoclonal antibody to IGF-I to the incubation medium, while the stimulation by IGF-I was completely blocked by anti-IGF-I. This suggests that the stimulation of glucose metabolism by GH in 3T3 adipocytes does not involve participation of IGF-I released into the incubation medium, although other systems may work this way (D'Ercole et al. 1984).

THE GROWTH HORMONE RECEPTOR IN 3T3 ADIPOCYTES

To probe more directly into the cellular mechanism of the action of GH, we have used the 3T3 cells to characterize the growth hormone receptor, and to examine whether the GH receptor, like those for several other growth-promoting peptides (e.g., Cohen et al.1980; Kasuga et al. 1982; Nishimura et al. 1982; Jacobs et al. 1983; Carter-Su and Pratt 1984), has ligand-promoted tyrosine kinase activity. Specifically, we investigated whether the GH receptor undergoes tyrosyl phosphorylation and whether GH stimulates the phosphorylation of its receptor in the 3T3-F442A cells. To this end, we used cross-linking agents in combination with a highly specific antibody to phosphorylated tyrosyl residues (Pang et al. 1985).

^{125}I-hGH was covalently cross-linked to receptors in intact 3T3-F442A fibroblasts in the presence or absence of excess unlabeled GH, and the cross-linked cells were solubilized and passed over a column of phosphotyrosyl-binding antibody immobilized on protein-A Sepharose. Figure 8 shows autoradiograms of SDS polyacrylamide gels of the solubilized GH-receptor complex before (starting material) and after immunoadsorption and elution from the column. Before immunoadsorption, the solubilized proteins contained a major labeled species with $M_r = 134,000$, which was not present when excess unlabeled GH was present. This is consistent with the GH receptor on the 3T3-F442A cells being similar to the GH receptor previously identified on rat adipocytes (Carter-Su et al. 1984; Gorin and Goodman 1984) and other cell types (Donner 1983; Asakawa et al. 1985; Hughes et al. 1983).

After immunoadsorption, the eluate from the antibody column contained a major $M_r = 134,000$ ^{125}I-GH receptor complex. The complex was absent when cells were incubated with excess unlabeled GH, consistent with this protein being the GH receptor. ^{125}I-GH by itself did not bind to the immobilized antibody. Furthermore, addition of O-phosphotyrosine to the solubilized protein before chromatography prevented the GH-receptor complex from binding to the antibody column, but O-phosphoserine and O-phosphothreonine did not prevent it. These data indicate that at least a portion of the $M_r = 134,000$ GH-receptor complex contains at least one phosphorylated tyrosine residue. A similar result was obtained when cross-linked GH receptor from the adipocyte form of 3T3-F442A cells was immunoadsorbed to and eluted from the antibody column.

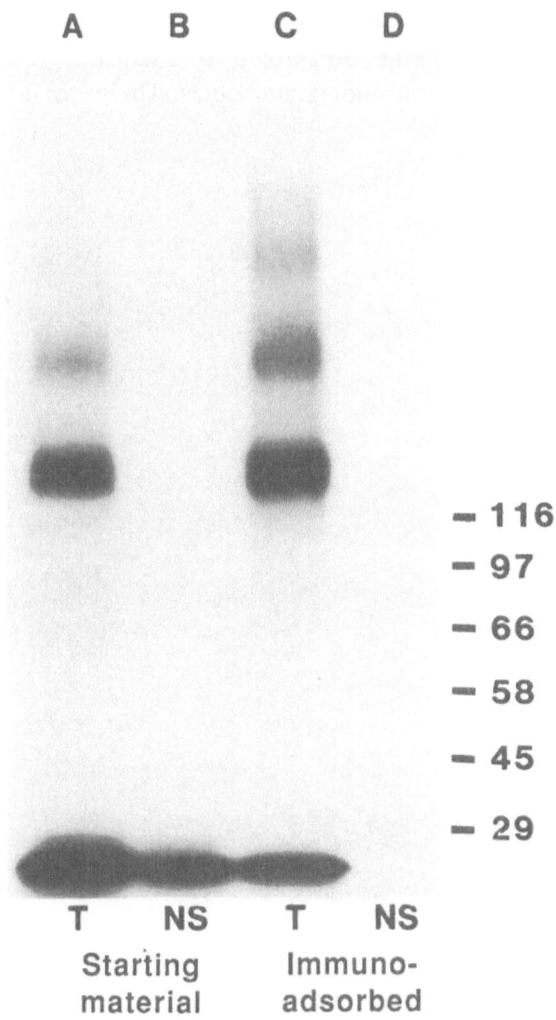

Fig. 8. Binding of ^{125}I-hGH affinity-labeled GH receptor to phosphotyrosyl binding antibody. 3T3-F442A fibroblasts incubated with ^{125}I-hGH for 1 h at 24°C without (T) or with (NS) an excess of unlabeled hGH. Cells were washed, incubated with disuccinimidyl suberate and solubilized with 0.1% triton. Solubilized proteins before chromatography were subjected to SDS-polyacrylamide gel electrophoresis in the presence of beta-mercaptoethanol (*lanes* A and B). Additional solubilized material was added to a phosphotyrosyl-binding antibody column and was eluted with p-nitrophenylphosphate. Column eluates were subjected to SDS-PAGE and autoradiography (*lanes* C and D). (Reproduced from Foster et al. [1988] by permission).

To examine GH promotion of phosphorylation, 3T3 fibroblasts were metaboli-
cally labeled with ^{32}P and then incubated with or without GH for 1 h. As shown
in Figure 9, GH stimulated the formation of a ^{32}P-labeled protein which bound to
immobilized phosphotyrosyl binding antibodies. The M_r of 114,000 obtained for

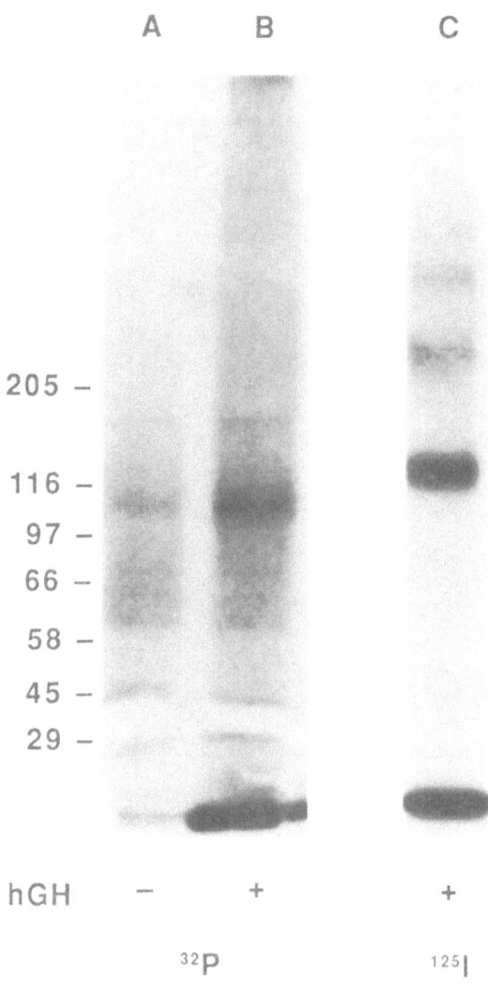

Fig. 9. Stimulation by GH of ^{32}P, incorporation into the GH receptor. 3T3-F442A
fibroblasts were incubated overnight with ^{32}P, in phosphate-free DMEM and were then
incubated for 1 h with or without met-hGH (29 ng/ml). Solubilized cell extracts were passed
over the phosphotyrosyl-binding antibody column, concentrated by TCA precipitation and
processed as described in Fig. 8 (*lanes* A and B). In a parallel experiment (*lane* C), cells
incubated with ^{125}I-GH for 1 h (29 ng/ml) were then subjected to cross-linking, passage over
and elution from the phosphotyrosyl-binding antibody column, and were concentrated and
processed as described. (Reproduced from Foster et al. [1988] by permission).

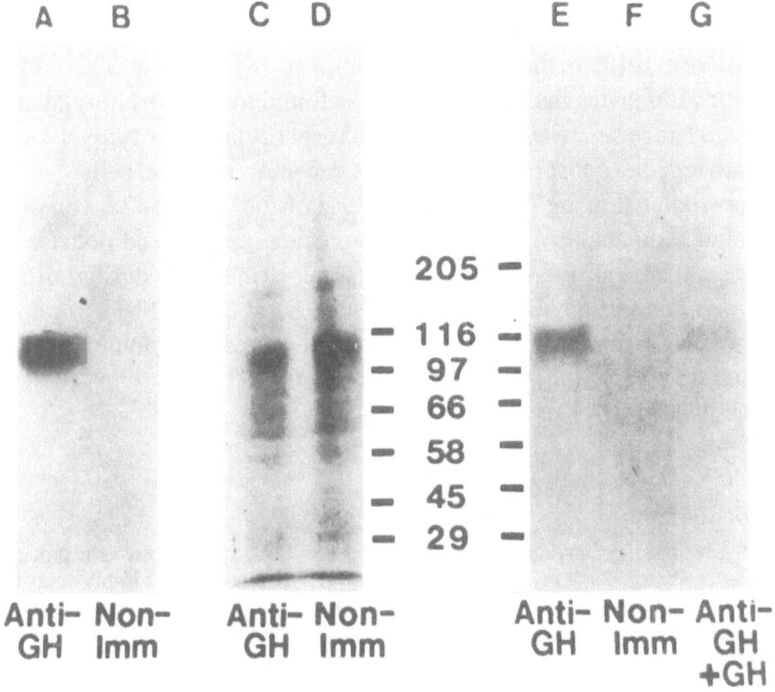

Fig. 10. Recognition of phosphorylated GH receptor by anti-hGH antibodies. 3T3-F442A fibroblasts were incubated with $^{32}P_i$, then with met-hGH (50 ng/ml), and were solubilized and adsorbed on the phosphotyrosyl-binding antibody column. Eluates were treated with anti-hGH antibody (*lanes* A, C, E) or nonimmune serum (*lanes* B, D, F) and hGH (1 µg/ml) (*lane* C). Immunoglobulin-bound proteins were then collected after incubation with protein-A Sepharose and subjected to SDS-PAGE under reducing conditions (*lanes* A, B, E–G). Aliquots of nonimmunoprecipitated material were also subjected to SDS-PAGE (lanes C and D). (Reproduced from Foster et al. [1988] by permission).

this protein is similar to that expected for noncross-linked GH receptor.This phosphoprotein could be immunoprecipitated with anti-GH antibody (Fig. 10), indicating that GH remained non covalently bound to the protein during adsorption to and elution from the immobilized phosphotyrosyl-binding antibody. Phosphoamino acid analysis of this $M_r=114,000$ phosphoprotein confirmed the presence of phosphotyrosyl residues. These observations provide strong evidence that the binding of GH to its receptor stimulates the phosphorylation of tyrosyl residues on the GH receptor.

Taken together, these studies indicate that 3T3-F442A adipocytes show the characteristic insulin-like and anti-insulin responses to GH. The responses occur equally well with hGH of pituitary or recombinant-DNA origin. GH was found to stimulate glucose uptake two to threefold in 5-15 min. This is one of the most rapid effects of GH reported, and its rapidity suggests that it is a direct effect of

GH on the 3T3 adipocytes. In contrast, after 24-48 h GH inhibited glucose uptake and metabolism, effects that were not mimicked by IGF-I. In 3T3-F442A preadipocytes and adipocytes the GH receptor was found to be phosphorylated on tyrosyl residues, GH receptor was found to be phosphorylated on tyrosyl residues, and GH stimulated the phosphorylation of its receptor in these cells.

These studies thus indicate the versatility of the 3T3-F442A cells for analysis of the cellular mechanisms for GH action. Since cultured adipocytes have been used for analysis of the regulation of gene expression during differentiation (Spiegelman et al. 1983; Doglio et al. 1986; Distel et al. 1987), we are now in a position to gain insight into possible changes in gene expression in the mechanisms of GH action.

Acknowledgments

Dr. H. Green kindly provided 3T3-F442A cells. We thank Dr. J. Kostyo for providing pituitary human GH, Genentech for providing methionyl human GH preparations, KabiVitrum for providing methionyl-IGF-I, Amgen for providing Thr^{59} IGF-I, Dr. J. Van Wyk for providing human IGF-I and monoclonal antibody against IGF-I, Dr. J. Florini for providing rIGF-II, Dr. R. Chance for providing porcine insulin, and the National Hormone and Pituitary Program for providing anti-hGH serum.

REFERENCES

Ader M, Agajanian T, Finegood DT, Bergman RN (1986) Recombinant deoxyribonucleic-derived 22-K and 20-K human growth hormone generate equivalent diabetogenic effects during chronic infusion in dogs. *Endocrinology* **120**: 725-731

Altszuler N (1974) Growth hormone and carbohydrate metabolism. In: Knobil E, Sawyer W (eds) *Handbook of Physiology* sec 7, vol 4 part 2, American Physiological Society, Bethesda, pp 233-252

Asakawa K, Grunberger G, McElduff A, Gorden P (1985) Polypeptide hormone receptor phosphorylation: is there a role in receptor-mediated endocytosis of human growth hormone? *Endocrinology* **117**: 631-637

Carter-Su C, Pratt WB (1984) Receptor phosphorylation. In: Conn PM (ed) *The Receptors.* Academic, New York, vol 1, pp 541-585

Carter-Su C, Schwartz J, Kikuchi G (1984) Identification of a high-affinity growth hormone receptor in rat adipocyte membranes. *J Biol Chem* **259**: 1099-1104

Cohen S, Carpenter G, King L jr (1980) Epidermal growth factor-receptor-protein kinase interactions. Co-purification of receptor and epidermal growth factor-enhanced phosphorylation activity. *J Biol Chem* **255**: 4834-4842

Daughaday WH (1980) Growth hormone and somatomedin. In: Daughaday WH (ed) *Endocrine Control of Growth.* Academic, New York, pp 1-21

Davidson MB (1987) Effect of growth hormone on carbohydrate and lipid metabolism. *Endocr Rev* **8**: 115-131

D'Ercole AJ, Stiles AD, Underwood LE (1980) Tissue concentrations of somatomedin C:

further evidence for multiple sites of synthesis and paracrine or autocrine mechanisms of action. *Proc Nat Acad Sci USA* **81**: 935-939

Doglio A, Dani C, Grimaldi P, Ailhaud G (1986) Growth hormone regulation of the expression of differentiation-dependent genes in preadipocyte Ob1771 cells. *Biochem J* **238**: 123-129

Donner DB (1983) Covalent coupling of human growth hormone to its receptor on rat hepatocytes. *J Biol Chem* **258**: 2736-2743

Distel RJ, Ro H-S, Rosen BS, Groves DL, Spiegelman BM (1987) Nucleoprotein complexes that regulate gene expression in adipocyte differentiation: direct participation of c-*fos*. *Cell* **49**: 835-844

Foster CM, Shafer JA, Rozsa FW, Wang XY, Lewis SD, Renken DA, Natale JE, Schwartz J., Carter-Su C (1988) Growth hormone promoted tyrosyl phosphorylation of growth hormone receptors in murine 3T3-F442A fibroblasts and adipocytes. *Biochemistry* **27**: 326-334

Gliemann J, Rees WD, Foley JA (1984) The fate of labelled glucose molecules in the rat adipocyte. Dependence on glucose concentration. *Biochim Biophys Acta* **804**: 68-76

Goodman HM (1968) Growth hormone and the metabolism of carbohydrate and lipid in adipose tissue. *Ann NY Acad Sci* **148**: 419-440

Gorin E, Goodman HM (1984) Covalent binding of growth hormone to surface receptors on rat adipocytes. *Endocrinology* **114**: 1279-1286

Green H, Kehinde O (1974) Sublines of mouse 3T3 cells that accumulate lipid. *Cell* **1**: 113-116

Green H, Morikawa M, Nixon T (1985) A dual effector theory of growth hormone action. *Differentiation* **29**: 195-198

Hughes JP, Simpson SA, Friesen HG (1983) Analysis of growth hormone and lactogenic binding sites cross-linked to iodinated human growth hormone. *Endocrinology* **112**: 1980-1985

Isaksson O, Edèn S, Jansson J-O (1985) Mode of action of pituitary growth hormone. *Annu Rev Physiol* **47**: 483-499

Jacobs S, Kull FC jr, Earp HS, Svoboda ME, Van Wyk JJ, Cuatrecasas P (1983) Somatomedin-C stimulates the phosphorylation of the beta-subunit of its own receptor. *J Biol Chem* **258**: 9581-9584

Kashiwagi A, Verso MA, Andrews J, Vasquez B, Reaven G, Foley JE (1983) *In vitro* insulin resistance of human adipocytes isolated from subjects with noninsulin-dependent diabetes mellitus. *J Clin Invest* **72**: 1246-1254

Kasuga M, Karlsson FA, Kahn CR (1982) Insulin stimulation of the 95,000-dalton subunit of its own receptor. *Science* **215**: 185-187

Kostyo JL, Cameron CM, Olson KC, Jones AJS, Pai R-C (1985) Biosynthetic 20-kilodalton methionyl human growth hormone has diabetogenic and insulin-like activities. *Proc Natl Acad Sci USA* **82**: 4250-4253

Maloff BL, Levine JH, Lockwood DH (1980) Direct effects of growth hormone on insulin action in rat adipose tissue maintained *in vitro*. *Endocrinology* **107**: 538-544

Morikawa M, Nixon T, Green H (1982) Growth hormone and the adipose conversion of 3T3 cells. *Cell* **29**: 783-789

Nishimura J, Huang JS, Duell TF (1982) Platelet-derived growth factor stimulates tyrosine-specific protein kinase activity in Swiss mouse 3T3 cell membranes. *Proc Natl Acad Sci USA* **79**: 4303-4307

Nixon TB, Green H (1984) Contribution of growth hormone to the adipogenic activity of serum. *Endocrinology* **114**: 527-532

Nyberg G, Smith U (1977) Human adipose tissue in culture VII. The long-term effect of growth hormone. *Horm Metab Res* **9**: 22-27

Pang DT, Sharma BR, Shafer JA (1985) Purification of the catalytically active phosphorylated form of insulin receptor kinase by affinity chromatography with O-phosphotyrosyl binding antibody. *Arch Biochem Biophys* **242**: 176-186

Schwartz J (1984) Growth hormone directly alters glucose utilization in 3T3 adipocytes. *Biochem Biophys Res Commun* **125**: 237-243

Schwartz J, Carter-Su (1988) Effects of growth hormone on glucose metabolism and glucose transport in 3T3-F442A cells: dependence on cell differentiation. *Endocrinology*. **122**: 2247-2256

Schwartz J, Foster CM (1986) Pituitary and recombinant deoxyribonucleic acid-derived human growth hormones alter glucose metabolism in 3T3 adipocytes. *J Clin Endocrinol Metab* **62**: 791-794

Schwartz J, Foster CM, Satin MS (1985) Growth hormone and insulin-like growth factors I and II produce distinct alterations in glucose metabolism in 3T3-F442A adipocytes. *Proc Natl Acad Sci USA* **82**: 8724-8728

Spiegelman BM, Frank M, Green H (1983) Molecular cloning of mRNA from 3T3 adipocytes. Regulation of mRNA content for glycerophosphate dehydrogenase and other differentiation-dependent proteins during adipocyte development. *J Biol Chem* **258**: 10083-10089

Van Wyk JJ, Underwood LE (1978) Somatomedins. In: Litwack G (ed) *Biochemical Actions of Hormones*, vol 5, pp 101-148

Walton PE, Etherton TD, Evock CM (1986) Antagonism of insulin activity in cultured pig adipose tissue by pituitary and recombinant porcine growth hormone. Potentiation by hydrocortisone. *Endocrinology* **118**: 2577-2581

SOMATOMEDINS

Advances in Growth Hormone and Growth Factor Research,
edited by E.E. Müller, D. Cocchi and V. Locatelli
Pythagora Press, Roma-Milano and Springer Verlag, Berlin-Heidelberg © 1989

Localization and regulation of IGF-I and IGF-II mRNA

M.A. Hynes[1], P.J. Brooks[2], J. English[2], J.J. Van Wyk[3] and P.K. Lund[2]

[1] *Howard Hughes Medical Institute Research Laboratory, Neurobiology and Behavior Center, Columbia University, New York, U.S.A.;* [2] *Department of Physiology and Curriculum in Neurobiology, University of North Carolina at Chapel Hill, U.S.A.; and* [3] *Department of Pediatrics, University of North Carolina at Chapel Hill, Chapel Hill, North Carolina, U.S.A.*

INTRODUCTION

Insulin-like growth factors I and II (IGF-I and -II) are peptide mitogens that exert a wide range of biological actions in many tissues and cell types. The actions of the IGFs include metabolic and differentiative effects as well as their capacity to stimulate cell proliferation. (Humbel 1984; Froesch et al. 1985; Van Wyk 1984). Traditionally the IGFs were considered as hormones that are transported in the circulation to act on target cells in an endocrine fashion (Humbel 1984; Froesch et al. 1985; Van Wyk et al. 1984). Based on studies of perfused liver (Schwander et al. 1983), liver derived cell lines (Moses et al. 1980), and primary cultures of liver cells (Richmond et al. 1985), the liver was considered the major source of serum IGFs. The IGFs do not appear to be stored in the liver to an appreciable extent, however, since the concentrations of IGFs are higher in blood perfusing the liver than in extracts of liver. More recently, cultured explants of many tissues in addition to liver have been found to secrete immunoreactive IGFs into media (D'Ercole et al. 1984). Extracts of multiple tissues in addition to liver also contain higher concentrations of immunoreactive IGF than can be attributed to concentrations in blood perfusing the tissues (D'Ercole· et al. 1984, 1986). These observations have raised the possibility that in addition to the endocrine actions of IGFs transported to target cells via the circulation, there may be paracrine or autocrine actions of IGFs synthesized locally in multiple tissues (D'Ercole et al. 1984, 1986).

Little is known about the precise sites of synthesis of the IGFs in different tissues or about the regulation of IGF-I and -II in different tissues in response to changes in endocrine status *in vivo*. Most information derives from studies of cells in culture or measurements of serum concentrations of these peptides *in vivo*.

Studies of the sites and regulation of synthesis of the IGFs in different tissues using traditional methods such as immunocytochemistry and radioimmunoassay have been difficult due to the low tissue concentrations of the IGFs.

Isolation of complementary DNAs (cDNAs) encoding IGF-I and -II from a number of mammalian species (Ullrich et al. 1977; Bell et al. 1984; Jansen et al. 1983, 1985; Dull et al. 1984; Soares et al. 1986) offers the opportunity to study IGF synthesis and its regulation by hybridization analyses of mRNAs encoding the peptides. In the present study, double stranded cDNAs encoding the IGFs and precursor peptides, as well as synthetic oligodeoxyribonucleotides (oligomers) corresponding to defined portions of characterized cDNAs were used to study the regulation and sites of synthesis of IGF mRNAs in selected rat tissues.

METHODS

Analyses of Rat IGF-I and IGF-II mRNAs by Northern Blot Hybridization

Polyadenylated RNA (poly A⁺ RNA) was isolated from rat tissues, size fractionated on an agarose gel, transferred to GENE SCREEN and hybridized with radioactive DNA probes by methods described in detail elsewhere (Lund et al. 1986). Poly A⁺ RNA was isolated from the liver of adult 250 g male Sprague-Dawley strain rats and from day 14 gestation whole rat embryos. Poly A⁺ RNA was isolated also from the liver and brain of rats of different growth hormone (GH) status. Poly A⁺ RNAs were extracted from the liver of 100 g untreated hypophysectomized rats, from hypophysectomized rats which had been given a single intraperitoneal (ip) injection of 200 μg human growth hormone (hGH, Crescormon Kabi) 4 h or 8 h before they were killed and from age-matched intact rats. Poly A⁺ RNAs were extracted from brain of 100g untreated hypophysectomized rats, from rats which had been given a single injection of 100ng of hGH into the lateral ventricle (icv injection) at 4 h or 8 h before they were killed and from intact age-matched rats.

In Situ Hybridization Histochemistry (ISHH) to Localize IGF-I and IGF-II mRNA

Animals used as a source of tissues for ISHH were anesthetized with sodium pentobarbital (6 mg /100 g i.m.) and perfused through the ascending aorta with phosphate buffered saline (PBS-0.15 M NaCl in 10 mM sodium phosphate pH 7.4) followed by 100 ml/100 g body weight of 4% paraformaldehyde and 0.2% glutaraldehyde in 0.1 M phosphate pH 7.4. Immediately after perfusion, tissues of interest were removed, immersed in fixative for a further 3h, and placed in PBS containing 30% sucrose for 12-18 h at 4°C. After sucrose infiltration, tissues were rapidly frozen onto cryostat chucks, and 8-15 micron sections were cut on a cryostat. Sections were thaw mounted onto gelatin-chrome alum subbed slides and stored at 4°C for 16 h to 4 days.

Hybridization and Autoradiography

Sections were treated with Proteinase K (25 µg/ml, Boehringer Mannheim) that had been previously self-digested at 37°C for 2 h to remove nucleases, rinsed in PBS, post-fixed in 0.2% glutaraldehyde for 10 min, incubated in a PBS-glycine buffer (0.1 M glycine in PBS pH 7.4) for 10 min and then washed in PBS containing 0.2% Triton X 100 for 1 h. The sections were then prehybridized in a buffer containing 30% deionized formamide, 6x SSC, 2x Denhardts, 10% dextran sulfate, 50 mM Tris-HCl pH 7.0, and 5 mM ethylenediamine-tetraacetate (EDTA) pH 7.0. Salmon sperm DNA was denatured by boiling for 2 min and added to IX hybridization buffer to give a final concentration of 50 µg/ml immediately before application. The sections were prehybridized for 2.5 h in humidified petri dishes at 37°C. After prehybridization, sections were dehydrated through 50%, 70%, 95%, and 100% ethanol, air-dried for 10-30 min and hybridized in the same buffer with the addition of 0.5-4 x 10^6 cpm/150 µl (0.5-4 pmol/150 µl) of ^{32}P-labeled probe at 37°C for 48 hrs. After hybridization, slides were washed successively for 1 h in 4x SSC at room temperature, for 5 h in 2x SSC at 37°C (with three changes of 1 h each) and in 0.1 x SSC for 10 min at room temperature, then dehydrated through an ascending series of ethanols each containing 0.33 M ammonium acetate, and air dried. Sections were exposed to X ray film at -100°C with intensifying screens for 16-18 h to provide an indication of the hybridization signal intensity then dipped in Kodak NTB 3 photoemulsion diluted 1:1 with distilled water, and exposed in sealed boxes for 5-21 days at 4°C. Slides were developed with Kodak D-19 developer, fixed, washed, air dried and stained with cresyl violet acetate. After staining, sections were dehydrated in ascending ethanol series and xylene, and coverslipped with permount. Hybridization signals were visualized with a Leitz photomicroscope with bright and dark field optics.

Hybridization Probes

Complementary DNA (cDNA) probes were labeled with ^{32}P by nick translation (Rigby et al. 1977). Oligomer probes were end-labeled with ^{32}P by use of polynu-

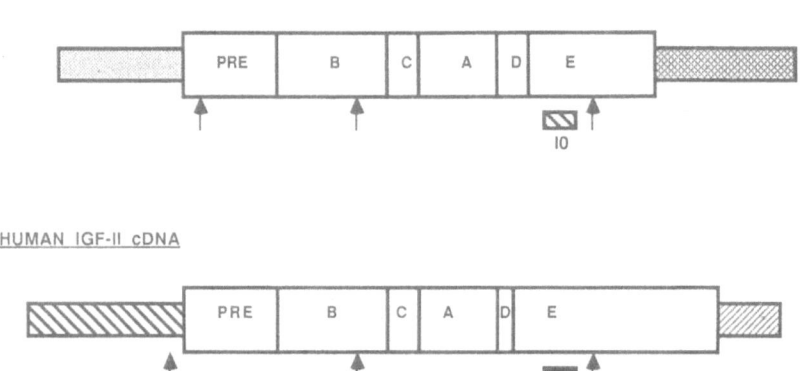

Fig. 1. Schematics of cDNAs encoding rat IGF-I (*top*; Casella et al. 1987) and human IGF-II (*bottom*; Jansen et al. 1985). Coding sequences for different domains of the IGF precursors — prepeptide (PRE), *B*, *C*, *A*, *D* and carboxyl-terminal precursor *E* domains — are indicated by the labeled *open bars*. Untranslated regions (UTs) are indicated by *shaded bars*. Locations of introns within the rat IGF-I gene and human IGF-II gene are indicated by *vertical arrows*. Locations of regions corresponding to the oligomer probes used in this study are indicated by numbered *boxes*. Regions of IGF-I and IGF-II mRNAs corresponding to these oligomers, #10 and #2, are conserved in all reported rat IGF-I mRNA types (Casella et al. 1987; Roberts et al. 1987) and all reported rat IGF-II mRNA types (Soares et al. 1986; Frunzio et al. 1986) respectively.

cleotide kinase followed by purification on 15-20% polyacrylamide gels containing 7 M urea (Maxam and Gilbert 1977). Specific probes utilized are summarized in Figures 1 and 5 and are described in the results section to facilitate data interpretation.

RESULTS

Multiple IGF-I and IGF-II mRNAs Each Contain Carboxyl-Terminal E Domain Coding Sequences

Complementary DNA (cDNA) probes encoding IGF-I and IGF-II were used in Northern blot hybridization analyses to investigate IGF-I and -II mRNAs in poly A⁺ RNAs from adult rat liver and whole rat embryos of day 14 gestation. As shown

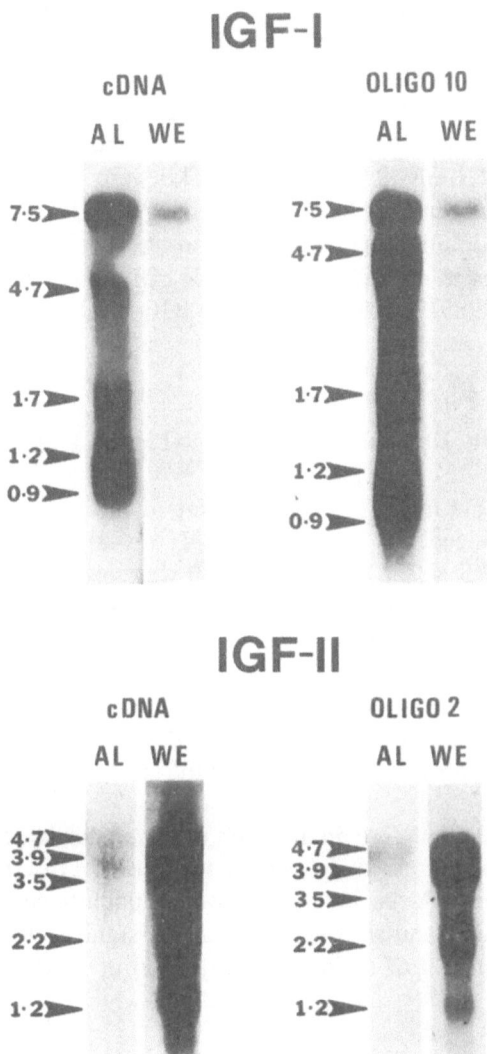

Fig. 2. Autoradiograms of Northern blots of poly A⁺ RNAs from adult rat liver and from whole embryo of gestational day 14 hybridized with IGF-I and IGF-II probes. *Top*, Shown on *left* are lanes from blots hybridized using a rat IGF-I cDNA as probe. Shown on *right* are lanes from replicate blots hybridized with synthetic oligomer # 10 (Fig. 1) that is complementary to carboxyl-terminal E domain coding sequences in the rat IGF-I cDNA (Fig. 1). *Bottom*, Shown on *left* are lanes from blots hybridized using a human IGF-II cDNA as probe. Shown on *right* are lanes from replicate blots hybridized with synthetic oligomer # 2 (Fig. 1) that is complementary to carboxyl-terminal E domain coding sequences in the human and the rat IGF-II cDNAs (Fig. 1).

in Fig. 2, using a rat IGF-I cDNA probe encoding the entire IGF-I precursor (Casella et al. 1987; Fig. 1), multiple IGF-I mRNAs of estimated sizes of 7.5, 4.7, 1.7, and 1.2-0.9 kb are detected in the whole rat embryo and in the adult rat liver. Each of these mRNAs also hybridized with synthetic oligomer #10 (Fig. 1) that is complementary to carboxyl-terminal E domain coding sequences in the rat IGF-I cDNA (Fig. 1). In cDNA probe encoding IGF-II (Figs. 1 and 5) multiple IGF-II mRNAs of estimated sizes 4.7, 3.9, 2.2, 1.75, and 1.2 kb were recognized in poly A⁺RNA from the embryo. In adult rat liver a hybridizing IGF-II mRNA of 4.7 kb is just discernible. Synthetic oligomer #2 corresponding to conserved carboxyl-terminal E domain sequences in rat and human IGF-II cDNAs (Figs. 1 and 5; Bell et al. 1985; Jansen et al. 1985) hybridized with the IGF-II mRNA species detected with the cDNA probe.

Growth Hormone Dependence of IGF-I and IGF-II mRNAs

Analyses of IGF-I mRNAs from the liver of intact, hypophysectomized rats and hypophysectomized rats treated with GH revealed that the multiple IGF-I mRNA species of 7.5, 4.7, 1.7, and 1.2-0.9 kb were each reduced in abundance in hypophysectomized rats compared with intact controls ($p < 0.01$) and were restored to normal abundance within 4 h of a single ip injection of hGH (Fig. 3). A similar pattern of GH dependence of IGF-I mRNAs was observed using either a cDNA probe encoding the entire IGF-I precursor or with oligomer #10 corresponding to carboxyl-terminal E domain sequences in the IGF-I cDNA (Fig. 3). There was no significant change in abundance of ubiquitin mRNAs in these samples of liver poly A⁺ mRNA.

Analyses of IGF-II mRNAs in the brain of intact rats, hypophysectomized rats, and hypophysectomized rats given a single icv injection of GH revealed that brain IGF-II mRNAs were reduced in abundance in hypophysectomized rats compared with controls (p<0.01), and showed a partial, but significant ($p < 0.05$) restoration of abundance within 8 h of a single icv injection of hGH (Fig. 4).

LOCALIZATION OF IGF mRNAs BY IN SITU HYBRIDIZATION HISTO-CHEMISTRY

Transgenic Mice Overexpressing Human IGF-I cDNA

Initial ISHH experiments were performed on tissues of transgenic mice overexpressing the human IGF-I cDNA to establish procedures for the localization of high abundance IGF-I mRNAs. Transgenic mice containing the human IGF-I cDNA (Bell et al. 1984) linked to the mouse metallothionein I promoter were provided by Dr. Ralph Brinster (University of Pennsylvania) and Drs. Richard

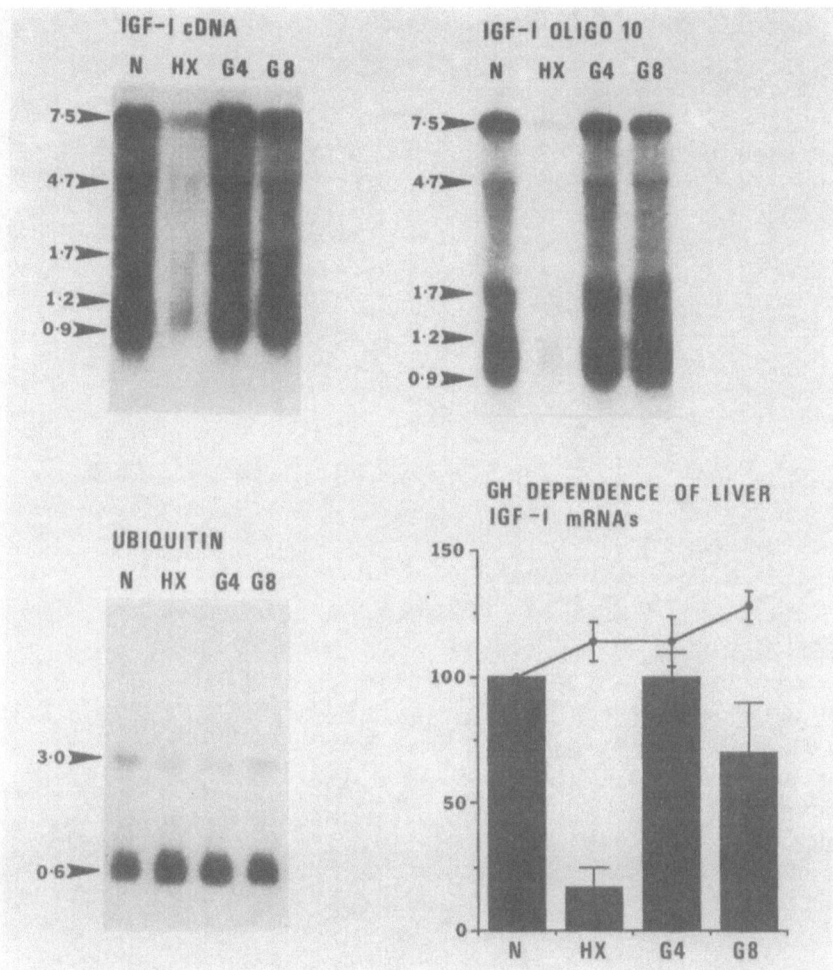

Fig. 3. Autoradiograms and quantitation of hybridization signals of Northern blots of poly A+ RNAs extracted from livers of normal (*N*), hypophysectomized (*HX*), and hypophysectomized rats 4 h (*G4*) or 8 h (*G8*) after ip injection of hGH. Fifteen micrograms poly A+ RNA is loaded into each lane. *Numbers* at the left of the blots are the estimated sizes of hybridizing mRNAs based on comparison with denatured HindIII fragments of lambda phage DNA. *Top left,* Blot hybridized with the rat IGF-I cDNA. *Top right,* Replicate blot to one on *left* except hybridization was with the IGF-I oligomer #10, specific for carboxyl-terminal trailer peptide sequences in the IGF-I cDNA. Bottom left, Rehybridization of blot shown in *top left* after stripping of IGF-I probe and rehybridization with a control ubiquitin probe. *Bottom right,* Abundance of IGF-I mRNAs and ubiquitin mRNAs. Histogram shows the abundance of IGF-II mRNAs from liver of normal (*N*), hypophysectomized (*HX*), and hypophysectomized rats 4 h (*G4*), or 8 h (*G8*) after icv injection of hGH. Mean abundance of the 0.6 kb ubiquitin mRNA is plotted for comparison. Abundance in each case is expressed as percentage of signal intensity observed in poly A+ RNAs from liver of normal rats. Each estimate of mRNA abundance in the mean ± SD of three different poly A+ RNA preparations from three different animals. Statistical comparisons of mRNA abundance are described in the text.

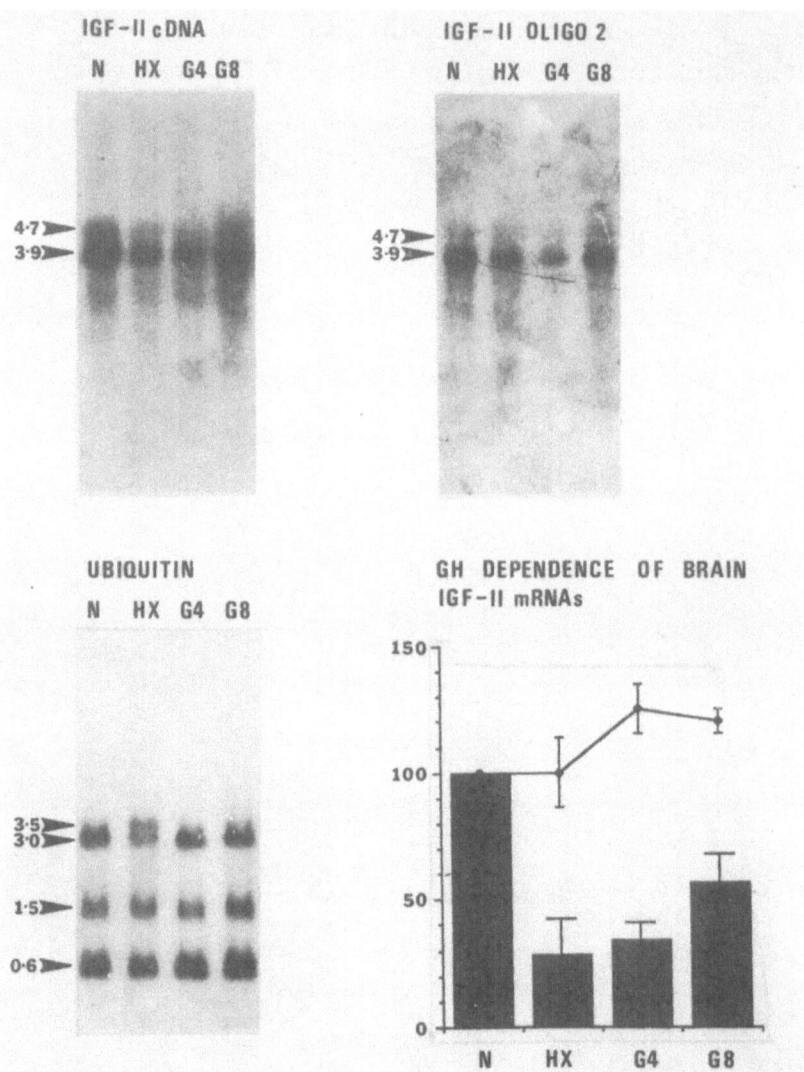

Fig. 4. Autoradiograms and quantitation of hybridization signals of Northern blots of poly A⁺ RNAs extracted from brains of normal (*N*), hypophysectomized (*HX*),and hypophysectomized rats 4 h (*G4*) or 8 h (*G8*) after icv injection of 200 ng hGH. Fifteen micrograms poly A⁺ RNA is loaded into each lane. *Numbers* at the left of the blots are the estimated sizes of hybridizing mRNAs based on comparison with denatured HindIII fragments of lambda phage DNA. *Top left*, Blot hybridized with the human IGF-II cDNA. *Top right*, Replicate blot to one on left except hybridization was with the IGF-II oligomer #2, specific for carboxyl-terminal trailer peptide sequences in the IGF-II cDNA. *Bottom left*, Rehybridization of blot shown in *top left* after stripping of IGF-II cDNA probe and re-hybridization with a control ubiquitin probe. *Bottom right*, Abundance of IGF-II mRNAs and ubiquitin mRNAs. Histogram shows the abundance of IGF-II mRNAs from whole brain of normal (*N*), hypophysectomized (*HX*), and hypophysectomized rats 4 h (*G4*), or 8 h (*G8*) after icv injection of hGH. Mean abundance of the 0.6 kb ubiquitin mRNA is plotted for comparison. Abundance in each case is expressed as percentage of signal intensity observed in poly A⁺ RNAs from normal brain. Each estimate of mRNA abundance in the mean ± SD of three different poly A⁺ RNA preparations from three different animals. Statistical comparisons of mRNA abundance were calculated according to Ott (1977) and are described in the text.

Palmiter and Lawrence Matthews (University of Washington). Preliminary studies in these animals had suggested high levels of expression of the human IGF-I transgene in the pancreas (Drs. Matthews and D'Ercole, University of North Carolina, personal communication).

Sections of liver, pancreas, small intestine and brain of transgenic mice and normal littermates were analyzed by hybridization with a 500 bp *Bam*HI/*Pst*I fragment of human IGF-I cDNA and with oligomer #4 complementary to the E

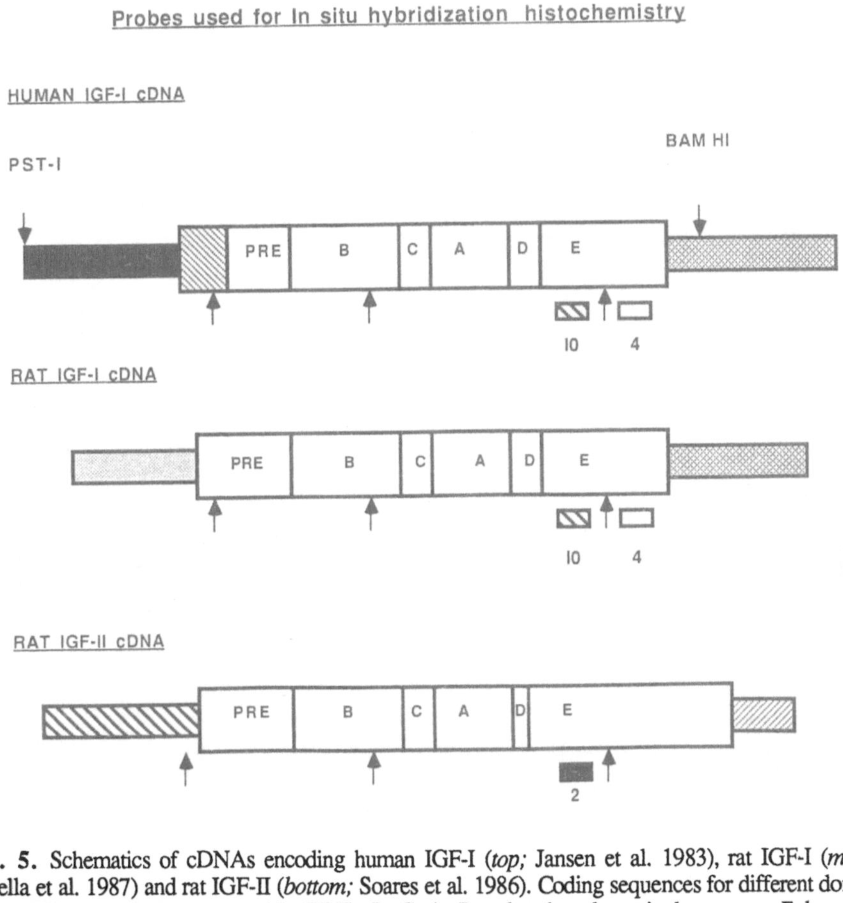

Fig. 5. Schematics of cDNAs encoding human IGF-I (*top;* Jansen et al. 1983), rat IGF-I (*middle;* Casella et al. 1987) and rat IGF-II (*bottom;* Soares et al. 1986). Coding sequences for different domains of the IGF precursors — prepeptide (PRE), *B*, *C*, *A*, *D* and carboxyl-terminal precursor *E* domains— are indicated by the labeled *open bars*. Untranslated regions (UTs) are indicated by *shaded bars*. Locations of introns within the IGF genes are indicated by *vertical arrows*. Locations of regions corresponding to the oligomer probes used in this study are indicated by numbered *boxes*. Regions of IGF-I and IGF-II mRNAs corresponding to these oligomers (#10 and #2) are conserved in all reported rat IGF-I mRNA types (Casella et al. 1987; Roberts et al. 1987) and all reported rat IGF-II mRNA types (Soares et al. 1986; Frunzio et al. 1986), respectively. Oligomer #4 shows 100% homology to sequences in the human IGF-I cDNA (Jansen et al. 1983) and has four mismatches with rat IGF-I (Casella et al. 1987).

domain coding sequences in human IGF-I cDNA (Fig. 5). As shown in Fig. 6, a
4 h X-ray film exposure of sections of normal and transgenic mouse tissues
hybridized with the *Bam*HI/*Pst*I fragment of human IGF-I cDNA revealed intense
hybridization of the probe to sections of transgenic mouse pancreas. A control
probe, the IGF-II oligomer #2, showed no detectable hybridization above background. Photomicrographs of sections of normal and transgenic mouse pancreas
are shown in Fig. 7. Normal pancreas showed no detectable hybridization above

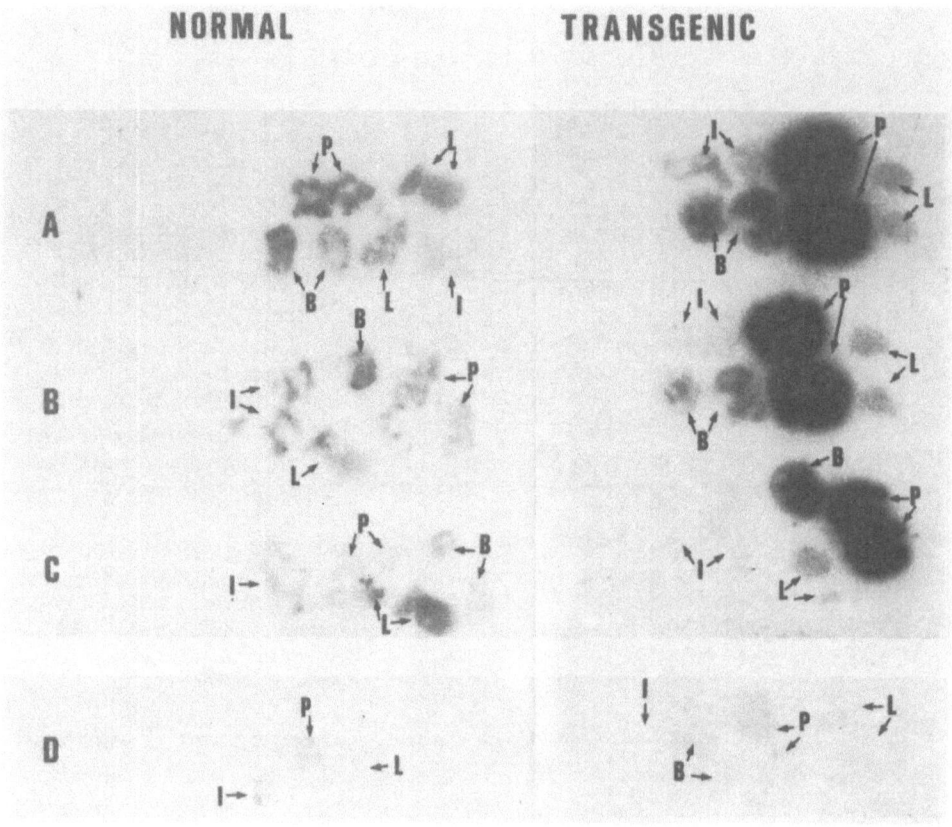

Fig. 6. ISHH labeling of human IGF-I mRNA in tissues of normal and transgenic mice.
Autoradiograms of slides bearing sections of normal (*A*) or transgenic (*B*) mouse liver (*L*),
pancreas (*P*), intestine (*I*) and brain (*B*) after incubation with 4 x 10^6 cpm per slide of ^{32}P-
labeled DNA probes. **A** and **B** IGF-I oligomer #4, **C** the 500 bp *Bam*HI/*Pst*I fragment of
the human IGF-I cDNA **D** IGF-II oligomer #2. Exposure time for the autoradiogram was
4 h at room temperature. The autoradiogram shows intense hybridization of IGF-I probes
to pancreas of transgenic animals. No detectable hybridization of the IGF-II oligomer to
sections of transgenic pancreas was observed.

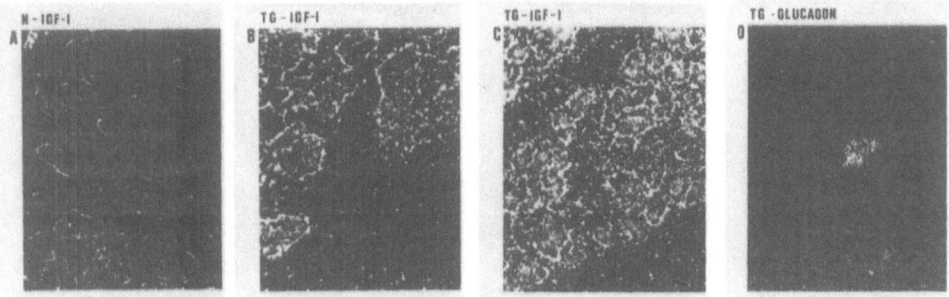

Fig. 7A-D. Photomicrographs of sections of pancreas from: **A** Normal mouse, hybridized with the IGF-I oligomer #4 (Fig. 5). Exposure time to photoemulsion was 12 h. **B** Transgenic mouse, same slide shown in A, hybridized with IGF-I oligomer #4 (Fig. 5). Exposure time to photoemulsion was 12 h. **C** Transgenic mouse, hybridized with *Bam*HI/*Pst*I fragment of the human IGF-I cDNA (Fig. 5). Exposure time to photoemulsion was 72 h. **D** Transgenic mouse, hybridized with a 20 bp oligomercomplementary to sequences within the rat proglucagon mRNA which code for glucagon-like peptide I (GLP-I) (Heinrich et al. 1984). Exposure time to photoemulsion was 12 h.

background with the human IGF-I *Bam*HI/*Pst*I cDNA fragment. In transgenic mouse pancreas intense hybridization was observed with both the human IGF-I cDNA fragment and the IGF-I oligomer #4. Hybridization was to exocrine pancreas with no hybridization above background in pancreatic islet tissue (Fig. 7). A control oligomer complementary to proglucagon mRNA (Heinrich et al. 1984) localized glucagon mRNAs in pancreatic islets of the transgenic mouse (Fig. 7).

Adult Rat Brain *IGF-II mRNA*

Previous studies indicated that the brain is the most abundant source of IGF-II mRNA in mature rats (Lund et al. 1986). Hybridization of coronal sections of rat brain with the human IGF-II cDNA probe and with the IGF-II oligomer #2 specific for carboxyl-terminal trailer peptide coding regions (Fig. 1) revealed specific hybridization to cells in the choroid plexus of the adult rat brain (Fig. 8). Specificity of localization of IGF-II mRNAs to rat choroid plexus was supported by five control procedures: (a) hybridization was symmetrical in the brain and seen over cells of the choroid plexus in each lateral ventricle; (b) hybridization to choroid plexus was seen using two different IGF-II probes: the IGF-II cDNA and oligomer #2; (c) a synthetic oligomer complementary to prosomatostatin mRNA did not hybridize to cells of the choroid plexus even though this probe showed positive hybridization to cells in the periventricular region of the hypothalamus; (d) hybridization to IGF-II mRNA in choroid plexus was completely blocked by the

IGF-II mRNA IN RAT CHOROID PLEXUS

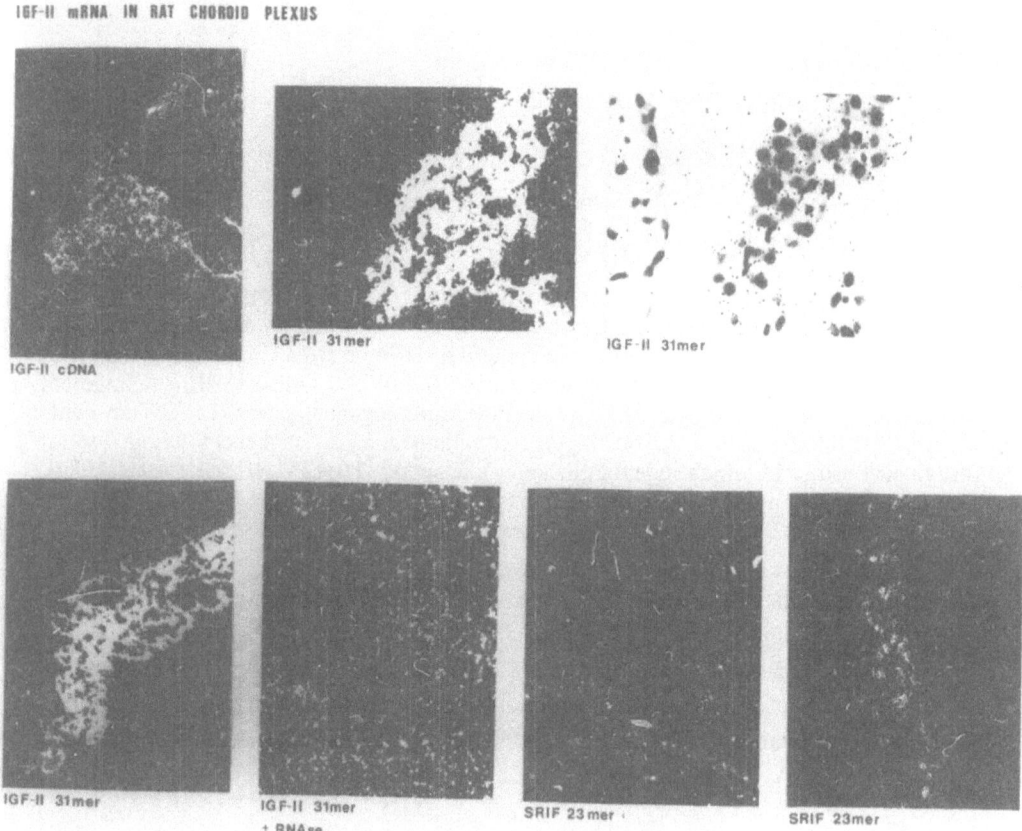

Fig. 8. *Top*, Low magnification dark field photomicrographs showing cells in choroid plexus of lateral ventricle of rat brain that hybridized with ^{32}P-labeled human IGF-II cDNA (*left*) or IGF-II oligomer #2 (*middle*) probe. On the *right* is a higher magnification bright field photomicrograph of a slide hybridized with the IGF-II oligomer #2. Accumulation of autoradiographic silver grains can be seen localized over cells in the chroid plexus. Over the fornix, density of silver grains was not higher than background hybridization. *Bottom*, Each dark field picture shows hybridization to slides bearing coronal sections from the same rat brain. The field of view in the first three pictures is over the lateral ventricle. In the fourth picture the field of view is the periventricular region of the hypothalamus. From left to right is (i) hybridization with ^{32}P-labeled IGF-II oligomer #2, (ii+iii) slides hybridized using identical conditions to the slide in (i) except that in (ii) the slide was pretreated with RNAase prior to hybridization and that the probe is a ^{32}P-labeled oligomer complementary to sequences in the rat somatostatin cDNA (Goodman et al. 1981), and (iv) photograph of the periventricular region of the hypothalamus showing positive hybridization of the somato-statin oligomer to cells in the periventricular region of the hypothalamus. Note that this photograph was from the same brain section as in iii.

addition of the nonlabeled homologous IGF-II probe before hybridization with the ^{32}P-labeled IGF-II probe (data not shown); (e) all hybridization to choroid plexus was eliminated by pretreatment of sections with RNase before hybridization (Fig. 8).

DISCUSSION

In previous studies we have detected multiple species of IGF-I and IGF-II mRNAs in fetal and adult rat tissues, including liver, lung, intestine, brain (Lund et al. 1986), pancreas (Hynes et al. 1987); and testies (Casella et al. 1987), but the precise structural relationships among the multiple mRNAs are yet to be defined. Recent characterization of rat IGF-I cDNAs and genes (Rotwein et al. 1986; Casella et al. 1987; Roberts et al. 1987) and of rat IGF-II genes (Soares et al. 1986; Frunzio et al. 1986) suggests that differential use of alternate promoters and/or splicing of exons corresponding to mRNA 5' untranslated regions as well as different polyadenylation sites account at least in part for IGF mRNA size heterogeneity.

In this study we aimed to develop procedures to allow localization of rat IGF-I and IGF-II mRNAs by in situ hybridization histochemistry (ISHH). Use of short single stranded oligomer probes for ISHH offers a number of advantages over longer double-stranded cDNA probes. The small size of oligomer probes facilitates penetration into tissue sections. Oligomer probes can be synthesized to defined regions of characterized mRNA and cDNA sequences that exclude regions of homology with related cDNAs that may compromise hybridization specificity. Use of oligomers corresponding to distinct regions in mRNA families that arise from the same gene may also be used to study particular mRNAs within the families. For ISHH studies in the rat, we initially identified oligomer probes that would hybridize with multiple IGF-I or IGF-II mRNAs previously identified using longer double-stranded cDNA probes (Lund et al. 1986). By Northern blot hybridization we established that oligomer #10, a 40-mer corresponding to carboxyl-terminal E domain sequences conserved among different reported rat IGF-I cDNAs (Casella et al. 1987; Roberts et al. 1987), hybridized with the same rat IGF-I mRNAs of 7.5, 4.7, 1.7, and 1.2-0.9 kb that were identified with a rat IGF-I cDNA probe (Casella et al. 1987). Furthermore, use of oligomer #10 and the rat IGF-I cDNA to study the GH dependence of rat liver IGF-I mRNA revealed a similar pattern of GH dependence with each probe. These findings suggest that each of the rat IGF-I mRNAs of 7.5, 4.7, 1.7, and 1.2-0.9 kb, which were identified with the cDNA probe, contains the carboxyl-terminal E domain coding sequences represented by oligomer #10.

Similarly, a human IGF-II cDNA and IGF-II oligomer #2, a 31- base- long oligomer complementary to carboxyl-terminal E domain coding sequences conserved in rat and human IGF-II mRNAs, each hybridized to IGF-II mRNAs of 4.7, 3.9, 2.2, and 1.2 kb in rat embryo suggesting that each mRNA contains E domain coding sequences represented by oligomer #2. Observations that a human IGF-II cDNA and IGF-II oligomer #2 hybridized to the same IGF-II mRNAs in rat brain poly A⁺ RNAs and revealed a similar pattern of GH dependence of rat brain IGF-II mRNAs, suggested that oligomer #2 would be of value as a probe for localization of rat brain IGF-II mRNAs by ISHH.

In situ hybridization histochemistry procedures for localization of IGF mRNAs were first tested on tissues of transgenic mice expressing the human IGF-I cDNA downstream from the mouse metallothionein promoter. Using as a probe either a double-stranded fragment of human IGF-I cDNA or an oligomer corresponding to E domain coding sequences in human IGF-I cDNA, intense hybridization compared with that in normal mice was seen in the exocrine pancreas of transgenic mice. These data were consistent with previous findings that the pancreas is a major site of expression of the human IGF-I cDNA transgene (Drs. Matthews, Palmiter and D'Ercole, personal communication) and suggest that ISHH procedures were appropriate for specific localization of high abundance IGF mRNAs in rodent tissues.

Analyses by ISHH of the cellular sites of synthesis of IGF-II mRNAs in the adult rat brain revealed hybridization of IGF-II oligomer #2 to cells of the choroid plexus and there was no detectable hybridization to other brain regions, cells or structures. These data suggest the choroid plexus is major site of IGF-II synthesis in the brain and a probable source of IGF-II secreted into cerebrospinal fluid, possibly for actions on target cells in other brain regions.

At present the cell type within the choroid plexus expressing IGF-II mRNA cannot be identified with the resolution of hybridization signals achieved. Uniform hybridization of the IGF-II probe over the entire choroid plexus suggests the epithelial cells or underlying stroma comprised of connective tissue as candidate cell types.

The lack of detectable hybridization of brain cell types other than choroid plexus with the IGF-II probe provides an indication that there is no, or only low level expression of IGF-II mRNAs in brain cells or regions other than choroid plexus. Negative in situ hybridization data should however, be interpreted with caution. Little or no information is currently available about IGF-II mRNA stability in different cell types and one alternate explanation for negative in situ hybridization data is the instability of IGF-II mRNAs in cell types other than those of choroid plexus. While control hybridizations indicated that somatostatin mRNAs are preserved in neuronal cells — using the in situ hybridization procedures employed — there is recent evidence to suggest short half lives of growth factor and oncogene mRNAs. It is also difficult to check that there is equal accessibility of mRNA for hybridization in all cell types and the negative in situ hybridization data could imply lack of access of the probe to hybridizable IGF-II mRNA. Clearly however the present results indicate that choroid plexus is a major site of synthesis of IGF-II mRNA in the adult rat brain.

The localization here of IGF-II mRNAs in the adult rat brain suggests ISHH will be a useful tool to study expression of IGF mRNAs at the cellular level. Further development of procedures should allow quantitation of expression of mRNAs in different cell types. Probes for ISHH used in this study were designed to hybridize either with all reported rat IGF-I or with all reported rat IGF-II mRNAs. Rat IGF-I cDNAs that differ in 5' untranslated regions and in coding sequence for carboxyl-terminal E domains (Casella et al. 1987; Roberts et al. 1987) have recently characterized. Insulin-like growth factor II mRNAs with different 5' untranslated regions have also been identified (Soares et al. 1986; Frunzio et al. 1986). Future use of oligomer probes that specifically recognize the different IGF-I and IGF-II mRNA types in ISHH will provide the opportunity to study whether there is cell-specific expression and/or regulation of the multiple IGF-I and IGF-II mRNAs and thereby provide insight into the biological significance of the multiple IGF-I and IGF-II mRNA species.

Acknowledgments

We would like to thank Eileen Hoyt and T. Hayes Woolen for excellent technical assistance, Drs. M. Jansen and J.L. Van Den Brande for provision of the human IGF-II cDNA probe, Drs. Ralph Brinster, Richard Behringer, Richard Palmiter and Lawrence Matthews for provision of transgenic mice and for sharing unpublished data about the transgenic mice, and Dr. A.J. D'Ercole for useful discussion and for sharing radioimmunoassay data on transgenic mice.

REFERENCES

Bell GI, Gerhard DS, Fong NM, Sanchez-Pescador R, Rall LB (1985) Isolation of the human insulin-like growth factor genes: insulin-like growth factor II and insulin genes are contiguous. *Proc Natl Acad Sci USA* **82**: 6450-6454

Bell GI, Merryweather JP, Sanchez-Pescador R, Stempien MM, Priestley L, Scott J, Rall LB (1984) Sequence of a cDNA clone encoding human preproinsulin -like growth factor II. *Nature* **310**: 775-777

Casella SJ, Smith EP, Van Wyk JJ, D'Ercole AJ, Hynes MA, Hoyt EC, Lund PK (1987) Isolation of rat testis cDNAs encoding an insulin-like growth factor I precursor. *DNA* **6**: 325-330

D'Ercole AJ, Stiles AD, Underwood LE (1984) Tissue concentrations of somatomedin C: further evidence for multiple sites of synthesis and paracrine or autocrine mechanisms of action. *Proc Natl Acad Sci USA* **81**: 935-939

D'Ercole AJ, Hill DJ, Strain A, Underwood LE (1986) Tissue and plasma somatomedin-C/insulin-like growth factor I concentrations in the human fetus during the first half of gestation. *Pediatric Res* **20**: 253-255

Dull TJ, Gray A, Hayflck JS, Ullrich A (1984) Insulin-like growth factor II precursor gene organization in relation to insulin gene family. *Nature* **311**: 777-781

Froesch ER, Schmid C, Schwander J, Zapf J (1985) Actions of insulin-like growth factors. *Annu Rev Physiol* **47**: 443-467

Frunzio R, Chiariotti L, Brown AL, Graham DE, Rechler MM, Bruni CB (1986) Structure and expression of the rat insulin-like growth factor II (rIGF-II) gene. rIGF-II RNAs are transcribed from two promoters. *J Biol Chem* **261**: 17138-17149

Goodman RH, Jacobs JW, Dee PC, Habener JF (1982) Somatostatin-28 encoded in a cloned cDNA obtained from a rat medullary thyroid carcinoma. *J Biol Chem* **257**: 1156-1159

Heinrich G, Gros P, Lund PK, Bentley RC, Habener JF (1984) Pre-proglucagon messenger ribonucleic acid: nucleotide and encoded amino acid sequences of the rat pancreatic complementary deoxyribonucleic acid. *Endocrinology* **115**: 2176-2181

Hynes MA, Van Wyk JJ, Brooks PJ, D'Ercole AJ, Jansen M, Lund PK (1987) Growth hormone dependence of somatomedin-C/insulin-like growth factor-I and insulin-like growth factor-II messenger ribonucleic acids. *Mol Endocrinol* **1**: 233-242

Humbel RE (1984) Insulin-like growth factors, somatomedins, and multiplication stimulating activity chemistry. In: Li CH (ed) *Hormonal Proteins and Peptides: Growth Factors*, vol 12. Academic, New York, pp 57-79

Lund PK, Moats-Staats BM, Hynes MA, D'Ercole AJ, Jansen M, Van Wyk JJ (1986) Somatomedin-C/insulin-like growth factor-I and insulin-like growth factor-II mRNAs in rat fetal and adult tissues. *J Biol Chem* **261**: 14539-14544

Jansen M, Van Schaik FMA, Richer AT, Bullock B, Woods DE, Gabbay KH, Nussbaum AL, Sussenbach JS, Van den Brande JL (1983) Sequence of cDNA encoding human insulin-like growth factor I precursor. *Nature* **306**: 609-611

Jansen M, Van Schaik FMA, von Tohl H, Van den Brande JL, Sussenbach JS (1985) Nucleotide sequences of cDNAs encoding precursors of human insulin-like growth factor II (IGF-II) and IGF-II variant. *FEBS Lett* **179**: 243-246

Maxam A, Gilbert W (1977) A new method for sequencing DNA. *Proc Natl Acad Sci USA* **74**: 560-564

Moses AC, Nissley SP, Short PA, Rechler MM, White RM, Knight AB, Higa OZ (1980) Increased levels of multiplication-stimulating activity, an insulin-like growth factor, in fetal rat serum. *Proc Natl Acad Sci USA* **77**: 3649-3653

Ott L (1977) In: Ott L (ed) An Introduction to Statistical Methods and Data Analysis. Duxbury, Belmont, pp 354-407

Richmond RA, Benedict MR, Florini JR, Toly BA (1985) Hormonal regulation of somatomedin secretion by fetal rat hepatocytes in primary culture. *Endocrinology* **116**: 180-188

Rigby PW, Diekmann M, Rhodes C, Berg P (1977) Labeling deoxyribonucleic acid to high specific activity *in vitro* by nick translation with DNA polymerase I. *J Mol Biol* **113**: 237-251

Roberts CT Jr, Lasky SR, Lowe WL, Seaman WT, LeRoith D (1987) Molecular cloning of rat insulin-like growth factor I complementary deoxyribounucleic acids: differential messenger ribonucleic acid processing and regulation by growth hormone in extrahepatic tissues. *Mol Cell Endocrinol* **1**: 243-248

Rotwein P (1986) Two insulin-like growth factor I messenger RNAs are expressed in human liver. *Proc Natl Acad Sci USA* **83**: 77-81

Schalch DS, Heinrich UE, Draznin B, Johnson CJ, Miller LL (1979) Role of the liver in regulating somatomedin activity: hormonal effects on the synthesis and release of insulin-like growth factor and its carrier protein by the isolated perfused rat liver. *Endocrinology* **104**: 1143-1151

Schwander JC, Hauri C, Zapf J, Froesch ER (1983) Synthesis and secretion of insulin-like growth factor and its binding protein by the perfused rat liver: dependence on growth hormone status. *Endocrinology* **113**: 297-305

Soares MB, Turken A, Ishii D, Mills L, Episkopou V, Cotter S, Zeitlin S, Efstradiatis A (1986) Rat insulin-like growth factor II gene. A single gene with two promoters expressing a multitranscript family. *J Mol Biol* **192**: 737-752

Ullrich A, Shire J, Chrigwin JM, Pictet R, Tischer E, Rutter WJ, Goodman HM (1977) Rat insulin genes: construction of plasmids containing the coding sequences. *Science* **196**: 1313-1319

Van Wyk JJ (1984) The somatomedins: biological actions and physiologic control mechanisms. In: Li CH (ed) *Hormonal Proteins and Peptides: Growth Factors,* vol 12. Academic, New York, pp 81-125

Advances in Growth Hormone and Growth Factor Research,
edited by E.E. Müller, D. Cocchi and V. Locatelli
Pythagora Press, Roma-Milano and Springer Verlag, Berlin-Heidelberg © 1989

Studies on somatomedin binding protein*

G. Pòvoa[1], K. Hall[2] and V. P. Collins[3]

[1] Centro Biomédico, UFES, ES, Brazil; [2] Department of Endocrinology, Karolinska Hospital, Stockholm; [3] Ludwig Institute for Cancer Research, Stockholm Branch, Sweden

Since the initial work on the purification of the human insulin-like growth factors (IGFs) it has been demonstrated that endogenous IGF activity is found in high molecular weight fractions in plasma. After acidification this activity can be recovered in a low molecular-weight form (Burgi et al. 1986; Daughaday and Kipnis 1966; Froesch et al. 1967; Van Wyk et al. 1969; Van den Brande et al. 1971; Hall 1972). Using labeled IGFs and excess unlabeled IGFs, the high molecular weight forms have been shown to be composed of IGFs and binding proteins (BP) with specific and saturable binding sites (Hintz et al. 1974). Further characterization of these BPs demonstrated the presence of at least two forms, with molecular weights of about 150 K and 35–45 K, respectively. It has also been shown that about 80% of the circulating IGFs in normal plasma are bound to the high-molecular-weight BP and 20% to the smaller BP. Less than 1% of IGFs circulate free in plasma (Hintz and Liu 1980; Hall et al. 1979; Furlanetto 1980; Copeland et al. 1980; White et al. 1981; Daughaday et al. 1982).

Acidification not only dissociates IGFs from the BPs but also breaks down irreversibly the 150-K form, which dissociates into an acid stable part and another unidentified component. The acid stable part of the 150-K BP has a molecular weight of about 50 K and retains the binding capacity for IGFs (Zapf et al. 1975; Furlanetto 1980). This acid stable component has been purified and a radioimmunoassay has been developed for its determination (Baxter et al. 1986). The antibodies recognize the 150-K BP in plasma but not the 35-K BP. Serum levels of the 150-K BP were found to be low at birth, to rise during childhood and puberty, and decline with increasing age. Thus, the age-related pattern of this BP is similar to that of IGF-I (Baxter and Martin 1986). Like the IGF-I levels, the levels of the 150-K BP seem growth hormone regulated with high levels in acromegaly and low levels in growth hormone deficiency. It has also been proposed that this 150-K

* This work was supported by grant no. 4224 from the Swedish Medical Research Council. G. Pòvoa has a scholarship from the Wenner-Gren Foundation.

BP could represent an oligomer composed of protomers with low molecular weights of about 24–28 K (Wilkins and D'Ercole 1985).

While the 150-K form is confined to the circulation, IGFs in body fluids other than plasma, as well as when isolated from cells, organs and amniotic fluid, are always associated with a 35-K BP (Cohen and Nissley 1979; Binoux et al. 1982a). This form of BP has also been demonstrated in media conditioned by several cell lines (Moses et al. 1983).

The BPs appear to be specific for the IGFs. The closely related polypeptides insulin and proinsulin do not interfere with the binding of IGFs. The BPs have affinity for both IGF-I and IGF-II but some differences in their affinity for each have been reported (Zapf et al. 1978; Hintz and Liu 1980; Binoux et al. 1982b). The physiological significance of the BPs is still unclear. The BPs prolong the half-life of the IGFs in the circulation and thereby ensure a supply to the tissues (Daughaday et al. 1968; Draznin et al. 1979). They have been proposed to protect the organism from the insulin action of the IGFs in the target organs for insulin (Oelz et al. 1970; Meuli et al. 1978) as well as to provide IGFs selectively to tissues capable of dissociating the complex (Hintz 1984). Addition of either the 50-K acid stable semipurified BP from plasma or even of unpure fractions isolated from amniotic fluid inhibits the binding of IGFs to receptors *in vitro* and thus prevents their biological activity (Zapf et al. 1979; Drop et al. 1979). However, it has recently been demonstrated that the 35-K BP purified from human amniotic fluid enhances the growth promoting effect of IGF-I on cultured fibroblasts (Elgin et al. 1986).

In this article we summarize what is currently known about the 35-K BP and present our own findings and ongoing work.

PURIFICATION AND CHARACTERIZATION OF THE 35-K BINDING PROTEIN

As has been shown by many research groups, the 35-K form of BP is always present wherever somatomedins/IGFs are released. Human amniotic fluid was found to be rich in this form of BP (Chochinow et al. 1977; Drop et al. 1979). The levels in amniotic fluid are highest around the 20th week of gestation, when the level is several times the plasma concentration, then decrease toward the end of pregnancy. This BP competes with human placenta membranes for labeled IGFs. Consequently, the placenta radioreceptor assay for IGFs could be used for the detection of the BP during the purification. Using amniotic fluid as starting material we have isolated a pure protein using a three-step procedure, namely ammonium sulphate precipitation, hydrophobic chromatography, and anion exchange chromatography (Pòvoa et al. 1984a). The molecular-weight of this protein was 32 K as determinad by SDS-PAGE under denaturing conditions and 3 K on gel

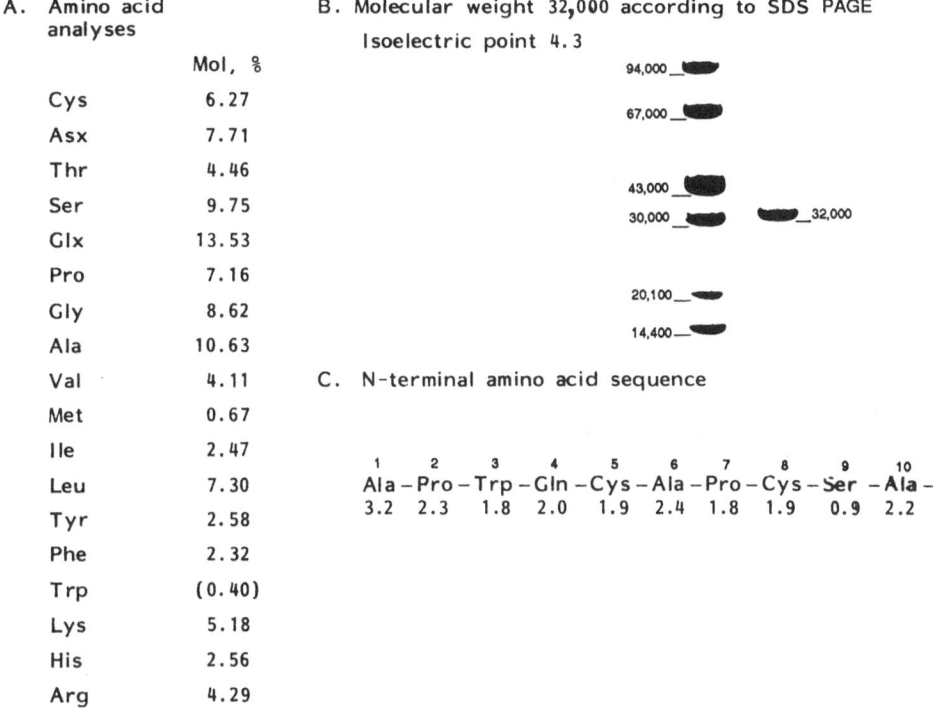

A. Amino acid analyses

	Mol, %
Cys	6.27
Asx	7.71
Thr	4.46
Ser	9.75
Glx	13.53
Pro	7.16
Gly	8.62
Ala	10.63
Val	4.11
Met	0.67
Ile	2.47
Leu	7.30
Tyr	2.58
Phe	2.32
Trp	(0.40)
Lys	5.18
His	2.56
Arg	4.29

B. Molecular weight 32,000 according to SDS PAGE

Isoelectric point 4.3

94,000
67,000
43,000
30,000 _____32,000
20,100
14,400

C. N-terminal amino acid sequence

1	2	3	4	5	6	7	8	9	10
Ala	Pro	Trp	Gln	Cys	Ala	Pro	Cys	Ser	Ala
3.2	2.3	1.8	2.0	1.9	2.4	1.8	1.9	0.9	2.2

Fig. 1. Structural data for the amniotic fluid IGFs-binding protein and SDS/polyacrylamide gel electrophoresis under denaturation conditions. Values (means of two preparations) as determined after hydrolysis with 6 M HCl/0.5% phenol for 24 h at 100°C. Cys determined at Cys(Cm) after carboxymethylation of the protein. Trp not determined but tentatively estimated from the acid hydrolysates. N-terminal amino acid sequence as determined by liquid-phase sequencer degradation of 4 nmol carboxymethylated BP. Cys analyzed as Cys(Cm). Values show nmol phenyl-thiohydantoins recovered (From Pòvoa et al. 1984a).

exclusion chromatography, and the isoelectric point was 4.3. The N-terminal amino acid sequence Ala-Pro-Trp-Gln-Cys-Ala-Pro-Cys-Ser-Ala disclosed that it was a previously uncharacterized protein. The amino acid composition also disclosed a high content of acidic/amidated residues (Fig. 1). The isolated 35-K binding protein bound both IGF-I and IGF-II with a calculated affinity constant of 10^{-9}. Later on, some groups, using either a modification of this method or other methods, purified and confirmed the first ten amino acids from the N-terminal amino acid sequence and the amino acid composition of this protein (Baxter et al. 1986; Elgin et al. 1986). A protein purified from placental tissues and termed PP12 was also shown to contain the same N-terminal amino acid sequence (Koistinen et al. 1986).

DEVELOPMENT OF A RADIOIMMUNOASSAY FOR 35-K BP AND ITS USE FOR IBP DETERMINATION IN SERUM

The BP isolated from human amniotic fluid was used to immunize rabbits and a radioimmunoassay with polyclonal antiserum was developed (Pòvoa et al. 1984b). It was shown that a 35-K BP from human serum, but not the 150-K BP, cross-reacted in this assay (Pòvoa et al. 1984b). In healthy adults a diurnal rhythm appears to exist, with high levels during the night (unpublished results).

The serum 35-K BP is age-dependent with high levels in cord blood, with a decline during childhood and the lowest levels in puberty. Thereafter, the levels increase during adult life, with higher levels in old age (unpublished results). This age- dependency shows a converse pattern to those of the 150-K BP and IGF-I levels, which are GH dependent. Furthermore, an inverse correlation was found between GH production and 35-K BP, with elevated levels in growth hormone deficiency and low levels in acromegaly (Fig. 2; Pòvoa et al. 1984b). In acromegaly,

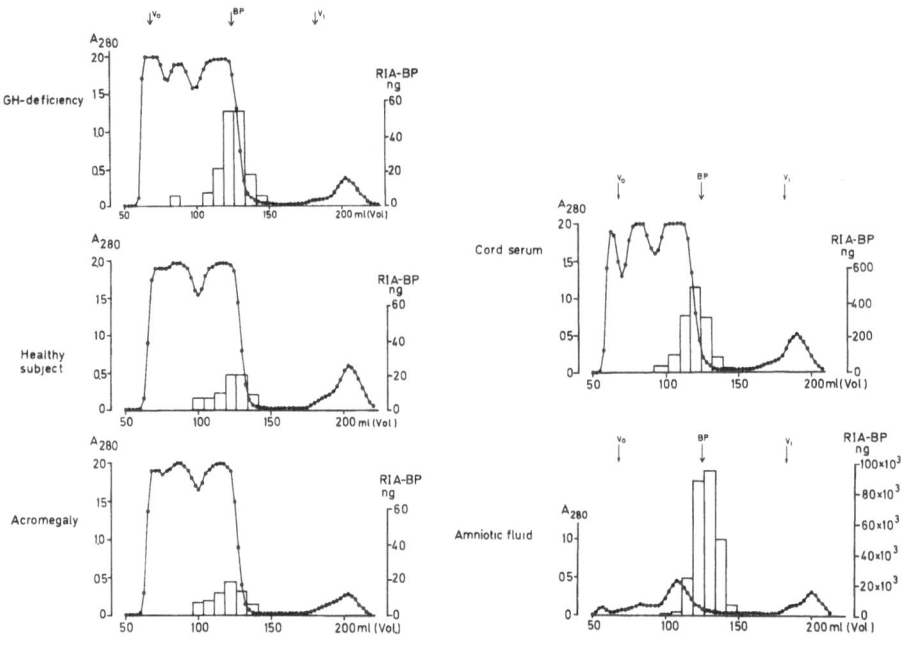

Fig. 2. Gel exclusion chromatography on a column of Sephadex G-200 (20 x 1.6 cm) at neutral conditions (0.05 M Tris, pH 7.5) of: *left* 3 ml serum from a patient with GH deficiency (*top*), a healthy subject (*middle*), and a patient with acromegaly (*bottom*). *right* 3 ml cord serum (*top*) and 3 ml of amniotic fluid containing 188 mg/ml of immunoreactive protein (*bottom*). *Bars* show immunoreactive binding protein (RIA-BP) in each fraction. *Arrows* indicate the elution volumes of V_o, V_i, and the pure amniotic fluid *BP* (From Pòvoa et al. 1984b).

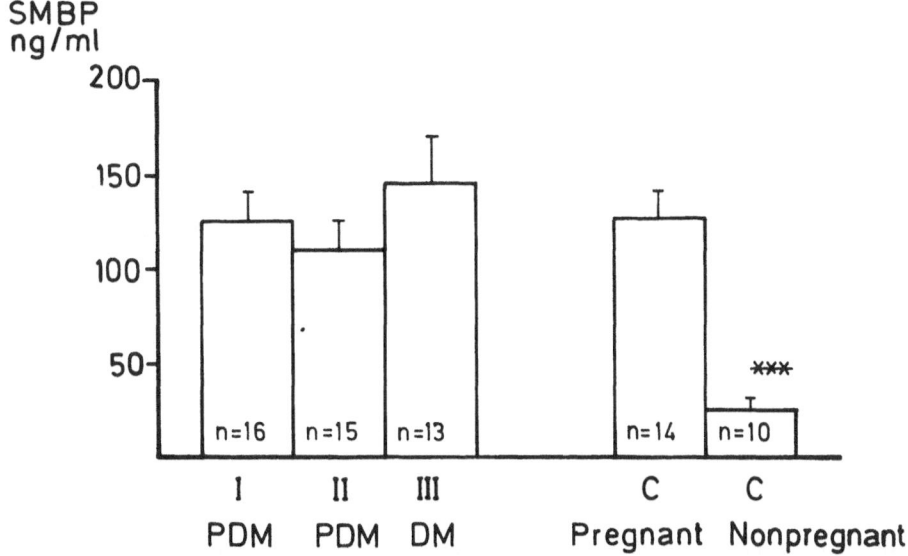

Fig. 3. Maternal serum levels of immunoreactive BP (x ± SEM) in women with pregnancy diabetes (*PDM*), treated with diet alone (*I*) or diet + insulin (*II*), and diabetes mellitus (*DM*) compared with those in pregnant and nonpregnant healthy women (C).

there was a significant negative correlation between 35-K BP and GH levels. However, no direct relationship seems to exist, because the levels of 35-K BP are also high in patients with anorexia nervosa or malnutrition, despite high growth hormone levels (unpublished results). Low levels were found in Cushing's syndrome.

There is an elevation of the serum level of 35-K BP throughout pregnancy, reaching a level four times that of nonpregnant individuals during the last trimester (Fig. 3; Hall et al. 1986). The serum levels declined after parturition with a half-life of about 24 h. The maternal serum 35-K BP pattern differs from that seen in the amniotic fluid, the latter having the highest levels around the 20th week. The sources of production are probably different. The administration of high levels of estrogens to tall girls has been found to cause a significant elevation during the treatment (Fig. 4; Pòvoa 1986).

POSSIBLE REGULATORY EFFECTS OF INSULIN ON THE SERUM 35-K BP

The factors that regulate the levels of 35-K BP are mainly unknown. We have found that in adolescents and adults with insulin-dependent diabetes mellitus

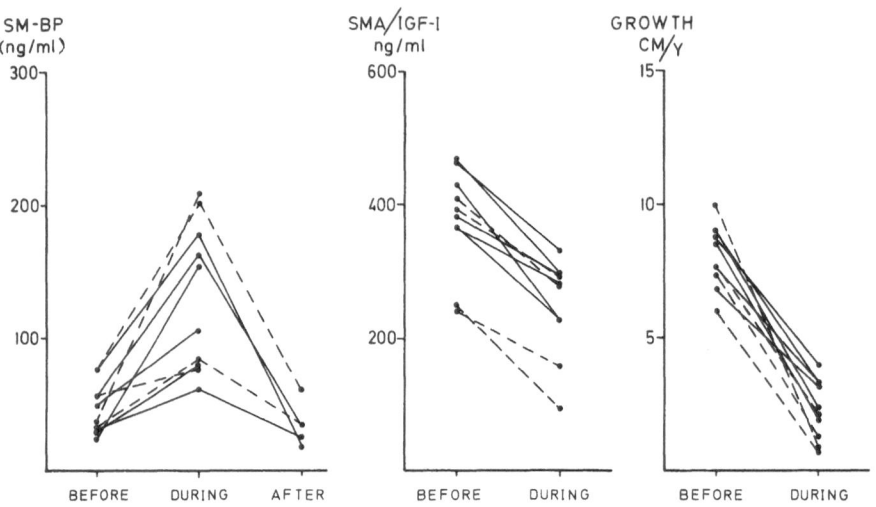

Fig. 4. Serum levels of SM-BP and IGF-I and the growth rate in tall girls before, during, and after treatment with 250 mg ethinylestradiol per day, over 1.5 – 2 years. At the start of treatment the skeletal age was <12.5 yr in six girls (●——●) and 13 – 13.5 yr in four girls (●----●) (From Pòvoa 1986).

(IDDM) and hyperglycemia the serum levels of 35-K BP were two- to three fold higher than in age-matched controls. However, the same was not true for patients with non-insulin-dependent diabetes mellitus (NIDDM) and a similar degree of hyperglycemia. This finding suggests that the elevation in the 35-K BP-levels could be related to insulin and not to the glucose levels (Hall et al., submitted for publication). Furthermore, improved glucose homeostasis in the group with IDDM by continuous insulin infusion normalized the 35-K BP-levels. In a group of adult diabetic patients the fasting serum levels of the 35-K BP showed a positive correlation with the glucose levels in IDDM, but no correlation was found in NIDDM (Brismar et al. 1987). On the other hand, the maternal 35-K BP levels during the last trimester in women with gestational diabetes as well as the cord blood levels, showed an inverse correlation with the C-peptide levels of cord blood (Hall et al. 1986). These findings taken together suggest a relationship between insulin and 35-K BP levels.

In order to further explore this hypothesis, hyperinsulinemia was induced in patients with IDDM by insulin infusion at a constant rate. Glucose was clamped after reaching normoglycemic levels, which were maintained by a Biostator-directed infusion of glucose for 2 h. The 35-K BP levels in plasma decreased with a calculated half-life of 60–90 min, independent of glucose levels (Brismar et al. 1987). Thus, insulin appears to regulate the 35-K BP levels directly and not through glucose regulation. It seems likely that insulin enhances the transport of the 35-K BP from the circulation to the target tissue.

HEP G2 CELL LINE AS A MODEL TO STUDY *IN VITRO* THE REGULATION OF THE 35-K BP

It had previously been demonstrated that the human hepatoma cell line Hep G2 produces a somatomedin BP with a molecular weight of about 35 K (Moses et al. 1983). Using our radioimmunoassay we detected a considerable amount of immunoreactive 35-K BP in medium conditioned by this cell line. This protein was isolated by immunoaffinity chromatography and proved to be identical to the 35-K BP from amniotic fluid with regard to size, charge, amino acid composition,

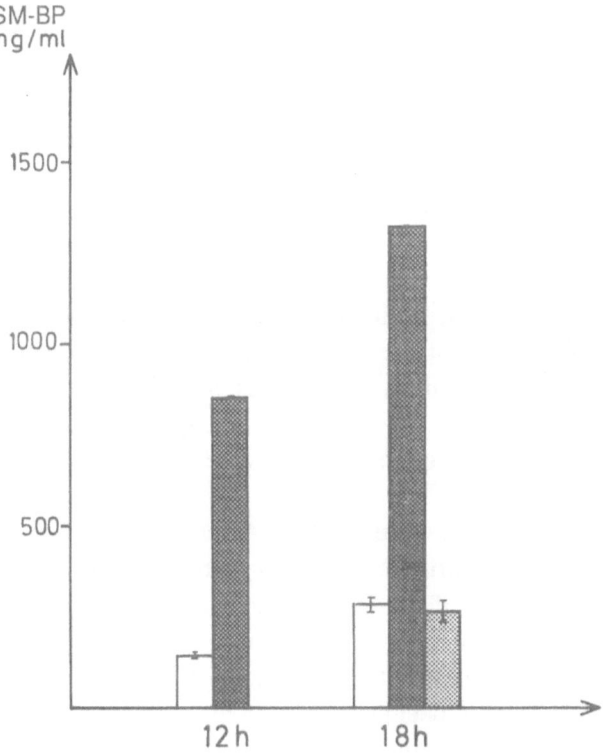

Fig. 5. Levels of immunoreactive SM-BP in medium conditioned by Hep G2 cells, 2 x 10^6 cells/ml. *Bars* show levels after 12 or 18 h of incubation with hCG preparation (500 U/ml) or pure hCG (200 mg/ml) compared with EMEM alone (From Pòvoa 1986).

and N-terminal amino acid sequence (Pòvoa et al. 1985). The Hep G2 cell line
has been used as a model to study the regulation of production of the 35-K BP.
The effects of different substances in physiological and pharmacological concentrations on the release of the binding protein were tested, with incubation periods
ranging from 6 h to 3 days. Hormones such as androgens, estrogens, insulin, growth
hormone, growth factors including IGFs, prolactin, placental lactogen, and pituitary gonadotropins did not influence the production of 35-K BP. Dexamethasone
in high doses had an inhibitory effect on production. Only a commercial preparation
of urinary human chorionic gonadotropin (hCG) elicited a time- and dose-related
stimulation (Fig. 5); highly purified hCG had no effect (Pòvoa 1986). It was
concluded that the stimulation was due to some contamination in the hCG
preparation. This substance is currently undergoing purification in our laboratory.

DEMONSTRATION OF THE 35-K BP IN HUMAN NORMAL TISSUES AND TUMORS

The radioimmunoassay for the amniotic fluid BP has been used in testing a
variety of cell lines, primary cell cultures and organ explants for the production
of this BP. It was found that decidual cells, the human breast-cancer cell lines MCF-
7 and MDA-MB-231 release immunoreactive BP in the conditioned media
(unpublished results). In a series of patients with primary liver cancers, very high
levels of serum 35-K BP were found in the majority of the cases (unpublished
results). The same was true of a series of human breast-cancer patients, irrespective
of the estrogen dependency of the tumor. Thus, in patients presenting with a tumor
producing high levels of the 35-K BP, monitoring serum levels of this BP may
be helpful in determining the course of the disease, i.e. recurrences may be preceded
by rising levels.

The polyclonal antisera were purified through a binding protein-Sepharose
column and the affinity-purified antibodies were used in immunohistochemical
studies of the distribution of the 35-K BP in a number of normal human tissues
and tumors as well as cell lines. The presence of immunoreactive BP could be
demonstrated in the cytoplasm of Hep G2 cells (Fig. 6a). The 35-K BP was also
present in the cytoplasm of decidual cells and the cytotrophoblast in the chorionic
villi (Fig. 6b). Nodules of primary liver cancer and normal hepatocytes were also
found to contain the 35-K BP (Fig. 6c); this was also the case in human breast
cancers (Fig. 6d), where the content varied greatly from case to case. Some of these
findings are similar to those reported for the placental protein called PP12 (Rutanen
et al. 1984).

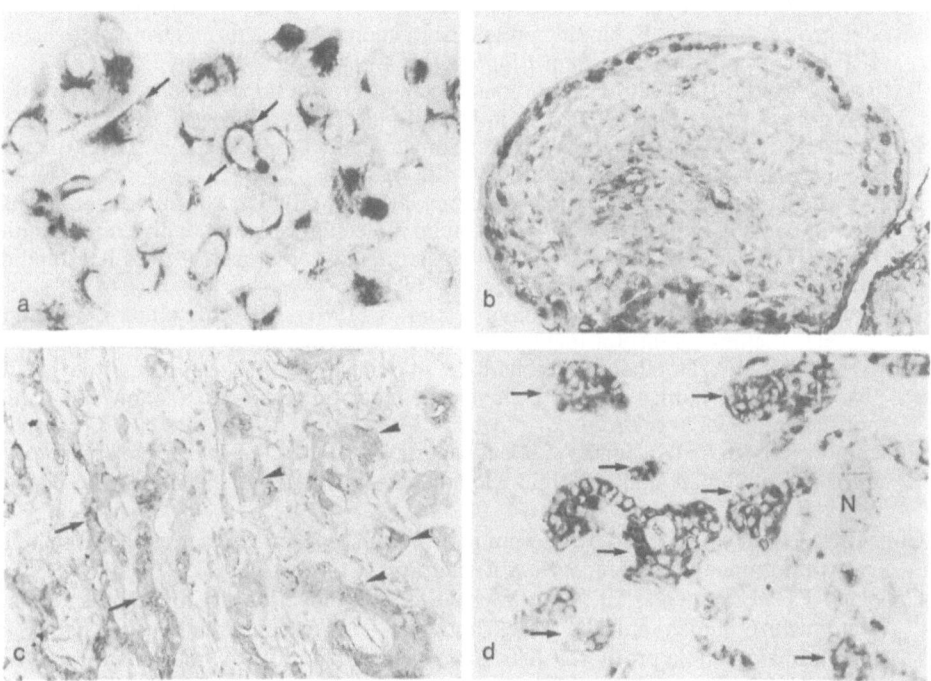

Fig. 6. a. Hep G2 cells studied immunocytochemically, using the affinity purified rabbit anti BP antiserum. Note the granular staining of the cytoplasm of the cells, mainly in the perinuclear area (*arrows*). The nucleus remains totally unstained. **b.** A chorionic villus studied in the same way as in Fig. 6a. Note the immunoreactivity mainly in the cytotrophoblast. **c.** The field shows normal hepatocytes (*arrows*) to the left and nodules of hepatocellular carcinoma to the right (*arrowheads*). Note that both cell types show immunoreactivity for the BP. Fibroblasts and other cell types in the connective tissue in the field are more or less negative. Nuclei in both normal and tumor cells are also negative. **d.** Multiple groups of breast-cancer cells (*arrows*), one of which is invading a nerve (*N*). Note the marked immunoreactivity in the tumor cells and the negative cell nuclei.

CONCLUSIONS

The levels of the 150-K form of the binding protein, which is restricted to the circulation, are growth-hormone regulated and the protein may function as a storage protein for IGFs. The 35-K form is widely distributed in many body fluids, may be a transport protein, and may play a role in regulating the availability of IGFs to the target cells. Insulin seems to be one regulator of this process. This protein is produced in many cell types, both normal and neoplastic.

REFERENCES

Baxter RC, Martin JL, Tyler MI, Howden MEH (1986) Growth hormone-dependent insulin-like growth factor (IGF) binding protein from human plasma differs from other human IGF-binding proteins. *Biochem Biophys Res Commun* **139**: 1256-1261

Baxter RC, Martin JL (1986) Radioimmunoassay of growth hormone-dependent insulin-like growth factor binding protein in human plasma. *J Clin Invest* **78**: 1504-1512

Binoux M, Lassarre C, Hardouin N (1982a) Somatomedin production by rat liver in organ culture. *Acta Endocrinol* (Copenh) **99**: 422-430

Binoux M, Hardouin N, Lassarre C, Hossenlopp P (1982b) Evidence for production by the liver of two IGF binding proteins with similar molecular weights but different affinities for IGF I and IGF II. Their relations with serum and cerebrospinal fluid IGF binding protein. *J Clin Endocrinol Metab* **55**: 600-602

Brismar K, Gutniak M, Pòvoa G, Werner S, Hall K (1987) The 35-K binding protein in diabetes mellitus is insulin regulated. *J Clin Endocrinol Metab* (submitted)

Burgi H, Müller WA, Humbel RE, Labhart A, Froesch ER (1966) Nonsuppressible insulin-like activity of human serum. I. Physiochemical properties, extraction and partial purification. *Biochem Biophys Acta* **121**: 349-359

Chochinow RH, Mariz IK, Hajek AS, Daughaday WH (1977) Characterization of a protein in mid-term human amniotic fluid which reacts in the somatomedin-C radioreceptor assay. *J Clin Endocrinol Metab* **44**: 902-908

Cohen KL, Nissley SP (1979) The serum half-life of somatomedin activity: evidence for growth hormone dependence. *Acta Endocrinol* (Copenh) **83**: 243-258

Copeland KC, Underwood LE, Van Wyk JJ (1980) Induction of immunoreactive somatomedin C human serum by growth hormone: dose-response relationships and effect on chromatographic profiles. *J Clin Endocrinol Metab* **50**: 690-697

Daughaday WH, Kipnis DM (1966) The growth-promoting and anti-insulin actions of somatotropin. *Recent Prog Horm Res* **22**: 49-93

Daughaday WH, Heins JN, Srivastava L, Hammer C (1968) Sulfation factor: studies of its removal from plasma and metabolic fate in cartilage. *J Lab Clin Med* **72**: 803-812

Daughaday WH, Ward AP, Goldberg AC, Trivedi B, Kapadia M (1982) Characterization of somatomedin binding in human serum by ultracentrifugation and gel filtration. *J Clin Endocrinol Metab* **55**: 916-921

Draznin B, Schalch DS, Heinrich UE, Schlueter R (1979) In: Giordano G, Van Wyk JJ, Minuto F (eds) *Somatomedins and Growth*. Academic, London, pp 149-161

Drop SLS, Valiquette G, Guyda HJ, Corvol MT, Posner BI (1979) Partial purification and characterization of a binding protein for insulin-like activity (ILAs) in human amniotic fluid: a possible inhibitor of insulin-like activity. *Acta Endocrinol* (Copenh) **90**: 505-518

Elgin RG, Busby WH Jr, Clemmons DR (1986) An insulin-like growth factor (IGF) binding protein enhances the biological response to IGF-I. *Proc Natl Acad Sci USA* **84**: 3254-3258

Froesch ER, Burgi H, Müller WA, Humbel RE, Jakob A, Labhart A (1967) Nonsuppressible insulin-like activity of human serum: purification, physiochemical and biological properties and its relation to total serum ILA. *Recent Prog Horm Res* **23**: 565-605

Furlanetto RW (1980) The somatomedin C binding protein: evidence for a heterologous subunit structure. *J Clin Endocrinol Metab* **51**: 12-19

Hall K (1972) Human somatomedin. *Acta Endocrinol* (Copenh) [Suppl 163]

Hall K, Brandt J, Enberg G, Fryklund L (1979) Immunorective somatomedin A in human serum. *J Clin Endocrinol Metab* **48**: 271-278

Hall K, Hansson U, Lundin G, Luthman M, Persson B, Pòvoa G, Stangenberg M, Öfverholm U (1986) Serum levels of somatomedins and somatomedin-binding protein in pregnant women with type I or gestational diabetes and their infants. *J Clin Endocrinol Metab* **63**: 1300-1306

Hintz RL (1984) Plasma forms of somatomedin and the binding protein phenomenon. *Clin Endocrinol Metab* **13**: 31-42

Hintz RL, Liu F (1980) Somatomedin plasma binding proteins. In: Pecile A, Müller EE (eds) *Growth Hormone and Other Biologically Active Peptides*. Excerpta Medica, Amsterdam, pp 133-143

Hintz RL, Orsini EM, Van Camp MG (1974) Interactions of somatomedin with plasma proteins. *Endocrinology* **94**: [Suppl A71]

Koistinen R, Kalkkinin N, Huhtala ML, Seppälä M, Bohn N, Rutanen EM (1986) Placental protein 12 is a decidual protein that binds somatomedin and has an identical N-terminal amino acid sequence with somatomedin-binding protein from human amniotic fluid. *Endocrinology* **118**: 1375-1378

Meuli C, Zapf J, Froesch ER (1978) NSILA-carrier protein abolishes the action of nonsuppressible insulin-like activity (NSILA-S) on perfused rat heart. *Diabetologia* **14**: 253-259

Moses AC, Freinkel AJ, Knowles BB, Aden DP (1983) Demonstration that a human hepatoma cell line produces a specific insulin-like growth factor carrier protein. *J Clin Endocrinol Metab* **56**: 1003-1008

Oelz O, Jakob A, Froesch ER (1970) Nonsuppressible insulin like activity (NSILA) of human serum. *Eur J Clin Invest* **1**: 48-53

Pòvoa G (1986) Low molecular form of somatomedin binding protein. Thesis, ISBN 91-7900-001-0

Pòvoa G, Enberg G, Jörnvall H, Hall K (1984a) Isolation and characterization of a somatomedin-binding protein from mid-term human amniotic fluid. *Eur J Biochem* **144**: 199-204

Pòvoa G, Roovete A, Hall K (1984b) Cross-reaction of serum somatomedin-binding protein in a radioimmunoassay developed for somatomedin-binding protein isolated from human amniotic fluid. *Acta Endocrinol* (Copenh) **107**: 563-570.

Pòvoa G, Isaksson M, Jörnvall H, Hall K (1985) The somatomedin-binding protein isolated from a human hepatoma cell line is identical to the human amniotic fluid somatomedin-binding protein. *Biochem Biophys Res Commun* **128**: 1071-1078

Rutanen EM, Wahlström T, Koistinen R, Sipponen P, Jalanko O, Seppälä M (1984) Protein 12 (PP12) in primãry liver cancer and cirrhosis. *Tumour Biol* **5**: 95-102

Spencer EM (1979) Synthesis by cultured hepatocytes of somatomedin and its binding protein. *FEBS Lett* **99**: 157-161

Van den Brande JL, Van Wyk JJ, Weaver RP, Mayberry HE (1971) Partial characterization of sulphation and thymidine factors in acromegalic plasma. *Acta Endocrinol* (Copenh) **66**: 65-81

Van Wyk JJ, Hall K, Weaver RP (1969) Partial purification of sulphation factor and thymidine factor from plasma. *Biochem Biophys Acta* **192**: 560-562

White RM, Nissley SP, Moses AC, Rechler MM, Johnsonbaugh RE (1981) The growth hormone dependence of a somatomedin- binding protein in human serum. *J Clin Endocrinol Metab* **53**: 49-57

Wilkins JR, D'Ercole AJ (1985) Affinity-labeled plasma somatomedin-C/insulin-like growth factor I binding proteins. Evidence of growth hormone dependence and subunit structure. *J Clin Invest* **75**: 1350-1358

Zapf J, Schoenle E, Froesch ER (1978) Insulin-like growth factors I and II: some biological actions and receptor-binding characteristics of two purified constituents of nonsuppressible insulin-like activity of human serum. *Eur J Biochem* **87**: 285-296

Zapf J, Schoenle E, Jagars G, Sand I, Grundwald J, Froesch ER (1979) Inhibition of the action of nonsuppressible insulin-like activity on isolated rat fat cells by binding to its carrier protein. *J Clin Invest* **63**: 1077-1084

Zapf J, Waldvogel M, Froesch ER (1975) Binding of nonsuppressible insulin-like activity to human serum. *Arch Biochem Biophys* **168**: 638-645

Advances in Growth Hormone and Growth Factor Research,
edited by E.E. Müller, D. Cocchi and V. Locatelli
Pythagora Press, Roma-Milano and Springer Verlag, Berlin-Heidelberg © 1989

Receptors for insulin-like growth factors I and II.*

R.G. ROSENFELD

Department of Pediatrics, Stanford University Medical Center, Stanford, California, U.S.A.

The somatomedins (SMs) comprise a family of growth hormone (GH)-dependent, insulin-like peptides with metabolic and mitogenic actions in a wide variety of cell lines (Phillips and Vassilopoulou-Sellin 1980). While their precise role in the regulation of cellular replication remains uncertain, a growing body of data supports the hypothesis that these peptides mediate many of the anabolic actions of GH (Phillips and Vassilopoulou-Sellin 1980; Schoenle et al. 1982). Two human SMs, insulin-like growth factors (IGF) I and II, have been purified, sequenced, and found to have 62% identity in amino acid positions (Rinderknecht and Humbel 1978a,b). Furthermore, these peptides have a striking structural homology to proinsulin. The sequences of cDNAs encoding for human pre-proIGF-I and -II have been elucidated and their respective 130 and 180 amino acid precursors predicted (Jansen et al. 1983; Bell et al. 1984; Dull et al. 1984). IGF-II has been mapped to the short arm of chromosome 11, tightly linked to both the insulin gene and the c-Ha-ras1 proto-oncogene (Brissenden et al. 1984; Tricoli et al. 1984). IGF-I maps to chromosome 12, which is evolutionarily related to chromosome 11 and carries the gene for the c-Ki-ras2 proto-oncogene. These striking homologies suggest evolutionary conservation of a common critical peptide skeleton and raise important questions concerning the precise role of each peptide and how cells mediate each peptide's biological activities. As their structural homology would suggest, IGF-I and -II have weak insulin-like metabolic effects, while insulin shares the mitogenic and growth-promoting actions of the IGFs (Rosenfeld and Hintz 1986). While competitive binding and affinity cross-linking studies have indicated the existence of specific receptors for each of the peptides, it has been difficult to discriminate among the functional roles of each

* This work was supported in part by NIH grants DK28229 and DK36054, as well as by the Diabetes Research and Education Foundation. Dr. Rosenfeld is the recipient of Research Cancer Development Award DK01275 from the NIH.

receptor because of (a) overlapping affinities of insulin and the IGFs for each other's receptors, (b) use of incompletely purified IGF preparations in both binding and biological studies, and (c) possible heterogeneity in binding properties of these receptors in different tissues and species.

STRUCTURE OF THE IGF RECEPTORS

The interrelationship of the insulin and IGF receptors has been clarified by attempts to purify and structurally characterize these membrane proteins. On the basis of affinity cross-linking studies, followed by polyacrylamide gel electrophoresis in sodium dodecyl sulfate (SDS-PAGE), both the insulin and IGF-I receptors have been found to be heterotetramers, composed of two $M_r = 135,000$ alpha-subunits and two $M_r = 90,000$ beta-subunits linked by interchain disulfide bonds (Kasuga et al. 1981; Chernausek et al. 1981; Massague and Czech 1982). Both subunits are glycosylated and exposed on the external surface of the cell. The alpha-subunit contains the binding site for insulin or IGF-I; the beta-subunit contains a transmembrane domain, an ATP-binding site, and a probable tyrosine autophosphorylation site (Kasuga et al. 1982; Jacobs et al. 1983b; Rubin et al. 1983; Zick et al. 1984). These findings are consistent with the observation that insulin and IGF-I can induce the phosphorylation of a tyrosine residue on the beta-subunit of their own receptors, as well as on exogenous protein substrates. Cloned cDNAs encoding for the entire 1370-1382 amino acid sequence of the human insulin receptor precursor have been identified and have been found to have significant sequence homologies to the IGF receptor and to the src family of tyrosine-specific protein kinases (Ullrich et al. 1985; Ebina et al. 1985). Nevertheless, the role(s) of phosphorylation in the biological actions of IGF-I is still not established.

More recently, Ullrich et al. (1986) determined the complete primary structure of the human IGF-I receptor from cloned cDNA. The deduced sequence was found to predict a 1367-amino acid receptor precursor, including a 30-amino acid signal peptide which is removed during translocation of the initial polypeptide chain. Cleavage of an Arg-Lys-Arg-Arg sequence results in the generation of an $M_r = 80,423$ alpha-subunit and an $M_r = 70,866$ beta-subunit. This compares with the respective $M_r = 135,000$ and $M_r = 90,000$ fully glycosylated subunits. As anticipated, extensive similarity with the insulin receptor was observed, including overall structure, subunit size, and primary sequence. However, while the insulin receptor has been localized on chromosome 19, the IGF-I receptor has been mapped to the distal band of the long arm of chromosome 15. These differences suggest that the insulin and IGF-I receptors are not only the products of distinct genes, but are probably also subject to different forms of regulatory signals.

Studies performed with IGF-II have indicated the existence of a structurally and immunologically distinct receptor (type II). Following affinity cross-linking and SDS-PAGE, this receptor migrates with an M_r = 220,000 in the unreduced state and 250,000 following reduction, suggesting that it is a single-chain polypeptide with internal disulfide bridges, which act to compact the molecule (Kasuga et al. 1981; Massague and Czech, 1982). Binding of ^{125}I-IGF-II to this receptor is preferentially inhibited by IGF-II > IGF-I , with no affinity for insulin, which appears to activate the appearance of type-II receptors on the cell surface (Oppenheimer et al. 1983; Wardzala et al. 1984). Although phosphorylation of the type-II receptor has been observed, this action could be mediated through the type-I receptor (Haskell et al. 1985). At this time, there is still no definitive evidence for the type-II receptor having intrinsic tyrosine kinase activity (Corvera et al. 1986).

Studies employing biosynthetic or cell-surface labeling have indicated that, in addition to the two major subunits (M_r = 135,000 and 95,000) of type-I receptors, there exist two higher molecular weight bands of M_r = 210,000 and 190,000 (Jacobs et al. 1983a). Pulse-chase studies demonstrate that the M_r = 190,000 band is the earliest labeled component, followed by the appearance of the M_r = 210,000 band and the alpha and beta subunits. Treatment of cells with monensin, which interferes with the biosynthesis of transmembrane and secretory proteins, blocks the disappearance of the M_r = 190,000 protein and the appearance of the mature alpha and beta subunits. Thus, the M_r = 190,000 component represents the high-mannose precursor form of the insulin or IGF-I receptor, which normally undergoes carbohydrate processing to produce the fully glycosylated M_r = 210,000 band and then undergoes proteolytic cleavage to generate the alpha and beta subunits. It is presumed that this M_r = 210,000 band represents the fully glycosylated form of the predicted M_r = 151,869 IGF-I receptor precursor. Jacobs et al. (1983a) have demonstrated similar maturational processing of the insulin and IGF-I receptors, although the respective precursors could be specifically immunoprecipitated by anti-insulin or anti-IGF-I receptor antibodies, confirming that the precursors for each receptor are distinct polypeptides.

Studies of the IGF-II receptor have indicated that the apparent molecular mass of this receptor in the absence of N-glycosylation is 232,000, and that glycosylation is required for the acquisition of binding activity (MacDonald and Czech 1985). The receptor is synthesized initially as an M_r = 245,000 precursor having 4-6 high mannose oligosaccharide side chains. Mannose removal and terminal sialylation converts this precursor to the M_r = 250,000 functional receptor. It is of note that recent immunohistochemical studies from our laboratory employing polyclonal and monoclonal antibodies to the type-II receptor have demonstrated high receptor concentrations in the Golgi complex, a major site of terminal glycosylation and proteolytic processing (Valentino et al. 1987).

Table 1 compares several of the structural and functional characteristics of the insulin and type-I and -II IGF receptors.

Table 1. Comparison of insulin and IGF receptors

	Insulin	Type-I IGF	Type-II IGF
M_r (unreduced)	>300,000	>300,000	240,000
M_r (reduced)	135,000	135,000	260,000
	90,000	90,000	
Subunits	2 alpha +	2 alpha +	none
	2 beta	2 beta	
Affinity	Ins > IGF-II	IGF-I > IGF-II	IGF-II >>
	> IGF-I	> Ins	IGF-I
Affinity for insulin	high	low	none
Glycosylation	+	+	+
Transmembrane	+	+	probable
Tyrosine kinase	+	+	−

FUNCTION(S) OF THE IGF RECEPTORS

As detailed above, structural analysis of the IGF receptors provided, for the first time, definitive proof of the existence of distinct receptors for insulin, IGF-I and IGF-II. Furthermore, by demonstrating the presence of tyrosine kinase activity on the insulin and type-I IGF receptor, these studies furnished some insight into potential mechanisms of action of the IGFs. However, despite these major additions to our understanding of the insulin and IGF receptors, several important fundamental questions have remained unanswered:

1. Can these receptors account for all of the biological action of insulin, IGF-I, and IGF-II?
2. Can the two identified IGF receptors account for all of the actions of growth hormone?
3. Which receptor(s) mediate each metabolic and mitogenic action of insulin, IGF-I and IGF-II?
4. What is the specific biological role of IGF-II and its receptor?

Answering these questions was greatly complicated by the overlapping affinities of each peptide for the other receptors (Rosenfeld and Hintz 1986). IGF-I and -II bind with weak affinity to the insulin receptor and are able to initiate insulin-like actions through this receptor. Similarly, insulin and IGF-II cross-react with the type-I IGF receptor. Although insulin does not appear to compete for occupancy of the type-II IGF receptor, variable affinity of IGF-I for this receptor has been reported, possibly due to contamination of natural IGF-I preparations with small amounts of IGF-II (Rosenfeld et al. 1987a). However, an approach to the complex issues listed above was provided by the discovery of naturally occurring anti-receptor antibodies (Flier et al. 1975). Incubation of rat adipocytes and human fibroblasts with Fab fragments derived from an anti-insulin receptor antibody resulted in a 30-fold rightward shift of the dose response for both insulin- and MSA-stimulated glucose oxidation, but there was no alteration in insulin- or MSA-stimulated DNA synthesis (King et al. 1980). It was concluded, therefore, that the insulin receptor mediates the metabolic actions of both insulin and IGF, while the IGF receptor(s) mediate the mitogenic actions. However, the generality of these conclusions has been questioned in light of the observation that the authors did not investigate the effect of this anti-receptor antibody on the type-I IGF receptor or on IGF-I action. Rosenfeld et al. (1981) have demonstrated that serum from a patient with insulin-resistant diabetes simultaneously inhibits both insulin and IGF-I binding, an observation which has been confirmed for the majority of naturally occurring anti-receptor antibodies (Jonas et al. 1982; Kasuga et al. 1983). Studies by Conover et al. (1985) have indicated that insulin stimulation of ^{3}H-thymidine incorporation in human fibroblasts is biphasic, with responses at insulin concentrations of 10-100 ng/ml apparently mediated through the insulin receptor, while responses at 1-100 µg/ml are mediated through the IGF-I receptor. These findings indicate that both the insulin and type-I IGF receptor are capable, at least in specific cell lines, of mediating the stimulation of cell replication. In some cell lines lacking IGF-I receptors, insulin appears to be capable of acting as a potent mitogen exclusively through its own receptor (Koontz and Iwahashi 1981). Similarly, Beguinot et al. (1985) have suggested that IGF-I and IGF-II stimulate glucose and amino acid uptake in L6 cells via interaction with their own receptors. Verspohl et al. (1984) reached the same conclusions concerning IGF stimulation of glycogen synthesis in HEP-G2 cells, indicating that IGF receptor(s) can mediate metabolic, as well as mitogenic actions.

These data, unfortunately, leave unanswered the question as to what the biological function of IGF-II is, and what role, if any, the type-II IGF receptor has in mediating these activities. Several studies have suggested a specific role for IGF-II in fetal growth, since plasma levels of IGF-II are relatively high in the fetus and newborn (particularly in the rat) (Moses et al. 1980), and since rat embryo fibroblasts synthesize large amounts of MSA (Adams et al. 1983). A second potential role for IGF-II is the regulation of cellular growth in the CNS, since specific receptors for IGF-II have been identified throughout the rat and human nervous systems, as well as within the pituitary (Sara et al. 1982, 1983; Goodyer et al. 1984; Rosenfeld et al. 1984). Furthermore, IGF-II, but not IGF-I, immunoreactivity has been identified in 24 distinct areas of the human brain, as well as in the cerebrospinal fluid (Haselbacher et al. 1985; Backstrom et al. 1984). Recio–Haselbacher Pinto et al. (1986) have reported that neurite outgrowth from chick embryo peripheral ganglion cells is stimulated by IGF-II at concentrations as low as 10 pM, supporting a specific neuritogenic role for IGF-II.

Virtually all data accumulated to date concerning the biologic function of the type-II receptor have been of the nature of comparative dose responses for IGF-II, IGF-I, and insulin.Studies by Heaton et al. (1980) in HTC rat hepatoma cells, by Schmid et al. (1983) in primary cultures of chicken skeletal muscle cells, by Janeczko and Etlinger, (1984) in chick myotubes, and by Beguinot et al. (1985) in L6 muscle cells have all supported a role for IGF-II mediated through the IGF-II receptor. Similarly, Verspohl and co-workers (1984) reported that IGF-I and -II stimulation of glycogen synthesis in HEP-G2 cells was not inhibited by a monoclonal antibody directed at the insulin receptor. More recently, Nishimoto et al. (1987a) have shown that IGF-II is significantly more potent than IGF-I in the stimulation of calcium influx in EGF-primed competent Balb/c 3T3 cells.

While the studies cited above support a role for the type-II receptor in mediating IGF-II action, other investigations have suggested that these effects may result from the binding of IGF-II to either insulin or type-I IGF receptors. Massague et al. (1982) have suggested that IGF-II stimulation of cell proliferation in H-35 rat hepatoma cells is through the insulin receptor. Employing the same cell line, Krett et al. (1987) have implicated the insulin receptor as the mediator of IGF-II-stimulated tyrosine aminotransferase, amino acid transport, and glycogen synthetase. Contrary to the report of Beguinot et al. (1985), Ewton and co-workers (1987) have provided data indicating that the type-II receptor in L6 myoblasts does not mediate amino acid uptake, cell proliferation and differentiation, or inhibition of protein degradation.

Initial studies employing antibodies against the insulin and IGF receptors did not support a role for the type-II receptor in mediating IGF-II action. In human fibroblasts, both Conover et al. (1986) and Furlanetto et al. (1987) showed that monoclonal antibodies against the type-I IGF receptor inhibited both IGF-I and IGF-II stimulation of DNA synthesis and cell replication. Similarly, IGF-II stimulation of amino acid transport in cultured human myotubes was inhibited by a monoclonal antibody against the type-I IGF receptor (Shimizu et al. 1986), and IGF-II stimulation of glucose uptake in TA1 mouse adipocytes was blocked by a monoclonal antibody against the insulin (and IGF-I) receptor kinase (Morgan and Roth, 1987). Employing a polyclonal antibody capable of inhibiting IGF-II binding, Mottola and Czech (1984) showed that pretreatment of H-35 cells did not affect IGF-II stimulation of DNA synthesis, supporting the concept that the actions of IGF-II in this cell line result from its weak affinity for the insulin receptor.

Three recent studies, however, strongly support a role for the type-II receptor in mediating IGF-II action:

1. Following up on the reports by Nishimoto et al. (1987a,b) of IGF-II stimulation of both calcium influx and DNA synthesis in EGF-primed competent Balb/c 3T3 cells, Kojima et al. (1988) investigated the effects of a polyclonal antibody capable of both immunoprecipitating and blocking the type-II IGF receptor. IGF-II was 50-fold more potent than IGF-I at stimulating both calcium

influx and thymidine incorporation, with an ED_{50} of 450 pM. Similarly, at receptor antibody concentrations of 100 µg/ml, calcium influx was stimulated four-fold and DNA synthesis ten-fold.

2. In prior studies, Verspohl et al. (1984) reported that IGF-II stimulation of glycogen synthesis in HEP-G2 cells was not inhibited by a monoclonal antibody directed against the insulin receptor. In a subsequent investigation, Hari et al. (1987) have reported that a polyclonal antibody directed against the human type-II receptor mimics the ability of IGF-II to stimulate glycogen synthesis in these cells.

3. Studies conducted in our laboratory, in collaboration with Dr. James Wyche, have focused on 18-54, SF cells, a rat tumor line capable of sustained growth under serum-free conditions. These cells are characterized by multiple forms of IGF-II mRNA, ranging from 1.1. to > 10 kb, with an abundant 4.4.-kb polyadenylated form and a nonpolyadenylated 1.1-kb form (James et al., submitted for publication). 18-54, SF cells secrete IGF-II into conditioned medium, attaining concentration as high as 10^{-8} M. When injected into athymic rats or mice, these cells produced large IGF-II-secreting tumors, which ultimately resulted in a tenfold rise in plasma IGF-II levels (Wilson et al. 1987). To explore whether IGF-II was functioning in an autocrine/paracrine manner in these cells and whether any such actions were mediated via the type-II receptor, the polyclonal antibody R-II-PAB-1, generated against the rat IGF-II receptor was employed (Rosenfeld et al. 1986, 1987b). At antibody concentrations of 100-400 µg/ml, R-II-PAB1 resulted in >80% inhibition of IGF-II binding, with no effect upon IGF-I binding. In densely plated cells, R-II-PAB1 resulted in a 70% decrease in cell number by day 5 and could block the mitogenic effect of exogenous IGF-II (Rosenfeld et al. 1987b). When R-II-PAB1 was employed in sparsely plated 18-54,SF cells, the major effect was a reduction in the number of cells per colony, rather than in the number of colonies formed. These data indicate that the major effect of this antibody against the type-II receptor was on the clonal proliferation of 18-54,SF cells, rather than on plating efficiency or cell viability.

These studies cannot be interpreted as definitive proof of the functional role(s) of the type-II IGF receptor. All three studies were performed with polyclonal antibodies, and one must consider the possibility that the observed effects were due to antibodies against other cellular antigens. Secondly, the cells employed were from unusual cell lines, and the resulting data cannot be generalized without considerable caution. On the other hand, recent immunohistochemical studies employing antibody against the type-II IGF receptor (R-II-PAB1) have indicated the presence of receptors in a wide variety of tissues and cells, where they are primarily localized to the Golgi complex and secondarily to the cell surface (Valentino et al. 1988; Rosenfeld and Pham 1987). Given the strikingly elevated levels of IGF-II and its mRNA in the rat fetus, and the demonstration by both binding studies and immunohistochemistry that type-II receptors are present in high concentrations in fetal tissues, a special role for IGF-II in the fetus and newborn has been hypothesized. Further investigations are clearly indicated to determine the role of the type-I and -II receptors in mediating these actions.

REFERENCES

Adams SO, Nissley SP, Handwerger S, Rechler MM (1983) Development patterns of insulin-like growth factor-I and -II synthesis and regulation in rat fibroblasts. *Nature* **302**: 150-153

Backström M, Hall K, Sara V (1984) Somatomedin levels in cerebrospinal fluid from adults with pituitary disorders. *Acta Endocrinol* (Copenh) **107**: 171-178

Beguinot F, Kahn CR, Moses AC, Smith RJ (1985) Distinct biologically active receptors for insulin, insulin-like growth factor-I, and insulin-like growth factor-II in cultured skeletal muscle cells. *J Biol Chem* **260**: 15892-15896

Bell GI, Merryweather JP, Sanchez-Pescador R, Stempien MM, Priestley L, Scott J, Rall LB (1984) Sequence of a cDNA clone encoding human preproinsulin-like growth factor-II. *Nature* **310**: 775-777

Brissenden JE, Ullrich A, Francke U (1984) Human chromosomal mapping of genes for insulin-like growth factors I and II and epidermal growth factor. *Nature* **310**: 781-784

Chernausek SD, Jacobs S, Van Wyk JJ (1981) Structural similarities between human receptors for somatomedin-C and insulin: analysis by affinity labeling. *Biochemistry* **20**: 7345-7350

Conover CA, Hintz RL, Rosenfeld RG (1985) Comparative effects of somatomedin-C and insulin on the metabolism and growth of cultured human fibroblasts. *J Cell Physiol* **122**: 133-141

Conover CA, Misra P, Hintz RL, Rosenfeld RG (1986) Effect of an anti-insulin-like growth factor-I receptor antibody on insulin-like growth factor-II stimulation of DNA synthesis in human fibroblasts. *Biochem Biophys Res Commun* **139**: 501-508

Corvera S, Whitehead RE, Mottola C, Czech MP (1986) The insulin-like growth factor-II receptor is phosphorylated by a tyrosine kinase in adipocyte plasma membranes *J Biol Chem* **261**: 7675-7679

Dull TJ, Gray A, Hayflick JS, Ullrich A (1984) Insulin-like growth factor-II precursor gene organization in relation to insulin gene family. *Nature* **310**: 777-781

Ebina Y, Ellis L, Jarnagin K, Edery M, Graf L, Clauser E, Ou J-H, Masiarz F, Kan YW, Goldfine ID, Roth RA, Rutter WJ (1985) The human insulin receptor cDNA: the structural basis for hormone-activated transmembrane signalling. *Cell* **40**: 747-758

Ewton DZ, Falen SL, Florini JR (1987) The type-II insulin-like growth factor (IGF) receptor has low affinity for IGF-I analogs: pleiotypic actions of IGFs in myoblasts are apparently mediated by the type-I receptor. *Endocrinology* **120**: 115-123

Flier JS, Kahn CR, Roth J, Bar RS (1975) Antibodies that impair insulin receptor binding in an unusual diabetic syndrome with severe insulin resistance. *Science* **190**: 63-65

Furlanetto RW, DiCarlo JN, Wisehart C (1987) The type-II insulin-like growth-factor receptor does not mediate deoxyribonucleic acid synthesis in human fibroblasts. *J Clin Endocrinol Metab* **64**: 1142-1149

Goodyer CG, De Stephano L, Lai WH, Guyda HJ, Posner BI (1984) Characterization of insulin-like growth-factor receptors in rat anterior pituitary, hypothalamus, and brain. *Endocrinology* **114**: 1187-1195

Hari J, Pierce SB, Morgan DO, Sara V, Smith MC, Roth RA (1987) The receptor for insulin-like growth factor-II mediates an insulin-like response. *EMBO J* **6**: 3367-3371

Haselbacher GK, Schwab ME, Pasi A, Humbel RE (1985) Insulin-like growth factor-II (IGF-II) in human brain: regional distribution of IGF-II and of higher-molecular-mass forms. *Proc Natl Acad Sci USA* **82**: 2153-2157

Haskell JF, Nissley SP, Rechler MM, Sasaki N, Greenstein L, Lee L (1985) Evidence for the phosphorylation of the type-II insulin-like growth-factor receptor in cultured cells. *Biochem Biophys Res Commun* **130**: 793-799

Heaton JH, Schilling EE, Gelehrter TD, Rechler MM, Spencer CJ, Nissley SP (1980) Induction of tyrosine aminotransferase and amino acid transport in rat hepatoma cells by insulin and the insulin-like growth factor, multiplication-stimulating activity. *Biochim Biophys Acta* **632**: 192-203

Jacobs S, Kull FC Jr, Cuatrecasas P (1983a) Monensin blocks the maturation of receptors for insulin and somatomedin-C: identification of receptor precursor. *Proc Natl Acad Sci USA* **80**: 1228-1231

Jacobs S, Kull FC Jr, Earp HS, Svoboda ME, Van Wyk JJ, Cuatrecasas P (1983b) Somatomedin-C stimulates the phosphorylation of the beta-subunit of its own receptor. *J Biol Chem* **258**: 9581-9584

Janeczko RA, Etlinger JD (1984) Inhibition of intracellular proteolysis in muscle cultures by multiplication-stimulating activity. *J Biol Chem* **259**: 6292-6297

Jansen M, van Schaik FMA, Ricker AT, Bullock B, Woods DE, Gabbay KH, Nussbaum AL, Sussenbach JS, Van den Brande JL (1983) Sequence of cDNA encoding human insulin-like growth factor-I precursor. *Nature* **306**: 609-611

Jonas HA, Baxter RC, Harrison LC (1982) Structural differences between insulin and somatomedin-C/insulin-like growth factor-I receptors revealed by autoantibodies to the insulin receptor. *Biochem Biophys Res Commun* **109**: 463-470

Kasuga M, Van Obberghen E, Nissley SP, Rechler MM (1981) Demonstration of two subtypes of insulin-like growth factor receptors by affinity cross-linking. *J Biol Chem* **256**: 5305-5308

Kasuga M., Karlsson FA, Kahn CR (1982) Insulin stimulates tyrosine phosphorylation of the 95,000-dalton subunit of its own receptor. *Science* **215**: 185-186

Kasuga M, Sasaki N, Kahn CR, Nissley SP, Rechler MM (1983) Anti-receptor antibodies as probes of insulin-like growth factor-receptor structure. *J Clin Invest* **72**: 1459-1469

King GL, Kahn CR, Rechler MM, Nissley SP (1980) Direct demonstration of separate receptors for growth and metabolic activities of insulin and multiplication-stimulating activity (an insulin-like growth factor) using antibodies to the insulin receptor. *J Clin Invest* **66**: 130-140

Kojima I, Nishimoto I, Iiri T, Ogata E, Rosenfeld RG (1988) Evidence that type-2 insulin-like growth factor receptor is coupled to a calcium-gating in Balb/c 3T3 cells. *Biochem Biophys Res Commun* **154**: 9-19

Koontz JW, Iwahashi M (1981) Insulin as a potent, specific growth factor in a rat hepatoma cell line. *Science* **211**: 947-949

Krett NL, Heaton JH, Gelehrter TD (1987) Mediation of insulin-like growth factor actions by the insulin receptor in H-35 rat hepatoma cells. *Endocrinology* **120**: 483-490

MacDonald RG, Czech MP (1985) Biosynthesis and processing of the type-II insulin-like growth factor receptor in H-35 hepatoma cells. *J Biol Chem* **260**: 11357-11365

Massague J, Czech MP (1982) The subunit structures of two distinct receptors for insulin-like growth factors I and II and their relationship to the insulin receptor. *J Biol Chem* **257**: 5038-5045

Massague J, Blinderman LA, Czech MP (1982) The high-affinity insulin receptor mediates growth stimulation in rat hepatoma cells. *J Biol Chem* **257**: 13958-13963

Morgan DO, Roth RA (1987) Acute insulin action requires insulin receptor kinase activity: introduction of an inhibitory monoclonal antibody into mammalian cells blocks the rapid effects of insulin. *Proc Natl Acad Sci USA* **84**: 41-45

Moses AC, Nissley SP, Short PA, Rechler MM, White RM, Knight AB, Higa OZ (1980) Increased levels of multiplication-stimulating activity, an insulin-like growth factor, in fetal rat serum. *Proc Natl Acad Sci USA* **77**: 3649-3653

Mottola C, Czech MP (1984) The type-II insulin-like growth factor receptor does not mediate increased DNA synthesis in H-35 hepatoma cells. *J Biol Chem* **259**: 12705-12713

Nishimoto I, Hata Y, Ogata E, Kojima I (1987a) Insulin-like growth factor-II stimulates calcium influx in competent Balb/c 3T3 cells primed with epidermal growth factor: characteristics of calcium influx and involvement of GTP-binding protein. *J Biol Chem* **262**: 12120-12126

Nishimoto I, Ohkuni Y, Ogata E, Kojima I (1987b) Insulin-like growth factor-II increases cytoplasmic free calcium in competent Balb/c 3T3 cells treated with epidermal growth factor. *Biochem Biophys Res Commun* **142**: 275-286

Oppenheimer CL, Pessin JE, Massague J, Gitomer W, Czech MP (1983) Insulin action rapidly modulates the apparent affinity of the insulin-like growth factor-II receptor. *J Biol Chem* **258**: 4824-4830

Phillips LS, Vassilopoulou-Sellin R (1980) Somatomedins. *N Engl J Med* **302**: 371-380, 438-446

Recio-Pinto E, Rechler MM, Ishii DN (1986) Effects of insulin, insulin-like growth factor-II, and nerve growth factor on neurite formation and survival in cultured sympathetic and sensory neurons. *J Neurosci* **6**: 1211-1219

Rinderknecht E, Humbel RE (1978a) The amino acid sequence of human insulin-like growth factor-I and its structural homology with proinsulin. *J Biol Chem* **253**: 2769-2776

Rinderknecht E, Humbel RE (1978b) Primary structure of human insulin-like growth factor-II. *FEBS Lett* **89**: 283-286

Rosenfeld RG, Baldwin D Jr, Dollar LA, Hintz RL, Olefsky JM, Rubenstein A (1981) Simultaneous inhibition of insulin and somatomedin-C binding to cultured IM-9 lymphocytes by naturally occurring anti-receptor antibodies. *Diabetes* **30**: 979-982

Rosenfeld RG, Ceda G, Wilson DM, Dollar LA, Hoffman AR (1984) Characterization of high-affinity receptors for insulin-like growth factor I and II on rat anterior pituitary cells. *Endocrinology* **114**: 1571-1575

Rosenfeld RG, Hintz RL (1986) Somatomedin receptors: structure, function and regulation. In: Conn PM (ed)*The Receptors*, vol III. Academic, Orlando, pp 281-329

Rosenfeld RG, Hodges D, Pham H, Lee PDK, Powell DR (1986) Purification of the insulin-like growth factor-II (IGF-II) receptor from an IGF-II-producing cell line and generation of an antibody which both immunoprecipitates and blocks the type-2 receptor. *Biochem Biophys Res Commun* **138**: 304-311

Rosenfeld RG, Pham H (1987) Production of monoclonal antibodies to the rat insulin-like growth factor-II (IGF-II) receptor. *Biochem Biophys Res Commun* **146**: 717-724

Rosenfeld RG, Conover CA, Hodges D, Lee PDK, Misra P, Hintz RL, Li CH (1987a) Heterogeneity of insulin-like growth factor-I affinity for the insulin-like growth factor-II receptor: comparison of natural, synthetic and recombinant DNA-derived insulin-like growth factor-I. *Biochem Biophys Res Commun* **143**: 199-205

Rosenfeld RG, Pham H, Han Z-Y, Shah R, Diaz G, Wyche JH (1987b) Demonstration of an autocrine role for insulin-like growth factor-II, mediated through the type-II receptor (submitted for publication)

Rubin JB, Shia MA, Pilch PF (1983) Stimulation of tyrosine-specific phosphorylation *in vitro* by insulin-like growth factor-I. *Nature* **305**: 438-440

Sara VR, Hall K, Von Holtz H, Humbel R, Sjogren B, Wettenberg L (1982) Evidence of the presence of specific receptors for insulin-like growth factors I (IGF-I) and 2 (IGF-2) and insulin throughout the adult human brain. *Neurosci Lett* **34**: 39-44

Sara VR, Hall K, Nisaki M, Fryklund L, Christensen N, Wettenberg L (1983) Ontogenesis of somatomedin and insulin receptors in the human fetus. *J Clin Invest* **71**: 1084-1094

Schmid C, Steiner T, Froesch ER (1983) Preferential enhancement of myoblast differentiation by insulin-like growth factors (IGF I and IGF II) in primary cultures of chick embryonic cells. *FEBS Lett* **161**: 117-121

Schoenle E, Zapf J, Humbel RE, Froesch ER (1982) Insulin-like growth factor II stimulates growth in hypophysectomized rats. *Nature* **296**: 252-253

Shimizu M, Webster C, Morgan DO, Blau HM, Roth RA (1986) Insulin and insulin-like growth factor receptor and responses in cultured human muscle cells. *Am J Physiol* **251**: E611-E615

Tricoli JV, Rall LR, Scott J, Bell GI, Shows TB (1984) Localization of insulin-like growth factor genes to human chromosomes 11 and 12. *Nature* **310**: 784-786

Ullrich A, Bell JR, Chen EY, Herrera R, Petruzzelli LM, Dull TJ, Gray A, Coussens L, Liao Y-C, Tsubokawa M, Mason A, Seeburg PH, Grunfeld C, Rosen OM, Ramachandran J (1985) Human insulin receptor and its relationship to the tyrosine kinase family of oncogenes. *Nature* **313**: 756-761

Ullrich A, Gray A, Tam AW, Yang-Feng T, Tsubokawa M, Collins C, Henzel W, Le Bon T, Kathuria S, Chen E, Jacobs S, Francke U, Ramachandran J, Fujita-Yamaguchi Y (1986) Insulin-like growth factor-I receptor primary structure: comparison with insulin receptor suggests structural determinants that define structural specificity. *EMBO J* **5**: 2503-2512

Valentino KL, Pham H, Ocrant I, Rosenfeld RG (1988) Distribution of insulin-like growth factor-II (IGF-II) receptor immunoreactivity in rat tissues. *Endocrinology* **122**: 2753-2763

Verspohl EJ, Roth RA, Vigneri R, Goldfine ID (1984) Dual regulation of glycogen metabolism by insulin and insulin-like growth factors in human hepatoma cells (HEP-G2). *J Clin Invest* **74**: 1436-1443

Wardzala LJ, Simpson IA, Rechler MM, Cushman SW (1984) Potential mechanism of the stimulatory action of insulin on insulin-like growth factor-II binding to the isolated rat adipose cell. *J Biol Chem* **259**: 8378-8383

Wilson DM, Thomas JA, Hamm TE Jr, Wyche J, Hintz RL, Rosenfeld RG (1987) Transplantation of insulin-like growth factor-II secreting tumors into nude rodents. *Endocrinology* **120**: 1896-1901

Zick Y, Sasaki N, Rees-Jones RW, Grunberger G, Nissley SP, Rechler MM (1984) Insulin-like growth factor-I (IGF-I) stimulates tyrosine kinase activity in purified receptors from a rat liver cell line. *Biochem Biophys Res Commun* **119**: 6-13

Advances in Growth Hormone and Growth Factor Research,
edited by E.E. Müller, D. Cocchi and V. Locatelli
Pythagora Press, Roma-Milano and Springer Verlag, Berlin-Heidelberg © 1989

In vivo actions of insulin-like growth factor-I*

J. Zapf[1], H.P. Guler[1], Ch. Schmid[1], A. Kurtz[2] and E.R. Froesch[1]

[1] *Metabolic Unit, Department of Medicine, University Hospital, Zürich, Switzerland; and* [2] *Institute of Physiology, University of Zürich, Zürich, Switzerland*

Insulin-like growth factors (IGFs) are structurally, phylogenetically and biologically closely related to insulin (Rinderknecht and Humbel 1978a, 1978b; Zapf et al. 1984). It may, therefore, not appear surprising that intravenous bolus injections of IGF-I or -II into rats cause the same effects as an intravenous injection of insulin (Zapf et al. 1986). In a more spectacular way this comparison has recently been carried out in man (Guler et al. 1987): When a bolus of 100 µg/kg body wt. of recombinant human IGF-I (rh IGF-I) is administered intravenously to normal human subjects a dramatic fall in the blood sugar is observed (Fig. 1). The lowest blood sugar levels are reached 30 min after the injection. The fall in the blood sugar is accompanied by symptoms of hypoglycemia, i.e., drowsiness, blurred vision, paleness of the face, sweating, trembling, and tachycardia, followed by hunger and fatigue. The hypoglycemic symptoms and the blood sugar curves are identical to those observed during a standard insulin tolerance test with an intravenous injection of 0.15 U/kg body wt. of insulin.

Although this hypoglycemic reaction represents a very impressive IGF effect, there is no evidence that endogenous IGFs are involved in the regulation of glucose homeostasis. Of the circulating endogenous IGFs, 92%–96% are tightly bound to specific serum carrier proteins (free IGF-I in normal human serum: 20–30 ng/ml, free IGF II: 7–15 ng/ml; J. Zapf, unpublished results). These complexed, "native" forms of IGF are inactive on insulin target tissues (Zapf et al. 1984). Although total endogenous IGF-I and -II concentrations in normal human serum do not differ from those of carrier protein-bound IGF-I and -II measured at various time points after the IGF I bolus, they never cause hypoglycemia. Therefore, the hypoglycemic reaction after iv IGF-I injection has been attributed to free IGF-I. As shown in Fig. 2, free IGF-I rose to over 300 ng/ml after 15 min when blood sugar levels had nearly

* This work was supported by grant no. 3.051-0.84 from the Swiss National Science Foundation.

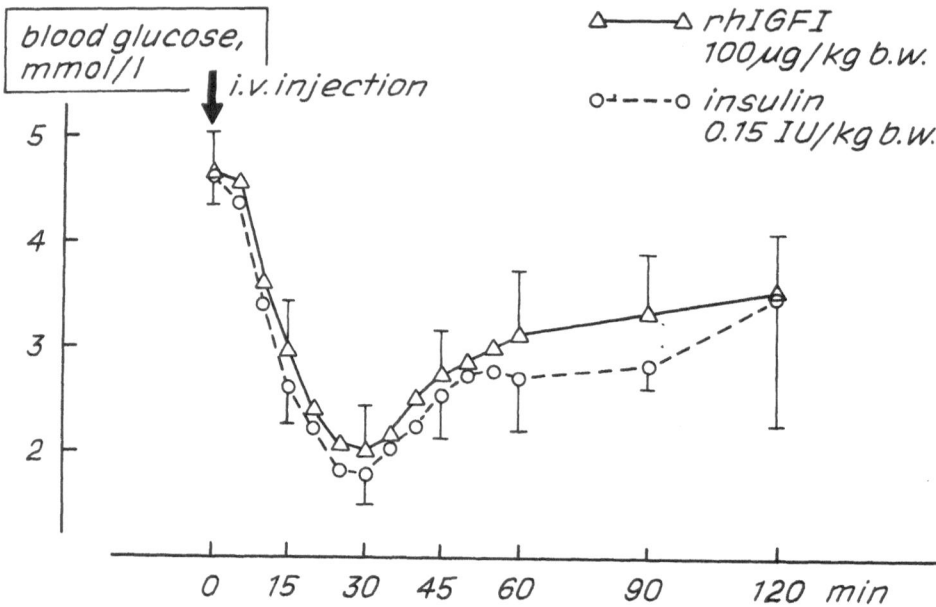

Fig. 1. Mean glucose levels (± SD) in eight healthy subjects after injection of 100 μg rh IGF-I or 0.15 IU insulin per kg body wt. (From Guler et al. 1987).

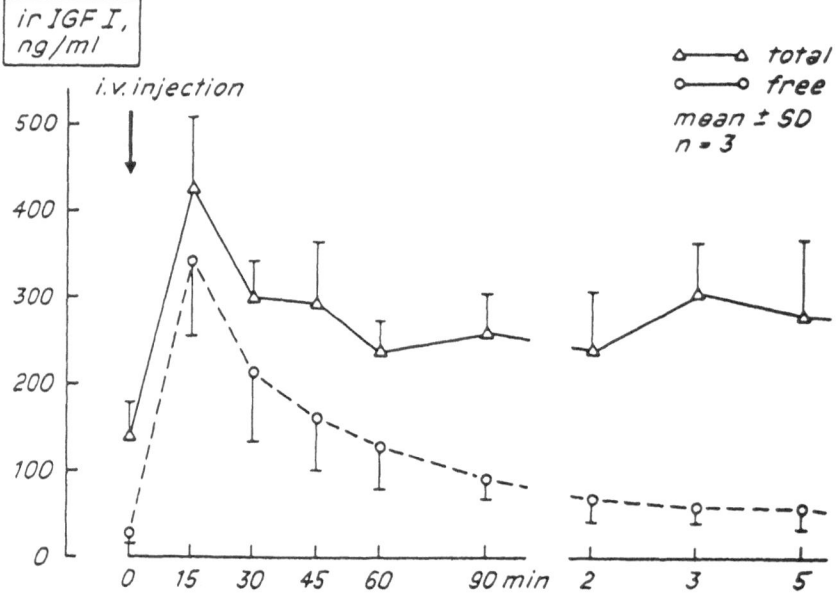

Fig. 2. Total and free radioimmunoassayable IGF-I in three healthy subjects after i.v. injection of 100 μg rh IGF-I/kg body wt. Total and free IGF-I were determined as described by Zapf et al. 1986. (From Guler et al. 1987).

reached their nadir. Thereafter, free IGF-I decreased with a half-life of 15 min, and bound IGF-I concomitantly increased while hypoglycemia subsided. Thus, acute insulin-like effects of IGF are observed only when its free form in serum increases so rapidly that it temporarily overrides the binding capacity of the IGF carrier proteins. The same conclusions have been derived from experiments in normal and hypophysectomized rats receiving intravenous bolus injections of IGF-I or -II (Zapf et al. 1986).

Under normal conditions, free IGF never rises to levels that are high enough to elicit acute insulin-like actions mediated via insulin receptors. Regulation of glucose homeostasis can therefore not be considered a physiological function of IGFs. Their biological significance *in vivo* has to be traced in a different direction and under different experimental conditions. Obviously, the latter have to be adjusted in such a way as to avoid rapid and drastic increases of free IGF and to allow for gradual equilibration with the circulating IGF carrier proteins.

The physiological role of IGF-I was postulated as early as 1972 in the somatomedin (somatomedin-C = IGF-I) hypothesis (Daughaday et al. 1972). According to it, growth hormone is the most important stimulus of IGF-I secretion by the liver, and IGF-I the major mediator of the effects of growth hormone on body growth. In addition to growth hormone, insulin and nutrition have been attributed important parts in the regulation of IGF-I levels and thus of growth (Chochinov and Daughaday 1976; Phillips and Unterman 1984). According to the somatomedin hypothesis growth arrest in endocrine conditions with low IGF-I serum levels should at least partly be reversed by raising serum IGF-I over a prolonged period of time. In order to test this we have used two experimental animal models that lack one or several of the above -mentioned growth regulators: the hypophysectomized and the insulin-deficient streptozotocin-diabetic rat. Rats stop growing after hypophysectomy, and due to growth hormone deficiency their serum IGF-I level decreases drastically. Similarly, growth of rats is arrested after the induction of severe insulin-deficient diabetes. Again, growth arrest is accompanied by a drop in growth hormone and endogenous IGF-I levels. However, in contrast to the hypophysectomized rat, who resumes growth under growth hormone treatment, growth hormone is ineffective in the diabetic rat (Young 1945; Scheiwiller et al. 1986).

EFFECTS OF RECOMBINANT HUMAN IGF-I IN HYPOPHYSECTOMIZED RATS

Bone and Organ Growth

In 1982, Schoenle et al. demonstrated that pure natural IGF-I infused subcutaneously into hypophysectomized rats over a period of 6 days stimulated gains in body weight and tibial epiphyseal width and increased the thymidine-incorporating activity of isolated costal cartilage. Similar results have been obtained with rh IGF-I in rats (Hizuka et al. 1987; Horiai et al. 1987; Guler et al. 1988) and in Snell dwarf mice (Van Buul-Offers et al. 1986). Although these results strongly supported the somatomedin hypothesis, the question of whether growth hormone might be important at one step of the regulation of longitudinal growth remained unanswered. As suggested by Isgaard et al. (1986), longitudinal bone growth is "the result of both cell differentiation, directly stimulated by growth hormone, and clonal expansion of cells in the proliferative layer of the growth plate due to the local production of growth factors." The authors continue to infer that "if this theory is correct, administration of IGF-I systemically or locally should only stimulate differentiated cells or cells committed to differentiate, but the effect of IGF-I should not be sustained due to the fact that these cells would have a limited proliferative capacity." According to these considerations, one could indeed argue that during a 6-day infusion with IGF-I enough target cells might still be available for IGF-I to cause a growth spurt of short duration. In order to test this argument it was important to infuse IGF-I into hypophysectomized rats for a longer period of time so that a growth hormone-sensitive differentiation step would become rate-limiting. In this case one would expect the effect of IGF-I to differ considerably from that of growth hormone.

A maximally effective dose of growth hormone (200 mU/day) and of rh IGF-I (300 µg/day) was therefore infused subcutaneously into hypophysectomized male rats (Tif RAI) over a period of 18 days. Control animals received a saline infusion. Total immunoreactive (ir) IGF-I serum levels at the end of the rh IGF-I infusion ranged between 440 and 540 ng/ml, free IGF-I between 20 and 40 ng/ml. In the saline-treated controls total irIGF was 12 ± 3 ng/ml, and no free IGF-I was measurable. The small free IGF-I steady-state concentration under the rh IGF-I infusion did not cause a drop in the blood sugar levels (6.55±1.1 mmol/l as compared to 6.72 ± 0.78 mmol/l in the controls).

As shown in Figure 3, body weight gain during the whole period of infusion did not differ significantly in the growth hormone and rh IGF-I-treated animals. This is even more convincingly reflected in the epiphyseal growth rate (Table 1) as determined by oxytetracycline staining (Hansson 1967; Hansson et al. 1972). Microscopic examination of the growth plates and the adjacent tissue showed a widening of the epiphysis with growth hormone similar to that with rh IGF-I (Table 1) and stacks of normally arranged chondrocytes and newly formed areas of trabecular bone (Guler et al. 1988). Preliminary morphometric studies of the epiphyseal cartilage after different infusion times did not reveal essential differences between growth hormone and rh IGF-I action (E. Hunziker and J. Zapf 1988): Whereas rh IGF-I caused a more rapid increase in the proliferative zone, maximal stimulation of both the proliferative and the hypertrophic zone appeared to be somewhat greater with growth hormone. Nevertheless, there is no evidence so far

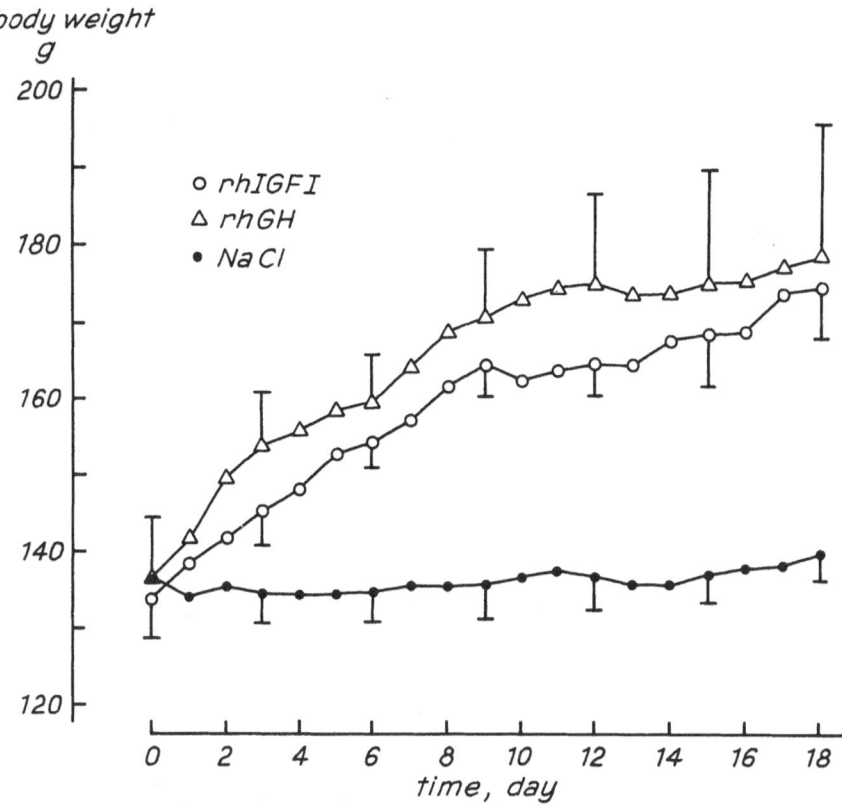

Fig. 3. Body weight gain in hypophysectomized rats infused for 18 days with saline, recombinant human growth hormone (200 mU/day), or rh IGF-I (300 µg/day) by subcutaneously implanted Alzet minipumps. All points are means ± SD from four rats per group (From Guler et al. 1988).

Table 1. Growth indices in hypophysectomized rats infused s.c. (Alzet miniosmotic pumps) for 18 days with saline, rh growth hormone, or rh IGF-I (Guler et al. 1988)

Treatment	Tibial epiphyseal width: µm (± SD)	Epiphyseal growth rate: µm/day (oxytetracycline marking)
Saline ($n = 4$)	188 ± 19	4.0 ± 0.7
rh Growth hormone 200 mU/d ($n = 4$)	321 ± 78 n.s.	52.5 ± 10.8 n.s.
rh IGF-I 300 µg/d ($n = 4$)	299 ± 47	37.1 ± 2.0

that IGF-I lacks any of the effects observed in growth hormone-treated hypophy-sectomized rats. On the contrary, the stimulatory effect of rh IGF-I on the weight of particular organs was even more pronounced (statistically significant) than that of growth hormone: during the 18 days of infusion with rh IGF-I the weight of the kidneys doubled and the weight of the thymus and the spleen increased threefold as compared with the hypophysectomized controls (Guler et al. 1988). Surpris-

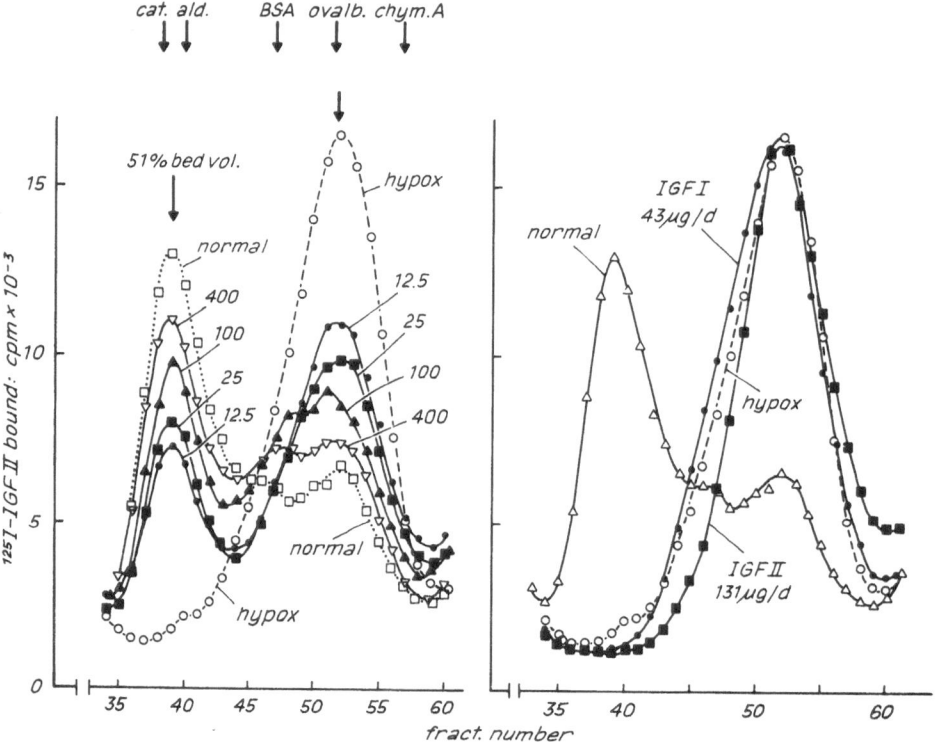

Fig. 4. Radiochromatographic patterns of sera from normal, hypophysectomized and IGF-I-, IGF-II-, or hGH-infused hypophysectomized rats after preequilibration of the sera with ^{125}I-IGF-II (24 h) and gel filtration on Sephadex G-200 at neutral pH. The numbers on the *left panel* give the mU/day of infused hGH; those on the *right panel* represent μg/day of infused IGF-I or -II. The methodology was the same as that described by Zapf et al. 1975. *cat*, catalase; *ald*, aldolase; *bsa*, bovine serum albumin; *ovalb*, ovalbumin; *chym A*, chymotrypsinogen A.

ingly, and in contrast to growth hormone, rh IGF-I did not significantly stimulate the weight of the gastrocnemius and soleus muscles.

Since growth hormone induces the synthesis of endogenous rat IGF-I, it is difficult to understand why IGF-I alone should have greater effects on some organs than growth hormone itself. One possible explanation may relate to differences in the pattern of the IGF carrier proteins in the treated animals: growth hormone induces the synthesis and secretion not only of endogenous IGF-I but also of the 150-K IGF carrier protein (Moses et al. 1976; Kaufmann et al. 1978; Schoenle et al. 1985) which is lacking in hypophysectomized rats. The 150-K carrier protein, however, does not appear to be induced by the infusion of rh IGF-I (Fig. 4; Schoenle et al. 1985). Instead, serum of hypophysectomized rats shows a pronounced increase of another IGF-binding protein with an apparent molecular weight of 40–50-K, as standardized by gel filtration on Sephadex G-200 (Fig. 4; Zapf et al. 1985a). Infused IGF-I equilibrates and circulates with this protein in serum, and only a small portion remains in the free form (see above). It appears that the affinity of IGF-I for the 40–50-K protein is smaller than that for the 150-K carrier (Kaufmann et al. 1977) and that the bioavailability of infused rh IGF-I for organs like the kidney, the spleen, or the thymus may be greater than that of endogenous IGF-I bound to the 150-K carrier protein. The contrary may be true for tissues like cartilage and bone. Thus, the binding proteins may function as vectors targeting IGF-I in a differential manner to different cells in the body. An important role of growth hormone would then consist of choosing the targets for IGF-I.

ERYTHROPOIESIS

The occurrence of anemia after hypophysectomy in rats has been known for many years (Meyer et al. 1937; Vollmer et al. 1939; Crafts 1941). At the same time hypoplasia of the bone marrow and a decrease in the number of reticulocytes in the peripheral blood have been observed (Overbeek 1936; Overbeek and Querido 1938) indicating that erythropoiesis in these animals is reduced. On the other hand, growth hormone has been reported to induce a marked increase in the number of reticulocytes in hypophysectomized rats (Meyer et al. 1940). According to the somatomedin hypothesis, this effect of growth hormone may well be mediated by insulin-like growth factor-I. In fact, insulin-like growth factor-I has been found to enhance erythropoiesis *in vitro* (Kurtz et al. 1982, 1985; Akahane et al. 1987). We therefore examined whether IGF-I stimulated erythropoiesis in hypophysectomized rats *in vivo*. Male Tif RAI rats (120–140 g body wt.) were infused for 6 days by means of subcutaneously implanted Alzet mini-pumps with rh IGF-I (120 μg/day) or with 28 mU/day of human growth hormone. As shown in Table 2, both hormones led to an increase in ^{59}Fe-incorporation into red blood cells and in the number of reticulocytes in the peripheral blood as compared with the saline-infused control animals. The hematocrit remained unchanged during growth hormone or rh IGF-I treatment. However, both hormones led to a significant increase in the immunoreactive erythropoietin concentrations after 6 days of infusion. Interestingly, stimulation of the erythropoietic indices preceded the rise

Table 2. Erythropoietic indices in hypophysectomized rats infused s.c. (Alzet miniosmotic pumps) for 6 days with saline, human growth hormone or rh IGF-I ($n=5$ for each group). (From Kurtz et al., in press)

Treatment	Reticulocytes (%) (mean ± SD)		^{59}Fe-incorporation into red blood cells (%): (mean ± SD)	
	After 4 days	After 6 days	Days 2-4	Days 4-6
Saline	4.3 ± 0.3	4.5 ± 0.4	25.0 ± 4.5	26.4 ± 5.5
Human growth hormone 28 mU/d	6.8 ± 0.6*	5.7 ± 0.3*	40.0 ± 5.6*	42.7 ± 3.6**
rh IGF-I 120 µg/d	5.6± 0.4*	7.7 ± 0.6*	47.2 ± 4.5***	46.4 ± 3.7**

*$P<0.05$ vs control; **$P<0.025$ vs control; ***$P<0.01$ vs control.

of the serum erythropoietin. A highly significant linear correlation was observed between the increase in body weight and the incorporation of ^{59}Fe into red blood cells.

These studies show that IGF-I stimulates erythropoiesis in hypophysectomized rats by a dual mechanism, i.e., directly and by enhancing erythropoietin production. Moreover, they demonstrate that IGF-I mediates the effects of growth hormone on erythropoiesis *in vivo* and is thus in line with the somatomedin hypothesis.

EFFECTS OF RH IGF-I IN DIABETIC RATS

Severely diabetic rats stop growing. They are insulin-deficient, the amplitude and duration of the growth hormone secretory episodes are markedly suppressed (Tannenbaum 1981), and serum IGF-I levels are low. Growth hormone is ineffective in restoring growth of these animals. Therefore, it was challenging to find out whether IGF-I treatment would be able to affect growth in the absence of growth hormone and insulin.

Infusion of 300 µg of rh IGF-I/day into severely diabetic male rats (Tif RAI) caused a significant increase in body weight, tibial epiphyseal width, and thymidine-incorporating activity of costal cartilage (Fig. 5) without attenuating the diabetic condition. Although the blood sugar levels were slightly reduced (probably

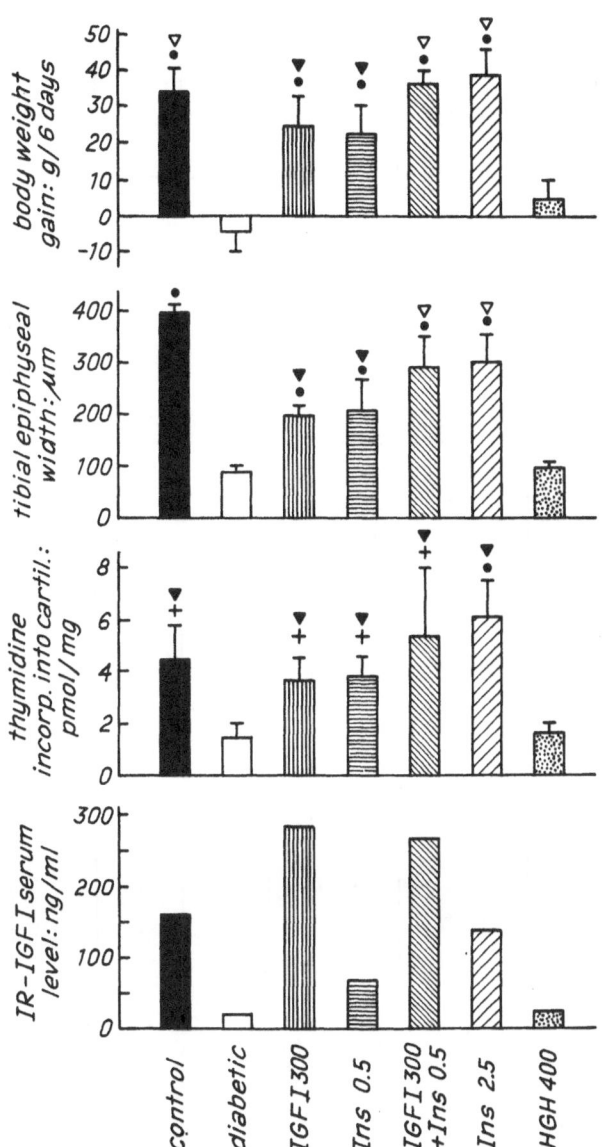

Fig. 5. Growth indices and ir IGF-I serum levels in normal, streptozotocin-diabetic, and treated diabetic rats. Hormones were infused for 6 days by subcutaneously implanted miniosmotic pumps in the following daily amounts: rh IGF-I, 300 µg/day; insulin, 0.5 and 2.5 U/day; a combination of 300 µg/day of rh IGF-I and 0.5 U/day of insulin; human growth hormone, 400 mU/day. Columns in the *upper three panels* represent mean values ($n = 5$–6), *brackets* standard deviations. ●, $P<0.001$ vs diabetic controls; +, $P<0.01$ vs diabetic controls; ▼, no significant difference against values designated by ▼; ▽, no significant difference against values designated with ▽; ▼ vs ▽, $P<0.05$ – 0.01. (Modified from Scheiwiller et al. 1986).

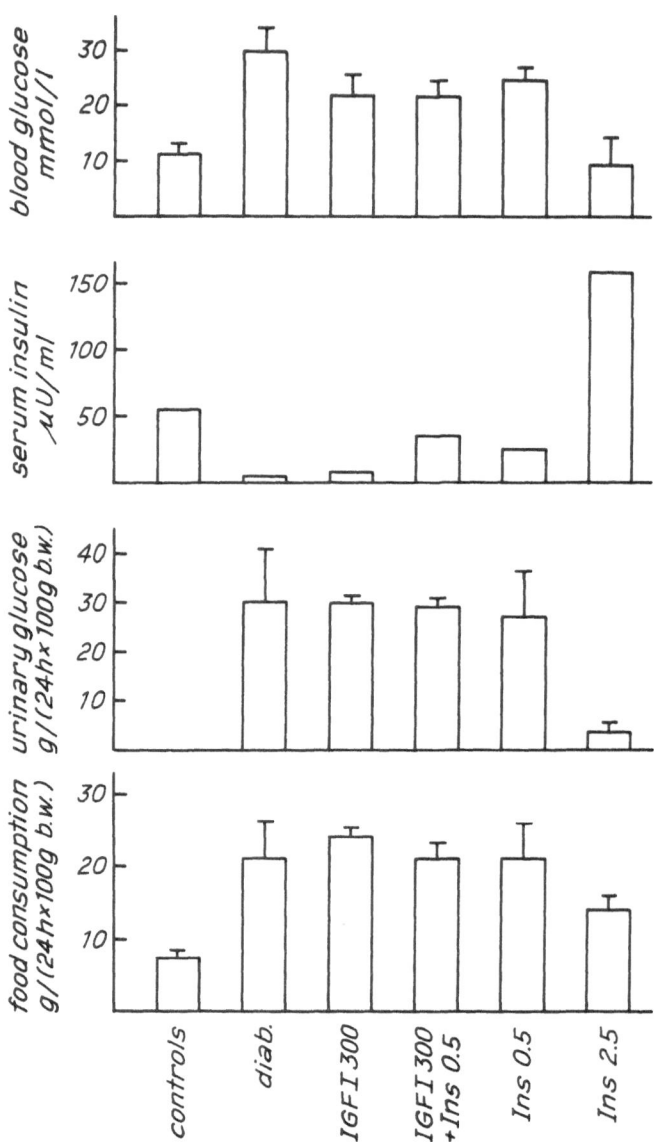

Fig. 6. Metabolic indices (blood glucose, serum insulin levels, urinary glucose excretion, and food consumption) in diabetic and in rh IGF-I- and insulin-infused diabetic rats. (From Scheiwiller et al 1986; Froesch and Zapf 1987).

due to a moderate rise of free IGF-I serum levels, which ranged between 40 and 50 ng/ml) as compared with those of the diabetic control rats, hyperglycemia persisted, and the extent of glycosuria, polyuria and polyphagia, was unaffected (Fig. 6). Much in contrast, the infusion of 2.5 U/day of insulin almost normalized the diabetic condition. Insulin caused a dose-dependent increase in the endogenous

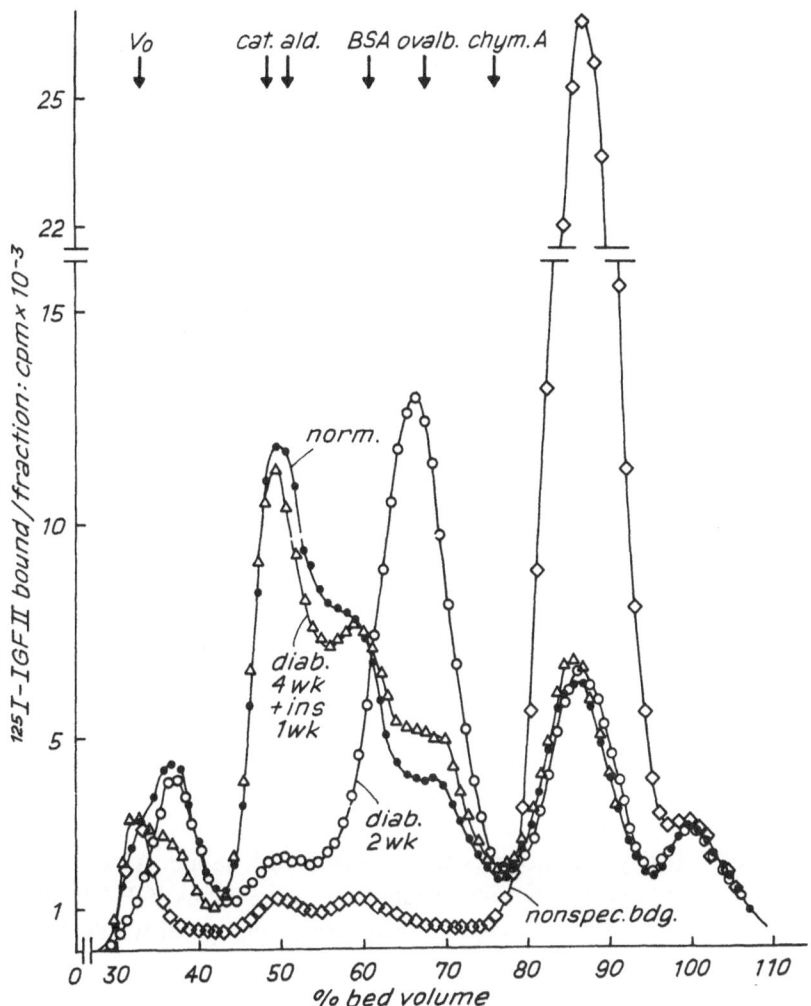

Fig. 7. Radiochromatographic patterns of sera from normal, streptozotocin-diabetic, and insulin-infused (2.5 U/day) streptozotocin-diabetic rats after preequilibration of the sera with ^{125}I-IGF-II (24 h) and gel filtration on Sephadex G-200 at neutral pH. The methodology was the same as that described by Zapf et al. 1975.

IGF-I level (Fig. 5). Concomitantly, insulin induced the 150-K IGF carrier protein, which was largely diminished in the diabetic controls (Fig. 7). Thus, the effect of insulin in the diabetic rat is similar to that of growth hormone in the hypophysec-tomized rat: both hormones induce endogenous IGF-I and the main IGF carrier protein (Zapf et al. 1985b).

Histological pictures of the epiphyseal cartilage and the adjacent trabecular structures of the diabetic rats treated with IGF-I or with insulin showed a marked restoration toward normal (Scheiwiller et al. 1986).

The fact that endogenous IGF-I levels in insulin-treated diabetic rats correlate significantly with the measured growth indices (Scheiwiller et al. 1986) suggests that insulin normalizes growth of diabetic rats mainly by raising endogenous IGF-I levels towards normal. Two mechanisms account for this: one is the restoration by insulin of growth hormone secretion (Robinson et al. 1987). However, since growth hormone itself is completely ineffective in raising endogenous IGF-I in the absence of insulin (Fig. 5) and thus is also ineffective on growth, insulin, in addition, must restore the impaired capacity of the liver to produce IGF-I and the 150-K IGF carrier protein in response to growth hormone (Zapf et al. 1985b).

Altogether, these results demonstrate that IGF-I acts on chondrocytes and bone cells *in vivo* and promotes growth. It displays these actions not only in the absence of growth hormone, as in the hypophysectomized rat, but also under conditions where, in addition to the lack of growth hormone, metabolism is severely deranged due to insulin deficiency and remains disturbed despite infusion of IGF-I.

DOES RH IGF-I AFFECT GROWTH PARAMETERS IN NORMAL RATS?

In order to determine whether normal growth could additionally be stimulated by increasing the steady-state level of IGF-I, normal male rats (130–135 g body wt.) were infused subcutaneously with 300 µg/day of rh IGF-I over 6 days. For comparison, three groups of rats received growth hormone at infusion rates of 25, 200, and 400 mU/day. Control rats were infused with saline. As shown in Table 3, growth hormone did not further increase the endogenous IGF-I level and had no additional effect on body weight gain or tibial epiphyseal width. Nor did the

Table 3. Growth parameters in normal male Tif-RAI rats (130-135 g) receiving a 6-day s.c. infusion (Alzet minipumps) of saline, growth hormone or rh IGF-I (from H.P. Guler, J. Zapf, E.R. Froesch, unpublished data)

Dose per 24 hours	n	Weight gain (g)		Tibial epiphyseal width (µm)		IGF-I serum level (ng/ml)	
NaCl	5	43.0	3.7	497	52	185	32
25 mU hGH	3	43.0	3.6	453	64	n.d.	
200 mU hGH	3	43.7	6.1	290	59	n.d.	
400 mU hGH	2	38.0	0.0	451	6	172	20
300 µg rh IGF-I	4	42.0	3.7	472	36	533	33

All values are means ± SD or range.
n.d., not determined.

rh IGF-I infusion stimulate these two growth indices, although it raised the circulating IGF-I level from 185 to 533 ng/ml. The animals did not become hypoglycemic.

It is likely not only that the duration of treatment was too short to demonstrate effects of growth hormone or rh IGF-I on normal growth, but also that in young rats skeletal growth is already so active that it would be difficult to stimulate it further. Our results fit the early observation of Ray et al. (1941) that 10 days of growth hormone treatment did not cause a significant increase in epiphyseal cartilage width in 54-day-old female rats. However, we suspect that under "appropriate" experimental conditions, i.e., (a) at a time when the growth curve flattens, (b) by increasing the infusion period, and (c) using female rats who grow slower than males, rh IGF-I would be able to accelerate normal growth. Surprisingly, Hizuka et al. (1987) reported a significant stimulatory effect of s.c. infused rh IGF-I (120 µg/day) on both body weight gain and tibial epiphyseal width in even younger normal male rats (initial body weight 55 g) after 7 days.

IN VIVO ACTION OF RH IGF-I INFUSIONS IN MAN

The conspicuous increase in kidney weight in hypophysectomized rats infused with rh IGF-I (see above) prompted us to investigate the effect of rh IGF-I during long-term infusion in two healthy adult subjects (both co-authors of this paper) on a standard diet. rh IGF-I was infused subcutaneously during 6 days at a dose of 20 µg/kg body wt/h (Guler et al., submitted for publication). While total IGF-I serum levels rose into a range between 600 and 900 ng/ml, IGF-II decreased concomitantly to 100 ng/ml with a half-life of 14–16 h. After the infusion had been stopped, IGF-II levels resumed at a much slower rate reaching half-maximal levels only after 54–64 h. Free IGF-I levels during the infusion rose from 20 to 60–80 ng/ml. Blood glucose levels remained normal.

Growth hormone secretion upon stimulation with growth hormone-releasing factor was markedly suppressed during IGF-I infusion and rebounded after the infusion was stopped. Similarly, nocturnal surges of growth hormone were blunted (decrease in the amplitude and in the number of spikes) during the IGF-I infusion. Both results suggest a negative feedback of high IGF-I serum concentrations on growth hormone secretion in man.

A pronounced effect of rh IGF-I was found on the glomerular filtration rate, which increased to 130% as estimated by creatinine clearance (Fig. 8). Plasma levels of creatinine, urea and uric acid decreased by 45% within 24–48 h after starting the infusion. All these indices remained decreased throughout the whole infusion period and reached preinfusion values 2–3 days after the infusion was stopped.

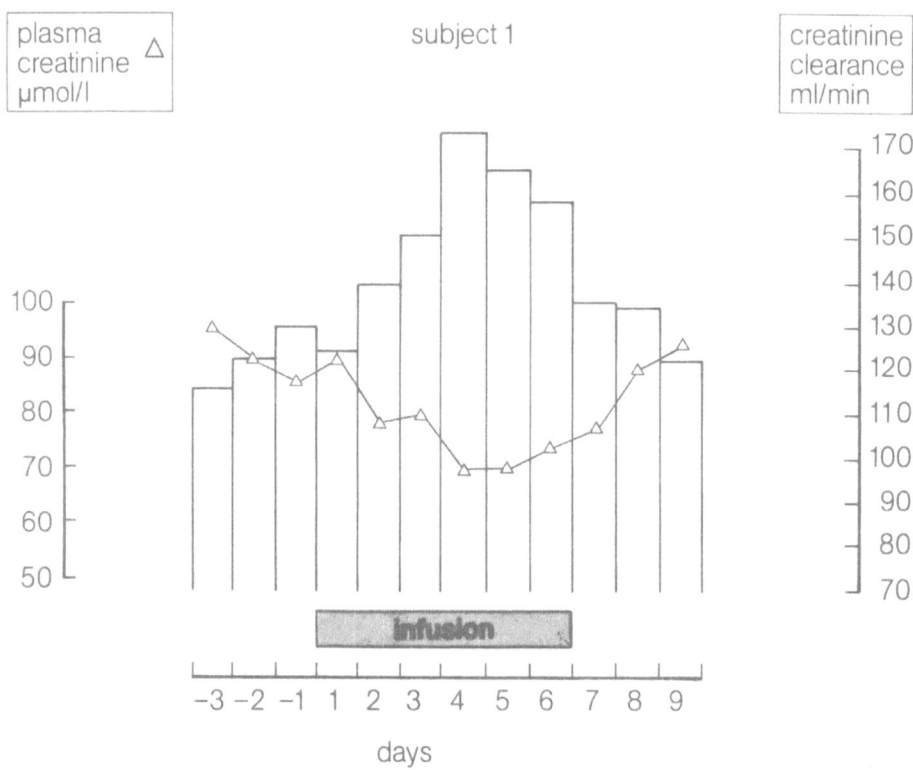

Fig. 8. Plasma levels of creatinine ($\triangle$-$\triangle$) and creatinine clearance (*bars*) in a healthy human adult (HPG) before, during, and after 6 days of a constant subcutaneous infusion of recombinant human IGF-I (infusion rate 20 µg/kg body wt./h). (From Guler et al., submitted for publication).

CONCLUSIONS AND REMARKS

The large amounts of IGF-I that circulate in serum suggest that IGF-I plays an important endocrine role. Here, we present data from two animal models where IGF-I, administered subcutaneously over a prolonged period of time increased body weight, tibial epiphyseal width and epiphyseal growth rate. IGF-I had an even more pronounced effect than growth hormone itself on particular organs such as the kidney, the spleen and the thymus. All of these data prove that IGF-I can act as an endocrine hormone. In addition, IGF-I may also function in a paracrine or autocrine manner. Thus, it is not only synthesized in the liver (Schwander et al. 1983), but also in cultured cells *in vitro* (Clemmons et al. 1981) and in a variety of tissues *in vivo* (D'Ercole et al. 1984) as documented by the presence of IGF-I messenger RNA (Mathews et al. 1986; Murphy et al. 1987). Which of these mechanisms is more important for which step in growth and differentiation cannot yet be decided. In any case, the endocrine and paracrine/autocrine mechanisms of action do not exclude each other, but may rather be complementary.

The finding that IGF-I has significantly greater effects than growth hormone itself on particular organs like the kidney or the spleen of hypophysectomized rats is unexpected, because growth hormone acts *via* IGF-I. One possible reason for these quantitative differences between IGF-I and growth hormone may be related to the way that IGF-I is presented to cells. During growth hormone treatment of hypophysectomized rats most of the newly synthesized endogenous IGF-I circulating in serum is bound to the simultaneously induced 150-K binding protein, whereas during IGF-I infusion most of the exogenously administered peptide circulates bound to the 40 to 50 -K IGF binding protein, to a small extent as free IGF-I. Thus, growth hormone may control cell- or tissue-specific targeting of IGF-I in the organism by regulating the synthesis of an IGF-I vector.

Most conspicuously, IGF-I can induce growth in severely diabetic rats without normalizing the diabetic condition. Since these animals are not only growth hormone-deficient, but also unresponsive to exogenous growth hormone as far as IGF-I generation and growth are concerned, this finding can be explained only by a direct stimulatory effect of IGF-I on cartilage and bone. This contention is also supported by findings *in vitro*, where IGF-I has been shown to act on human chondrocytes (Vetter et al. 1986) and rat osteoblasts more potently than insulin (Schmid et al. 1983, 1984; Ernst and Froesch 1987). This difference in growth-promoting potency in vitro may also explain why "physiological" doses of insulin have no growth-promoting effect on hypophysectomized animals who cannot synthesize IGF-I due to the absence of growth hormone. Because of fatal hypoglycemia it is extremely difficult to keep hypophysectomized rats alive under the infusion of pharmacological doses of insulin. However, weak growth effects of large insulin doses can be demonstrated in animals that survive the infusion (Zapf et al. 1987). In this case, one may assume that insulin, despite its low affinity for the type-I IGF receptor, reaches serum levels high enough to allow substantial cross-reaction, resulting in a noticeable growth effect.

In diabetic rats insulin treatment normalizes growth hormone secretion and growth hormone receptor responsiveness and thus raises IGF-I levels, leading to resumption of growth. It is extremely unlikely that infused IGF-I, like insulin, normalizes growth hormone secretion and growth hormone receptor responsiveness because it does not restore the metabolic disorder, which seems to be a prerequisite for normalization of the above two functions. Apart from this, one would rather expect an inhibitory action of high IGF-I concentrations on growth hormone secretion (Abe et al. 1983).

Investigation of the *in vivo* effects of IGF-I on man is just beginning. New and unexpected findings may result from such studies. The first treatment of a Laron dwarf by IGF-I does not seem to be far away. It does not appear unlikely that other routes of therapeutic use of IGF-I may open up in the future.

Acknowledgments

We thank E. Futo, Ch. Hauri, H. Häsler, I. Giger, and M. Waldvogel for their expert technical assistance and M. Salman for devoted secretarial work.

REFERENCES

Abe H, Molitch ME, Van Wyk JJ, Underwood LE (1983) Human growth hormone and somatomedin-C suppress the spontaneous release of growth hormone in unanesthetized rats. *Endocrinology* **113**: 1319-1324

Akahane K, Tojo A, Urabe A, Takaku F (1987) Pure erythropoietic colony and burst formations in serum-free culture and their enhancement by insulin-like growth factor-I. *Exp Hematol* **15**: 797-802

Chochinov RH, Daughaday WH (1976) Current concepts of somatomedin and other biologically related growth factors. *Diabetes* **25**: 994-1007

Clemmons DR, Underwood LE, Van Wyk JJ (1981) Hormonal control of immunoreactive somatomedin production by cultured human fibroblasts. *J Clin Invest* **67**: 10-19

Crafts RC (1941) The effect of endocrines on the formed elements of the blood. *Endocrinology* **29**: 606-618

Daughaday WH, Hall K, Raben MS, Salmon WD Jr, Van den Brande LJ, Van Wyk JJ (1972) Somatomedin: proposed designation for sulfation factor. *Nature* **235**: 107

D'Ercole AJ, Stiles AD, Underwood LE (1984) Tissue concentration of somatomedin C: further evidence for multiple sites of synthesis and paracrine/autocrine mechanisms of action. *Proc Natl Acad Sci USA* **81**: 935-939

Ernst M, Froesch ER (1987) Osteoblast-like cells in a serum-free methylcellulose medium form colonies: effects of insulin and insulin-like growth factor I. *Calcif Tissue Int* **40**: 27-34

Froesch ER, Zapf J (1987) Insulin, IGF-I and growth in diabetic rats. *Nature* **326**: 549

Guler HP, Zapf J, Froesch ER (1987) Short-term metabolic effects of recombinant human insulin-like growth factor I in healthy adults. *N Engl J Med* **317**: 137-140

Guler HP, Zapf J, Scheiwiller E, Froesch ER (1988) Recombinant human insulin-like growth factor I stimulates growth and has specific effects on organ size in hypophysectomized rats. *Proc Natl Acad Sci* (USA) **85**: 4889-4894

Hansson LI (1967) Daily growth in length of diaphysis measured by oxytetracycline in rabbit normally and after medullary plugging. *Acta Orthop Scand* [Suppl] **101**: 34-84

Hansson LI, Menander-Sellman K, Stenström A, Thorngren K-G (1972) Rate of normal longitudinal bone growth in the rat. *Calcif Tissue Res* **10**: 238-251

Hizuka N, Takano K, Asakawa K, Miyakawa M, Tanaka I, Horikawa R, Hasegawa S, Mikasa Y, Saito S, Shibasaki T, Shizume K (1987) *In vivo* effects of insulin-like growth factor-I in rats. *Endocrinol Jpn* **34**: 115-121

Horiai H, Asada T, Seki J, Motoyama Y, Ogawa T, Suzuki S, Niwa M, Ono T, Shibayama F, Kikuchi H (1987) Growth-promoting effect of recombinant insulin-like growth factor-I/somatomedin C in hypophysectomized rats. *Endocrinol Jpn* **34**: 164

Isgaard J, Nilsson A, Lindahl A, Jansson J-O, Isaksson OGP (1986) Effects of local administration of GH and IGF-I on longitudinal bone growth in rats. *Am J Physiol* **250**: E367-E372

Kaufmann U, Zapf J, Torretti B, Froesch ER (1977) Demonstration of a specific serum carrier protein of nonsuppressible insulin-like activity *in vivo. J Clin Endocrinol Metab* **44**: 160-166

Kaufmann U, Zapf J, Froesch ER (1978) Growth hormone dependence of nonsuppressible insulin-like activity (NSILA) and of NSILA carrier protein in rats. *Acta Endocrinol* (Copenh) **87**: 716-727

Kurtz A, Jelkmann W, Bauer C (1982) A new candidate for the regulation of enithropoiesis: insulin-like growth factor *I. FEBS Lett* **149**: 105-108

Kurtz A, Härtl W, Jelkmann W, Zapf J, Bauer C (1985) Activity in fetal bovine serum that stimulates erythroid colony formation in fetal mouse livers is insulin-like growth factor I. *J Clin Invest* **76**: 1643-1648

Kurtz A, Zapf J, Eckardt KU, Clemmuns G, Froesch ER, Bauer C. Insulin-like growth factor I stimulates erithropoiesis in hypophysectomized rats. *Proc Natl Acad Sci USA,* in press

Mathews LS, Norstedt G, Palmiter RD (1986) Regulation of insulin-like growth factor I gene expression by growth hormone. *Proc Natl Acad Sci USA* **83**: 9343-9347

Meyer OO, Stewart GE, Thewlis EW, Rush HP (1937) The hypophysis and hematopoiesis. *Fol Haemat* **57**: 99-109

Meyer OO, Thewlis EW, Rusch HP (1940) The hypophysis and hemopoiesis. *Endocrinology* **28**: 932-944

Moses A, Nissley SP, Cohen KL, Rechler MM (1976) Specific binding of a somatomedin-like polypeptide in rat serum depends on growth hormone. *Nature* **263**: 137-140

Murphy LJ, Bell GI, Friesen HG (1987) Tissue distribution of insulin-like growth factor-I and -II messenger ribonucleic acid in the adult rat. *Endocrinology* **120**: 1279-1282

Overbeek GA (1936) Reticulocytes in normal and hypophysectomised rats. *Arch Int Pharmacodyn Ther* **54**: 340-348

Overbeek GA, Querido A (1938) Hypophysis and blood-picture. II. *Arch Int Pharmacodyn Ther* **60**: 105-114

Phillips LS, Untermann TG (1984) Somatomedin activity in disorders of nutrition and metabolism. In: Daughaday WH (ed) *Clinics in Endocrinology and Metabolism: Tissue Growth Factors,* vol 13. Saunders, London, Philadelphia, Toronto, pp 145-189

Ray RD, Evans HM, Becks H (1941) Effect of the pituitary growth hormone on the epiphyseal disc of the tibia of the rat. *Am J Pathol* **18**: 509-528

Rinderknecht E, Humbel RE (1978a) The amino acid sequence of human insulin-like growth factor I and its structural homology with proinsulin. *J Biol Chem* **253**: 2769-2776

Rinderknecht E, Humbel RE (1978b) Primary structure of human insulin-like growth factor II. *FEBS Lett* **89**: 283-286

Robinson ICAF, Clark RG, Carlsson LMS (1987) Insulin, IGF-I and growth in diabetic rats. *Nature* **326**: 549

Scheiwiller E, Guler HP, Merryweather J, Scandella C, Maerki W, Zapf J, Froesch ER (1986) Growth restoration of insulin-deficient diabetic rats by recombinant human insulin-like growth factor-I. *Nature* **323**: 169-171

Schmid C, Steiner T, Froesch ER (1983) Insulin-like growth factors stimulate synthesis of nucleic acids and glycogen in cultured calvaria cells. *Calcif Tissue Int* **35**: 578-585

Schmid C, Steiner T, Froesch ER (1984) Insulin-like growth factor-I supports differentiation of cultured osteoblast-like cells. *FEBS Lett* **173**: 48-52

Schoenle E, Zapf J, Humbel RE, Froesch ER (1982) Insulin-like growth factor-I stimulates growth in hypophysectomized rats. *Nature* **296**: 252-253

Schoenle E, Zapf J, Hauri C, Steiner T, Froesch ER (1985) Comparison of *in vivo* effects of insulin-like growth factors I and II and of growth hormone in hypophysectomized rats. *Acta Endocrinol* (Copenh) **108**: 167-174

Schwander JC, Hauri C, Zapf J, Froesch ER (1983) Synthesis and secretion of insulin-like growth factor and its binding protein by the perfused rat liver: dependence on growth hormone status. *Endocrinology* **113**: 297-305

Tannenbaum GS (1981) Growth hormone secretory dynamics in streptozotocin diabetes: evidence for a role of endogenous circulating somatostatin. *Endocrinology* **108**: 76-82

Van Buul-Offers S, Ueda I, Van den Brande JL (1986) Biosynthetic somatomedin C (SM-C/IGF I) increases the length and weight of Snell dwarf mice. *Pediatr Res* **20**: 825-827

Vetter U, Zapf J, Heit W, Helbing G, Heinze E, Froesch ER (1986) Human fetal and adult chondrocytes. Effect of insulin-like growth factors I and II, insulin, and growth hormone on clonal growth. *J Clin Invest* **77**: 1903-1908

Vollmer EP, Gordon AS, Levenstein I, Charipper HA (1939) Effects of hypophysectomy upon the blood picture of the rat. *Endocrinology* **25**: 970-977

Young FG (1945) Growth and diabetes in normal animals treated with pituitary (anterior lobe) diabetogenic extract. *Biochem J* **39**: 515-536

Zapf J, Waldvogel M, Froesch ER (1975) Binding of nonsuppressible insulin-like activity to human serum: evidence for a carrier protein. *Arch Biochem Biophys* **168**: 638-645

Zapf J, Schmid C, Froesch ER (1984) Biological and immunological properties of insulin-like growth factors I and II. In: Daughaday WH (ed) *Clinics in Endocrinology and Metabolism: Tissue Growth Factors*,vol. 13. Saunders, London, Philadelphia, Toronto pp 3-30

Zapf J, Schoenle E, Froesch ER (1985a) *In vivo* effects of the insulin-like growth factors in the hypophysectomized rat: comparison with human growth hormone and the possible role of the specific IGF carrier proteins. In: *Growth Factors in Biology and Medicine,* Ciba Foundation Symposium 116. Pitman, London, pp 169-187

Zapf J, Pescatore P, Scheiwiller E, Froesch ER (1985b) Growth hormone, insulin and insulin-like growth factors: interdependent regulation and functional cooperativity. *J Endocr* **107** [Suppl] abstract 10

Zapf J, Hauri C, Waldvogel M, Froesch ER (1986) Acute metabolic effects and half-lives of intravenously administered insulin-like growth factors I and II in normal and hypophysectomized rats. *J Clin Invest* **77**: 1768-1775

Zapf J, Scheiwiller E, Guler HP, Froesch ER (1987) Insulin and insulin-like growth factor I: Comparative aspects of their in vivo actions on growth and glucose homeostasis. *Endocrinol Jpn* **34** (Suppl. 1): 123-129

GROWTH FACTORS

Advances in Growth Hormone and Growth Factor Research,
edited by E.E. Müller, D. Cocchi and V. Locatelli
Pythagora Press, Roma-Milano and Springer Verlag, Berlin-Heidelberg © 1989

The structure and physiology of epidermal growth factor and its receptor

R.W. DONALDSON, S. NISHIBE and G. CARPENTER

Department of Biochemistry, Vanderbilt University School of Medicine, Nashville, Tennessee, U.S.A.

INTRODUCTION

Epidermal growth factor (EGF), a small polypeptide of 53 amino acid residues and a molecular mass of approximately 6000 daltons, was identified and isolated from the mouse submaxillary gland in 1962 (Cohen). Since then, the physiological relevance of EGF and the mechanism by which it stimulates normal cell proliferation have been areas of intense investigation, and a great deal is known about the structure of both EGF and its receptor. It is known that EGF interacts with a plasma membrane receptor in sensitive cells and initiates a number of cellular responses that can be roughly divided into early and late events. The "early" events involve changes in membrane properties, such as protein phosphorylation and nutrient transport, while the "late" events include macromolecular synthesis, which requires nuclear activity and therefore some communication between the cell surface and the nucleus. The mechanism(s) by which these responses are initiated, however, is not yet understood. It is the purpose of this review to summarize recent developments of EGF and EGF receptor-related research.

STRUCTURE OF EGF

The mouse EGF (m-EGF) contains 53 amino acid residues and disulfide bonds between residues 5 and 20, 14 and 31, and 33 and 42 produce three disulfide loops in the secondary structure of the molecule (Fig. 1a) (Savage et al. 1972, 1973). The disulfide bonds in m-EGF are required for biological activity (Taylor et al. 1972). In the EGF molecules isolated from other species and in the EGF-like

* This work was supported by NIH grants CA43720 and CA24071.

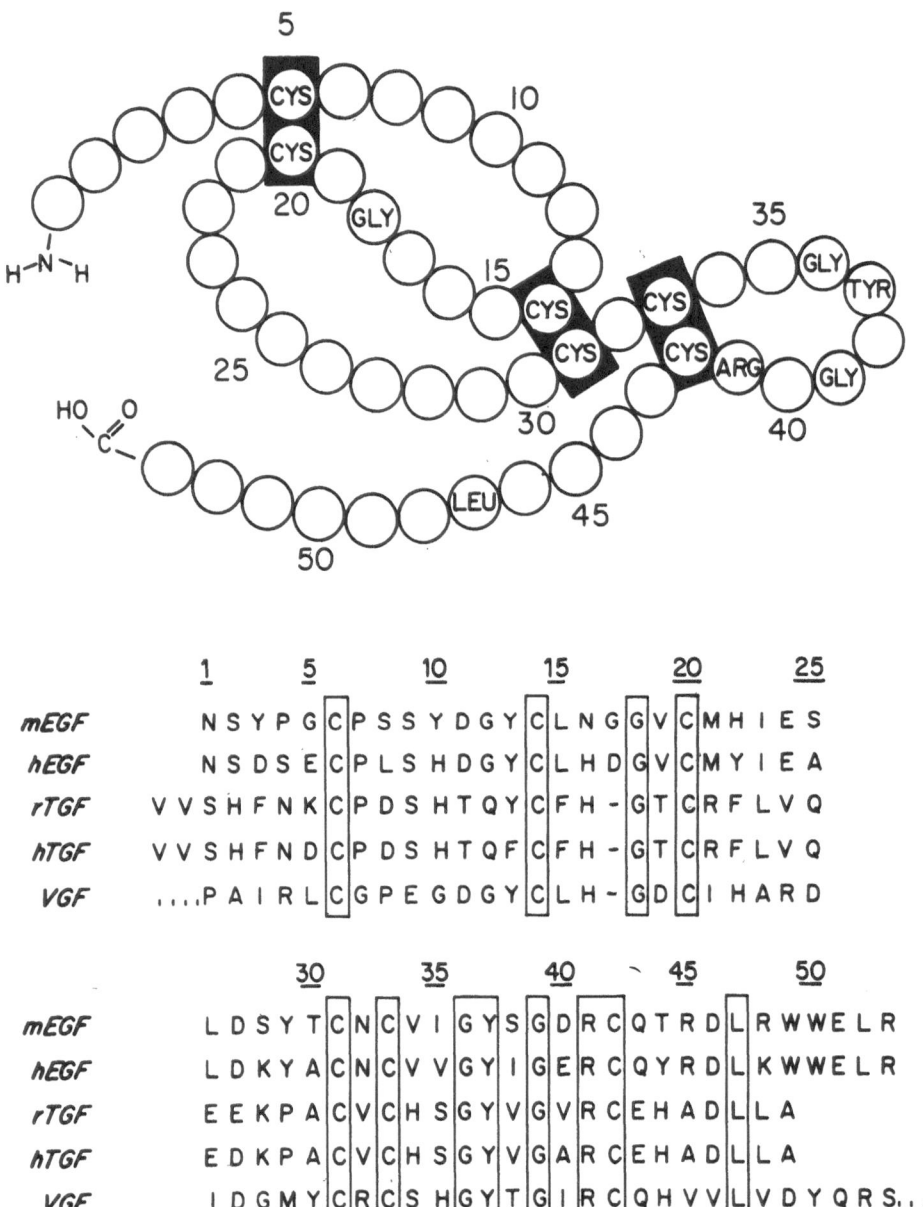

Fig. 1. a Representation of strictly conserved amino acid residues in the EGF family of growth factors and the positioning of disulfide bonds as derived from data on the mouse-derived EGF. b Amino acid sequences of growth factors related to EGF. The sequences of the individual polypeptides are taken from references cited in the text and are shown in comparison with the sequence of mouse EGF (From Carpenter and Zendegui 1986b).

polypeptides, the location of these cysteine residues, and presumably the placement of disulfide bonds, is the most conserved structural characteristic (Fig. 1b).

Human EGF (h-EGF) was identified (Starkey et al. 1975) and isolated (Cohen and Carpenter, 1975) from urine in 1975. In the same year, Gregory (1975) published the sequence of β-urogastrone, isolated from human urine as an inhibitor of gastric acid secretion. The human urogastrone sequence was very similar to that of m-EGF, and purified urogastrone preparations had the mitogenic biologic activity of EGF. It is now accepted tha β-urogastrone is, in fact, the human form of EGF and that EGF has activities, such as the pharmacologic inhibition of gastric acid secretion, that do not seem related to the mitogenic action of the growth factor.

During the mid-1980s the structures of other EGF-like molecules have been reported. Transforming growth factor (TGF) type α has been isolated from rat and human material (Marquardt et al. 1983, 1984; Derynck et al. 1984) and the sequences have been determined (Fig. 1b). These proteins are structurally analogous to the EGF sequence, compete with [^{125}I]EGF in radioreceptor assays, and produce EGF-like biological responses in sensitive cells. Transforming growth factor type α seems to be produced, unlike EGF as a result of cell transformation, but TGFa is also found in fetal fluids. Also, the TGFs are antigenically distinct from EGF.

In late 1984 and early 1985, three reports (Blomquist et al. 1984; Brown et al. 1985; Reisner, 1985) appeared that described significant sequence homology between EGF and a protein product of the vaccinia virus, a member of the poxvirus family. Although the vaccinia virus is the causative agent of cowpox, a disease not involving increased cell proliferation, other members of the pox family do produce abnormal cellular proliferation. The function of the vaccinia EGF-like protein (VGF) is not known, but its structure (Fig. 1b) is deduced from nucleic acid sequences of the virus genome (Blomquist et al. 1984; Brown et al. 1985; Reisner, 1985) and direct sequence analysis of the protein (Stroobant et al. 1985). Partially purified VGF from the media of vaccinia-infected cells has EGF-like activities (Twardizik et al. 1985).

PRODUCTION AND DISTRIBUTION OF h-EGF

The site(s) of EGF synthesis and/or storage in the human is not known. It has been reported tha there is evidence to indicate the presence of somewhat elevated levels of EGF in two human tissues the submandibular glands and Brunner's glands (Elden et al. 1978; Heitz et al. 1978). Hirata and Orth (1979) reported the concentrations of EGF in human tissues as follows: submandibular gland, 1.3 (ng/ g wet tissue); duodenum, 2.3; pancreas 2.8; jejunum, 4.8; thyroid, 5.3; and kidney, 5.5. The quantities of EGF detected in these human tissues are exceedingly low compared with the daily urinary excretion of the growth factor (50-60 μg/24 h), and their physiological significance is therefore not clear.

Human EGF has been detected in various body fluids (summarized in Carpenter, 1985). EGF is present in most human body fluids at concentrations ranging from 1 to 800 ng/ml. Changes in these levels have not as yet been correlated with any pathological processes. Interestingly, the concentration of EGF in the urine of newborns is low and increases about fivefold in the first 2 years of life (Mattila et al. 1985). This change represents the largest described alteration of EGF levels in vivo; however, its physiological meaning is not known.

The concentration of EGF in serum or plasma has been difficult to determine accurately. Most authors seem to agree that circulating levels of EGF are quite low -less than 1 ng/ml plasma. Based on a sensitive bioassay (which also detects TGF-α activity), we estimate that commercial fetal calf serum contains less than 300 pg of EGF activity per ml (Carpenter and Zendegui, 1986a). Cell culture medium with 10% serum therefore contains less than 30 pg/ml, far less than the 1 ng/ml at which most cultured cells respond to EGF, suggesting that EGF is unlikely to contribute significantly to the mitogenic activity of serum in most cell culture media. Furthermore, if EGF serves a mitogenic function in vivo, then the low concentration of this factor in serum suggests that the circulatory system is unlikely to be an important source for target cell populations. The more physiologically relevant source is likely to be localized production, leading to stimulation of neighboring cells in a paracrine fashion. It is also unlikely that filtration and/ or concentration of EGF from plasma accounts for the relatively high EGF levels in fluids such as urine, milk, and saliva.

BIOLOGICAL EFFECTS OF EGF

EGF has been demonstrated to elicit significant biologic responses in intact animals, organ cultures, and cell culture systems (Carpenter and Cohen 1979; Carpenter 1981). The administration of EGF to animals produces a number of dramatic effects associated with enhanced proliferation of epithelial tissues. In some cases, for example in skin, this stimulation of cell proliferation leads to a more rapid rate of differentiation (Cohen and Elliot 1963), while in other tissues, such as the trachea (Sundell et al. 1980), a metaplastic effect is produced.

Recently, it was reported that EGF induced different responses in the same cells or in different cells of the same organ, under varying conditions. For example, organ culture studies of tooth morphogenesis indicated that EGF stimulates cell proliferation in the dental epithelium but inhibits proliferation in the dental mesenchyma (Partanen et al. 1985). Interestingly, dissociated cells of the dental mesenchyma were stimulated by EGF. Also, EGF can have stimulatory or inhibitory effects on different cell populations within the same organ. In the wool follicle, EGF was found to stimulate mitosis of peripheral cells of the acini but to inhibit mitosis in the bulb cells (Moore et al. 1985). These findings suggest that the effect of EGF on these cells may be modulated by the interaction of the target cells with nearby cells in the tissue, i.e., that the cells may respond positively or negatively to EGF, depending on the associations with nearby cells and on the constraints imposed by tissue organization.

The A-431 human squamous cell carcinoma line is a valuable tool in research on the EGF receptor, as it contains EGF receptor levels which are at least 20-fold higher than those found in most other cell lines (summarized in Stoscheck and Carpenter 1983). EGF inhibits the growth of these cells (Barnes 1982; Gill and Lazar 1981), an effect which is not understood but which is generally viewed as an anomaly of little physiological relevance and ascribed to the excessive number of receptors present on the cell surface. Recent studies (Cowley et al. 1986; Yamamoto et al. 1986; Kamata et al. 1986) show that carcinomas of epidermal and epithelial origins (including the A-431 line), as well as glioblastomas (Libermann et al. 1985), quite frequently have increased EGF receptor number. Receptor number in these cell lines often correlates directly with EGF inhibition of proliferation. The inhibitory effect of EGF on the growth of these epithelial-derived lines suggests that in some circumstances EGF may act as a signal for differentiation rather than proliferaton. For example, EGF-treated A-431 cells show an arrest in G1 (MacLeod et al. 1986) and increased levels of involucrin, a precursor for cornified envelope formation in keratinization (Rosdy et al. 1986). On the other hand, human esophageal carcinoma cells, which can have fewer EGF receptors that the corresponding normal esophageal keratinocytes, proliferate markedly in response to EGF addition (Banks-Schlegel and Quintero 1986).

A number of other nonmitogenic effects of EGF have been reported. These include increased prolactin secretion by the rat pituitary tumor cell line GH3, accompanied by inhibition of cell proliferation (Schonbrunn et al. 1980) and increased chorionic gonadotropin secretion by choriocarcinoma cells (Benveniste et al. 1978).

Considerably more information is available concerning the biochemical action of EGF on isolated cell populations grown in culture. EGF receptors were found on most nonhematopoietic cell types, and nearly all receptor-positive cells respond to EGF. The most frequently observed biologic response of cultured cells to EGF is mitogenesis, with a few exceptions in which EGF increases the synthesis and secretion of peptide hormones without any measurable mitogenic response.

DISTRIBUTION OF THE EGF RECEPTOR

Experiments performed in the mid-1970s provided the basis for the identification of a plasma membrane receptor for EGF. These early studies demonstrated that cultured cells responded to the mitogenic signal provided by EGF, and that radiolabeled EGF bound in a specific, saturable manner and with high affinity to the surface of these responsive cells (Hollenberg and Cuatrecasas 1973; Armelin 1973). Almost all cell types, with the exception of hematopoietic cells, express the EGF receptor at levels that vary widely, from 20,000 to 200,000 receptor per cell in nontransformed populations. In these cells there is no relationship between receptor number and the responsiveness of cells to EGF.

Several cell lines have been reported in which the EGF receptor is overexpressed at levels of approximately 1×10^6 receptors per cell (Haigler et al. 1978; Kamata et al. 1986). Nearly all of the receptor-overexpressing cell lines are derived from squamous carcinomas. Altered expression of the EGF receptor has been reported for a variety of human tumor carcinoma tissues including breast (Sainsbury et al. 1985), liver (Kaneko et al. 1985), bladder (Neal et al. 1985), pancreas (Korc et al. 1986), glioblastomas (Liberman et al. 1985), sarcomas (Gusterson et al. 1985), and lung (Hendler and Ozanne 1984), but most frequently squamous carcinomas (Cowley et al. 1984). In some instances of overexpression the EGF receptor gene is amplified, but in other cases the gene is not amplified and increased receptor mRNA may be due to either transcriptional or post-transcriptional mechanisms.

A key element in elucidating the biochemistry of the EGF receptor has been the use of an EGF receptor-overproducing cell line, designated A-431 (Fabricant et al. 1977). This cell line, derived from a human vulva epidermoid carcinoma, binds approximately 2.5×10^6 molecules of [^{125}I]EGF per cell (Haigler et al. 1978), and the EGF receptor is estimated to consitute nearly 0.2% of the total cell protein (Stoscheck and Carpenter 1984). The intact EGF receptor from A-431 cell was first purified to near homogeneity by the use of affinity chromatography (Cohen et al. 1982b) and shown to have an apparent molecular mass of 170,000. A similar EGF receptor has also been isolated in active form from mouse liver (Cohen et al. 1982a) and human placenta (Pike et al. 1984).

STRUCTURAL DOMAINS OF THE EGF RECEPTOR AND THE RELATED erb B PROTEINS

EGF was demonstrated to activate a protein kinase *in vitro* in 1978 (Carpenter et al. 1978), which was later shown to be specific for tyrosine residues (Ushiro and Cohen 1980). Although several lines of biochemical and immunological evidence suggested that the kinase activity was intrinsic to the receptor molecule, it was the cloning and sequencing of the EGF receptor which provided hard evidence that the receptor molecule possessed tyrosine kinase activity.

Proteolytic fragments of purified A-431 cell EGF receptor were sequenced by

Downward et al. (Downward et al. 1984). Comparison of these partial receptor sequences with known protein sequences showed that a suprisingly high level of homology existed between the EGF receptor and the v-erb B oncogene product of the avian erythroblastosis virus (Yamamoto et al. 1983). Comparison of the complete h-EGF receptor sequence (Ullrich et al. 1984) with the v-erb B protein indicated that a large part of the receptor was highly homologous and that this conserved region contained a 250-amino acid sequence shared by all tyrosine kinases. It was later shown that induction of erythroblastosis by the avian leukosis virus involved transcriptional activation of c-erb sequences (Nilson et al. 1985). A schematic outline of the structures of the EGF receptor, the v-erb B protein, and the activated c-erb B protein are shown in Fig. 2. All three proteins contain one amino acid transmembrane sequence. The 622-amino acid-long external domain of the EGF receptor contains the amino terminus and 12 canonical sites where N-linked glycosylaton might occur (consensus sequences of ASN-X-SER/THR). The external domain contains a large proportion of cysteine residues (9%), which appear in two clusters. As this cysteine clustering has been observed in the external domain of other receptors, it may be related to the function of ligand binding.

The internal domain of the v-erb B protein (which shows substantial homology with EGF receptor residues 694-937) is truncated, and the ligand binding site is therefore lost. There is also a truncation at the carboxyl terminal of the intracellular domain, which includes the primary sites of EGF receptor autophosphorylation. These discoveries have led to predictions that removal of the ligand binding domain and autophosphorylation sites of the EGF receptor would be sufficient to convert the receptor to a transforming protein whose tyrosine kinase activity is constitutively activated. Three groups have recently demonstrated the presence of auto-phosphorylation activity in the v-erb B molecule (Decker 1985; Gilmore et al. 1985; Kris et al. 1985), but kinase activity has not been observed. Interestingly, and in contrast with the v-erb B molecule, the activated c-erb B protein contains the entire carboxyl terminal region, analogous to the autophosphorylation sites of the EGF receptor. An analysis of AEV-related viruses of differing oncogenic potential indicates that deletions in the carboxyl terminal allow a greater potential to transform other cell types (Gannett et al. 1986).

The cytoplasmic domain of the EGF receptor has been the subject of intense interest for two reasons: the tyrosine kinase activity encoded in this region is considered the primary effector system in the transmembrane signaling process, and similar activity is present in several other growth factor receptors as well as in several oncogene protein products (Hunter and Cooper 1985). The 542-amino acid region comprising the cytoplasmic domain is separated from the external domain by a transmembrane sequence comprised of 26, mostly hydrophobic amino acids (Ullrich et al. 1984). Immediately adjacent to this transmembrane sequence there is, at the cytoplasmic interface, a 13-residue sequence that is highly enriched in basic amino acids. It is probable that this sequence functions to stop receptor extrusion into the endoplasmic reticulum following translation of the mRNA. Interest in a functional role for this sequence in mature receptor functioning has been elicited by the presence of a phosphorylation site for protein kinase-C, at threonine 654, which lies in the midst of this very basic region that encompasses amino acids 648-661 (Cochet et al. 1984).

There is a region of approximately 250 amino acids which shows sequence homology with other members of the tyrosine kinase family and shares some homology with the A-kinase catalytic subunit as well (Shoji et al. 1981; Barker and Dayhoff 1982). This homology is shared by a variety of viral oncogene protein products, some which have been shown to demonstrate protein kinase activity: *v-src*, *v-yes*, *v-fgr*, *v-fms*, *v-fps*, *v-fes*, *v-able*, and *v-ros*; and some which have not: *v-mil*, *v-raf*, and *v-mos* (Hunter and Cooper 1985).

POST-TRANSLATIONAL MODIFICATIONS

Several sites of post-translational modification in the cytoplasmic domain of the EGF receptor have been mapped (Fig. 2). In general, these residues are similarly positioned in the v-*erb* B protein, although the modified residues per se have not been isolated from the viral protein. The primary sites of autophosphorylation map to tyrosine residues at positions 1173, 1148, and 1058 (Downward et al. 1984). The penultimate tyrosine, which appears to be the most highly phosphorylated (residue 1173), is missing from the v-*erb* B molecule due to a truncation at the C-terminal.

Treatment of the EGF receptor with calpain (a calcium-dependent protease) releases a 20,000-dalton fragment which contains the major autophosphorylation sites (Gates and King 1985). The remaining 150,000-dalton receptor fragment retains autophosphorylation activity, indicating that additional sites can be utilized. Tyrosine 845 is homologous to the pp60v-s^{rc} kinase major autophosphorylation site 416 (Hunter and Cooper 1985) and may be a candidate for autophosphorylation in the 150,000-dalton receptor species.

It has been known for several years that the EGF receptor contains, in addition to phosphotyrosine, significant levels of phosphothreonine and phosphoserine (Hunter and Cooper 1980; Carlin and Knowles 1982). These nontyrosine phosphorylations are carried out by protein kinases other than the receptor itself. Protein kinase-C, activated by the phorbol ester TPA in intact cells or added *in vitro* to EGF receptor preparations, is able to catalyze phosphorylation of the EGF receptor. Phosphopeptide maps indicate several sites in which phosphorylation of the EGF receptor by protein kinase-C occurs, but threonine 654 is the predominant site, and phosphorylation at this site appears to attenuate receptor function (Hunter et al. 1984; Davis and Czech 1985, 1986). Site-directed mutagenesis of threonine

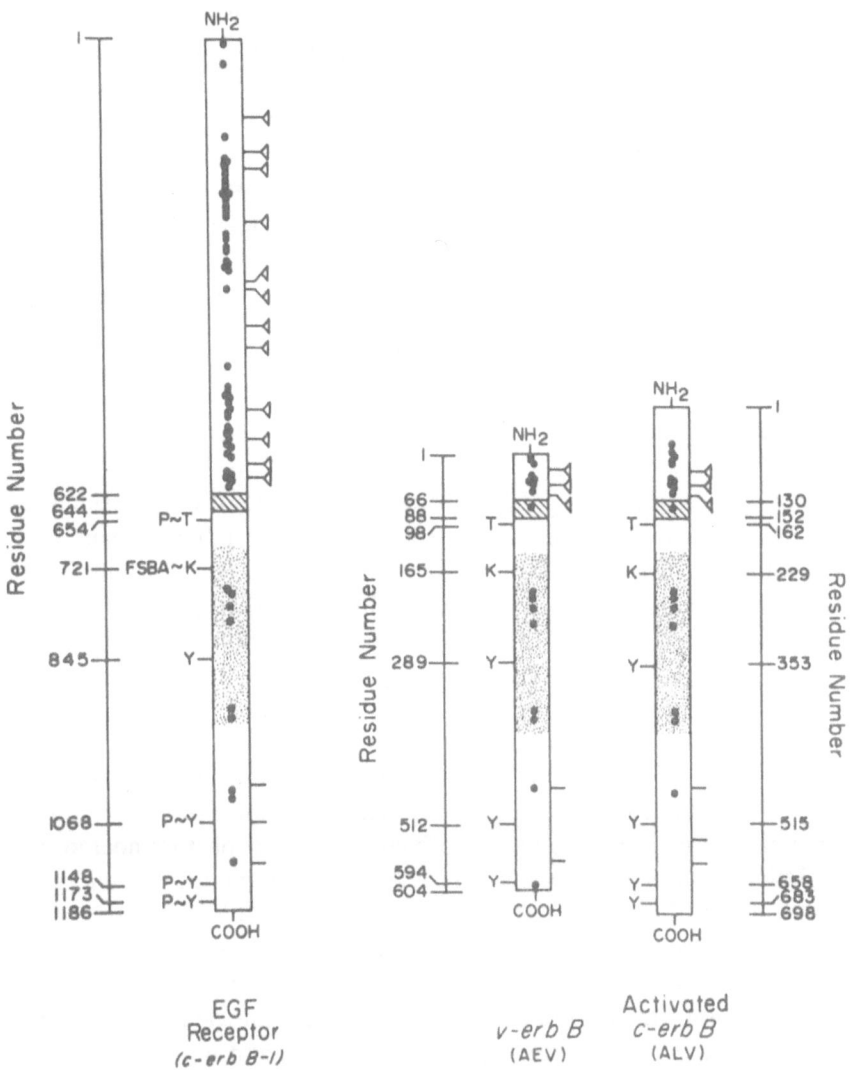

Fig. 2. Schematic representation of the structures of the EGF receptor from A-431 cells and the v-*erb* B protein. The amino acid sequence is based on the data of Ullrich et al. (1984) for the EGF receptor and Yamamoto et al. (1984) for the v-*erb* B protein. ●, cysteine residues in the primary sequence. All canonical sites for potential N-linked glycosylation are indicated (–) and those that are likely to actually be utilized are designated (-◁). The predicted cross-membrane sequences are identified by *cross-hatching. Stippled areas* are sequences similar to *src. Y*, tyrosine residues. Known sites of autophosphorylation are designated by *P~Y* (Downward et al. 1984). The C-kinase-induced threonine phosphorylation (Hunter et al. 1984; Davis and Czech 1985) is shown as *P~T*, and the lysine residue labeled with an ATP analogue (Russo et al. 1985) is designated *FSBA~K*.

654 to alanine yielded alanine-654 receptors which bound [^{125}I]EGF equally well in the presence or absence of TPA (Lin et al. 1986). It is possible that transmodulation of EGF receptors by other growth factors, such as PDGF, or by EGF itself may be due to diacylglycerol-mediated protein kinase-C activation. Treatment of A-431 cells with EGF results in a 70% reduction in tyrosine phosphorylation of exogenous protein substrates (Chinkers and Garbers 1986). Similarly, such treatments lead to activation of protein kinase-C and phosphorylation of a threonine-654 EGF receptor peptide (King and Cooper 1986; Whiteley and Glaser 1986). Protein kinase-C therefore seems to be a critical intermediate in the EGF system itself and, through heterologous growth factors, to regulate EGF receptor activity.

TYROSINE KINASE ACTIVITY AND RECEPTOR FUNCTION

Several recent experiments point to the essential nature of the tyrosine kinase activity in EGF receptor function. Inactivation of the ATP-binding domain by site-directed substitution of Lys 721 with methionine yielded a receptor with inactive kinase activity despite the ability to effectively bind [^{125}I]EGF (Chen et al. 1987). Cells possessing this site-specific mutation failed to exhibit several typical responses to EGF, such as increased doubling time, activation of c-*fos* transcription, Ca^{2+} mobilization, and internalization of occupied receptors.

Similar results were obtained when four amino acids were inserted into the EGF receptor after amino acid 708 in an attempt to alter the nearby ATP-binding domain (Livneh et al. 1978). Although this modified receptor bound [^{125}I]EGF, it lacked the ability to phosphorylate artificial substrates or to undergo autophosphorylation. It similary appeared to lack the ability to induce phosphorylation of ribosomal subunit S6, which occurs through the activation of an S6 kinase, or to induce DNA synthesis as measured by [^{3}H]thymidine incorporation. In contrast to the results obtained with the site-directed mutant for Lys 721, it was reported that in this insertion mutant occupied EGF receptors were rapidly internalized.

Results have been reported for similar mutants of the insulin receptor. When alanine wad substituted for Lys 1018 in the ATP-binding domain of the β subunit, the resulting receptor lacked tyrosine kinase activity and failed to mediate multiple post-receptor actions, including internalization of occupied receptors, as seen with the altered EGF receptor (Chou et al. 1987; Russell et al. 1987). From this it must be concluded that intrinsic tyrosine kinase activity in receptor molecules is essential in transmitting the signal generated by ligand binding and probably in the signals necessary for internalization of ligand: receptor complexes.

Just as the significance of the autophosphorylation event remains unclear, so does the understanding of EGF receptor-mediated tytosine phosphorylation of exogenous substrates. Although several proteins have been shown to be phosphorylated *in vivo* or *in vitro*, the identity of some and the functional significance of tyrosine phosphorylation of all are unknown.

CONCLUDING REMARKS

Since the initial report of EGF in 1962 (Cohen 1962), a substantial amount of information concerning the structure and function of both EGF and its receptor has been assembled. The convergence of viral oncology with the growth factor field has emphasized the concept of common pathways in growth control and has provided a way to observe numerous variations on a theme. The next major chapter in this story will concern the nature of the signal which is transmitted to the nucleus and how it functions to induce DNA synthesis in the cell.

Acknowledgment

The authors thank Susan Haever for typing the manuscript.

REFERENCES

Armelin HA (1973) Pituitary extracts and steroid hormones in the control of 3T3 cell growth. *Proc Natl Acad Sci USA* **70**: 2702-2706

Banks-Schlegel SP, Quintero J (1986) Human esophageal carcinoma cells have fewer, but higher-affinity epidermal growth factor receptors. *J Biol Chem* **261**: 4359-4362

Barker WC, Dayhoff MO (1982) Viral *src* gene products are related to the catalytic chain of mammalian cAMP-dependent protein kinase. *Proc Natl Acad Sci USA* **79**: 2836-2839

Barnes DW (1982) Epidermal growth factor inhibits growth of A-431 human epidermoid carcinoma in serum-free cell culture. *J Cell Biol* **93**: 1-4

Benveniste R, Speeg KV Jr, Carpenter G, Cohen S, Lindner J, Rabinowitz D (1978) Epidermal growth factor stimulates secretion of human chorionic gonadotropin by cultured human choriocarcinoma cells. *J Clin Endocrinol Metab* **46**: 162-172

Blomquist MC, Hunt LT, Barker WC (1984) Vaccinia virus 19-kilodalton protein: relationship to several mammalian proteins including two growth factors. *Proc Natl Acad Sci USA* **81**: 7363-7367

Brown JP, Twardzic DR, Marquardt H, Todaro GJ (1985) Vaccinia virus encodes a polypeptide homologous to epidermal growth factor and transforming growth factor. *Nature* **313**: 491-492

Carlin CR, Knowles BB (1982) Identity of human epidermal growth factor (EGF) receptor with glycoprotein SA-7: evidence for differential phosphorylation of the two components of the EGF receptor from A431 cells. *Proc Natl Acad Sci USA* **79**: 5026-5030

Carpenter G (1981) Epidermal growth factor. *Handbook Exp Pharmacol* **57**: 89-132

Carpenter G (1985) Epidermal growth factor: biology and receptor metabolism. *J Cell Sci* [Suppl. 3]: 1-9

Carpenter G, Cohen S (1979) Epidermal growth factor. *Annu Rev Biochem* **48**: 193-216

Carpenter G, Zendegui J (1986) A biological assay for epidermal growth factor/urogastrone and related polypeptides. *Anal Biochem* **153**: 279-282

Carpenter G, Zendegui J (1987) Epidermal growth factor, its receptor, and related proteins. *Exp Cell Res* **164**:1-10

Carpenter G, King L Jr, Cohen S (1978) Epidermal growth factor stimulates phosphorylation in membrane preparations *in vitro. Nature* **276**: 409-410

Carpenter G, King L, Cohen S (1979) Rapid enhancement of protein phosphorylation in A-431 cell membrane preparations by epidermal growth factor. *J Biol Chem* **254**: 193-216

Chen WS, Lazar CS, Poenie M, Tsien RY, Gill G, Rosenfeld MG (1987) Requirement for intrinsic protein tyrosine kinase in the immediate and late actions of the EGF receptor. *Nature* **328**: 820-823

Chinkers M, Garbers DL (1986) Suppression of protein tyrosine kinase activity of the epidermal growth factor receptor by epidermal growth factor. *J Biol Chem* **261**: 11073-11078

Chou C-K, Dull TJ, Russell DS, Gherzi R, Lebwohl D, Ullrich A, Rosen OM (1987) Human insulin receptors mutated at the ATP-binding site lack protein tyrosine kinase activity and fail to mediate post-receptor effects of insulin. *J Biol Chem* **262**: 1842-1847

Cochet C, Gill GN, Meisenhelder J, Cooper JA, Hunter T (1984) Protein kinase-C phosphorylates the epidermal growth factor receptor and reduces its epidermal growth factor-stimulated tyrosine protein kinase activity. *J Biol Chem* **259**: 2553-2558

Cohen S (1962) Isolation of a mouse submaxillary gland protein accelerating incisor eruption and eyelid opening in the newborn animal. *J Biol Chem* **237**: 1555-1562

Cohen S, Carpenter G (1975) Human epidermal growth factor: isolation and chemical and biological properties. *Proc Natl Acad Sci USA* **72**: 1317-1321

Cohen S, Elliot GA (1963) The stimulation of epidermal keratinization by a protein isolated from the submaxillary gland of the mouse. *J Invest Dermatol* **40**: 1-5

Cohen S, Fava RA, Sawyer ST (1982a) Purification and characterization of epidermal growth factor receptor/protein kinase from mouse liver. *Proc Natl Acad Sci USA* **79**: 6237-6241

Cohen S, Ushiro H, Stoscheck C, Chinkers M (1982b) A native 170,000 epidermal growth factor receptor kinase complex from shed membrane vesicles. *J Biol Chem* **257**: 1523-1531

Cowley G, Smith JA, Gusterson B, Hendler F, Ozanne B (1984) The amount of EGF receptor is elevated on squamous cell carcinomas. *Cancer Cells* **1**: 5-10

Cowley GP, Smith JA, Gusterson BA (1986) Increased EGF receptors on human squamous carcinoma cell lines. *Br J Cancer* **53**: 223-229

Davis RJ, Czech MP (1985) Tumor-promoting phorbol diesters cause the phosphorylation of the epidermal growth factor receptor in normal human fibroblasts at threonine-654. *Proc Natl Acad Sci USA* **82**: 1974-1978

Davis RJ, Czech MP (1986) Inhibition of the apparent affinity of the epidermal growth factor receptor caused by phorbol diesters correlates with phosphorylation of threonine-654 but not other sites on the receptor. *Biochem J* **233**: 435-441

Decker SJ (1985) Phosphorylation of the *erb* B gene product from an avian erythroblastosis virus-transformed chick fibroblast cell line. *J Biol Chem* **260**: 2003-2006

Derynck R, Roberts AB, Winkler ME, Chen EY, Goeddel DV (1984) Human transforming growth factor-α: precursor structure and expression in *E. coli*. *Cell* **38**: 287-297

Downward J, Parker P, Waterfield MD (1984) Autophosphorylation sites on the EGF receptor. *Nature* **311**: 483-485

Elder JB, Williams G, Lacey E, Gregory H (1978) Cellular localisation of human urogastrone/epidermal growth factor. *Nature* **271**: 466-467

Fabricant RN, DeLarco JE, Todaro GJ (1977) Nerve growth factor receptors on human melanoma cells in culture. *Proc Natl Acad Sci USA* **74**: 565-569

Gannett DC, Tracy SE, Robinson HL (1986) Differences in sequences encoding the carboxyl-terminal domain of the epidermal growth factor receptor correlate with differences in the disease potential of viral *erb* B genes. *Proc Natl Acad Sci USA* **83**: 6053-6057

Gates RE, King LE Jr (1982) Calcium facilitates endogenous proteolysis of the EGF receptor-kinase. *Mol Cell Endocrinol* **27**: 325-327

Gill GN, Lazar CS (1981) Increased phosphotyrosine content and inhibition of proliferation in EGF-treated A431 cells. *Nature* **293**: 305-307

Gilmore T, DeClue JE, Martin GS (1985) Protein phosphorylation of tyrosine is induced by the v-*erb* B gene product *in vivo* and *in vitro*. *Cell* **40**: 609-618

Gregory H (1975) Isolation and structure of urogastrone and its relationship to epidermal growth factor. *Nature* **257**: 325-327

Gusterson B, Cowley G, McIlhinney J, Ozanne B, Fisher C, Reeves B (1985) Evidence for increased epidermal growth factor receptors in human sarcomas. *Int J Cancer* **36**: 689-693

Haigler H, Ash JF, Singer SJ, Cohen S (1978) Visualization by fluorescence of the binding and internalization of epidermal growth factor in human carcinoma cells A-431. *Proc Natl Acad Sci USA* **75**: 3317-3321

Heitz PU, Kaspar M, Van Noorden S, Polak JM, Gregory H, Pearse AGE (1978) Immunohistochemical localization of urogastrone in human duodenal and submandibular glands. *Gut* **19**: 408-413

Hendler FJ, Ozanne B (1984) Human squamous cell lung cancers express increased epidermal growth factor receptors. *J Clin Invest* **74**: 647-651

Hirata Y, Orth DN (1979) Epidermal growth factor (urogastrone) in human tissues. *J Clin Endocrinol Metab* **48**: 667-672

Hollenberg MD, Cuatrecasas P (1973) Epidermal growth factor: receptors in human fibroblasts and modulation of action by cholera toxin. *Proc Natl Acad Sci USA* **70**: 2964-2968

Hunter T, Cooper JA (1981) Epidermal growth factor induces rapid tyrosine phosphorylation of proteins in A-431 human tumor cells. *Cell* **24**: 660-669

Hunter T, Cooper JA (1985) Protein-tyrosine-kinases. *Annu Rev Biochem* **54**: 897-930

Hunter T, Ling N, Cooper JA (1984) Protein kinase-C phosphorylation of the EGF receptor at a threonine residue close to the cytoplasmic face of the plasma membrane. *Nature* **311**: 480-483

Kamata N, Chida K, Rikimaru K, Horikoshi M, Enomoto S, Kuroki T (1986) Growth-inhibitory effects of epidermal growth factor and over-expression of its receptor on human squamous cell carcinomas in culture. *Cancer Res* **46**: 1648-1653

Kaneko Y, Shibuya M, Nakayama T, Hayashida N, Toda G, Endo Y, Oka H, Oda T (1985) Hypomethylation of c-*myc* and epidermal growth factor receptor genes in human hepatocellular carcinomas and fetal liver. *Jpn J Cancer Res* **76**: 1136-1140

King CS, Cooper JA (1986) Effects of protein kinase-C activation after epidermal growth factor binding on epidermal growth factor receptor phosphorylation. *J Biol Chem* **261**: 10073-10078

Korc M, Meltzer P, Trent J (1986) Enhanced expression of epidermal growth factor receptor correlates with alterations of chromosome 7 in human pancreatic cancer. *Proc Natl Acad Sci USA* **83**: 5141-5144

Kris RM, Lax I, Gullick W, Waterfield MD, Ullrich A, Fridkin M, Schlessinger J (1985) Antibodies against a synthetic peptide as a probe for the kinase activity of the avian EGF receptor and v-*erb* B protein. *Cell* **40**: 619-625

Libermann TA, Nusbaum HR, Razon N, Kris R, Lax I, Soreq H, Whittle N, Waterfield MD, Ullrich A, Schlessinger J (1985) Amplification, enhanced expression and possible rearrangement of EGF receptor gene in primary human brain tumors of glial origin. *Nature* **313**: 144-147

Lin CR, Chen WS, Lazar CS, Carpenter CD, Gill GN, Evans RM, Rosenfeld MG (1986) Protein kinase-C phosphorylation at Thr 654 of the unoccupied EGF receptor and EGF binding regulate functional receptor loss by independent mechanisms. *Cell* **44**: 839-848

Livneh E, Reiss N, Berent E, Ullrich A, Schlessinger J (1987) An insertional mutant of epidermal growth factor receptor allows dissection of diverse receptor functions. *EMBO J* **6**: 2669-2676

MacLeod CL, Luk A, Castagnola J, Cronin M, Vendelson J (1986) EGF induces cell cycle arrest of A431 human epidermoid carcinoma cells *J Cell Physiol.* **127**: 175-182

Marquardt H, Hunkapiller MW, Hood LE, Twardzik DR, DeLarco JE, Stephenson JR, Todaro GJ (1983) Transforming growth factors produced by retrovirus-transformed rodent fibroblasts and human melanoma cells: amino acid sequence homology with epidermal growth factor. *Proc Natl Acad Sci USA* **80**: 4684-4688

Marquardt H, Hunkapiller MW, Hood LE, Todaro GJ (1984) Rat transforming growth factor type 1: structure and relation to epidermal growth factor. *Science* **223**: 1079-1082

Mattila A-L, Perheentupa J, Pesonen K, Viinikka L (1985) Epidermal growth factor in human urine from birth to puberty. *J Clin Endocrinol Metab* **61**: 997-1000

Moore GPM, Panaretto BA, Carter NB (1985) Epidermal hyperplasia and wool follicle regression in sheep infused with epidermal growth factor. *J Invest Dermatol* **84**: 172-175

Neal DE, Marsh C, Bennett MK, Abel PD, Hall RR, Sainsbury JRC, Harris AL (1985) Epidermal-growth-factor receptors in human bladder cancer: comparison of invasive and superficial tumours. *Lancet* **i**: 366-368

Nilson TW, Maroney PA, Goodwin RG, Rottman FM, Crittendon LB, Raines MA, Kung HJ (1985) c-*erb* B activation in ALV-induced erythroblastosis: novel RNA processing and promoter insertion result in expression of an amino-truncated EGF receptor. *Cell* **41**: 719-726

Partanen A-M, Ekblom P, Thesleff I (1985) Epidermal growth factor inhibits morphogenesis and cell differentiation in cultured mouse embryonic teeth. *Dev Biol* **111**: 84-94

Pike LJ, Kuenzel EA, Casnelli JE, Krebs EG (1984) A comparison of the insulin - and epidermal growth factor - stimulated protein kinase from human placenta. *J Biol Chem* **259**: 9913-9921

Reisner AH (1985) Similarity between the vaccinia virus 19K early protein and epidermal growth factor. *Nature* **313**: 801-803

Rosdy M, Bernard BA, Schmidt R, Darmon M (1986) Incomplete epidermal differentiation of A-431 epidermoid carcinoma cells. *In Vitro Cell Devl Biol* **22**: 295-300

Russell DS, Gherzi R, Johnson EL, Chou C-K, Rosen OM (1987) The protein-tyrosine kinase activity of the insulin receptor is necessary for insulin-mediated receptor down-regulation. *J Biol Chem* **262**: 11833-11840

Russo MW, Lukas TJ, Cohen S, Staros JV (1985) Identification of residues in the nucleotide binding site of the epidermal growth factor receptor/kinase. *J Biol Chem* **260**: 5205-5208

Sainsbury JRC, Sherbert GV, Farndon JR, Harris AL (1985) Epidermal growth factor receptors and oestrogen receptors in human breast cancer. *Lancet* **i**: 364-366

Savage CR Jr, Inagami T, Cohen S (1972) The primary structure of epidermal growth factor. *J Biol Chem* **247**: 7612-7621

Savage CR Jr, Hash JH, Cohen S (1973) Epidermal growth factor: location of disulfide bonds. *J Biol Chem* **248**: 7669-7672

Schonbrunn A, Krasnoff M, Westendorf JM, Tashjian AH (1980) Epidermal growth factor and thyrotropin-releasing hormone act similarly on a clonal pituitary cell strain. Modulation of hormone production and inhibition of cell proliferation. *J Cell Biol* **85**: 786-797

Shoji S, Parmelee DC, Wade RD, Kimar S, Ericsson LH, Walsh KA, Neurath H, Long GL, Demaille JG, Fisher EH, Titani K (1981) Complete amino acid sequence of the catalytic subunit of bovine cardiac muscle cyclic AMP-dependent protein kinase. *Proc Natl Acad Sci USA* **78**: 848-851

Starkey RH, Cohen S, Orth DN (1975) Epidermal growth factor: identification of a new hormone in human urine. *Science* **189**: 800-802

Stoscheck CM, Carpenter G (1983) Biology of the A-431 cell: a useful organism for hormone research. *J Cell Biochem* **23**: 191-202

Stoscheck CM, Carpenter G (1984) Characterization of the metabolic turnover of epidermal growth factor receptor protein in A-431 cells. *J Cell Physiol* **120**: 296-302

Stroobant P, Rice AP, Gullick WJ, Cheng DJ, Kerr IM, Waterfield MD (1985) Purification and characterization of vaccinia virus growth factor. *Cell* **42**: 383-393

Sundell HW, Gray ME, Serenius FS, Escobedo MB, Stahlman MT (1980) Effects of epidermal growth factor on lung maturation in fetal lambs. *Am J Pathol* **100**: 707-726

Taylor JM, Mitchell WM, Cohen S (1972) Epidermal growth factor: physical and chemical properties. *J Biol Chem* **247**: 5928-5934

Twardzik DR, Brown JP, Ranchalis JE, Todaro GJ, Moss B (1985) Vaccinia virus-infected cells release a novel polypeptide functionally related to transforming and epidermal growth factors. *Proc Natl Acad Sci USA* **82**: 5300-5304

Ullrich A, Coussens L, Hayflick JS, Dull TJ, Gray A, Tam AW, Lee J, Yarden Y, Libermann L, Schlessinger J, Downward J, Mayes ELV, Wittle N, Waterfield MD, Seeburg PH (1984) Human epidermal growth factor receptor cDNA sequences and aberrant expression of the amplified gene in A431 epidermoid carcinoma cells. *Nature* **309**: 418-425

Ushiro H, Cohen S (1980) Identification of a phosphotyrosine as a product of epidermal growth factor-activated protein kinase in A-431 cell membranes. *J Biol Chem* **255**: 8363-8365

Whiteley B, Glaser L (1986) Epidermal growth factor (EGF) promotes phosphorylation at threonine-654 of the EGF receptor: possible role of protein kinase-C in homologous regulation of the EGF receptor. *J Cell Biol* **103**: 1355-1362

Yamamoto T, Nishida T, Miyajima N, Kawai S, Ooi T, Toyoshima K (1983) The *erb*-B gene of avian erythroblastosis virus is a member of the *src* gene family. *Cell* **35**: 71-78

Yamamoto T, Kamata N, Kawano H, Shimizu S, Kuroki T, Toyoshima K, Rikimaru K,

Nomura N, Ishizaki R, Pastan I, Gamou S, Shimizu N (1986) High incidence of amplification of the epidermal growth factor receptor gene in human squamous carcinoma cell lines. *Cancer Res* **46**: 414-416

Advances in Growth Hormone and Growth Factor Research,
edited by E.E. Müller, D. Cocchi and V. Locatelli
Pythagora Press, Roma-Milano and Springer Verlag, Berlin-Heidelberg © 1989

Platelet-derived growth factor – Structural and functional aspects of the A-chain gene

C. Betsholtz[1], F. Rorsman[1], M. Bywater[1], C.-H. Heldin[2] and B. Westermark[1]

[1] *Department of Pathology, University Hospital, Uppsala, Sweden; and* [2] *Ludwig Institute for Cancer Research, Biomedical Center, Uppsala, Sweden*

Platelet-derived growth factor (PDGF) is one of the major mitogens present in serum for fibroblasts, smooth muscle cells, and glia cells (for recent reviews see Heldin et al. 1985; Ross et al. 1986). Its cellular effects include the activation of protein phosphorylation, ion transport, membrane ruffling, expression of specific genes (e.g. c-*fos*, c-*myc*), synthesis of specific proteins including extracellular matrix components, and directed cell migration (for original references see Heldin et al. 1985; Ross et al. 1986). PDGF is normally stored in the platelet α-granules and released at sites of vascular injury. This, in conjunction with its cellular effects, has suggested that PDGF participates in wound healing by acting specifically on connective tissue cells. In recent years, PDGF expression has been demonstrated in several normal cell types; i.e., endothelial cells (DiCorleto and Bowen-Pope 1983; Barrett et al. 1984), macrophages (Shimokado et al. 1985; Martinet et al. 1986), cytotrophoblasts of the placenta (Goustin et al. 1985), vascular smooth muscle cells (Seifert et al. 1984; Nilsson et al. 1985), and foreskin fibroblasts (Paulsson et al. 1987). In all instances, expression of one or both of the PDGF genes (see below) was seen only in an activated cellular phenotype; expression was induced following lipopolysaccharide activation of macrophages or mitogen stimulation of smooth muscle cells or fibroblasts and ceased in differentiating endothelial cells and cytotrophoblasts (fusing to the multinucleated non-proliferating syncytiotrophoblast layer). This suggests that several sources of PDGF may be utilized in tissue-repair processes and that additional normal functions of PDGF, such as the organization of extraembryonal tissues, have to be considered.

PDGF has also been implied in various pathological processes such as atherosclerosis, myelofibrosis, and neoplasia. Ideas about its involvement in initiation and/or progression of tumorigenesis stem from the observation in 1980 that human osteosarcoma cells produce a PDGF-like growth factor (Heldin et al. 1980). Since these cells are derived from a normal cell supposed to carry receptors for PDGF, an autocrine loop driving the proliferation of the tumor cells could be envisioned. That these cells indeed expressed PDGF receptors was later confirmed (Betsholtz et al. 1984). However, attempts to interfere with the neoplastic phenotype by interrupting the autocrine loop were unsuccessful (Betsholtz et al. 1984). Numerous examples of the production of PDGF-like growth factors by a wide variety of human tumor cell lines have since been reported (reviewed in Heldin et al. 1986a). In tumors such as glioma, there is frequently co-expression of PDGF-like growth factors and PDGF receptors, but also in these cases evidence for autocrine growth control is lacking (Nistér et al. 1988).

The idea that the uncontrolled expression of PDGF in a receptor-bearing cell could nevertheless be important in the initiation of the tumorigenic process is supported by the biological features of simian sarcoma virus (SSV). This primate retrovirus has transduced the gene for one of the constituent polypeptide chain of PDGF, the B-chain (see below), and thereby attained transforming properties (Devare et al. 1983; Waterfield et al. 1983; Doolittle et al. 1983). The transduced B-chain gene (v-sis) encodes a structurally and functionally normal PDGF B-chain that exerts the transforming function by interacting with PDGF receptors at the external side of the plasma membrane (Robbins et al. 1983; Johnsson et al. 1985). Thus, acute transformation by SSV *in vitro* is the result of the establishment of an autocrine loop involving the production and response to a PDGF-like growth factor (reviewed in Westermark et al. 1987). Consequently, the transformed phenotype in this case is, in contrast to the situations in established cell lines from human tumors, inhibitable by PDGF antagonists such as PDGF antibodies or suramin (Johnsson et al. 1985; Betsholtz et al. 1986a). Transformation by SSV *in vitro* is in all respects indistinguishable from chronic stimulation by PDGF (Johnsson et al. 1986). Infected normal human fibroblasts do not acquire a fully transformed phenotype, for example they do not become immortalized. Therefore, several questions regarding the transforming properties of SSV *in vivo* remain to be answered. Injection of SSV into the brain of newborn marmosets gives rise to malignant glioblastomas at a relatively high incidence (Wolfe et al. 1971). Apparently, the autocrine loop established in the SSV-transformed cell *in vivo* may become complemented and perhaps overriden by subsequent changes in the cellular genotype. How this occurs, and which the relevant changes are, remain essential questions in investigating the role of the PDGF genes in the development of glioma.

STRUCTURE OF PDGF AND PDGF-LIKE GROWTH FACTORS

PDGF is a 30-K protein composed of two related but distinct polypeptide chains, A and B, of approximately the same size (Johnsson et al. 1982), linked by disulphide bridges. The native human platelet PDGF molecule is a heterodimer between one A- and one B-chain (Hammacher et al. 1988). However, several of the characterized PDGF-like growth factors have been shown to be homodimers of either of the chains. Porcine PDGF (Stroobant and Waterfield 1984) and the SSV oncogene product (Robbins et al. 1983) are B-chain homodimers, whereas growth factors purified from the conditioned medium of human osteosarcoma (Heldin et al. 1986b), melanoma (Westermark et al. 1986) and glioblastoma (Nistér et al. 1988) cell lines are homodimers of A-chains. The presence of two different polypeptide chains that may be assembled into three different dimeric forms raises questions about possible differences in function. We have recently addressed these issues by transfecting cloned PDGF A- and B-chain cDNAs under transcriptional control by a retroviral LTR into PDGF receptor-bearing cells and studied the effects on cellular transformation. The present review focuses on these data, as well as on the structural organization of the human PDGF A-chain locus in comparison with the PDGF B-chain locus.

THE PDGF A-CHAIN cDNA AND GENE

The nucleotide sequence of PDGF A-chain cDNA, isolated from a human glioma cDNA library, predicts that the A-chain, in analogy with the B-chain, is synthesized as a prepropeptide (Betsholtz et al. 1986b). Thus, it contains a signal sequence directing transport of the nascent polypeptide across the endoplasmic reticulum membrane, as well as an N-terminal propeptide with unknown function. C-terminal processing may occur as well, but putative processing sites have not been identified. The mature A-chain contains a site for N-linked glycosylation, but whether this is utilized is not known. A comparison between the amino acid sequences for the predicted A- and B-chain precursors is shown in Fig. 1. The overall amino acid sequence homology between the PDGF precursor molecules is about 50%. All eight cysteine residues within the mature chains participate in disulphide bonding and are conserved, indicating similarities in tertiary structure.

Several laboratories have performed a structural characterization of the PDGF B-chain/c-sis gene (Josephs et al. 1984; Johnsson et al. 1984; Chiu et al. 1984). We recently performed a structural characterization of the PDGF A-chain gene (Rorsman et al. 1988). The overall structure of the PDGF A-chain gene in comparison with the PDGF B-chain gene, their mRNAs, protein products, and assembly into dimeric molecules are shown in Fig. 2. Both genes have seven exons and span some 20 kb pairs of genomic DNA. The exon/intron arrangement is also very similar; notably, the exons appear to divide the protein precursor into different functional domains. In both genes exon 1 encodes the signal sequence and exons 2 and 3 the N-terminal propeptide. The N-terminal proteolytic processing site occurs after a stretch of basic amino acids (see Fig. 1), the codons for which are located close to the 3' end of exon 3 in both genes. Consequently, the mature chains

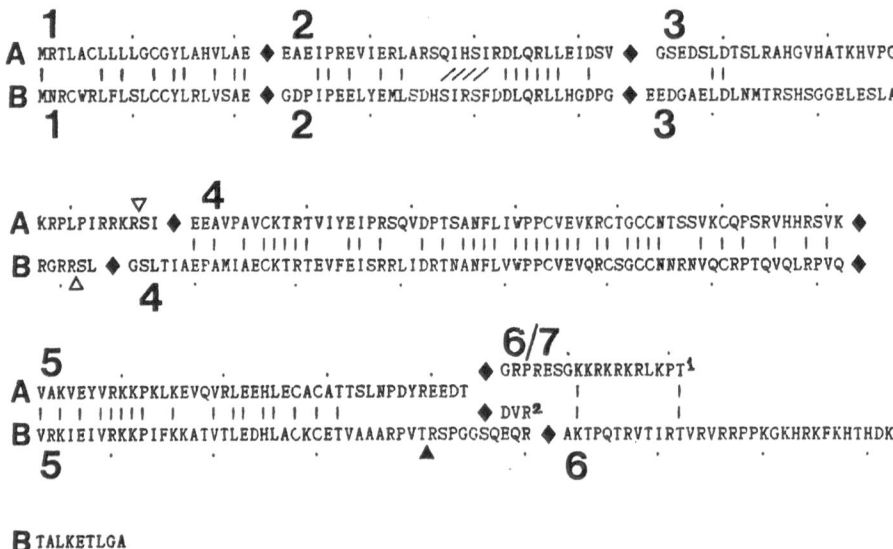

Fig. 1. Comparison of the primary structures of the PDGF A- and B-chain precursors. The one-letter amino acid code is used. *Vertical lines* indicate homology. The number of the exon encoding a particular part of each polypeptide is shown, as well as the location of the splice junctions. The two alternative A-chain C-terminals, depending on whether exon 6 is utilized or not, are indicated. *Open triangles* mark the N-terminal processing sites. The *filled triangle* marks a possible C-terminal processing site in the B-chain precursor.

are almost entirely encoded by exons 4 and 5; only the first few amino acids derive from exon 3. The B-chain C-terminal sequence encoded by exon 6 is probably removed in the mature B-chain. There is circumstantial evidence that the C-terminal processing site follows the threonine residue at position 190 in the B-chain precursor (Fig. 1; Johnsson et al. 1984). The B-chain exon 7 contains only an untranslated sequence. In the A-chain gene, exon 6 appears to be alternatively utilized, probably because of differential splicing. This was detected through the characterization of two types of cDNA clones, with or without exon 6 sequences present, in a glioma cDNA library (Betsholtz et al. 1986b). With exon 6 present, a sequence 15 amino acids longer and extremely basic is predicted. In this case, exon 7 is untranslated. In the transcript lacking exon 6, exon 7 substitutes the last three codons. The shorter A-chain transcript appears to be the more common one in the glioma cell line. Using the excised exon 6 as a probe, it was evident that other tumor cell lines also contained both types of transcript; however, the relative amount of the longer transcript (containing exon 6) was lower in other cell lines compared with the glioma (Rorsman et al. 1988). In addition, the longer A-chain

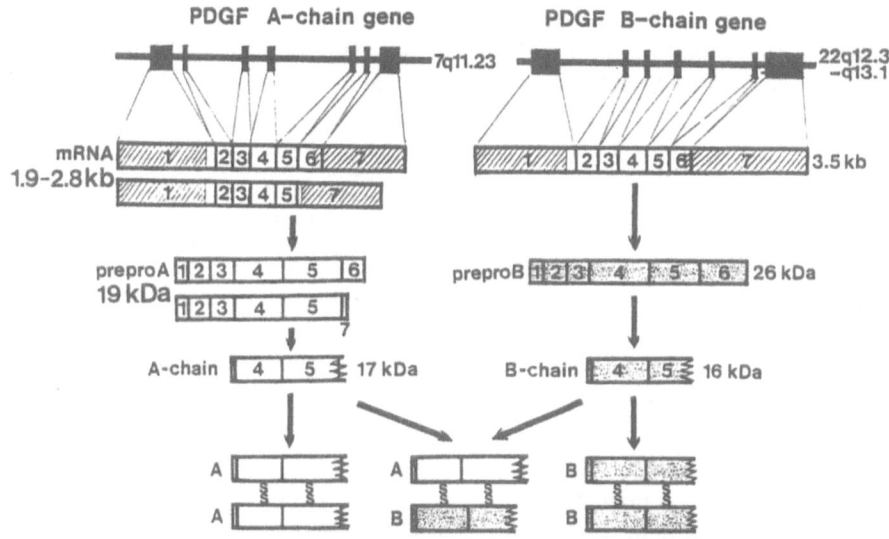

Fig. 2. The two PDGF genes and their products. The approximate outline of the two genes and their chromosomal localization are given at the top of the figure. *Boxes* represent exons. In the mRNAs, the *numbers* indicate the coding by each exon. The *open parts* represent translated sequences, whereas *hatched parts* indicated nontranslated 5' and 3' sequences. The approximate sizes of the transcripts are given (*kb* = kilobase pairs). The differential splicing of the A-chain mRNA precursor is illustrated; a minor fraction of the transcripts contains exon 6. The prepro A/B-chain and the protein coding by each exon, as well as the processing of each precursor, are shown in the middle part of the figure. The approximate sizes of the mature chains are given (*kDa* = kiloDaltons). The various dimeric forms of the two chains are shown at the bottom. All three types exist in nature and possess different biological activities (see the text).

transcript was not detected in normal endothelial cells (Collins et al. 1987). Thus, the alternative splicing involving exon 6 may to some extent be cell-type specific; whether it has any biological relevance remains to be determined.

The B-chain (c-*sis*) mRNA has been identified as a single 3.5-kb species (Rao et al. 1986). The A-chain mRNA is heterogeneous; three species ranging between 1.9 and 2.8 kb have been identified using Northern blot analysis (Betsholtz et al. 1986b). These differences cannot be explained by differential splicing involving exon 6, since it is too small (69 bp) and the excised and radiolabeled exon 6 hybridizes with all three A-chain mRNAs. Only one promoter-like element has been identified upstream of exon 1 in the A-chain gene (Rorsman et al. 1988). However, one cannot exclude an alternative promoter further upstream associated with an additional exon. It is perhaps more likely to assume that the size differences occur in the 3' end, but this has been difficult to define due to the lack of any isolated polyadenylated cDNA clones. However, several alternative polyadenylation signals are found in the 3' end of the gene (Rorsman et al. 1988), and the alternative utilization of these may explain why three different-size A-chain mRNAs are present in human cells actively expressing the A-chain gene.

FUNCTIONAL ANALYSIS OF THE PDGF A– AND B–CHAINS BY GENE TRANSFECTION

Both PDGF genes are highly conserved in the mammalian and chicken genomes (F. Rorsman and C. Betsholtz unpublished manuscript). This indicates specific and indispensible functions associated with each of their protein products. Our group has taken several approaches to reveal differences in function between the two chains (in reality, between the three types of dimers formed [Fig. 1]). One strategy has been to transfect cells with A- and B-chain cDNAs under control of strong promoters and to study the transforming functon of such constructs. In previous studies, both cDNA and genomic clones corresponding to the PDGF B-chain/c-*sis* under viral promoter control have been shown to be transforming for NIH/3T3 cells in a manner similar to v-*sis* (Clarke et al. 1984; Gazit et al. 1984; Chiu et al. 1984). In order to compare the transforming activities of the two PDGF genes, we constructed retrovirus vectors containing full-length cDNA clones corresponding to PDGF-A and PDGF-B, as well as a gene conferring resistance to G-418, and transfected these into retrovirus packaging cell lines to generate stocks of helper-free recombinant virus. Subsequently, rat and human fibroblasts were infected, G-418 selected, and analyzed with regard to *in vitro* properties related to transformation. The result demonstrated a significant difference between the two genes. In spite of similar levels of mRNA expressed from the various recombinant viruses and the production of intact homodimers of A- and B-chains, respectively, in the infected rat and human cells, only the cells infected by the B-chain virus attained a transformed phenotype comparable to that seen in SSV-transformation (Bywater et al. 1988). This included changed cellular morphology, focus-forming ability, growth to higher saturation densities in monolayer, growth in low serum concentrations, and growth in soft agar. The cells infected by the A-chain virus were morphologically normal, and no discrete foci were detected. However, the cells grew to high saturation densities, proliferated in low serum concentrations, and showed an increased ability for anchorage-independent growth, although to a lesser extent than cells infected by the B-chain virus (Bywater et al. 1988). Similar differences in transforming abilities between the PDGF A- and B-chains have been found using vectors utilizing the mouse metallothionein promotor and NIH/3T3 cells as targets (Beckmann et al. 1988). In the latter study, however, it was also demonstrated that a dramatic overexpression of the A-chain resulted in a weak focus-forming effect.

CONCLUSIONS AND SPECULATIONS

Several explanations for the different results obtained with the A- and B-chains virus may be considered. Accumulating evidence, however, suggests that the different biological effects of A- and B-chain homodimers is the major determinant. A PDGF-like molecule purified from glioma cells that has the structure of a PDGF A-chain homodimer (Nistér et al. 1988) and recombinant A-chain homodimers expressed in CHO cells and yeast (Östman et al., submitted for publication) are less potent as mitogens compared with heterodimeric PDGF or B-chain homodimers, which in turn are equipotent. A-chain homodimers also lack several other biological features of PDGF, such as the ability to induce directed cell migration and cytoskeletal reorganization (Nistér et al. 1988). Obviously, the A-chain homodimer might be associated with other functions, the nature of which remains to be disclosed.

One function of the A-chain might be to confer secretory properties to the heterodimer. Even if both the A- and the B-chain precursors possess signal sequences and enter the secretory pathway, the BB-homodimer has been shown to remain associated with cellular membranes to a large extent (Robbins et al. 1985), whereas the A-chain homodimer is readily secreted (Betsholtz et al. 1986b).

Human tumor cell lines express the genes for the PDGF A- and B-chains at a relatively high frequency (Eva et al. 1982; Betsholtz et al. 1986b; Perez et al. 1987; Söderdahl et al. 1988; Nistér et al. 1988), but they do not appear to be coordinately expressed. Some lines express only A-chain mRNA, others only B-chain mRNA, and some express both types of mRNA. Thus, the genes, which reside on different chomosomes, the A-chain gene on 7 (Betsholtz et al. 1986b) and the B-chain gene on 22 (Swan et al. 1982; Dalla Favera et al. 1982), are differentially regulated. This is compatible with different functions associated with the two genes. In relation to the assumed role for PDGF in certain normal and pathological conditions with cell proliferation, further functional characterization of the members of the PDGF family is highly warranted.

Note added in proof

Since the submission of this manuscript two classes of PDGF receptors with different affinity for the various dimeric PDGF forms have been reported (Heldin et al. 1988; Hart et al. 1988). The previously known PDGF receptor (called type B PDGF receptor), characterized biochemically and by cDNA cloning (Yarden et al. 1986; Claesson-Welsh et al. 1988), binds PDGF-BB with high affinity and PDGF-AB with lower affinity but does not bind PDGF-AA (Heldin et al. 1988; Claesson-Welsh et al. 1988). The second PDGF receptor (type A PDGF receptor) binds all three PDGF dimers with high affinity (Heldin et al. 1988; Hart et al. 1988). Thus, the different effects of PDGF-BB and PDGF-AA in autocrine systems and following exogenous administrations (Nistér et al. 1988; Bywater et al. 1988; Beckman et al. 1988) may relate to the fact that they in part act via different receptors.

REFERENCES

Barrett TB, Gajdusek CM, Schwartz SM, McDougall JK, Benditt EP (1984) Expression of the *sis* gene by endothelial cells in culture and *in vivo*. *Proc Natl Acad Sci USA* **81**: 6772-6774

Beckman MP, Betsholtz C, Heldin C–H, Westermark B, Di Marco E, Di Fiore PP, Robbins KC, Aaronson SA (1988) Human PDGF-A and PDGF-B chains differ in their biological properties and transforming potential. *Science*, in press

Betsholtz C, Westermark B, Ek B, Heldin C-H (1984) Co-expression of a PDGF-like growth factor and PDGF receptors in a human osteosarcoma cell line: implications for autocrine receptor activation. *Cell* **39**: 447-457

Betsholtz C, Johnsson A, Heldin C-H, Westermark B (1986a) Efficient reversion of SSV-transformation and inhibition of growth factor-induced mitogenesis by suramin. *Proc Natl Acad Sci USA* **83**: 6440-6444

Betsholtz C, Johnsson A, Heldin C-H, Westermark B, Lind P, Urdea MS, Shows TB, Philpott K, Mellor A, Knott TJ, Scott J (1986b) cDNA sequence and chromosomal localization of human platelet-derived growth factor A-chain and its expression in tumour cell lines. *Nature* **320**: 695-699

Bywater M, Rorsman F, Bongeam-Rudloff E, Mark G, Hammacher A, Heldin C-H, Westermark B, Betsholtz C (1988) Expression of recombinant platelet-derived growth factor A- and B- chain homodimers in Rat-1 cells and human fibroblasts reveals differences in protein processing and effects. *Mol Cell Biol* **8**: 2753-2762

Chiu I-M, Reddy EP, Givol D, Robbins KC, Tronick SR, Aaronson SA (1984) Nucleotide sequence analysis identifies the human c-*sis* proto-oncogene as a structural gene for platelet-derived growth factor. *Cell* **37**: 123-129

Claesson-Welsh L, Eriksson A, Morèn A, Severinsson L, Ek B, Östman A, Heldin C-H (1988) cDNA cloning and expression of a human platelet-derived growth factor (PDGF) receptor specific for B-chain-containing PDGF molecules. *Mol Cell Biol* **8**: 3476-3486

Clarks MF, Westin E, Schmidt D, Josephs SF, Ratner L, Wong-Staal F, Gallo RC, Reitz MS (1984) Trasformation of NIH 3T3 cells by a human c-*sis* cDNA clone. *Nature* **308**: 464-467

Collins T, Bonthron TD, Orkin SH (1987) Alternative RNA splicing affects function of encoded platelet-derived growth factor A-chain. *Nature* **328**: 621-623

Dalla Favera R, Gallo RC, Giallongo A, Croce CM (1982) Chromosomal localization of the human homolog (c-*sis*) of the simian sarcoma virus *onc* gene. *Science* **218**: 686-688

Devare SG, Reddy EP, Law JD, Robbins KC, Aaronson SA (1983) Nucleotide sequence of the simian sarcoma virus genome: demonstration that its acquired cellular sequences encode the transforming gene product p28[sis]. *Proc Natl Acad Sci USA* **80**: 731-735

DiCorleto PE, Bowen-Pope DF (1983) Cultured endothelial cells produce a platelet-derived growth factor-like protein. *Proc Natl Acad Sci USA* **80**: 1919-1923

Doolittle RF, Hunkapiller MW, Hood LE, Devare SG, Robbins KC, Aaronson SA, Antoniades HN (1983) Simian sarcoma virus *onc* gene, v-*sis*, is derived from the gene (or genes) encoding a platelet-derived growth factor. *Science* **221**: 275-277

Eva A, Robbins KC, Andersen PR, Srinavasan A, Tronick SR, Reddy EP, Ellmore NW, Galen AT, Lautenberg JA, Papas TS, Westin EH, Wong-Staal F, Gallo RC, Aaronson SA (1982) Cellular genes analogous to retroviral *onc* genes are transcribed in human tumor cells. *Nature* **295**: 116-119

Gazit A, Igarashi H, Chiu I-M, Srinivasan A, Yaniv A, Tronik SR, Robbins KC, Aaronson

SA (1984) Expression of the normal *sis*/PDGF-2 coding sequence induces cellular transformation. *Cell* **39**: 89-97

Goustin AS, Betsholtz C, Pfeifer-Ohlsson S, Persson H, Rydnert J, Bywater M, Holmgren G, Heldin C-H, Westermark B, Ohlsson R (1985) Co-expression of the *sis* and *myc* proto-oncogenes in human placenta suggests autocrine control of trophoblast growth. *Cell* **41**: 301-312

Hart CE, Forstrom JV, Kelly JD, Seifert RA, Smith RA, Ross R, Murray MJ, Bowen-Pope DF (1988) Two classes of PDGF receptors recognize different forms of PDGF. *Science* **240**: 1529-1531

Heldin C-H, Westermark B, Wasteson Å (1980) Chemical and biological properties of a growth factor from human cultured osteosarcoma cells: resemblance with platelet-derived growth factor. *J Cell Physiol* **105**: 235-246

Heldin C-H, Wasteson Å, Westermark B (1985) Platelet-derived growth factor. *Mol Cell Endocrinol* **39**: 169-187

Heldin C-H, Betsholtz C, Johnsson A, Westermark B (1986a) Role of PDGF-like growth factors in malignant transformation. *Cancer Rev* **2**: 34-47

Heldin C-H, Johnsson A, Wennergren S, Wernestedt C, Betsholtz C, Westermark B (1986b) A human osteosarcoma cell line secretes a growth factor structurally related to a homodimer of PDGF A-chains. *Nature* **319**: 511-514

Heldin C-H, Bäckström G, Östman A, Hammacher A, Rönnstrand L, Rubin K, Nistér M, Westermark B (1988) Binding of different dimeric forms of PDGF to human fibroblasts: evidence for two separate receptor types. *EMBO J* **7**: 1387-1393

Hommacher A, Hellman U, Johnsson A, Östman A, Gunnarsson K, Westermarle B, Wasterson Å, Heldin C-H (1988) PDGF purified from human platelets is a heterodimer of A and one B chain. *J Biol Chem,* in press

Johnsson A, Heldin C-H, Westermark B, Wasteson Å (1982) Platelet-derived growth factor: identification of constituent polypeptide chains. *Biochem Biophys Res Commun* **104**: 66-74

Johnsson A, Heldin C-H, Wasteson Å, Westermark B, Deuel TF, Huang JS, Seeburg PH, Gray A, Ullrich A, Scrace G, Stroobant P, Waterfield MD (1984) The c-*sis* gene encodes a precursor of the B-chain of platelet-derived growth factor. *EMBO J* **3**: 921-928

Johnsson A, Betsholtz C, Heldin C-H, Westermark B (1985) Antibodies against platelet-derived growth factor inhibit acute transformation by simian sarcoma virus. *Nature* **317**: 438-440

Johnsson A, Betsholtz C, Heldin C-H, Westermark B (1986) The phenotypic characteristics of simian sarcoma virus-transformed human fibroblasts suggest that the v-*sis* product solely acts as a PDGF receptor agonist in cell transformation. *EMBO J* **5**: 1535-1541

Josephs SF, Ratner L, Clarke MF, Westin EH, Reitz MS, Wong-Staal F (1984) Transforming potential of human c-*sis* nucleotide sequences encoding platelet-derived growth factor. *Science* **225**: 636-639

Martinet Y, Bitterman PB, Mornex J-F, Grotendorst G, Martin GT, Crystal RG (1986) Activated human monocytes express the c-*sis* proto-oncogene and release a mediator showing PDGF-like activity. *Nature* **319**: 158-160

Nistér M, Hammacher A, Mellström K, Siegbahn A, Rönnstrand L, Westermark B, Heldin C-H (1988) A glioma-derived PDGF A chain homodimer has different functional activities from a PDGF AB heterodimer purified from human platelets. *Cell* **52**: 791-799

Nilsson J, Sjölund M, Palmberg L, Thyberg J, Heldin C-H (1985) Arterial smooth muscle cells in primary culture produce a platelet-derived growth factor-like protein. *Proc Natl Acad Sci USA* **82**: 4418-4422

Paulsson Y, Hammacher A, Heldin C-H, Westermark B (1987) Possible positive autocrine feedback in the prereplicative phase of human fibroblasts. *Nature* **328**: 715-717

Perez R, Betsholtz C, Westermark B, Heldin C-H (1987) Frequent expression of growth factors for mesenchymal cells in human mammary carcinoma cell lines. *Cancer Res* **47**: 3425-3429

Rao CD, Igarashi H, Chiu I-M, Robbins KC, Aaronson SA (1986) Structure and sequence of the human c-*sis*/platelet-derived growth factor 2 (SIS/PDGF2) transcriptional unit. *Proc Natl Acad Sci USA* **83**: 2392-2396

Robbins KC, Antoniades HN, Devare SG, Hunkapiller MW, Aaronson SA (1983) Structural and immunological similarities between simian sarcoma virus gene product(s) and human platelet-derived growth factor. *Nature* **305**: 605-608

Robbins KC, Leal F, Pierce JH, Aaronson SA (1985) The v-*sis*/PDGF-2 transforming gene product localizes to cell membranes but is not a secretory protein. *EMBO J* **4**: 1783-1792

Rorsman F, Bywater M, Knott TJ Scott J, Betsholtz C (1988) Structural characterization of the human platelet-derived growth factor A-chain cDNA and gene: alternative exon usage predicts two different precursor proteins. *Mol Cell Biol* **8**: 571-577

Ross R, Raines EW, Bowen-Pope DF (1986) The biology of platelet-derived growth factor. *Cell* **46**: 155-169

Seifert RA, Schwartz SM, Bowen-Pope DF (1984) Developmentally regulated production of platelet-derived growth factor-like molecules. *Nature* **311**: 669-671

Shimokado K, Raines EW, Madtes DK, Barrett TB, Benditt EP, Ross R (1985) A significant part of macrophage-derived growth factor consists of at least two forms of PDGF. *Cell* **43**: 277-286

Söderdahl G, Betsholtz C, Johansson A, Nilsson K, Bergh J (1988) Differential expression of platelet-derived growth factor and transforming growth factor genes in small- and non-small cell human lung cancer cell lines. *Int J Cancer* **41**: 636-641

Stroobant P, Waterfield MD (1984) Purification and properties of porcine platelet-derived growth factor. *EMBO J* **2**: 2963-2967

Swan DC, McBride OW, Robbins KC, Keithley DA, Reddy EP, Aaronson SA (1982) Chromosomal mapping of the simian sarcoma virus *onc* gene analogue in human cells. *Proc Natl Acad Sci USA* **79**: 4691-4695

Waterfield MD, Scrace GT, Whittle N, Stroobant P, Johnsson A, Wasteson Å, Westermark B, Heldin C-H, Huang JS, Deuel TF (1983) Platelet-derived growth factor is structurally related to the putative transforming protein p28^{is} of simian sarcoma virus. *Nature* **304**: 35-39

Westermark B, Johnsson A, Paulsson Y, Betsholtz C, Heldin C-H, Herlyn M, Rodeck U, Koprowski H (1986) Human melanoma cell lines of primary and metastatic origin express the genes encoding the constitutive chains of PDGF and produce a PDGF-like growth factor. *Proc Natl Acad Sci USA* **83**: 7197-7200

Westermark B, Betsholtz C, Johnsson A, Heldin C-H (1987) Acute transformation by simian sarcoma virus is mediated by an externalized PDGF-like growth factor. In: Kjeldgaard NO, Forchhammer J (eds) *Viral Carcinogenesis*. Alfred Benzon Symposium 24, Munksgaard, Copenhagen, pp 445-454

Wolfe LG, Deinhart F, Thiele GH, Rabin H, Kawakami T, Bustad LK (1971) Induction of tumors in marmoset monkeys by simian sarcoma virus, type 1 (Lagothrix): a preliminary report. *JNCI* **47**: 1115-1120

Yarden Y, Escobedo JA, Kuang W-J, Yang-Feng TL, Daniel TO, Tremble PM, Chen EY, Ando ME, Harkins RN, Francke U, Fried VA, Ulrich A, Williams LT (1986) Structure of the receptor for platelet-derived growth factors helps define a family of closely related growth factor receptor. *Nature* **323**: 226-232

Advances in Growth Hormone and Growth Factor Research,
edited by E.E. Müller, D. Cocchi and V. Locatelli
Pythagora Press, Roma-Milano and Springer Verlag, Berlin-Heidelberg © 1989

Biological effects of transforming growth factors

H.L. Moses[1], J. Keski-Oja[2], R.M. Lyons[1], N.J. Sipes[1], C.C. Bascom[1] and R.J. Coffey Jr[1]

[1] Department of Cell Biology, Vanderbilt University School of Medicine, Nashville, Tennessee, U.S.A.; and [2] Department of Virology, University of Helsinki, Helsinki, Finland

Transforming growth factors (TGFs) were originally defined by their biological effects on fibroblastic cells (for review see Goustin et al. 1986). These effects included induction of morphological transformation in monolayer culture and stimulation of colony formation in soft agar. While the early studies with TGFα were somewhat misleading with respect to the function of these factors, they did lead to the purification and cloning of two important growth-regulatory molecules, TGFα and TGFβ. Interestingly, one of these factors (TGFα) is a potent mitogen for a wide variety of cell types, while the other (TGFβ) is the most potent growth-inhibitory polypeptide known for most cell types (Goustin et al. 1986).

TGFα AND ITS RECEPTOR

TGFα was originally described by De Larco and Todaro (1978), who used the term "sarcoma growth factor". It was later shown that the preparation called sarcoma growth factor contained both TGFα and TGFβ (Anzano et al. 1983). TGFα is a 50-amino acid molecule that has some sequence and significant structural homology to epidermal growth factor (EGF) (Marquardt et al. 1984). TGFα binds to the EGF receptor and apparently mediates all of its effects through this EGF-receptor binding; there is no evidence for a TGFα receptor distinct from the EGF receptor. The cell culture effects of TGFα are virtually identical to those of EGF (Anzano et al. 1983). There are some quantitative differences between the biological effects of TGFα and EGF in organ culture and *in vivo* assays, with TGFα tending to be more potent (for review see Derynck, 1986).

TGFα was originally shown to be produced by murine sarcoma virus-transformed mouse 3T3 cells, but not by the nontransformed parent cells (DeLarco and Todaro 1978), and was later found in medium conditioned by human carcinoma cells in culture (Todaro et al. 1980). Subsequently, TGFα was identified in embryonic tissue (Twardzik et al. 1982), resulting in the widespread view that TGFα was an embryonic molecule inappropriately expressed in some cancer cells. It was in conjunction with the discovery of TGFα that the autocrine hypothesis was first published as an explanation for the excessive growth that occurs in neoplastic cells (Sporn and Todaro 1980).

TGFβ AND ITS RECEPTOR

While the confusing terminology would imply otherwise, TGFβ bears little relationship to TGFα. TGFβ was first described as a growth-stimulatory molecule by its ability to induce soft agar colony formation of mouse embryo-derived, fibroblastic AKR-2B cells (Moses et al. 1981) and shortly thereafter by its biological effects in combination with EGF on rat fibroblastic NRK cells (Roberts et al. 1981). Although originally described as being produced by neoplastically transformed cells (Moses et al. 1981; Roberts et al. 1981), it is now known that TGFβ is a highly ubiquitous molecule, and it has been purified from several normal tissues (for review see Roberts et al. 1983). Platelets, which give rise to the TGFβ found in serum (Childs et al. 1982), are the most abundant source for purification of TGFβ (Assoian et al. 1983). TGFβ is released by cells in culture and by platelets in a latent form that is irreversibly activated by acid treatment (Lawrence et al. 1984). The intact, active TGFβ molecule has a molecular weight of 25 K and is composed of two apparently identical disulfide-linked subunits of 12 K (Assoian et al. 1983). The gene for human TGFβ has been cloned, and the amino acid sequence deduced from the cDNA sequence indicates a subunit of 112 amino acids (Derynck et al. 1985). These studies further suggest a precursor encoded in a 390-residue open reading frame where each subunit is encoded by residues 279-390. The precursor is processed by proteolytic cleavage to yield the active molecule (Gentry et al. 1987). The murine TGFβ has also been cloned, and comparison with the human sequence shows an exceptionally high degree of evolutionary conservation (Derynck et al. 1986). A second TGFβ (called TGFβ2 to distinguish it from the originally described TGFβ, now called TGFβ1) has been identified in porcine platelets and bovine bone (Cheifetz et al. 1987). TGFβ2 apparently binds to the same receptor as TGFβ1 and has virtually identical biological activities (Cheifetz et al. 1987). An apparently identical molecule has been identified as an immuno-suppressive agent produced by glioma cells (Wrann et al. 1987). Other molecules with structural and some sequence homology to TGFβ have been purified or identified by gene cloning and cDNA sequencing. These include Müllerian inhibiting substance (Cate et al. 1986), inhibins (and their β-chain dimers, activins) (Mason et al. 1985), and the *Drosophila* decapentaplegic gene (Padgett et al. 1987). Müllerian inhibiting substance and inhibins/activins have receptors that are distinct from the TGFβ receptor (Ying et al. 1986; Coughlin et al. 1987).

TGFβ has its own specific cell membrane receptors which, like the TGFβ molecule itself, are highly ubiquitous (Tucker et al. 1984a). Specific binding of [^{125}I]-TGFβ to various mesenchymal and epithelial cells in primary and secondary cultures and continuous cell lines, both normal and neoplastic, has been reported. The dissociation constants reported have ranged from 25 to 140 pM and receptor number per cell from 10,000 to 40,000 (Frolik et al. 1984; Tucker et al. 1984a; Massague and Like 1985). The TGFβ receptor is apparently quite different from other growth factor receptors (Massague 1985). Recently, at least two types of receptors for TGFβ have been proposed on the basis of chemical cross-linking studies (Cheifetz et al. 1987). No kinase or other enzymatic activities have been reported for the TGFβ receptor thus far.

TGFβ has been reported to have numerous, diverse biological activities. It is mitogenic only for fibroblastic and selected other mesenchymal cells (Moses et al. 1985). TGFβ is a potent stimulator of extracellular matrix production by increasing synthesis of matrix components (Ignotz and Massague 1986; Ignotz et al. 1987; Raghow et al. 1987) and by diminishing matrix degradation through stimulating protease inhibitor (Laiho et al. 1986) and decreasing protease production (Matrisian et al. 1986; Edwards et al. 1987). TGFβ's potent chemotactic effect on fibroblasts (Postlethwaite et al. 1987) is another action that may contribute to its ability to stimulate connective tissue formation (Roberts et al. 1986). TGFβ also inhibits differentiation of adipocytes and myoblasts (Massague 1987).

We have demonstrated that the growth inhibitor originally described by Holley et al. (1978) from African green monkey (BSC-1) cells is similar, if not identical, to human platelet-derived TGFβ (Tucker et al. 1984b). Holley et al. (1978) had suggested that autocrine inhibition by the growth inhibitor was important in the growth regulation of the BSC-1 cells. In addition to demonstrating apparent identity, it was further shown that human platelet-derived TGFβ is a highly potent inhibitor of the BSC-1 and CCL-64 (mink lung) epithelial cells. This led to studies of inhibitory effects on a variety of normal cells, demonstrating that TGFβ is the most potent growth-inhibitory polypeptide known for a wide variety of cell types including epithelial, lymphoid, and myeloid cells (Moses and Leof 1986).

AUTOCRINE CONTROL OF KERATINOCYTE PROLIFERATION BY TGFs

In studies of the role of TGFs on normal epithelial cells we have utilized secondary cultures of human skin keratinocytes grown in low-calcium, serum-free medium (Shipley et al. 1986). These cells require EGF/TGFα for proliferation and retain the ability to differentiate under high-calcium conditions. Recent studies from our laboratory have demonstrated that TGFα is produced by normal neonatal and adult keratinocytes, both *in vitro* and *in vivo* (Coffey et al. 1987). It was further shown that TGFα expression in these cells was dependent upon the presence of EGF, and that both EGF and TGFα induced significant levels of TGFα mRNA; EGF was demonstrated to stimulate the release of TGFα protein into the culture medium. The autoinduction of TGFα may be a mechanism of signal amplification for fine-tuning of the proliferative response. These studies further demonstrate that TGFα production does occur in normal adult epithelial cells that are also capable of responding to the factor, suggesting the possibility of normal autocrine regulation of cell proliferation. These findings do not exclude the possibility that abnormal autocrine stimulation by TGFα could be involved in neoplastic transformation. Unregulated production of TGFα could be important in this process. With the normal skin keratinocytes, the production of TGFα is clearly regulated, since the cells produce very little TGFα when grown in the absence of EGF for 48 h (Coffey et al. 1987).

We have also demonstrated that TGFβ is a potent growth inhibitor for secondary cultures of human foreskin keratinocytes (Moses et al. 1985; Shipley et al. 1986). The keratinocytes were found to be reversibly inhibited in their growth by TGFβ, with the majority of cells blocked in the G_1 phase of the cell cycle. Half-maximal inhibition was obtained at 12 pM TGFβ. There was no induction of any of several differentiation markers examined, indicating that the mechanism of growth inhibition is not through induction of terminal differentiation. The keratinocytes were also demonstrated to synthesize and release TGFβ into the medium, with confluent cultures producing as much as 80 pM per 24 h (Shipley et al. 1986). However, all of the detectable TGFβ released was in a latent form, detectable only after acid treatment of the medium, similar to observations with many other cell types. Whether the latent TGFβ activates spontaneously or can be activated by the cells, with the active material subsequently binding to cell surface receptors, is not known. However, since the keratinocytes have receptors for TGFβ (Shipley et al. 1986), are capable of responding to the factor, and secrete relatively large quantities into conditioned medium, the possibility of negative autocrine regulation by TGFβ in keratinocytes must be entertained as a viable possibility.

PROPOSED MODEL FOR AUTOCRINE REGULATION BY TGFs

The data summarized above demonstrate that control of keratinocyte proliferation may involve both positive and negative polypeptide regulators that bind to cell surface membrane receptors. The keratinocytes produce the same peptides that stimulate or inhibit their proliferation. TGFα stimulates proliferation and the keratinocytes produce TGFα This production is autoregulated. TGFβ reversibly inhibits proliferation of keratinocytes, which also secrete this factor. It is hypothesized that autocrine regulation by both stimulators and inhibitors of proliferation is an important physiologic mechanism of regulation of keratinocytes proliferation. A similar mechanism may be involved in the regulation of proliferation in other cell types as well, and may involve peptide growth factors and growth inhibitors other than TGFa and TGFβ. The presence of opposing regulatory pathways should allow for a more precise control of the very important process of cell proliferation than a single on/off stimulatory pathway provided by the growth factors. This suggests that the growth-inhibitory polypeptides may play as important a role in the control of cell proliferation as the growth -stimulatory factors do.

MECHANISMS OF GROWTH INHIBITION BY TGFβ

The mechanisms by which TGFβ inhibits cell proliferation are largely unknown. The growth-inhibitory effects of TGFβ do not appear to be secondary to cytotoxicity, since the inhibition is reversible in most circumstances. Induction of terminal differentiation, although reported in one cell type (Masui et al. 1986), does not appear to be a general phenomenon and does not occur in human foreskin keratinocytes (Shipley et al. 1986). It seems likely that TGFβ is primarily a growth inhibitor and that stimulation of fibroblastic cells is fortuitous through the induction of c-*sis* and autocrine activity by PDGF, which is the direct mitogen (Leof et al. 1986). TGFβ does not appear to interfere with growth factor-receptor interactions or with transduction of the growth factor signal (Like and Massague 1986; Coughlin et al. 1987). TGFβ like the interferons and tumor necrosis factor, does cause a reduction in c-*myc* expression (Ferdandez-Pol et al. 1987; Takehara et al. 1987). It is hypothesized that TGFβ, through some unknown signal transduction mechanism, selectively reduces the expression of a gene or genes necessary for proliferation.

CHANGES IN AUTOCRINE REGULATION IN NEOPLASTIC TRANSFORMATION

Alterations in the TGFα autocrine stimulatory pathway in neoplastic transformation have been proposed (Todaro et al. 1980; Coffey et al. 1987) and could result in an increased proliferative potential. The role of TGFβ in neoplastic transformation of epithelial cells is probably very different from that involved in fibroblastic cells, where autocrine stimulation may occur (Keski-Oja et al. 1987). We have demonstrated that a squamous carcinoma cell line has lost the inhibitory response to TGFβ exhibited by normal keratinocytes (Shipley et al. 1986), and similar results have been obtained by Masui et al. (1986) in bronchial-derived squamous carcinoma cell lines. The loss of the normal inhibitory response to TGFβ in epithelial cells could also result in an enhanced proliferative potential and thereby account for part of the neoplastic phenotype. Loss of the ability to activate the latent TGFβ released by most cells, including platelets, could also result in loss of the normal inhibitory effects of TGFβ.

REFERENCES

Anzano MA, Roberts AB, Smith JM, Sporn MB, DeLarco JE (1983) Sarcoma growth factor from conditioned medium of virally transformed cells is composed of both type-alpha and type-beta transforming growth factors. *Proc Natl Acad Sci USA* **80**: 6264-6268

Assoian RK, Komoriya A, Meyers CA, Miller DM, Sporn MB (1983) Transforming growth factor-beta in human platelets. Identification of a major storage site, purification, and characterization. *J Biol Chem* **258**: 7155-7160

Cate RL, Mattaliano RJ, Hession C, Tizard R, Farber NM, Cheung A, Ninfa EG, Frey AZ, Gash DJ, Chow EP, Fisher A, Bertonis JM, Torres G, Wallner BP, Ramachandran KL, Ragin RC, Managanaro TF, MacLaughlin DT, Donahoe PK (1986) Isolation of the bovine and human genes for Müllerian inhibiting substance and expression of the human gene in animal cells. *Cells* **45**: 685-698

Cheifetz S, Weatherbee JA, Tsang ML-S, Anderson JK, Mole JE, Lucas R, Massague J (1987) The transforming growth factor-beta system, a complex pattern of cross-reactive ligands and receptors. *Cells* **48**: 409-415

Childs CB, Proper JA, Tucker RF, Moses HL (1982) Serum contains platelet-derived transforming growth factor. *Proc Natl Acad Sci USA* **79**: 5312-5316

Coffey RJ, Derynck R, Wilcox JN, Bringman TS, Goustin S, Moses HL, Pittelkow MR (1987) Production and auto-induction of transforming growth factor-beta in human keratinocytes. *Nature* **328**: 817-820

Coughlin JP, Donahoe PK, Budzik GP, MacLaughlin DT (1987) Müllerian inhibiting substance blocks autophosphorylation of the EGF receptor by inhibiting tyrosine kinase. *Mol Cell Endocrinol* **49**: 75-86

DeLarco JE, Todaro GJ (1978) Growth factors from murine sarcoma virus-transformed cells. *Proc Natl Acad Sci USA* **75**: 4001-4005

Derynck R (1986) Transforming growth factor-alpha: structure and biological activities. *J Cell Biochem* **32**: 293-304

Derynck R, Jarrett JA, Chen EY, Eaton DH, Bell JR, Assoian RK, Roberts AB, Sporn MB, Goeddel DV (1985) Human transforming growth factor-beta complementary DNA sequence and expression in normal and transformed cells. *Nature* **316**: 701-705

Derynck R, Jarrett JA, Chen EY, Goeddel DV (1986) The murine transforming growth factor-beta precursor. *J Biol Chem* **261**: 4377-4379

Edwards DR, Murphy G, Reynolds JJ, Whitham SE, Docherty JP, Angel P, Heath JK (1987)

Transforming growth factor beta modulates the expression of collagenase and metal-loproteinase inhibitor. *EMBO J* **6**: 1899-1904

Fernandez-Pol JA, Talked VD, Klos DJ, Hamilton PD (1987) Suppression of the EGF-dependent induction of c-*myc* proto-oncogene expression by transforming growth factor-beta in a human breast carcinoma cell line. *Biochem Biophys Res Commun* **144**: 1197-1205

Frolik CA, Wakefield LM, Smith DM, Sporn MB (1984) Characterization of a membrane receptor for transforming growth factor-beta in normal rat kidney fibroblasts. *J Biol Chem* **259**: 10995-11000

Gentry LE, Webb NR, Lim GJ, Brunner AM, Ranchalis JE, Twardzik DR, Lioubin MN, Marquardt H, Purchio AF (1987) Type-1 transforming growth factor-beta: amplified expression and secretion of mature and precursor polypeptides in chinese hamster ovary cells. *Mol Cell Biol* **7**: 4318-4327

Goustin AS, Leof EB, Shipley GD, Moses HL (1986) Perspective in cancer research: growth factors and cancer. *Cancer Res* **46**: 1015-1029

Holley RW, Armour R, Baldwin JH (1978) Density-dependent regulation of growth of BSC-1 cells in cell culture: growth inhibitors formed by the cells. *Proc Natl Acad Sci USA* **75**: 1864-1866

Ignotz RA, Massague J (1986) Transforming growth factor-beta stimulates the expression of fibronectin and collagen and their incorporation into the extracellular matrix. *J Biol Chem* **261**: 4337-4345

Ignotz RA, Endo R, Massague J (1987) Regulation of fibronectin and type-I collagen mRNA levels by transforming growth factor-beta. *J Biol Chem* **262**: 6443-6446

Keski-Oja J, Lyons RM, Moses HL (1987) Immunodetection and modulation of cellular growth with antibodies against native transforming growth factor-beta. *Cancer Res* **47**: 6451-6458

Laiho M, Saksela O, Andreasen PA, Keski-Oja J (1986) Enhanced production and extracel-lular deposition of the endothelial-type plasminogen activator inhibitor in cultured human lung fibroblasts by transforming growth factor-beta. *J Cell Biol* **103**: 2403-2410

Lawrence DA, Pircher R, Kryceve-Martinerie C, Jullien P (1984) Normal embryo fibrob-lasts release trasforming growth factors in a latent form. *J Cell Physiol* **121**: 184-188

Leof EB, Proper JA, Goustin AS, Shipley GD, DiCorleto E, Moses HL (1986) Induction of c-*sis* mRNA and activity similar to platelet-derived growth factor by transforming growth factor-beta: a proposed model for indirect mitogenesis involving autocrine activity. *Proc Natl Acad Sci USA* **83**: 2453-2457

Like B, Massague J (1986) The antiproliferative effect of type beta transforming growth factor occurs at a level distal from receptors for growth-activating factors. *J Biol Chem* **261**: 13426-13429

Marquardt H, Hunkapiller MW, Hood LE, Todaro GJ (1984) Rat transforming growth factor type 1: structure and relation to epidermal growth factor. *Science* **223**: 1079-1082

Mason AJ, Hayflick JS, Ling N, Esch F, Ueno N, Ying Y, Guillemin R, Niall H, Seeburg PH (1985) Complementary DNA sequences of ovarian follicular fluid inhibin show precursor structure and homology with transforming growth factor-beta. *Nature* **318**: 659-663

Massague J (1985) Subunit structure of a high-affinity receptor for type-beta transforming growth factor. Evidence for a disulfide-linked glycosylated receptor complex. *J Biol Chem* **260**: 7059-7066

Massague J (1987) The TGF-beta family of growth and differentiation factors. *Cell* **49**: 437-438

Massague J, Like B (1985) Cellular receptors for type-beta transforming growth factor. Ligand binding and affinity labeling in human and rodent cell lines. *J Biol Chem* **260**: 2636-2645

Masui T, Wakefield LM, Lechner JF, La Veck MA, Sporn B, Harris CC (1986) Type-beta transforming growth factor is the primary differentiation-inducing serum factor for normal human bronchial epithelial cells. *Proc Natl Acad Sci USA* **83**: 2438-2442

Matrisian LM, Leroy P, Ruhlmann C, Gesnel M-C, Breathnach R (1986) Isolation of the oncogene and epidermal growth factor-induced transin gene: complex control in rat fibroblasts. *Mol Cell Biol* **6**: 1679-1686

Moses HL, Leof EB (1986) Transforming growth factor-beta. In: Kahn P, Graf T (eds) *Oncogene and Growth Control.* Springer, Berlin, Heidelberg, New York, pp. 51-57

Moses HL, Branum EB, Proper JA, Robinson RA (1981) Transforming growth factor production by chemically transformed cells. *Cancer Res* **41**: 2842-2848

Moses HL, Tucker RF, Leof EB, Coffey RJJ, Halper J, Shipley GD (1985) Type-beta transforming growth factor is a growth stimulator and a growth inhibitor. In: Feramisco J, Ozanne B, Stiles C (eds) *Growth Factors and Transformation. Cancer Cells,* vol 3. Cold Spring Harbor Press, Cold Spring Harbor, pp 65-71

Padgett RW, St Johnston RD, Gelbart WM (1987) A transcript from a *Drosophila*-pattern gene predicts a protein homologous to the transforming growth factor-beta family. *Nature* **325**: 81-84

Postlethwaite AE, Keski-Oja J, Moses HL, Kang AH (1987) Stimulation of the chemotactic migration of human fibroblasts by transforming growth factor-beta. *J Exp Med* **165**: 251-256

Raghow R, Postlethwaite AE, Keski-Oja J, Moses HL, Kang H (1987) Transforming growth factor-beta increases steady-state levels of type-I procollagen and fibronectin mRNA posttranscriptionally in cultured human dermal fibroblasts. *J Clin Invest* **79**: 1285-1288

Roberts AB, Anzano MA, Lamb LC, Smith JM, Sporn MB (1981) New class of transforming growth factors potentiated by epidermal growth factor: isolation from non-neoplastic tissue. *Proc Natl Acad Sci USA* **78**: 5339-5343

Roberts AB, Frolik CA, Anzano MA, Sporn MB (1983) Transforming growth factors from neoplastic and non-neoplastic tissues. *Fed Proc* **42**: 2621-2626

Roberts AB, Sporn MB, Assoian RK, Smith JM, Roche NS, Wakefield LM, Heine UI, Liotta LA, Falanga V, Kehrl H, Fauci AS (1986) Transforming growth factor type beta: rapid induction of fibrosis and angiogenesis *in vivo* and stimulation of collagen formation *in vitro. Proc Natl Acad Sci USA* **83**: 4167-4171

Shipley GD, Pittelkow MR, Wille JJJ, Scott RE, Moses L (1986) Reversible inhibition of normal human prokeratinocyte proliferation by type-beta transforming growth factor-growth inhibitor in serum-free medium. *Cancer Res* **46**: 2068-2071

Sporn MB, Todaro GJ (1980) Autocrine secretion and malignant transformation of cells. *N Engl J Med* **303**: 878-880

Takehara K, LeRoy EC, Grotendorst GR (1987) TGF-beta inhibition of endothelial cell proliferation: alteration of EGF binding and EGF-induced growth-regulatory (competence) gene expression. *Cell* **49**: 415-422

Todaro GJ, Fryling C, DeLarco JE (1980) Transforming growth factors produced by certain human tumor cells: polypeptides that interact with epidermal growth factor receptors. *Proc Natl Acad Sci USA* **77**: 5258-5262

Tucker RF, Branum EL, Shipley GD, Ryan RJ, Moses HL (1984a) Specific binding to cultured cells of ^{125}I-labeled transforming growth factor type-beta from human platelets.

Proc Natl Acad Sci USA **81**: 6757-6761

Tucker RF, Shipley GD, Moses HL, Holley RW (1984b) Growth inhibitor from BSC-1 cells closely related to the platelet type-beta transforming growth factor. *Science* **226**: 705-707

Twardzik DR, Ranchalis JE, Todaro GJ (1982) Mouse embryonic transforming growth factors related to those isolated from tumor cells. *Cancer Res* **42**: 590-593

Wrann M, Bodmer S, de Martin R, Siepl C, Hofer-Warbinek R, Frei K, Hofer E, Fontana A (1987) T cell suppressor factor from human glioblastoma cells is a 12.5-kd protein closely related to transforming growth factor-beta. *EMBO J* **6**: 1633-1636

Ying S-Y, Becker A, Ling N, Ueno N, Guillemin R (1986) Inhibin and beta-type transforming growth factor (TGF-beta) have opposite modulating effects on the follicle-stimulating hormone (FSH)-induced aromatase activity of cultured rat granulosa cells. *Biochem Biophys Res Commun* **136**: 969-975

NEURAL REGULATION OF GROWTH
HORMONE SECRETION

Advances in Growth Hormone and Growth Factor Research,
edited by E.E. Müller, D. Cocchi and V. Locatelli
Pythagora Press, Roma-Milano and Springer Verlag, Berlin-Heidelberg © 1989

Ontogeny of growth hormone-releasing factor and its role in fetal and neonatal growth*

W.B. Wehrenberg[1] and R. C. Gaillard[2]

[1]*Department of Health Sciences, University of Wisconsin - Milwaukee, Milwauke, Wisconsin, U.S.A.; and* [2]*Clinique Medicale and Division d'Endocrinologie Hôpital Cantonal Universitaire de Genève, Geneva, Switzerland*

The role of growth hormone (GH) in controlling postnatal growth is unquestioned (Underwood and Van Wyk 1981). The secretion of this hormone is in turn mediated by hypothalamic growth hormone-releasing factor (GRF; Guillemin et al. 1982; Rivier et al. 1982) and growth hormone-releasing inhibiting factor, also known as somatostatin (Brazeau et al. 1973). Under normal conditions the release of GH is episodic in nature. The variations in plasma GH concentrations cannot be explained by corresponding hypothalamic release of somatostatin, since infusion of antiserum raised against somatostatin does not abolish the pulses of GH secretion (Ferland et al. 1976; Terry and Martin 1981). It has now been demonstrated that the administration of antiserum raised against GRF abruptly abolishes GH pulses (Wehrenberg et al. 1982; Chihara et al. 1984). Furthemore, direct evidence suggests that GRF is released by the hypothalamus in an episodic fashion (Plotsky and Vale 1985). Thus, while somatostatin plays a role in modulating GH secretion, it appears that GRF is responsible for elevated GH concentrations and its pulsatile pattern of secretion. In light of this relationship and the clear role GH has in somatic growth, we became interested in the role of GRF in regulating somatic growth of the fetus and neonate. This review covers our current knowledge of the ontogeny of GRF and experiments we have conducted to evaluate its role in fetal and neonatal growth in the rat.

* This research was supported by the University of Wisconsin Foundation/Shaw Research Fund, and the Swiss National Research Foundation, grant no. 3.814-084.

ONTOGENY OF GROWTH HORMONE-RELEASING FACTOR

In Man

Bresson et al. (1984) reported the presence of GRF immunoreactivity in both perikarya and nerve processes in the infundibular (arcuate) nucleus of the human hypothalamus as early as the 18th week of gestation. Bloch and colleagues (1984) reported a slightly later onset for the appearance of GRF staining, with immunoreactive cells appearing around the 29th week of fetal life and cell fibers around the 31st week. Initially, the cells show a typical neuroblastic aspect. They are quite small (10 μm in diameter), without processes and a large nucleus. In older fetuses and neonates, the cells are located in the same area but with more intense immunoreactive staining. Branched processes are observed at 9 weeks postnatally. The late appearance of GRF in the human fetal hypothalamus is in contrast to the earlier appearance of the other hypophysiotropic factors; gonadotropin-releasing hormone (GnRH) is detected as early as the 11th week of gestation (Bloch 1978; Bugnon et al. 1978), somatostatin during the 14th week (Bugnon et al. 1977), and corticotropin-releasing factor during the 16th week (Bugnon et al. 1982). All of these systems appear to be fully mature, with established connections in the median eminence as early as the 16th week of gestation, long before the GRF system is established.

In the Rat

Numerous investigators have reported on the ontogeny of GRF in the rat. Ishikawa et al. (1986) and Daikoku et al. (1985) observed GRF immunoreactivity as early as day 18 of fetal life and De Gennaro et al. (1986) from day 20 of gestation on. GRF was observed only in nerve terminals and fibers during fetal life. Perikarya staining for GRF were not observed until days 1-2 of age. Radioimmunoassay of hypothalamic GRF has confirmed these results. GRF is undetectable on day 17 of gestation but is increased to 30-65 pg/hypothalamus during days 18-20 (Jansson et al. 1987). Both the histochemical and radioimmunoassay data indicate that GRF concentrations increase following birth. In contrast to human beings, GRF appears to exhibit parallel development to the other hypophysiotropic hormones in the rat. The first histochemical appearance of GnRH is on day 18 of gestation (Kawano et al. 1980), and somatostatin appears on day 17 (Daikoku et al. 1983).

The ontogeny of GRF in the rat, coupled with the fact that direct vascular connections between the hypothalamus and the pituitary are established by day 18 of gestation, suggest that GRF can reach the fetal pituitary. We have shown that pituitary somatotrophs obtained from 18-day-old fetuses are responsive to GRF stimulation (Baird et al. 1984) and that GRF can be isolated from the rat placenta (Baird et al. 1985). These data suggest that the high GH concentrations observed during the later stages of gestation in the fetus are regulated by hypothalamic GRF. Furthermore, we have observed that GH secretion in the 1-day-old rat is GRF dependent (Wehrenberg 1986). Since the neuroendocrine system in the rat at birth is less developed than that of human beings, the dependence of GH secretion on GRF implies that a similar situation may exist in man during the later stages of gestation. We have conducted a series of experiments in the rat to evaluate the possible role of GRF and GH in fetal and neonatal growth.

HYPOTHALAMIC REGULATION OF MATERNAL GROWTH HORMONE SECRETION DURING PREGNANCY AND ITS INVOLVEMENT IN FETAL GROWTH

The episodic secretion of GH is evident throughout pregnancy in the rat, with the intervals between spontaneous pulses similar to those in nonpregnant animals (Saunders et al. 1976; Klindt et al. 1981). However, the concentrations of GH increase dramatically on days 18-21 of pregnancy, with both trough and peak values increasing. On the day of parturition there is a distinct increase in GH concentrations which coincides with the onset of delivery. We have sought to evaluate the mechanisms by which the hypothalamus regulates maternal growth hormone secretion during pregnancy and at parturition. In light of the significant role GH has in controlling postnatal growth, we have also investigated the role of maternal GH secretion on fetal development and viability.

Female rats were paired with males and checked daily for mating. Day O of pregnancy was designated the day sperm was observed in a vaginal lavage. On day 10 of gestation, the pregnant females were outfitted with a chronic indwelling venous catheter while anesthetized with sodium pentobarbital (50 mg/kg, ip). Details of the construction of the catheters and the surgical procedures used are described elsewhere (Wehrenberg et al. 1984). Following surgery, the animals were housed individually in isolation chambers. On days 17 through 22 of gestation the catheters were flushed and the females received a 0.2 ml iv injection of normal rabbit serum (NRS) or antiserum raised against GRF (GRF-ab) at 8 a.m. Blood sampling was initiated at 9 a.m., and blood samples (0.25 ml) were drawn at 20-min intervals for 3 h. The samples were, centrifuged, plasma was taken for the measurement of GH by RIA, and the red blood cells were resuspended in saline and returned to the animals. Due to the fact that animals gave birth on different days of gestation, the data have been normalized by expressing results in an antedated fashion from the day of parturition. Nonpregnant females and females 4-6, 2 and 1 day prior to parturition and on the day of parturition were studied. At least six females were studied in each treatment group. The antisera against

GRF were prepared by immunizing rabbits with a mixture of synthetic rat GRF and methylated bovine serum albumin in Freund's adjuvant as previously described (Benoit et al. 1982). The antisera are highly specific for rat GRF and are directed toward the carboxyl terminal of the molecule. Further details of the antisera have been published elsewhere (Wehrenberg 1986). GH concentrations were determined by radioimmunoassay using reagents provided by the NIH, except for the first antiserum, which was kindly provided by Dr. Y. Sinha (Sinha 1972b). The reference standard was NIH RP-2. Within and between assay variations averaged 8% and 14% respectively. Data are expressed as the mean ±SEM. Significant treatment effects were detected by analysis of variance, with consideration given for repeated measures; differences between means were then detected by the Newman-Keuls method (Winer 1971).

The results show that plasma GH concentrations demonstrated a pulsatile pattern of secretion in the normal and pregnant females treated with NRS (Figs.1-5). Females studied 4-6 days prior to parturition had plasma GH concentrations in the 20-60 ng/ml range, with clear evidence of episodic peaks (Fig. 2). Animals studied 2 days prior to parturition also demonstrated pulsatile GH secretion. The

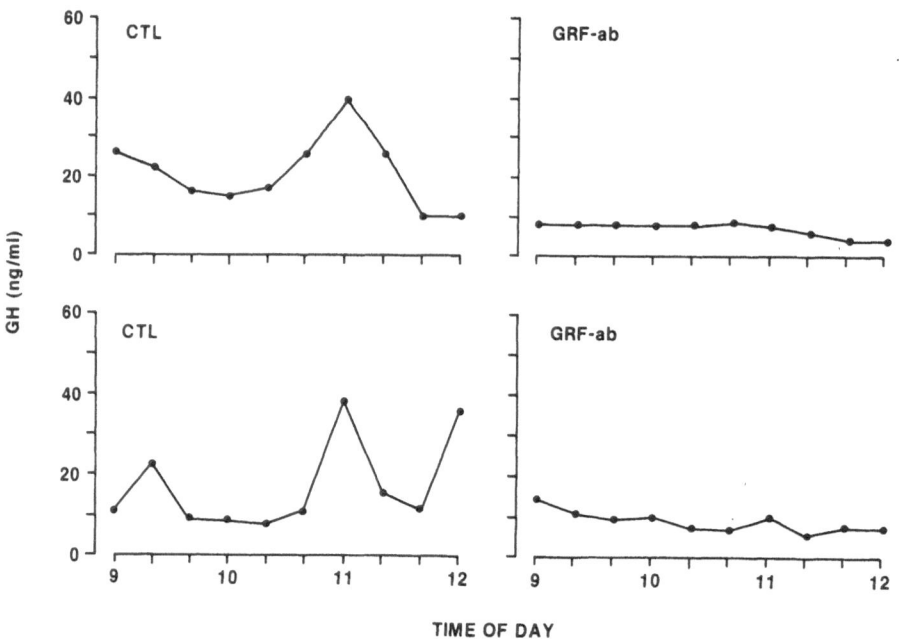

Fig. 1. Plasma GH concentrations in normal female rats. Control (*CTL*) females received 0.2 ml normal rabbit serum iv 1 h before the initiation of blood sampling, while treated females received 0.2 ml antiserum raised against growth hormone-releasing factor (GRF-ab).

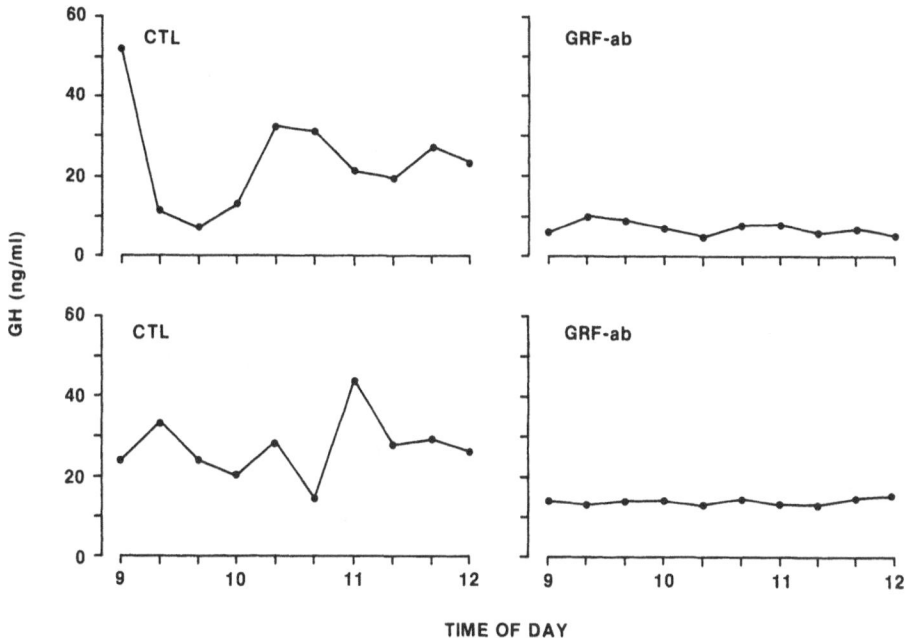

Fig. 2. Plasma GH concentrations in pregnant rats 4-6 days prior to parturition. Control (*CTL*) females received 0.2 ml normal rabbit serum iv 1 h before the initiation of blood sampling, while treated females received 0.2 ml antiserum raised against growth hormone-releasing factor (GRF-ab).

pattern of secretion was similar to that observed in females studied 4-6 days prior to parturition, although there was a slight tendency for the mean GH concentrations to be higher (Fig. 3 and Table 1). One day prior to parturition, the GH concentrations in NRS-treated animals were still pulsatile in nature, but the overall values were significantly higher (P<0.05) than those observed 4-6 days prior to parturition (Fig. 4 and Table 1). Plasma GH concentrations on the day of parturition were quite variable. It was difficult to assess pulsatile GH secretion at this time, since a large increase in GH concentrations occurred at the time of the first birth (Fig. 5). The overall GH concentrations were significantly higher (P<0.05) than those observed 4-6 days prior to parturition.

In contrast to the pulsatile nature of GH secretion and the increasing GH concentrations observed as gestation progressed in the NRS-treated animals, the GH concentrations observed in the GRF-ab treated females were low and constant over time. Figures 1 through 5 reflect examples of the patterns and concentrations of GH observed in normal and pregnant females treated with the GRF-ab. In general, the GH concentrations averaged less than 10 ng/ml, and there was little evidence of a pulsatile pattern of secretion. The data indicate that GRF-ab-treated

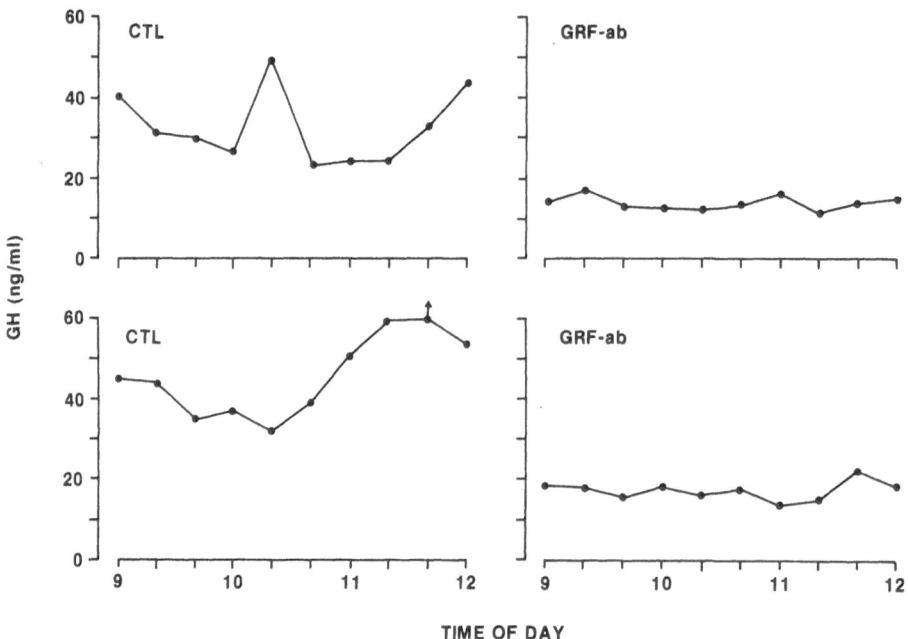

Fig. 3. Plasma GH concentrations in pregnant rats 2 days prior to parturition. Control (*CTL*) females received 0.2 ml normal rabbit serum iv 1h before the initiation of blood sampling, while treated females received 0.2 ml antiserum raised against growth hormone-releasing factor (GRF-ab). *Arrow* associated with data point indicates GH concentrations greater than 60 ng/ml.

Table 1. Plasma GH concentrations in pregnant female rats treated with normal rabbit serum (NRS) or antiserum raised against GRF (GRF-ab) on various days of gestation

Treatment	GH(ng/ml) concentration days prior to parturiton			
	4–6	2	1	0
NRS	18.5 ± 1.3	30.1 ± 2.6	42.5 ± 4.5[a,b]	35.4 ± 4.9[a,b]
GRF-ab	11.1 ± 0.6	19.1 ± 0.8	14.3 ± 0.8	17.9 ± 1.6

All values expressed as mean ± SEM.
[a] $P<0.05$ NRS days 4–6 vs NRS day 1 and day 0.
[b] $P<0.05$ NRS vs GRF-ab on the same day.

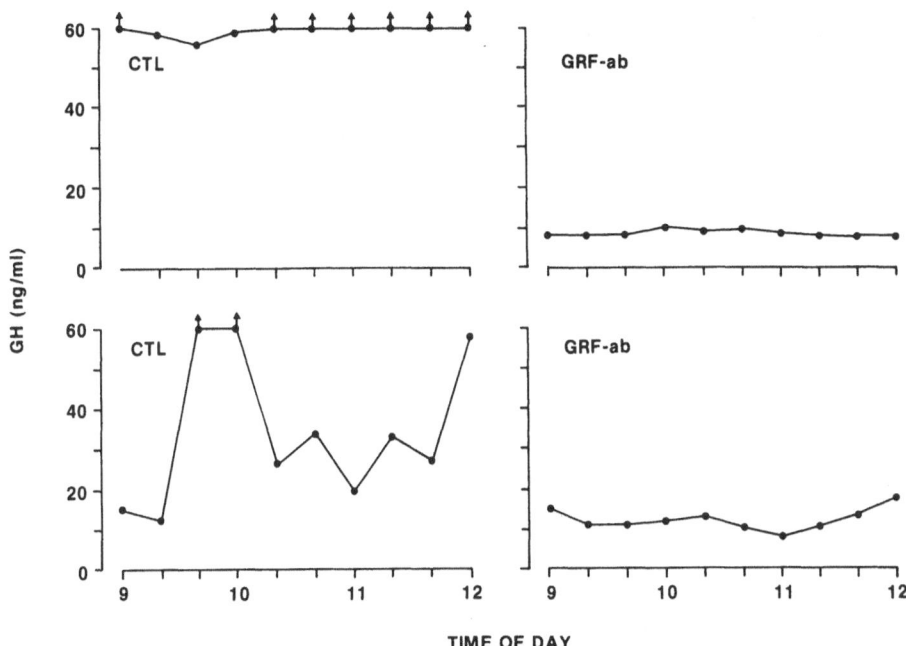

Fig. 4. Plasma GH concentrations in pregnant rats 1 day prior to parturition. Control (*CTL*) females received 0.2 ml normal rabbit serum iv 1 h before the initiation of blood sampling, while treated females received 0.2 ml antiserum raised against growth hormone-releasing factor (GRF-ab). *Arrows* associated with data points indicate GH concentrations greater than 60 ng/ml.

animals showed a slight increase in GH concentrations on the day of parturition. This appeared to coincide with the time of the first birth. The litter size of the GRF-ab treated and control females was similar (Table 2). In spite of the remarkable difference in GH concentrations between the two groups of pregnant females, the onset of parturition and the birth weight of the pups born were also not different. The pups born to the control females and GRF-ab-treated females averaged 5.5-5.7 g, regardless of treatment (Table 2).

The dynamics of GH secretion in normal rats studied throughout the estrus cycle and in pregnant rats have been reported previously (Saunders et al. 1976; Klindt et al. 1981). Our studies have confirmed the earlier reports that GH concentrations increase during the later stages of pregnancy in the rat. Mean GH concentrations increased 11 ng/ml per day from days 18 trough 21 of gestation in control animals. The present studies also investigated the mechanisms of hypothalamic regulation of episodic GH secretion in the female rat, particularly during the last 3-4 days of gestation, when GH concentrations increase dramatically. The present data clearly show that, regardless of reproductive condition, the

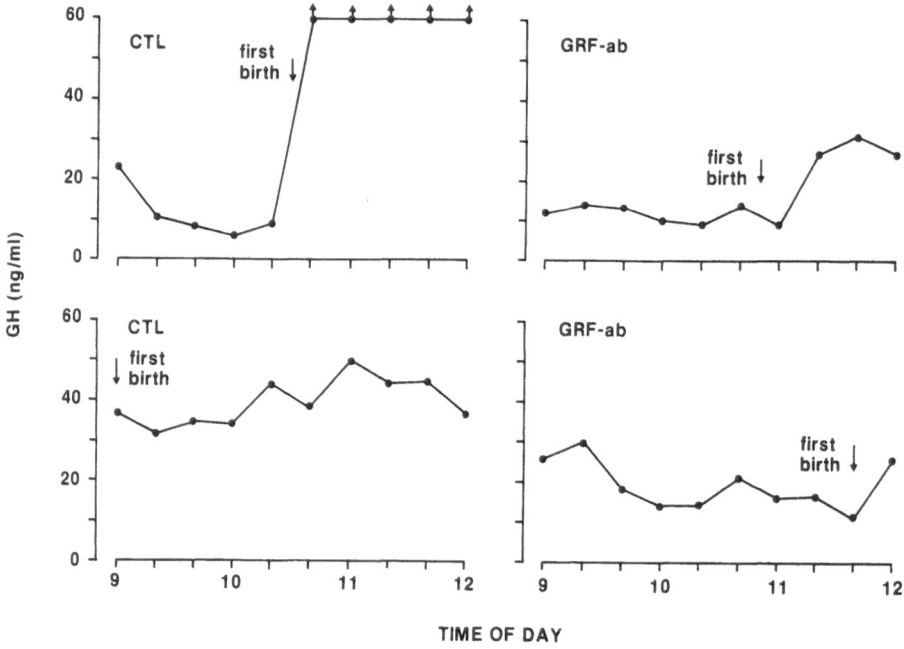

Fig. 5. Plasma GH concentrations in pregnant rats on the day of parturition. Control (*CTL*) females received 0.2 ml normal rabbit serum iv 1 h before the initiation of blood sampling, while treated females received 0.2 ml antiserum raised against growth hormone-releasing factor (GRF-ab). *Arrows* associated with data points indicate GH concentrations greater than 60 ng/ml. The time of the *first birth* is indicated.

Table 2. Vital statistics of pups born from female rats treated with normal rabbit serum (NRS) or antiserum raised against GRF (GRF-ab) on days 16 through 21 of gestation

Treatment	Length of gestation	Litter size	Birth weight	Plasma GH concentrations (ng/ml)[a]
NRS	21.0 ± 0.3	13.4 ± 0.9	5.55 ± 0.06	44.9 ± 3.1*
GRF-ab	21.5 ± 0.4	11.8 ± 0.7	5.73 ± 0.07	27.0 ± 3.7

All values expressed as mean ± SEM.
[a] Neonatal plasma GH concentrations determined from samples taken 24 h after birth.
* $P<0.01$ NRS- vs GRF-ab-treated.

administration of GRF-ab inhibits the episodic release of GH by the pituitary in the female rat. This results in plasma GH concentrations which are low to non-detectable. The inhibition of maternal GH secretion in GRF-ab treated animals during late gestation and parturition does not alter the length of the gestation, the litter size, or the weight of the pups as compared with control treated animals. These results lead us to conclude that maternal GH plays an insignificant role in the somatic development of the fetus and in the events leading to normal parturition.

HYPOTHALAMIC REGULATION OF MATERNAL GROWTH HORMONE AND PROLACTIN SECRETION DURING LACTATION AND THEIR IN-VOLVEMENT IN NEONATAL GROWTH

The importance of lactation in neonatal growth is well recognized, yet the hypothalamic regulation of the hormones involved is poorly understood. The suckling stimulus in lactating rats is known to cause a rapid increase in serum GH and prolactin concentrations (Saunders et al. 1976; Miki et al. 1981; Riskind et al. 1984). There is evidence to suggest that the neuroendocrine mechanisms regulating the release of these two hormones may have characteristics in common. First, GH and prolactin have evolved from one common precursor, and although the hormones are distinct, there is the possibility that under certain circumstances the neuroendocrine secretory mechanisms may have remained common. Second, a common level of control, i.e., the endogenous opioid peptide system, has been suggested for the suckling-induced increase of both GH and prolactin (Miki et al. 1981; Riskind et al. 1984; Selmanoff and Gregerson 1986). The present studies were undertaken to determine if the neuroendocrine mechanisms regulating the increase in GH and prolactin are common or distinct.

Female rats were paired with males. During late gestation the pregnant females were transferred to individual cages with nesting material. On the first day post-partum, litter size was adjusted to eight pups per mother. On day 7 or 8 post-partum, females were outfitted with an indwelling venous catheter while anesthetized with ether. The females were returned to their litters approximately 15 min following the termination of surgery and all animals were nursing within 1 h. On day 9 or 10 post-partum, the pups were removed from their mothers at 8 a.m. At 11a.m. the mothers were injected iv with 0.2 ml of NRS or GRF-ab. A blood sample (0.4 ml) was drawn at 2 p.m. to determine baseline hormone concentrations. This was followed by the return of the pups to their respective mothers. The pups had been weighed immediately prior to their being returned. Subsequent blood samples were drawn 10, 20, 30, and 60 min following the onset of suckling. Suckling was defined as at least five of eight pups attached to the teats. In most instances, suckling occurred within 10 min of the pups being returned to their mothers. Blood samples, radioimmunoassays and data analysis were performed as described earlier.

In the NRS-treated mothers suckling induced a significant and sustained rise in serum prolactin concentrations as compared with concentrations observed prior to the onset of suckling. Maximum values exceeded 200 ng/ml and concentrations remained high for the entire sampling period (Table 3). Administration of the GRF-ab did not affect the suckling-induced changes in prolactin concentrations. Both the absolute concentrations of prolactin and its time course of release were unaltered as compared with the NRS-treated animals. Suckling caused a rapid and transient increase in serum GH in NRS-treated rats, with values rising over 2.5 fold within 10 min of the onset of suckling. Thereafter, concentrations declined, reaching baseline by 60 min. In distinct contrast, GH did not increase during lactation in females treated with the GRF-ab (Table 3). Furthemore, the baseline values for GH in the GRF-ab treated females were lower than those in the NRS-treated ones. Regardless of the treatment, all pups gained approximately 0.6 g during the 1h period of suckling.

The present study confirms previous reports that suckling induces a rapid rise in serum GH concentrations and a slower and more sustained rise in prolactin. More importantly, it also demonstrates that the neuroendocrine pathways regulating GH secretion during lactation express their actions through GRF. They further show that these mechanisms are distinct and can operate independently from those regulating prolactin secretion. This is consistent with the evidence now accumulating which suggests that GRF is the final common denominator regulating GH secretion. Based on studies using similar passive-immunization techniques, Miki et al. (1984) and Cella et al. (1987) have shown that alpha-adrenergic stimulation of GH secretion is mediated via GRF. Likewise, it appears opiates and opioid peptides also mediate their effects via GRF (Miki et al. 1984; Wehrenberg et al. 1985).

The GRF-ab treatment did not affect the suckling-induced increase in prolactin. These results demonstrate that the rise in prolactin and GH which occur during lactation can be dissociated. Previous studies have shown that the rise in GH and prolactin are both interrupted by naloxone treatment (Miki et al. 1981; Riskind et al. 1984; Selmanoff and Gregerson 1986). These observations, coupled with the present ones, suggest that the involvement of the endogenous opiod system in lactation occurs prior to involvement of the hypothalamic GRF system. It is possible that the endogenous opioid peptides are not required to mediate the increases in prolactin during lactation, since Riskind et al. (1984) have shown that the prolactin rise during lactation can occur without changes in peripheral plasma β-endorphin. However, this interpretation is tempered by the fact that changes in hormone concentrations at a peripheral level may not accurately reflect changes in the central nervous system. Furthemore, there appear to be different receptors mediating the GH and prolactin responses to opioids (Spiegel et al. 1982).

The fact that the pups of the antibody-treated and NRS-treated mothers did

Table 3. Serum GH and prolactin concentrations in mothers passively immunized with normal rabbit serum (NRS-Controls) or GRF antiserum (GRF-ab) in response to a 1 h suckling stimulus

Hormone	Treatment	Time from onset of suckling (min)					
		0	10	20	30	60	Mean value
GH	Controls	3.8 ± 0.3	9.9 ± 1.4	8.3 ± 2.6	5.1 ± 0.7	3.9 ± 0.3	6.2 ± 1.4
	GRF-ab	2.7 ± 0.2	3.3 ± 0.2	2.7 ± 0.1	2.6 ± 0.1	2.5 ± 0.1	2.8 ± 0.7*
Prolactin	Controls	3 ± 1	42 ± 12	108 ± 24	159 ± 26	214 ± 10	88 ± 12
	GRF-ab	3 ± 1	23 ± 10	57 ± 18	108 ± 21	199 ± 16	65 ± 11

Data expressed as mean ± SEM.
* Significantly different ($P<0.05$) from NRS-treated animals.

not differ in weight gain during the 1 h period of suckling serves as a good indicator that maternal nursing behavior was not affected by the treatment. It can also be stated that acute suppression of GH secretion by GRF-ab treatment did not alter lactational performance in the mothers. However, this does not imply that normal GH concentrations are not required for normal lactation. Sinha and colleagues (1972a) have clearly shown that chronic GH antiserum treatment in mice inhibits the body-weight gain of suckling pups.

Acknowledgments

We thank Prof. A. F. Mueller for his support in conducting this reserarch. Portions of this work were perfomed while W.B. Wehrenberg was an Invited Professor of the Department of Medicine, Hôpital Cantonal Universitaire de Genève, Genève, Switzerland. We thank J. Wehrenberg, S. Corder, D. Turnill, and M. Giacomini for their technical assistance in performing this work.

REFERENCES

Baird A, Wehrenberg WB, Ling N (1984) Ontogeny of the response to growth hormone-releasing factor. *Regul Peptides* **10**: 23-28

Baird A, Wehrenberg WB, Böhlen P, Ling N (1985) Immunoreactive and biologically active growth hormone-releasing factor in the rat placenta. *Endocrinology* **117**: 1598-1601

Benoit R, Böhlen P, Ling N, Brisken A, Esch F, Brazeau P, Ying S, Guillemin R (1982) Presence of somatostatin-28-(1-12) in hypothalamus and pancreas. *Proc Natl Acad Sci USA* **79**: 917-921

Bloch B (1978) Les neurones producteurs de LH-RH chez l'homme au cours de la vie foetale. Etude cyto-immunologique en microscopie optique et en microscopie electronique. *Int Ann Sci Univ France Comte* **5**: 1-216

Bloch B, Gaillard RC, Brazeau P, Lin HD, Ling N (1984) Topographical and ontogenetic study of the neurons producing growth hormone-releasing factor in human hypothalamus. *Regul Peptides* **8**: 21-31

Brazeau P, Vale W, Burgus R, Ling N, Butcher M, Rivier J, Guillemin R (1973) Hypothalamic polypeptide that inhibits the secretion of immunoreactive pituitary growth hormone. *Science* **179**: 77-79

Bresson JL, Clacequin MC, Fellmann D, Bugnon C, (1984) Ontogeny of the neuroglandular system revealed with hpGRF-44 antibodies in human hypothalamus. *Neuroendocrinology* **39**: 68-73

Bugnon C, Bloch B, Lenys D, Fellmann D (1978) Cytoimmunological study of LH-RH neurons in humans during fetal life. In: *Brain-Endocrine Interaction*. 3rd International Symposium, Wurzburg, Germany, 1977. III. Neural hormone and reproduction. Karger, Basel, pp 183-196

Bugnon C, Fellmann D, Bloch B (1977) Immunocytochemical study of the ontogenesis of the hypothalamic somatostatin-containing neurons in the human fetus. *Cell Tissue Res* **183**: 319-328

Bugnon C, Fellmann D, Bresson JL, Clavequin MC (1982) Etude immunocytochimique de l'ontogenese du systeme neuroglandulaire a CRF chez l'homme. *CR Acad Sci* [III] **294**: 491-494

Cella SG, Locatelli V, De Gennaro V, Wehrenberg WB, Müller EE (1987) Pharmacological manipulations of α-adrenoceptors in the infant rat and effects on growth hormone secretion. Study of the underlying mechanisms of action. *Endocrinology* **120**: 1639-1643

Chihara K, Minamitani N, Kaji H, Kodama H, Kita T, Fujita T (1984) Noradrenergic modulation of human pancreatic growth hormone-releasing factor (hpGHRF 1-44) induced growth hormone release in conscious male rabbits: involvement of endogenous somatostatin. *Endocrinology* **114**: 1402-1406

Daikoku S, Hisano S, Kawano H, Okamura Y, Tsuruo Y (1983) Ontogenetic studies on the topographical heterogeneity of somatostatin-containing neurons in rat hypothalamus. *Cell Tissue Res* **233**: 347-357

Daikoku S, Kawano H, Nogushi M, Toluzen M, Chihara K, Saito H, Shibasaki T (1985) Ontogenetic appearance of immunoreactive GHRH-containing neurons in the rat hypothalamus. *Regul Peptides* **8**: 21-31

De Gennaro V, Redaelli M, Locatelli V, Cella SG, Fumagalli G, Wehrenberg WB, Müller EE (1986) Ontogeny of growth hormone-relaeasing factor in the rat hypothalamus. *Neuroendocrinology* **44**: 59-64

Ferland L, Labrie F, Jobin M, Arimura A, Schally AV (1976) Physiological role of somatostatin in the control of growth hormone and thyrotropin secretion. *Biochem Biophys Res Commun* **68**: 149-156

Guillemin R, Brazeau P, Böhlen P, Esch F, Ling N, Wehrenberg WB (1982) Growth hormone-releasing factor from a human pancreatic tumor that caused acromegaly. *Science* **218**: 585-587

Ishikawa K, Katakami H, Jansson JO, Frohman LA (1986) Ontogenesis of growth hormone-releasing hormone neurons in the rat hypothalamus. *Neuroendocrinology* **43**: 537-542

Jansson JO, Ischikawa K, Katakami H, Frohman LA (1987) Pre-and postnatal developmental changes in hypothalamic content of rat growth hormone-releasing factor. *Endocrinology* **120**: 525-530

Kawano H, Watanabe YG, Daikoku S (1980) Light-and electron-microscopic observations on the apperarance of immunoreactive LHRH in perinatal rat hypothalamus. *Cell Tissue Res* **213**: 465-474

Klindt J, Robertson MC, Friesen HG (1981) Secretion of placental lactogen, growth hormone, and prolactin in late pregnant rats. *Endocrinology* **109**: 1492-1495

Miki N, Ono M, Shizume K (1984) Evidence that opiatergic and alpha-adrenergic mechanisms stimulate rat growth hormone release via growth hormone-releasing factor. *Endocrinology* **114**: 1950-1952

Miki N, Sonntag WE, Forman LJ, Meites J (1981) Suppression by naloxone of rise in plasma growth hormone and prolactin induced by suckling. *Proc Soc Exp Biol Med* **168**: 330-333

Plotsky PM, Vale W (1985) Patterns of growth hormone-releasing factor and somatostatin secretion into the hypophyseal-portal circulation of the rat. *Science* **230**: 461-463

Riskind PN, Millard WJ, Martin JB (1984) Opiate modulation of the anterior pituitary hormone response during suckling in the rat. *Endocrinology* **114**: 1231-1237

Rivier J, Spiess J, Thorner M, Vale W (1982) Characterization of a growth hormone-releasing factor from a human pancreatic islet tumour. *Nature* **300**: 276-278

Saunders A, Terry LC, Audet J, Brazeau P, Martin JB (1976) Dynamic studies of growth hormone and prolactin secretion in the female rat. *Neuroendocrinology* **21**: 193-203

Selmanoff M, Gregerson KA (1986) Suckling-induced prolactin release is suppressed by naloxone and stimulated by β-endorphin. *Neuroendocrinology* **42**: 255-259

Sinha YN, Lewis UJ, Vanderlaan WP (1972) Effects of administering antisera to mouse growth hormone and prolactin on gain in litter weight and on mammary nucleic acid content of lactating C3H mice. *J Endocr* **55**: 31-40

Sinha YN, Selby FW, Lewis UJ, Vanderlaan WP (1972) Studies of GH secretion in mice by a homologous radioimmunoassay for mouse GH. *Endocrinology* **91**: 784-792

Spiegel K, Kourides IA, Pasternak GW (1982) Prolactin and growth hormone release by morphine in the rat: different receptor mechanism. *Science* **217**: 745-747

Terry LC, Martin JB (1981) The effects of lateral hypothalamic-medial forebrain stimulation and somatostatin antiserum on pulsatile growth hormone secretion in freely behaving rats: evidence for a dual regulatory mechanism. *Endocrinology* **109**: 622-627

Underwood LE, Van Wyk JJ (1981) Hormones in normal and aberrant growth. In: Williams RH (ed) *Textbook of Endocrinology*. Saunders, Philadelphia, pp 1149-1191

Wehrenberg WB (1986) The role of growth hormone-releasing factor and somatostatin on somatic growth in rats. *Endocrinology* **118**: 489-494

Wehrenberg WB, Bloch B, Ling N (1985) Pituitary secretion of growth hormone in response to opioid peptides and opiates is mediated through growth hormone-releasing factor. *Neuroendocrinology* **41**: 13-22

Wehrenberg WB, Brazeau P, Ling N (1985) Pituitary growth hormone response in rats during a 24-hour infusion of growth hormone-releasing factor. *Endocrinology* **114**: 1613-1616

Wehrenberg WB, Brazeau P, Luben R, Böhlen P, Guillemin R (1982) Inhibition of the pulsatile secretion of growth hormone by monoclonal antibodies to the hypothalamic growth hormone-releasing factor (GRF). *Endocrinology* **111**: 2147-2148

Winer BJ (1971) Multifactor experiments having repeated measures on the same elements. In: Winer BJ (ed) *Statistical Principles in Experimental Design, 2nd edn.*, McGraw-Hill, New York, pp 514-603

Advances in Growth Hormone and Growth Factor Research,
edited by E.E. Müller, D. Cocchi and V. Locatelli
Pythagora Press, Roma-Milano and Springer Verlag, Berlin-Heidelberg © 1989

Structure and expression of growth hormone- releasing hormone (GHRH) genes

K. E. Mayo

Department of Biochemistry, Molecular Biology, and Cell Biology, Northwestern University, Evanston, Illinois, U.S.A.

The neurohumoral regulation of growth hormone secretion is mediated in part by two hypothalamic peptides, somatostatin and growth hormone-releasing hormone (GHRH). Although somatostatin, a peptide possessing inhibitory activity for growth hormone release, was one of the first of the hypothalamic releasing or release-inhibiting factors to be chemically characterized (Brazeau et al. 1973), a positively acting growth hormone-releasing factor was not unequivocally identified until 1982 (Guillemin et al. 1982; Rivier et al. 1982). The initial isolation of GHRH was facilitated by the existence of human tumors ectopically producing growth hormone-releasing activity (Frohman and Szabo 1981; Cronin et al. 1982), and it was from two tumors of this type that human GHRH was characterized as colinear peptides of 40 or 44 amino acids. Subsequently, the identity of this structure with that of GHRH of human hypothalamic origin (Ling et al. 1984) and the similarity of this structure with that of GHRH of rat hypothalamic origin (Spiess et al. 1983) were verified. The biology of GHRH action has been the topic of several recent reviews (Wehrenberg et al. 1985; Frohman and Jansson 1986; Gelato and Merriam 1986), and is further considered in other chapters within this volume.

Molecular cloning of the precursors to hypothalamic peptides such as GHRH promises to be an effective approach to understanding their biosynthesis and its neural and hormonal control. Toward this goal, we, and others, have isolated cDNA clones specific for GHRH. Initially, two pancreatic tumors from which the GHRH peptide was purified served as an mRNA source for the construction of libraries from which cDNAs encoding the human GHRH precursor were isolated (Mayo et al. 1983; Gubler et al. 1983). Subsequently, we isolated a rat GHRH cDNA directly from a hypothalamic library (Mayo et al. 1985a). Sequence analyses of these clones has revealed the existence and structures of proteins that serve as precursors to the mature, biologically active GHRH peptides. In addition to providing structural information, these cDNAs are useful as probes for the measurement of GHRH mRNA, both in the hypothalamus and in the periphery.

In this short article, I will review our current knowledge concerning the structure of the GHRH gene both in man and in rodents. As an example of the application of GHRH probes to understanding GHRH biosynthesis, recent experiments involving hormonal feedback control of GHRH mRNA will be discussed. Finally, the use of transgenic animals expressing recombinant GHRH genes to study the physiological actions of GHRH will be described.

STRUCTURE OF THE GHRH PRECURSOR AND GENE

Using oligonucleotide and antibody screening of a cDNA expression library, a cDNA clone encoding human GHRH was isolated from a pancreatic tumor (Mayo et al. 1983). Use of this cDNA to screen human genomic libraries resulted in the elucidation of the complete structure of the human GHRH gene (Mayo et al. 1985b). The structures of the human gene, cDNA, and GHRH precursor protein are schematically presented in Figure 1. Like most small peptides, GHRH is generated by the proteolytic processing of a larger precursor protein. Two forms of this precursor are predicted from the sequence of cDNA clones; those cDNAs isolated in our laboratory encode a 108-amino acid protein, while cDNAs encoding either a 107- or a 108-amino acid precursor have been reported by others (Gubler et al. 1983). The difference in these proteins is the presence or absence of a serine residue at position 103, which falls outside of the GHRH-encoding domain. It appears that both of the major forms of human GHRH ($1-40_{OH}$ and $1-44_{NH2}$) are generated from this precursor by cleavage and amidation reactions that follow established processing rules (Docherty and Steiner 1982; Loh and Gainer 1983). A second predicted peptide of 29 or 30 amino acids is found within the precursor, C-terminal to GHRH. Immunohistochemical evidence suggests that this peptide is C-terminally amidated and is co-released into portal capillaries with GHRH (Bloch et al. 1986), but no function has as yet been ascribed to this peptide.

The GHRH gene is probably present in a single copy in the human genome, and has been mapped to human chromosome 20 (Mayo et al. 1985b; Riddell et al. 1985). The gene includes five exons spanning about 10 kilobases of DNA, and there is a clear segregation of functional domains of the GHRH precursor into the five exons of the gene (Fig. 1). DNA sequencing has indicated that the two forms of the GHRH precursor (107 and 108 amino acids) most likely arise by alternative processing of a single primary transcript, in that differential usage of two splice-acceptor sites at the beginning of exon 5 would result in the inclusion or exclusion of the nucleotides encoding serine 103. Whether this differential processing is

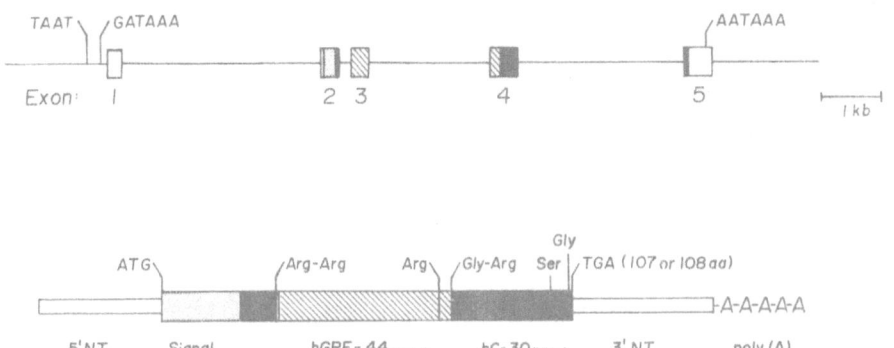

Fig. 1. Structure of the human GHRH gene and cDNA. The *top line* shows the structure of the gene, with exons represented as *boxes* and consensus sequences involved in transcription and polyadenylation indicated. The *bottom line* shows the structure of the mRNA as deduced from cDNA clones. The two major forms of human GHRH (*GRF*) are indicated, along with the amino acid residues involved in their proteolytic processing from the precursor. *hC* is the carboxyl- terminal peptide thought to be generated from the precursor along with GHRH. Glycine 103, present in some but not all cDNA clones, is also indicated.

found in the hypothalamus as well as in the pancreas, whether it is regulated, and whether it has physiological importance all remain to be determined.

The rat GHRH precursor and gene are closely related to their human counterparts, and their structures are shown schematically in Fig. 2. A single species of rat GHRH of 43 amino acids with a free C-terminus is predicted from the cDNA sequence, in agreement with results from characterization of the protein (Spiess et al. 1983). Although most of the 104-amino acid rat GHRH precursor protein is similar to the human precursor, there is an interesting divergence of the two in their C-terminal domains, such that the last 13 residues contain no identities. Comparison of the rat and human gene sequences suggests that this is again due to differential splicing of the two transcripts between exons 4 and 5, such that the 3' end of the two mRNAs is read in a different translational reading frame.

Comparison of DNA sequences between the rat and human GHRH genes indicates that the 5'-flanking regions of the two genes are highly conserved. This presumably reflects a functional importance of this region in directing expression of the GHRH gene. However, when the human GHRH promoter fused to a marker gene was introduced into transgenic mice, thymic- rather than hypothalamic-specific expression resulted, suggesting that this region alone is not sufficient to direct appropriate tissue-specific expression of the GHRH gene (Botteri et al. 1987).

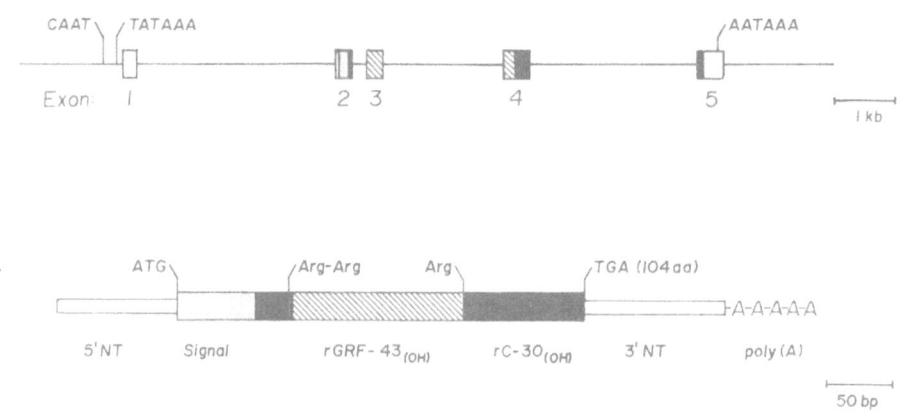

Fig. 2. Structure of the rat GHRH gene and cDNA. The *top line* shows the rat gene and the *bottom line* the mRNA as deduced from partial rat cDNA clones and from similarity to full-length human cDNAs. Items indicated are as in the legend to Figure 1.

EXPRESSION OF GHRH mRNA IN RAT HYPOTHALAMUS

With the advent of cDNA probes for GHRH, it became possible to examine the expression of the gene in the mammalian hypothalamus. As a model, we initially focused on the hormonal feedback regulation of GHRH biosynthesis that occurs during growth hormone deficiency produced by hypophysectomy in the male rat. Fig. 3 is a Northern RNA blot of medial-basal hypothalamic RNA from control or 7-day hypophysectomized rats, probed with a rat GHRH cDNA. There is a sixfold increase in GHRH mRNA following hypophysectomy, suggesting a loss of negative feedback to the hypothalamus by growth hormone or one of its induced products.

In the above experiment, pooled groups of six animals were studied to facilitate detection of the very low abundance of GHRH mRNA. To examine GHRH mRNA in individual animals, and to determine where GHRH mRNA is being expressed in the brain, we utilized *in situ* hybridization techniques (Young 1986; Schwartz and Costa 1986). Using radiolabeled anti-sense RNA probes specific for rat GHRH, we were able to localize GHRH mRNA predominantly to the arcuate nucleus of the rat hypothalamus (Mayo et al. 1986), in agreement with the known immunohistochemical distribution of the GHRH peptide (Merchenthaler et al. 1984; Sawch-

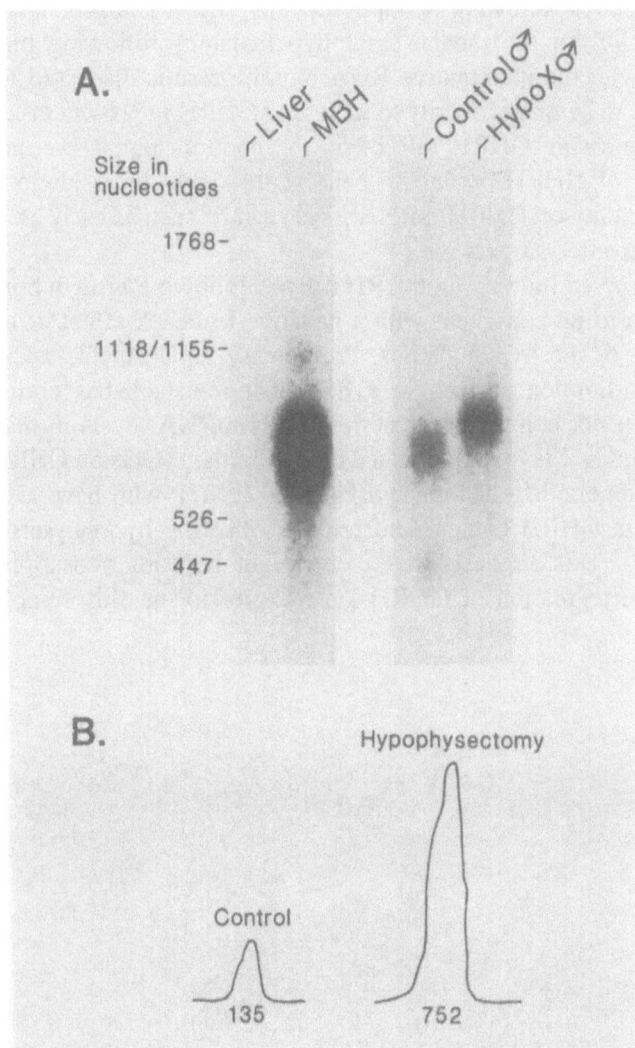

Fig. 3. Induction of GHRH mRNA in rat hypothalamus in response to hypophysectomy. Panel *A* is an autoradiogram of a Northern RNA blot (Meinkoth and Wahl 1984). *Liver* and *MBH* (medial basal hypothalamus) are negative and positive controls for GHRH mRNA, while *Control* and *Hypox* indicate the experimental RNAs from 200-g male rats. Each lane contains 5 µg poly(A)$^+$ RNA, and the blot is probed with a rat GHRH cDNA. Panel *B* is a densitometric scan of the two experimental lanes from A, with the relative densities indicated.

enko et al. 1985). Figure 4 is an example of such an *in situ* hybridization analysis of GHRH mRNA in the male rat brain, comparing a control animal with one examined 7 days following hypophysectomy. There is a pronounced induction of GHRH mRNA in the medial-basal hypothalamus following pituitary removal, consistent with the quantitative RNA blotting results discussed above. Although this appears to be due primarily to increased GHRH mRNA in the same population of cells expressing GHRH mRNA in the control animal, we cannot rule out a recruitment of GHRH-producing cells at this level of resolution. The effect of hypophysectomy on GHRH mRNA levels can be seen as early as 3 days or as late as 3 months after surgery.

The observed increase in GHRH mRNA following growth hormone depletion would seem to be consistent with a negative-feedback effect of growth hormone at the level of the hypothalamus. Preliminary replacement experiments indicate that growth hormone, given in a dose that re-establishes normal growth, can partially, but not completely, restore GHRH mRNA levels in this model system (D. Kulik and K. Mayo, unpublished data). Similar effects on GHRH mRNA levels have been reported by Chomczynski et al. (1987), who have also demonstrated a decrease in GHRH content and release following hypophysectomy (Katakami et al. 1987). This suggests that there might be both transcriptional and post-transcription regulation of GHRH gene expression in this complex system.

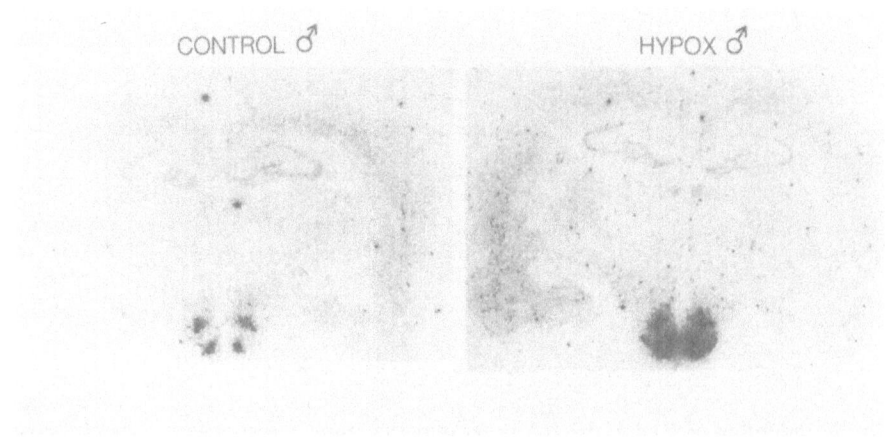

Fig. 4. In situ hybridization analysis of GHRH mRNA in the rat hypothalamus. Fixed and frozen 20 μm sections from 200-g male rats (control and 7 days following hypophysectomy [HYPOX]) were hybridized on microscope slides to a rat GHRH anti-sense ^{32}P-RNA probe, washed, and exposed to autoradiographic film (Woodruff et al. 1987).

TRANSGENIC MICE PRODUCING GHRH

To further examine the role of GHRH in animal growth, we have generated lines of transgenic mice that express supraphysiological levels of GHRH. Transgenic animals containing experimentally introduced genes provide unique model systems for studying many aspects of gene expression and regulation (Palmiter and Brinster 1985). Transgenic animal systems have been widely exploited for investigating the control of animal growth (Hammer et al. 1985a), and animals transgenic for growth hormone fusion genes (Palmiter et al. 1982) and somatostatin fusion genes (Low et al. 1985) have been generated. Following the isolation of GHRH cDNA and genomic clones, we utilized such an approach to determine whether excessive production of GHRH could lead to elevated growth hormone secretion and increased growth in animals. For these experiments, a new gene including the promoter and regulatory regions of a mouse metallothionein-I gene (Durnam et al. 1980) fused to a human GHRH minigene was constructed (referred to as MT-GRF). This fusion gene was introduced into the germ line of animals by microinjection into fertilized mouse eggs (Hammer et al. 1985b). The process used to generate and analyze transgenic mice is schematically reviewed in Fig. 5.

Many of the mice transgenic for the MT-GRF fusion gene have immuno-reactive human GHRH in serum (up to 200 ng/ml), have highly elevated serum growth hormone levels (up to 1000 ng/ml), and grow to be abnormally large (Hammer et al. 1985b). Both the MT-GRF fusion gene and the large phenotype are stably inherited by the offspring of these animals as a Mendelian trait. Figure 6 shows one of the MT-GRF animals compared with an age- and sex- matched control littermate; although independently derived MT-GRF founder animals have shown variable increases in growth rate and in eventual size, many of these mice reach about twice normal size as adults. In a complementary series of experiments, antibodies against GHRH have been shown to inhibit somatic growth in rats (Wehrenberg 1986), again indicating the importance of GHRH in the regulation of animal growth.

Metallothionein gene expression is strongly inducible by heavy metals (Durnam and Palmiter 1981); however, we have been unable to detect any effect of dietary zinc on growth of the MT-GRF animals, suggesting that factors other than GHRH are rate limiting for growth. The metallothionein sequences do seem to play an important role in determining which tissues express the MT- GRF transgene. Most of the MT-GRF animals were found to express human GHRH mRNA in the liver, a major site of metallothionein synthesis, as well as in other peripheral tissues. Figure 7 is an *in situ* hybridization analysis showing GHRH mRNA in the liver of one of these transgenic animals. In general, tissues that express the metallothionein gene also seem to express the MT-GRF fusion gene (Hammer et al. 1985b); however, there are marked contrasts in expression among

INTRODUCTION OF CLONED GENES INTO MOUSE EMBRYOS

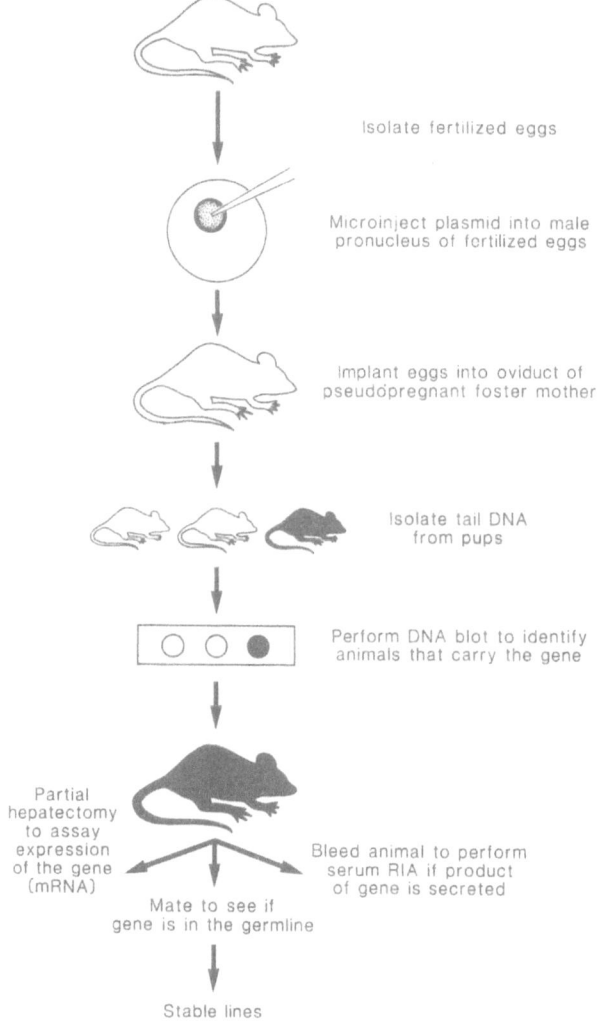

Isolate fertilized eggs

Microinject plasmid into male pronucleus of fertilized eggs

Implant eggs into oviduct of pseudopregnant foster mother

Isolate tail DNA from pups

Perform DNA blot to identify animals that carry the gene

Partial hepatectomy to assay expression of the gene (mRNA)

Bleed animal to perform serum RIA if product of gene is secreted

Mate to see if gene is in the germline

Stable lines

Fig. 5. The production of transgenic mice by microinjection of cloned genes into fertilized mouse eggs.

the different lines of MT-GRF animals. One major site of GHRH expression in these mice is the pituitary, and Figure 7 also shows the localization of GHRH mRNA in this tissue. Expression of a metallothionein- somatostatin fusion gene in the pituitary has been observed, and in these animals somatostatin was shown to be localized in gonadotrophs (Low et al. 1986). In the MT-GRF animals, expression of GHRH within the pituitary appears to be confined to somatotrophs, suggesting that sequences other than those in the metallothionein promoter are important for this specificity (Mayo et al. 1988).

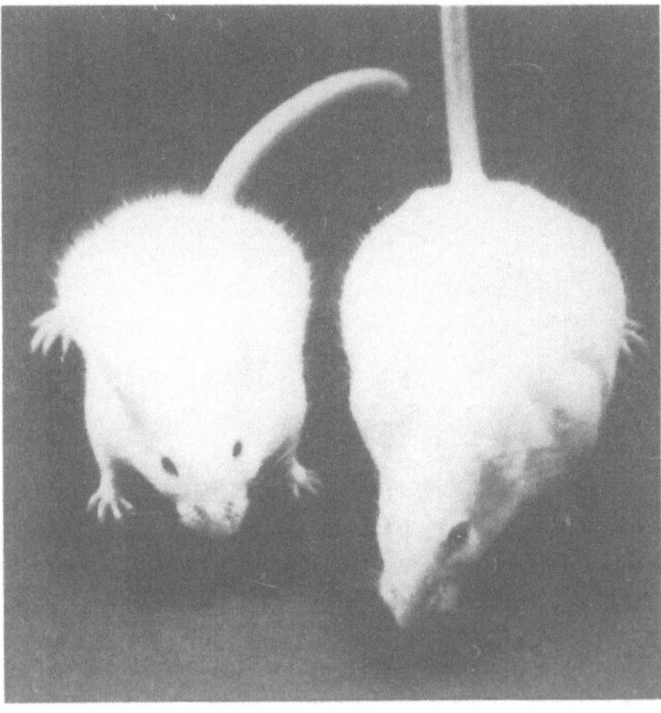

Fig. 6. Transgenic mouse expressing human GHRH. A second generation animal (*right*) of the 803-4 founder line of MT-GRF transgenic mice is compared with an age- and sex-matched littermate (*left*) that does not have the transgene. (Photograph courtesy of R. Hammer and R. Brinster, University of Pennsylvania).

Although many of the internal organs of these animals are enlarged (see Fig. 7), the pituitary gland has been found to be grossly enlarged. An example of this is shown in Figure 8, which compares the pituitary from a transgenic and a control animal. This increase in pituitary size seems to be due primarily to hyperplasia of the somatotrophs (Mayo et al. 1988), although hypertrophy of cells also appears to be a contributing factor (Hammer et al. 1985a). This suggests that pituitary somatotrophs might continuously proliferate in response to the chronically elevated GHRH levels in these mice and is in agreement with observations of somatotroph hyperplasia in patients with GHRH- secreting tumors (Thorner et al. 1982; Asa et al. 1984), as well as with recent results indicating a direct mitogenic activity of GHRH on somatotrophs in primary culture (Billestrup et al. 1986). In addition to stimulating growth hormone secretion, GHRH has been shown to increase transcription of the growth hormone gene (Barinaga et al. 1983; Gick et al. 1984). Observations in several systems now suggest yet another potential function for GHRH as a specific growth factor for its target cell, the pituitary somatotroph.

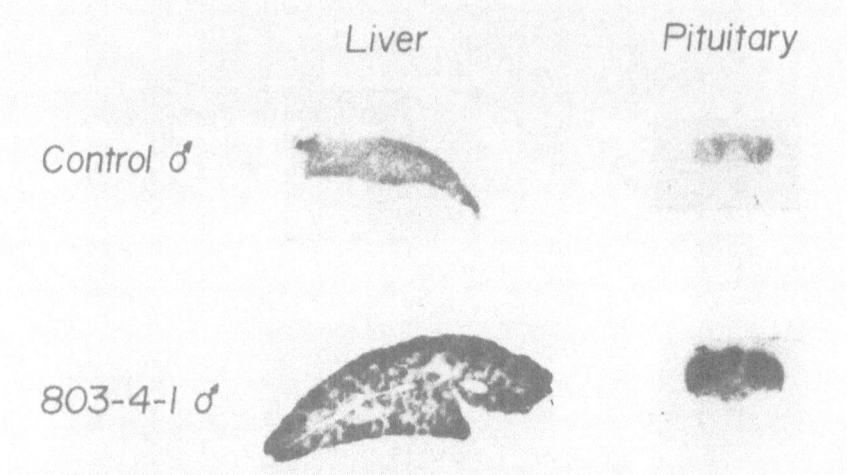

Fig. 7. Detection of human GHRH mRNA in transgenic mouse tissues. Liver and pituitary sections from a control and an MT-GRF transgenic animal were assayed for human GHRH mRNA using in situ hybridization, as described in the legend to Figure 4.

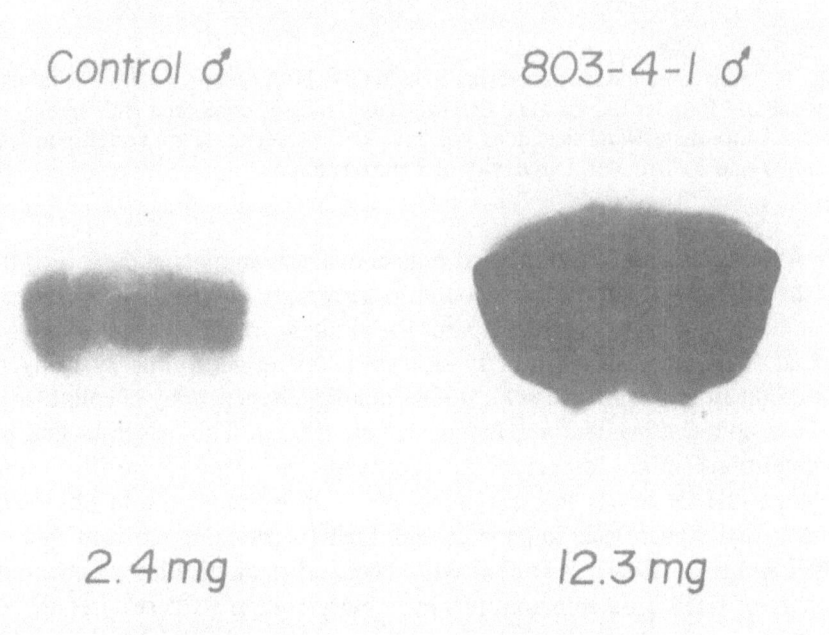

Fig. 8. Effect of chronic GHRH production on pituitary development in MT- GRF transgenic mice. The *left* pituitary is from a control animal weighing 23 g, the *right* from an MT-GRF transgenic animal expressing human GHRH and weighing 46 g.

CONCLUSIONS

The isolation and characterization of the GHRH genes has provided valuable structural insights into GHRH biosynthesis, revealing the complete structure of the GHRH precursor proteins in several species. More importantly, the availability of probes that specifically detect GHRH mRNA in these species will allow questions concerning the mechanism by which GHRH gene expression is regulated to be addressed. Although hormonal feedback has been considered here as one example, it is likely that a wide variety of neural, hormonal, and developmental signals will affect GHRH synthesis and secretion. In the absence of permanent cell lines that express GHRH, it will be necessary to develop defined environments in which GHRH gene expression and regulation can be examined, such as transfected cell lines and transgenic animals.

The importance of GHRH for normal somatic growth has been strongly indicated by a variety of clinical observations and experimental results, including those reviewed here using a transgenic animal system. The expression of foreign genes using transgenic animal approaches has often resulted in the identification of novel and unexpected consequences of transgene expression, such as the interesting effect on pituitary development observed in the MT-GRF mice. In addition to showing increased growth, the MT-GRF transgenic mice provide an intriguing model in which to investigate the effects of chronic GHRH production on other hormones involved in growth regulation, in which to examine the tissue specificity of transgene expression, and in which to study its developmental consequences.

REFERENCES

Asa SL, Scheithauer BW, Bilbao JM, Horvath E, Ryan N, Kovacs K, Randall RV, Laws ER, Singer W, Linfoot JA, Thorner MO, Vale W (1984) A case for hypothalamic acromegaly: a clinicopathological study of six patients with hypothalamic gangliocytomas producing growth hormone-releasing factor. *J Clin Endocrinol Metab* 58: 798-803

Barinaga M, Yamamoto G, Rivier C, Vale W, Evans R, Rosenfeld MG (1983) Transcriptional regulation of growth hormone gene expression by growth hormone-releasing factor. *Nature* 306: 84-85

Billestrup N, Swanson L, Vale W (1986) Growth hormone-releasing factor stimulates proliferation of somatotrophs *in vitro*. *Proc Natl Acad Sci USA* 83: 6854-6857

Bloch B, Baird A, Ling N, Guillemin R (1986) Immunohistochemical evidence that growth hormone-releasing factor neurons contain an amidated peptide derived from cleavage of the carboxyl-terminal end of the GRF precursor. *Endocrinology* 118: 156-162

Botteri F, van der Putten H, Wong D, Sauvage C, Evans R (1987) Unexpected thymic hyperplasia in transgenic mice harboring a neuronal promoter fused with simian virus 40 large-T antigen. *Mol Cell Biol* **7**: 3178-3184

Brazeau P, Vale W, Burgus R, Ling N, Butcher M, Rivier J, Guillemin R (1973) Hypothalamic polypeptide that inhibits the secretion of immunoreactive pituitary growth hormone. *Science* **179**: 77-79

Chomczynski P, Downs TR, Frohman LA (1987) Feedback regulation of growth hormone (GH)-releasing hormone by GH in rat hypothalamus. 69th Annual Meeting of the Endocrine Society, p 175 (abstract)

Cronin MJ, Rogol AD, Dabney LG, Thorner MO (1982) Selective growth hormone and cyclic AMP-stimulating activity is present in a human pancreatic islet cell tumor. *J Clin Endocrinol Metab* **55**: 381-383

Docherty K, Steiner DF (1982) Post-translational proteolysis in polypeptide hormone biosynthesis. *Annu Rev Physiol* **44**: 625-638

Durnam DM, Palmiter RD (1981) Transcriptional regulation of the mouse metallothionein-I gene by heavy metals. *J Biol Chem* **256**: 5712-5716

Durnam DM, Perrin F, Gannon F, Palmiter RD (1980) Isolation and characterization of the mouse metallothionein-I gene. *Proc Natl Acad Sci USA* **77**: 6511-6515

Frohman LA, Szabo M (1981) Ectopic production of growth hormone- releasing factor by carcinoid and pancreatic islet tumors associated with acromegaly. In: *Physiopathology of Endocrine Diseases and Mechanisms of Hormone Action*. Liss, New York, pp 259-271

Frohman LA, Jannson, J-O (1986) Growth hormone-releasing hormone. *Endocr Rev* **7**: 223-253

Gelato MC, Merriam GR (1986) Growth hormone releasing hormone. *Annu Rev Physiol* **48**: 569-591

Gick GG, Zeytin F, Brazeau P, Ling NC, Esch F, Bancroft FC (1984) Growth hormone-releasing factor regulates growth hormone mRNA in primary cultures of rat pituitary cells. *Proc Natl Acad Sci USA* **81**: 1553-1555

Gubler U, Monahan JJ, Lomedico PT, Bhatt RS, Collier KJ, Hoffman BJ, Böhlen P, Esch F, Ling N, Zeytin F, Brazeau P, Poonian MS, Gage LP (1983) Cloning and sequence analysis of cDNA for the precursor of human growth hormone-releasing factor, somatocrinin. *Proc Natl Acad Sci USA* **80**: 4311-4315

Guillemin R, Brazeau P, Böhlen P, Esch F, Ling, N, Wehrenberg WB (1982) Growth hormone-releasing factor from a human pancreatic tumor that caused acromegaly. *Science* **218**: 585-587

Hammer RE, Brinster RL, Palmiter RD (1985a) Use of gene transfer to increase animal growth. *Cold Spring Harbor Symp Quant Biol* **50**: 379-387

Hammer RE, Brinster RL, Rosenfeld MG, Evans RM, Mayo KE (1985b) Expression of human growth hormone-releasing factor in transgenic mice results in increased somatic growth. *Nature* **315**: 413-416

Katakami H, Downs TR, Frohman LA (1987) Effect of hypophysectomy on hypothalamic growth hormone-releasing factor content and release in the rat. *Endocrinology* **120**: 1079-1082

Ling N, Esch F, Böhlen P, Brazeau P, Wehrenberg WB, Guillemin R (1984) Isolation, primary structure and synthesis of human hypothalamic somatocrinin: growth hormone-releasing factor. *Proc Natl Acad Sci USA* **81**: 4302-4306

Loh YP, Gainer H (1983) Biosynthesis and processing of neuropeptides. In: Krieger DT, Brownstein MJ, Martin JB (eds) *Brain Peptides*.Wiley, New York, pp 79-116

Low MJ, Hammer RE, Goodman RH, Habener JF, Palmiter RD, Brinster RL (1985) Tissue-specific post-transcriptional processing of pre-prosomatostatin encoded by a metal-lothionein-somatostatin fusion gene. *Cell* **41**: 211-219

Low MJ, Lechan RM, Hammer RE, Brinster RL, Habener JF, Mandel G, Goodman RH (1986) Gonadotroph-specific expression of metallothionein fusion genes in pituitaries of transgenic mice. *Science* **231**: 1002-1004

Mayo KE, Vale W, Rivier J, Rosenfeld MG, Evans RM (1983) Expression-cloning and sequence of a cDNA encoding growth hormone-releasing factor. *Nature* **306**: 86-88

Mayo KE, Cerelli G, Rosenfeld GM, Evans RM (1985a) Characterization of cDNA and genomic clones encoding the precursor to rat hypothalamic growth hormone-releasing factor. *Nature* **314**: 464-467

Mayo KE, Cerelli G, Lebo R, Bruce BD, Rosenfeld MG, Evans RM (1985b) Gene encoding human growth hormone-releasing factor precursor: structure, sequence, and chromosomal assignment. *Proc Natl Acad Sci USA* **82**: 63-67

Mayo KE, Evans RM, Rosenfeld MG (1986) Genes encoding neuroendocrine peptides: strategies toward their identification and characterization. *Annu Rev Physiol* **48**: 431-446

Mayo KE, Hammer RE, Swanson LW, Brinster RL, Rosenfeld MG, Evans RM (1988) Dramatic pituitary hyperplasia in transgenic mice expressing a human growth hormone-releasing factor gene. *Mol Endocrinol* **2**: 606-612

Meinkoth J, Wahl G (1984) Hybridization to nucleic acids immobilized on solid supports. *Anal Biochem* **138**: 267-284

Merchenthaler I, Vigh S, Schally AV, Petrusz P (1984) Immunocytochemical localization of growth hormone-releasing factor in rat hypothalamus. *Endocrinology* **114**: 1082-1085

Palmiter RD, Brinster RL (1985) Transgenic mice. *Cell* **41**: 343-345

Palmiter RD, Brinster RL, Hammer RE, Trumbauer ME, Rosenfeld MG, Birnbirg NC, Evans RM (1982) Dramatic growth of mice that develop from eggs microinjected with metallothionein-growth hormone fusion genes. *Nature* **300**: 611-615

Riddell DC, Mallonee R, Phillips JA, Parks JS, Sexton LA, Hamerton JL (1985) Chromosomal assignment of human sequences encoding arginine vasopressin-neurophysin II and growth hormone releasing factor. *Somatic Cell Mol Genet* **11**: 189-195

Rivier J, Spiess J, Thorner M, Vale W (1982) Characterization of a growth hormone-releasing factor from a human pancreatic islet tumour. *Nature* **300**: 276-278

Sawchenko PE, Swanson LW, Rivier J, Vale WW (1985) The distribution of growth hormone-releasing factor (GRF) immunoreactivity in the central nervous system of the rat: an immunohistochemical study using antisera directed against rat hypothalamic GRF. *J Comp Neurol* **237**: 100-115

Schwartz JP, Costa E (1986) Hybridization approaches to the study of neuropeptides. *Annu Rev Neurosci* **9**: 277-304

Spiess J, Rivier J, Vale W (1983) Characterization of rat hypothalamic growth hormone-releasing factor. *Nature* **303**: 532-535

Thorner MO, Perryman RL, Cronin MJ, Rogol AD, Draznin M, Johanson A, Vale W, Horvath E, Kovacs K (1982) Somatotroph hyperplasia. Successful treatment of acromegaly by removal of a pancreatic islet tumor secreting a growth hormone – releasing factor. *J Clin Invest* **70**: 965-977

Wehrenberg WB, Baird A, Zeytin F, Esch F, Böhlen P, Ling N, Ying SY, Guillemin R (1985) Physiological studies with somatocrinin, a growth hormone- releasing factor. *Annu Rev Pharmacol Toxicol* **25**: 463-483

Wehrenberg WB (1986) The role of growth hormone-releasing factor and somatostatin on somatic growth in rats. *Endocrinology* **118**: 489-494

Woodruff TK, Meunier H, Jones PBC, Hseuh AJW, Mayo KE (1987) Rat inhibin: molecular cloning of α- and β-subunit cDNAs and expression in the ovary. *Mol Endocrinol* **1**: 561-568

Young WS III (1986) In situ hybridization histochemistry and the study of the nervous system. *Trends Neurosci* **9**: 549-551

Advances in Growth Hormone and Growth Factor Research,
edited by E.E. Müller, D. Cocchi and V. Locatelli
Pythagora Press, Roma-Milano and Springer Verlag, Berlin-Heidelberg © 1989

Multifactorial regulation of growth hormone

C. Berthelier, P. Bertrand, M.T. Bluet-Pajot, H. Clauser, D. Durand, A. Enjalbert, J. Epelbaum, E. Rerat and C. Kordon

Unitè de Neuroendocrinologie Institut National de la Santè et de la Recherche Medicale, Inserm U. 159, Centre Paul Broca, Paris, France

Two sets of data lead us to an in-depth reappraisal of the nature of growth hormone control. The first derives from pharmacological studies which have shown that somatotrophs can be affected by many more neurotransmitters and neuropeptides than only a GH-inhibitory factor (SRIF) and a GH-releasing factor (GHRH), as originally postulated. Even if these are still believed to have a major role in the regulation of the hormone, GH cells can respond to at least six other factors.

The second series of results concerns the hypothalamic level of regulation and was produced by modern techniques of functional neuroanatomy. These have permitted mapping of peptide-producing hypothalamic neurons and, in some cases, identification of their postsynaptic targets. Coincidence of proven synaptic contacts and of *in vitro* peptide effects can be considered a fair presumption of functional interaction.

In the present chapter, we describe the organization of hypothalamic neuronal networks involved in GH control, list direct transmitter and peptide actions on somatotrophs, and review the mechanisms by which GH cells integrate the various signals which they receive from the hypothalamus.

HYPOTHALAMIC ASPECTS OF GH CONTROL

The best-documented hypothalamic interactions governing GH secretion are, of course, those which affect the activity of GHRH and SRIF neurons.

A major component of GH regulation is provided by noradrenergic afferents to the arcuate nucleus (Palkovits et al. 1980), a structure which contains most GHRH cell bodies (Sawchenko et al. 1985). A positive action of norepinephrine

(NE) on GH release was described some time ago (Collu et al. 1972). It accounts for the inhibition of pulsatile (Durand et al. 1978) or induced GH secretion (Bluet-Pajot and Schaub 1978) observed after administration of α-noradrenergic antagonists, as well as for restoration of GH pulses by clonidine, an $α_2$-agonist, after pharmacological depletion of brain catecholamines (Durand et al. 1978). Although synaptic contacts between catecholamine projections and GHRH-containing cell bodies have not been formally characterized as yet, it seems very likely that this effect is mediated by noradrenergic fibers originating from the A1 mesencephalic nucleus and innervating the arcuate nucleus by the ventral hypothalamic noradrenergic bundle (Palkovits 1981).

Complementary effects of noradrenergic neurons on SRIF release have also been postulated. A direct action of NE on SRIF release from the median eminence could not be documented; however, NE triggers the release of SRIF from the anterior periventricular hypothalamus (Epelbaum et al. 1981), where most SRIF cell bodies are located (Elde and Parsons 1975). This could involve an action on dendritic release; increased SRIF concentrations at that level could in turn shut off the activity of SRIF neurons and hence reduce SRIF inhibition of GH (Epelbaum et al. 1981). An autocrine, ultrashort SRIF-SRIF negative fredback has indeed been demonstrated in the periventricular hypothalamus (Peterfreund and Vale 1984; Epelbaum 1986). In addition, synapses interconnecting SRIF neurons have been observed in that region by ultrastructural methods (Alonso et al. 1985). SRIF neurons also receive a local Gabaergic innervation (Willoughby et al. 1987).

Another important input to SRIF neurons comes from the paraventricular nucleus. Fibers originating from paraventricular cell bodies and containing CRF project to the vicinity of somatostatin cell bodies in the periventricular area (Swanson et al. 1983). In parallel, CRF was shown to stimulate the release of somatostatin, both *in vitro* (Peterfreund and Vale 1983) and upon intraventricular infusion (Ono et al. 1984). In addition, other axons from the paraventricular nucleus also project to the periventricular area (Conrad and Pfaff 1976). They may account for the effect of VIP on SRIF. In contrast to CRF, that peptide is inhibitory to somatostatin release (Epelbaum et al. 1979; Vijayan et al. 1979). The VIP-SRIF interaction was shown to be dose dependent, and its specificity could be characterized by using active VIP analogs (Epelbaum et al. 1979).

Within the periventricular system itself, neurotensin neurons are stimulatory to somatostatin (Shimatsu et al. 1982).

β-endorphin-containing nerve endings coming from the arcuate nucleus are also important for SRIF and GHRH control. The opioid peptide was shown to inhibit the release of somatostatin from hypothalamic slices (Drouva et al. 1980). The effect is dose dependent and can be blocked by opioid antagonists (Drouva et al. 1981).

Conversely, periventricular fibers innervate the arcuate nucleus (Conrad and Pfaff 1976). They may convey SRIF influences, which are believed to play an important role in synchronizing GHRH pulsatile activation (Plotsky and Vale

1985). SRIF can also inhibit the release of TRH (Hirooka et al. 1978; Tapia-Arancibia et al. 1984).

Finally, neuronal interactions within the arcuate nucleus are also involved in GHRH regulation. Arcuate β-endorphin neurons establish numerous contacts with neighboring cells. Experiments with intraventricular administration of morphine or opioid peptides, combined with passive immunization against SRIF (Ferland et al. 1977) or against GHRH (Wehrenberg et al. 1985), suggest that GHRH activation is more important than SRIF inhibition in mediating opioid-induced release of growth hormone, but no direct evidence of GHRH stimulation by opioids has been provided so far. In addition, another arcuate peptide, galanin, which appears to colocalize in dopamine as well as in GHRH neurons (Melander et al. 1986), stimulates GH upon intraventricular infusion (Ottlecz et al. 1985). The effect does not involve a direct action of galanin on somatotrophs and is thus believed to be mediated by GHRH neurons (Ottlecz et al. 1985) via epinephrine mediation (see Camanni et al., this volume). NPY also colocalizes in some GHRH neurons (Ciofi and Tramu 1986).

Intrahypothalamic peptide interactions affecting GH secretion are represented in Figure 1. Their functional relevance for growth hormone regulation will be discussed in the next section.

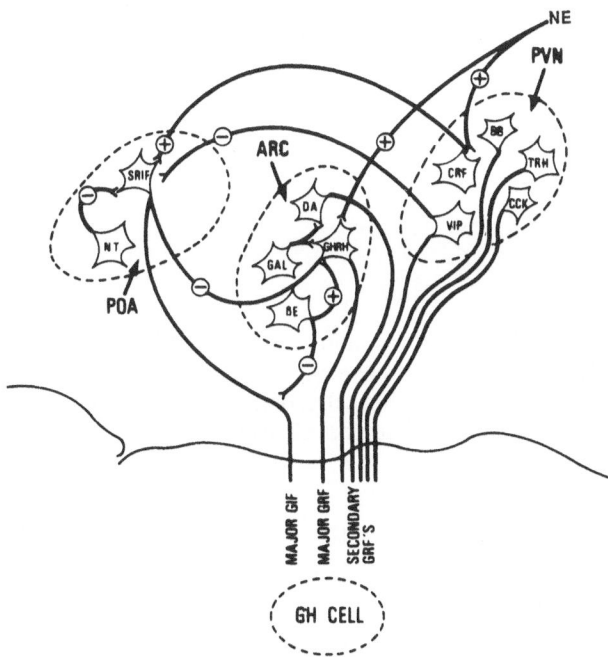

Fig. 1. Hypothalamic neuronal networks involved in GH control. ARC, arcuate nucleus; BB, bombesin; BE, β-endorphin; CCK, cholecystokinin; CRF, corticotropin-releasing hormone; DA, dopamine; GAL, galanin; GHRH, growth hormone-releasing hormone; NE, norepinephine; NT, neurotensin; POA, preoptic area; PVN, paraventricular nucleus; SRIF, somatostatin; TRH, thyrotropin-releasing hormone; VIP, vasoactive intestinal polypeptide.

Among the regulatory components, a peculiar neuroanatomical organization of four neuronal cell types seems to account for the pulsatile pattern of GH secretion. As shown in Figure 2, SRIF, GHRH, and OPI neurons are organized as loops and exhibit redundant, reciprocal effects. The figure is a hypothetical representation, but it is based on neuroanatomical as well as neuropharmacological data. These loops are located entirely within the hypothalamus, an observation consistent with the finding that hypothalamic deafferentation does not disrupt episodic secretion of GH (Willoughby et al. 1977). Extrinsic inputs, in particular

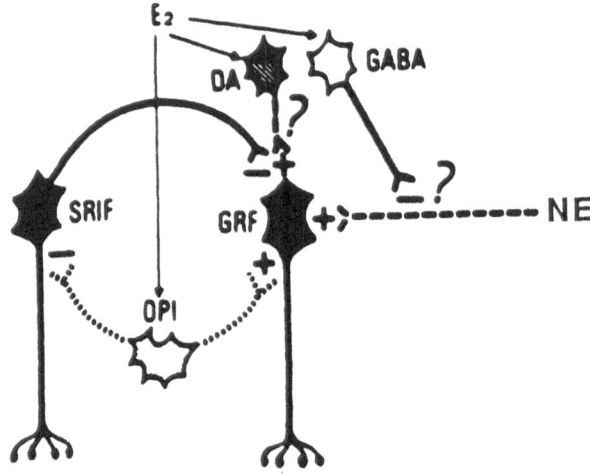

Fig. 2. Hypothalamic "loops" operating as pacemakers for the episodic control of GH. Opioid (*OPI*), *SRIF*, and *GRF* neurons exhibit redundant interactions, whereas, norepinephrine (*NE*) inputs and, more hypothetically, dopamine (*DA*) and *GABA* neurons modulate the activity of the loop. Several neuronal elements of the loop are targets of estradiol (E_2).

NE-containing projections from the mesencephalon, can affect at least two components of the loop. A similar circuitry, in which various elements of the loop can reinforce each other and induce "resonance" effects, has also been postulated to act as a primary oscillator of LH episodic secretion (Kordon et al. 1987). In addition, the β-endorphin component of the loop has the capacity to express receptors for estrogens (Morrell et al. 1985; Jirikowski et al. 1986); secretion of the peptide is influenced by the steroid (Wardlaw et al. 1982). This feature could account for the action of estrogen, which increases the frequency of episodic GH release.

MULTIFACTORIAL CONTROL OF SOMATOTROPHS

A large number of hypothalamic peptides released into the hypothalamo-hypophyseal portal system can directly affect somatotrophs. With the exception of SRIF, these all seem stimulatory to GH release. Like GHRH itself, three of them belong to the "glucagon peptide superfamily": VIP (Denef et al. 1985; Bluet-Pajot et al. 1987), PHI (Vigh and Schally 1984), and glucagon (Takahara et al. 1978) itself. But several other peptides, almost all produced in neurons of the paraventricular nucleus, have a similár effect: these are cholecystokinin (Vijayan et al. 1979), TRH (Bluet-Pajot et al. 1986), bombesin (Bicknell and Chapman 1983), and neurotensin (Maeda and Frohman 1978). In addition, recent data have shown that dopamine (DA) is also able to induce a transient stimulation of GH release from pituitary cells *in vitro* (Serri et al. 1987).

The functional relevance of these actions is still very uncertain, because in most cases, lack of appropriate antagonists has thus far not permitted testing the role of the endogenous peptides.

Another characteristic of peptide effects on somatotrophs is their variability according to endocrine conditions. GH secretion is highly susceptible to thyroid hormones, but this seems due to their predominant role in initiating transcription of the GH gene, rather than to an effect mediated by peptide receptors. *In vivo* GH stimulation by TRH was reported to occur only under circumstances in which the pituitary is deprived of hypothalamic signals, as after lesion of the median eminence (Müller et al. 1977; Bluet-Pajot et al. 1986). Similar "atypical" responses to GH have been observed in human beings under several pathological conditions such as acromegaly (Irie and Tsushima 1972) or depression (Maeda et al. 1975). The absence of such responses in intact rats (Müller et al. 1977) or in healthy human subjects (Gonzalez-Barcena et al. 1973) suggests that, under normal conditions, other signals received by somatotrophs blunt or antagonize the TRH effect. VIP has been proposed as one such factor (Bluet-Pajot et al. 1987). Interestingly, the effects of VIP and TRH, two peptides able to stimulate GH in *vitro* when added separately to incubated pituitaries (Bluet-Pajot et al. 1986, 1987), are not additive or even antagonistic on GH release when added together (Denef et al. 1985), whereas they are additive on lactotrophs (Fig. 3).

Why does GH respond to TRH or to VIP only *in vitro*, or under conditions in which hypothalamo-hypophyseal connections have been disrupted? Theoretically, this could be accounted for by the presence of endogenous SRIF, or alternatively, of endogenous DA in intact animals. In this sense, paradoxical or pathological GH responses ·in functional tests using TRH could thus point to a hypothetical disruption in the supply of these neuropeptides to the pituitary. Dopamine, which induces a transient stimulation of GH release from perifused pituitary cells, has a paradoxical, inhibitory effect on GHRH-induced GH secretion (Serri et al. 1987). This inhibitory action, however, does not account for the

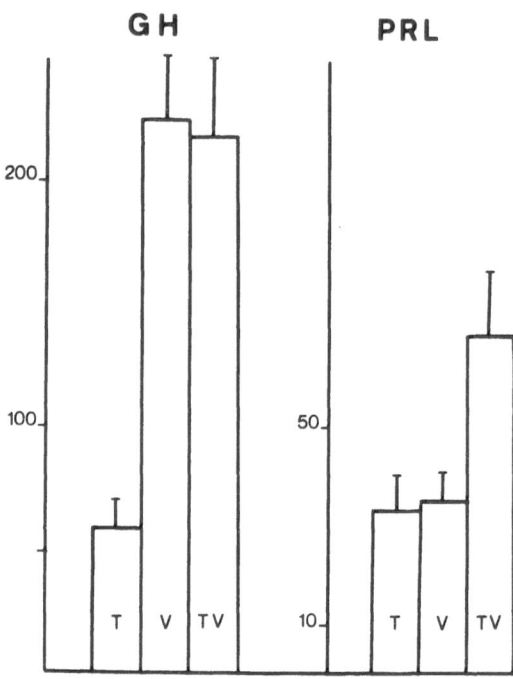

Fig. 3. Effect of equimolar (10^{-8}M) concentrations of TRH and VIP, added alone or in combination to a perfusate of pituitary fragments, on GH and prolactin (PRL) secretion. T, TRH; V, VIP; TV, TRH and VIP

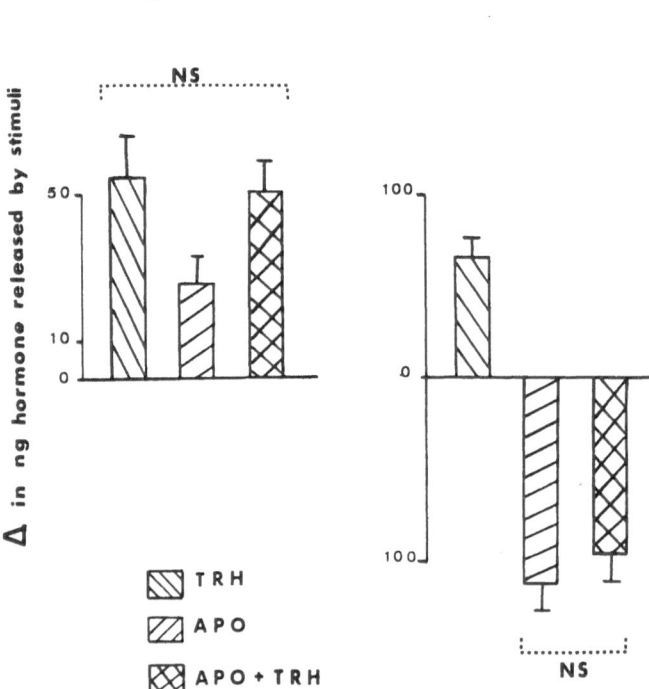

Fig. 4. *GH* and prolactin (*PRL*) responses from superfused rat pituitary fragments *in vitro* to *TRH* (10^{-8} *M*), apomorphine (*APO*) (10^{-7} *M*) or TRH plus apomorphine (*APO-TRH*). Δ represents the total amount of additional GH or prolactin released as a result of drug addition to the medium (difference between induced and control release). Values are the mean ± SEM of five determinations.

increased GH responses to VIP or TRH in lesioned animals, for treatment with DA antagonists does not elicit TRH responsiveness in intact animals, nor does DA itself block TRH-induced GH responses from perifused pituitaries (Fig. 4).

RECEPTOR AND TRANSDUCTION MECHANISMS

The capacity of peptides to reinforce or to antagonize one another's action or the existence of inconstant somatotropic responses to a given hypothalamic factor stress the importance of reception and transduction mechanisms for modulation of GH control.

A major difficulty in analyzing these mechanisms is that of localizing receptors or second messengers present on GH cells. Most available data on receptor and coupling processes were obtained with heterogeneous cell culture or membrane homogenates; hence, they do not really address the question. We recently attempted to re-evaluate the problem by stydying populations of dispersed pituitary cells separated by unit gravity sedimentation. When these were too scarce to characterize coupling mechanisms under appropriate conditions, "additivity" or "nonadditivity" assessment of second messengers was achieved on the whole cell population. This method relies on the assumption that whenever two receptors coupled to the same second messenger induce nonadditive or only partially additive effects, it can be concluded that they overlap on the same cell type.

GHRH receptors seem to be located only on GH cells (Wehrenberg et al. 1984), whereas SRIF receptors are present on lactotrophs and on thyreotrophs as well (Epelbaum et al. 1987). In human pituitary tumors, SRIF receptors have been shown to correlate with pathological GH secretion (Moyse et al. 1984). GHRH receptors are positively coupled with adenylate cyclase (Cronin et al. 1982; Bilezikjian and Vale 1983), while SRIF is inhibitory to the enzyme (Bilezikjian and Vale 1983; Epelbaum et al. 1987; Chneiweiss et al. 1987). The presence of DA receptors on somatotrophs could not be confirmed; if there are any, their concentration on this cell type was probably too small and, under our experimental conditions, GH cells were insufficiently separated from prolactin cells, a major source of DA receptors. But autoradiographic studies are indicative of DA binding on somatotrophs (Rosenzweig et al. 1982). In addition, VIP and TRH receptors are also present on somatotrophs. As in other tissues, VIP receptors on somatotrophs appear coupled to adenylate cyclase, as indicated by incomplete additivity of cyclase stimulation by GHRH and VIP (Fig. 5). At the concentrations of VIP used in that study, a direct interaction of the peptide with GHRH receptors, as described in other systems (Waelbroeck et al. 1985), seems unlikely.

In an attempt to determine whether adenylate cyclase modulation did indeed

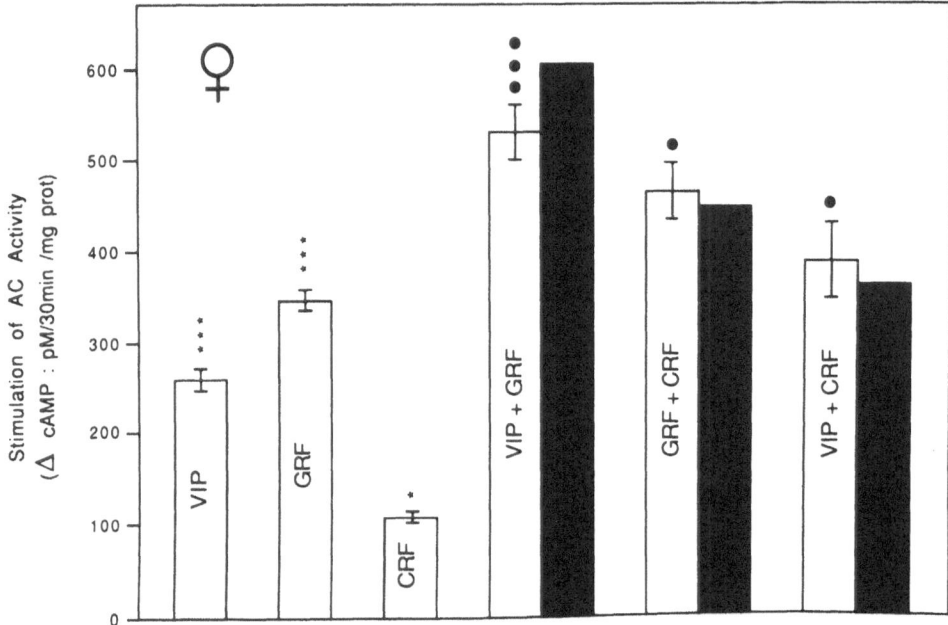

Fig. 5. Additivity of the effects of *VIP, GRF,* and *CRF* on adenylate cyclase activity of female rat adenohypophysis. *Black bars* represent the theoretical additivity of the effect of paired neurohormones. Results (mean ± SEM of three individual experiments) are compared by analysis of variance: *, $P < .05$ and ***, $P < .001$ versus basal activity (167 ± 84 pM/30 min/mg of protein); ●, $P < .05$ and ●●●, $P < .001$ versus the major effect of neurohormones tested separately. Total additivity of the effect of CRF with that of VIP or GRF indicates that CRF receptors are localized on cell types not affected by VIP or GRF.

account for GHRH and SRIF actions on GH secretion, pertussis toxin was used (a bacterial toxin which uncouples receptor-mediated adenylate cyclase inhibition). Under these conditions, cAMP production was no longer inhibited by SRIF, as expected, GH inhibition, however, was only partially blocked. This discrepancy does not depend upon the presence of SRIF receptor subtypes which could affect cAMP and other processes in an independent manner, since only one category of binding sites has been described in the pituitary (Epelbaum et al. 1987; Fig. 6). It rather suggests that, in addition to cAMP, other transduction mechanisms are involved in controlling receptor-dependent inhibition of GH.

So far, no effect of SRIF on phospholipase-C has been found. SRIF (Wehrenberg et al. 1984) actions independent of cAMP are also unlikely to involve phospholipase-A2 and the arachidonic cascade. Inhibition of the enzyme phospholipase-A2 by a competitive analogue of arachidonic acid, ETYA, inhibits basal release of GH by pituitary cells but does not interfere with the actions of GHRH (Fig. 7) or SRIF (Fig. 8) on secretion of the hormone.

In contrast, inhibition of ionic channels, in particular Ca^{2+}-dependent K^+

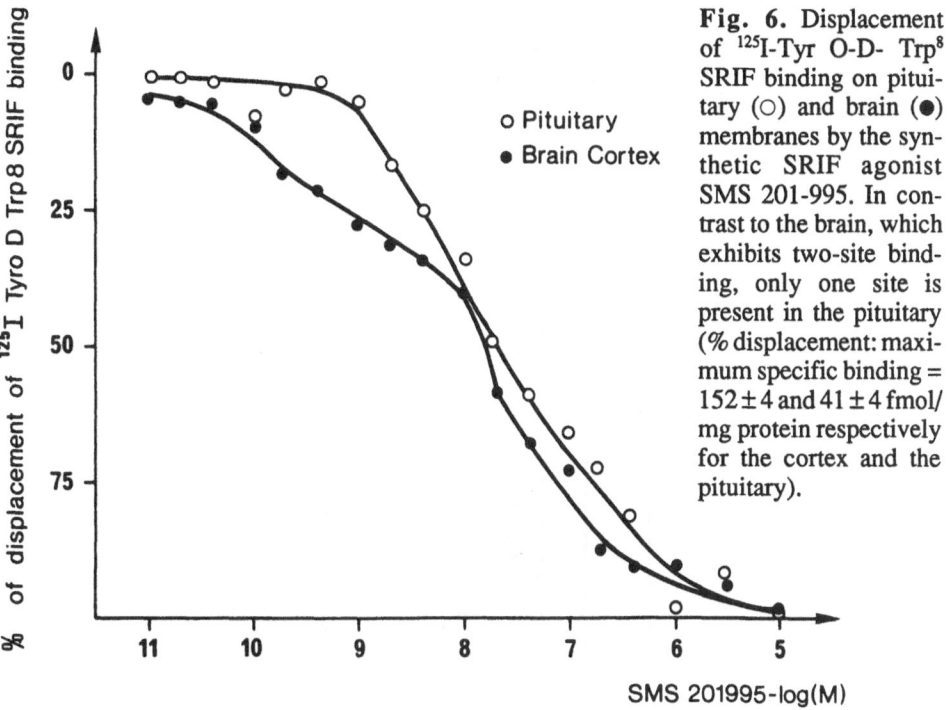

Fig. 6. Displacement of ^{125}I-Tyr O-D- Trp^8 SRIF binding on pituitary (○) and brain (●) membranes by the synthetic SRIF agonist SMS 201-995. In contrast to the brain, which exhibits two-site binding, only one site is present in the pituitary (% displacement: maximum specific binding = 152 ± 4 and 41 ± 4 fmol/mg protein respectively for the cortex and the pituitary).

Fig. 7. Combined effects of ETYA (eicosa - 5, 8, 11, 14, tetraynoic acid), an inhibitor of arachidonate metabolism ($10^{-5} M$), and GRF ($10^{-8} M$) on GH secretion by cultured adenohypophyseal cells: the experiment reveals preferential inhibition of basal (- 56%) in comparison with stimulated (- 34%) secretion.

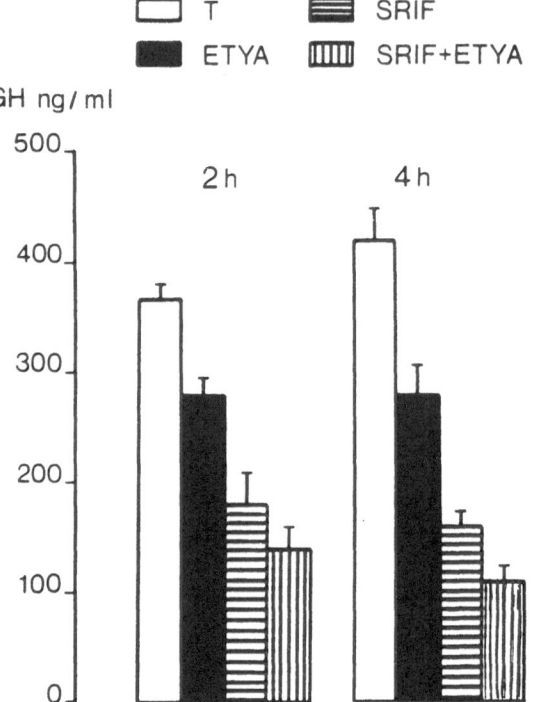

Fig. 8. Additivity of ETYA (10^{-5} M)- and SRIF (10^{-8} M)- induced inhibition of GH secretion by cultured adenohypophyseal cells. Note the increase of inhibition with increasing incubation periods.

channels, has been reported to follow application of the peptide (Koch and Schonbrunn 1986); the effect could be a direct one, but it is more likely to involve mediation by coupling proteins which could link SRIF receptors to the channel.

CONCLUSIONS

Among factors which modulate growth hormone release directly or indirectly, two (CRF and opioids) are predominantly involved in stress responses. But their overall effect seems paradoxical: the SRIF-releasing action of CRF fits well with the frequent occurrence of decreased plasma GH levels during stress in rodents (Mounier et al. 1983). In contrast, opioids tend to inhibit SRIF and to activate GHRH release, two actions which should rather result in elevated GH secretion. This dual control of GH by stress peptides could account for the species differences observed in the somatotropic response to stress: whereas GH inhibition is a major neuroendocrine parameter of stress in rodents (Mounier et al. 1983), plasma levels of growth hormone increase in primates under comparable conditions (Brown and Reichlin 1972). The differential GH response would thus reflect a different balance in positive and negative influences between those species. This interspecific "flexibility" contrasts with the constancy of other hormonal responses to stress, such as those of prolactin and ACTH, which are always stimulated, irrespective of the species. Species differences in GH responsiveness to stress are possibly reinforced by the fact that other peptides originating from the paraventricular nucleus, a structure primarily concerned with processing stress information (Doris 1984; Weiner et al. 1988), stimulate GH, i.e., exhibit a converse action to that of CRF.

The complex signaling of hypothalamic transmitters to the pituitary is suggestive of a "coded" regulation system. In that sense, the overall response of somatotrophs, as that of other pituitary cell types, depends upon signal combinations rather than of upon additive effects of individual signals. Such "coding" could explain why GH and prolactin cells can respond differentially to hypothalamic stimulation in spite of exhibiting several common neuropeptide receptors.

REFERENCES

Alonso G, Tapia-Arancibia L, Assenmacher I (1985) Electron-microscopic immunocytochemical study of somatostatin neurons in the periventricular nucleus of the rat hypothalamus with special references to their relationships with homologous neuronal processes. *Neuroscience* **16**: 297-306

Bicknell RJ, Chapman C (1983) Bombesin stimulates growth hormone secretion from cultured bovine pituitary cells. *Neuroendocrinology* **36**: 33-38

Bilezikjian LM, Vale WW (1983) Stimulation of the adenosine 3', 5'-monophosphate production by growth hormone-releasing factor and its inhibition by somatostatin in anterior pituitary cells *in vitro. Endocrinology* **113**: 1726-1731

Bluet-Pajot MT, Schaub C (1978) Effects of inhibitors of catecholamine synthesis and catecholamine agonists on morphine- and hypoglycemia-induced release of growth hormone in the rat. *J Endocr* **76**: 365-366

Bluet-Pajot MT, Durand D, Drouva S, Mounier F, Pessac M, Kordon C (1986) Further evidence that thyrotropin-releasing hormone participates in the regulation of growth hormone secretion. *Neuroendocrinology* **44**: 70-75

Bluet-Pajot MT, Mounnier F, Léonard JF, Kordon C, Durand D (1987) Vasoactive intestinal peptide induces a transient release of growth hormone in the rat. *Peptides* **8**: 35-38

Brown GM, Reichlin S (1972) Psychologic and neural regulation of GH secretion. *Psychosom Med* **34**: 45-61

Chneiweiss H, Bertrand P, Epelbaum J, Kordon C, Glowinski J, Prémont J Enjalbert A (1987) Somatostatin receptors on cortical neurons and adenohypophysis: comparison between specific binding and adenylate cyclase inhibition. *Eur J Pharmacol* **138**: 249-255

Ciofi P, Tramu G (1986) Nouveau système neuronique tubéro-extra-infundibulaire caractérisé par la co-existence de G/CCK, NPY et hGRF. *Ann Endocrinol (Paris)* **47**: 21N

Collu R, Fraschini F, Visconti P, Martini L (1972) Adrenergic and serotoninergic control of GH secretion in adult male rats. *Endocrinology* **90**: 1231-1237

Conrad LC, Pfaff DW (1976) Efferents from medial basal forebrain and hypothalamus in the rat. An autoradiographic study of the anterior hypothalamus. *J Comp Neurol* **169**: 221-262

Cronin MJ, Rogol AD, Dabney LG Thorner MO (1982) Selective growth hormone and adenosine 3'-5'-mono phosphate-stimulating activity is present on a human pancreatic islet cell tumor. *J Clin Endocrinol Metab* **55**: 381-383

Denef C, Schramme C, Baes M (1985) Stimulation of growth hormone release by vasoactive intestinal peptide and peptide PHI in rat anterior pituitary reaggregates. *Neuroendocrinology* **40**: 88-91

Doris PA (1984) Vasopressin and central integrative processes. *Neuroendocrinology* **38**: 75-85

Drouva SV, Epelbaum J, Tapia-Arancibia L, Laplante E, Kordon C (1980) Metenkephalin inhibition of K^+ induced LHRH and SRIF release from rat mediobasal hypothalamic slices. *Eur J Pharmacol* **61**: 411-412

Drouva SV, Epelbaum J, Tapia-Arancibia L, Laplante E, Kordon C (1981) Opiate receptors modulate LHRH and SRIF release from mediobasal hypothalamic neurons. *Neuroendocrinology* **32**: 163-167

Durand D, Martin JB, Brazeau P (1978) Evidence for a role of α-adrenergic mechanisms in regulation of episodic growth hormone secretion in the rat. *Endocrinology* **100**: 722-728

Elde RP, Parson JA (1975) Immunocytochemical localization of somatostatin in cell bodies of the rat hypothalamus. *Am J Anat* **144**: 541-548

Epelbaum J (1986) Somatostatin in the central nervous system: physiology and pathological modifications. *Prog Neurobiol* **27**: 63-100

Epelbaum J, Tapia-Arancibia L, Besson J, Rotsztejn WH, Kordon C (1979) Vasoactive intestinal peptide inhibits release of somatostatin from hypothalamic slices *in vitro*. *Eur J Pharmacol* **58**: 493-495

Epelbaum J, Tapia-Arancibia L, Kordon C (1981) Noradrenaline stimulates somatostatin release from incubated slices of the amygdala and the hypothalamic preoptic area. *Brain Res* **215**: 393-397

Epelbaum J, Enjalbert A, Krantic S, Musset F, Bertrand P, Rasolonjanahary R, Shu C, Kordon C (1987) Somatostatin receptors on pituitary somatotrophs, thyreotrophs and lactotrophs: pharmacological evidence for loose coupling to adenylate cyclase. *Endocrinology* **121**: 2177-2185

Ferland L, Labrie F, Arimura K, Schally AV (1977) Stimulated release of hypothalamic growth hormone releasing activity by morphine and pentobarbital. *Mol Cell Endocrinol* **6**: 247-252

Gonzalez-Barcena D, Kastin AJ, Schalch DS, Torrez-Zamora M, Perez-Pasten E, Kato A, Schally AV (1973) Response to thyrotropin-releasing hormone in patients with renal failure and after infusion in normal men. *J Clin Endocrinol Metab* **36**: 117-120

Hirooka Y, Hollander CS, Suzuki S, Ferdinand F, Juan S (1978) Somatostatin inhibits release of thyrotropin-releasing factor from organ cultures of rat hypothalamus. *PNAS* **75**: 4509-4513

Irie M, Tsushima T (1972) Increase of serum growth hormone concentration following thyrotropin-releasing hormone injection in patients with acromegaly or gigantism. *J Clin Endocrinol Metab* **35**: 97-100

Jirikowski GF, Merchenthaler I, Rieger GE, Stumpf WE (1986) Estradiol target sites immunoreactive for B-endorphin in the arcuate nuclei of rat and mouse hypothalamus. *Neurosci Lett* **65**: 121-126

Koch D, Schonbrunn A (1986) A transmembrane K^+ gradient is required for somatostatin to decrease intracellular free Ca^{2+} and inhibit hormone release via a cAMP-independent mechanism. 16th Ann. Meeting of Neuroscience (Abstr)

Kordon C, Rotten D, Durand D, Bluet-Pajot MT (1987) Neuroendocrine control of episodic hormone secretion. In: Wagner TOF, Filicori M (eds) *Episodic Hormone Secretion: from Basic Science to Clinical Application*. pp. 25-36

Maeda K, Frohman LA (1978) Dissociation of systemic and central effects of neurotensin on the secretion of growth hormone, prolactin and thyrotropin. *Endocrinology* **103**: 1903-1909

Maeda K, Kato Y, Ohgo S, Chihara K, Yoshimoto Y, Yamaguchi N, Kuromaru S, Imura H (1975) GH and prolactin release after injection of TRH in patients with depression. *J Clin Endocrinol Metab* **40**: 501-505

Melander T, Hökfelt T, Rökeaus A, Cuello AL, Oertel WH, Verhofstad A, Goldstein M (1986) Coexistence of galanin-like immunoreactivity with catecholamines, 5-hydroxytryptamine, Gaba and neuropeptides in the rat CNS. *J Neurosci* **6**: 3640-3654

Morrell DI, McGinty JF, Pfaff DW (1985) A subset of β-endorphin- or dynorphin-containing neurons in the medial basal hypothalamus accumulates estradiol. *Neuroendocrinology* **41**: 417-426

Mounier F, Bluet-Pajot MT, Durand D, Schaub C (1983) Effects of acute stress on growth hormone release induced by morphine and clonidine in the rat. *Neuroendocr Lett* **5**: 221-226

Moyse E, Le Dafniet M, Epelbaum J, Pagesy P, Peillon F, Kordon C, Enjalbert A (1984) Somatostatin receptors in human growth hormone and prolactin-secreting pituitary adenomas. *J Clin Endocrinol Metab* **61**: 98-103

Müller EE, Panerai AE, Cocchi D, Gil-Ad I, Rossi GL, Olgiati VR (1977) Growth hormone releasing activity of thyrotropin-releasing hormone in rats with hypothalamic lesions. *Endocrinology* **100**: 1663-1667

Ono N, Lumpkin MD, Samson WK, Mc Donald JK, Mc Cann SM (1984) Intrahypothalamic action of corticotropin-releasing factor (CRF) to inhibit growth hormone and LH release in the rat. *Life Sci* **35**: 1117-1123

Ottlecz AW, Samson K, McCann SM (1985) The effects of gastric inhibitory polypeptide (GIP) on the release of anterior pituitary hormones. *Peptides* **6**: 115-119

Palkovits M (1981) Catecholamines in the hypothalamus: an anatomical review. *Neuroendocrinology* **33**: 123-128

Palkovits M, Zaborsky L, Feminger A, Mezey E, Fekete MK, Herman JP, Kanyicska B, Szabo D (1980) Noradrenergic innervation of the rat hypothalamus: experimental biochemical and electron-microscopic studies. *Brain Res* **191**: 161-171

Peterfreund R, Vale W (1983) Ovine corticotropin-releasing factor stimulates somatostatin secretion from cultured brain cells. *Endocrinology* **112**: 1275-1279

Peterfreund R, Vale W (1984) Somatostatin analogs inhibit somatostatin secretion from cultured hypothalamic cells. *Neuroendocrinology* **39**: 397-402

Plotsky PM, Vale W (1985) Patterns of growth hormone-releasing factor and somatostatin secretion into the hypophyseal portal circulation of the rat. *Science* **230**: 461-463

Rosenzweig LJ, Kanwar YS (1982) Dopamine internalization by and intracellular distribution within prolactin cells and somatotrophs of the rat anterior pituitary as determined by quantitative electron-microscopic autoradiography. *Endocrinology* **111**: 1817-1824

Sawchenko PE, Swanson LW, Rivier J, Vale W (1985) The distribution of growth hormone-releasing factor (GRF) immunoreactivity in the central nervous system of the rat: an immunohystochemical study using antisera directed against rat hypothalamic GRF. *J Comp Neurol* **237**: 100-115

Serri O, Deslauriers N, Brazeau P (1987) Dual action of dopamine on growth hormone release *in vitro*. *Neuroendocrinology* **45**: 363-367

Shimatsu A, Kato Y, Matsushita N, Katakami H, Yanaihara N, Imura H (1982) Effects of glucagon, neurotensin and vasoactive intestinal polypeptide on somatostatin release from perifused rat hypothalamus, *Endocrinology* **110**: 2113-2117

Swanson LW, Sawchenko PE, Rivier J, Vale W (1983) Organization of ovine corticotropin-releasing factor immunoreactive cells and fibers in the rat brain: an immunocytochemical study. *Neuroendocrinology* **36**: 165-186

Takahara J, Yunoki S, Yamauchi J, Yakushi J, Hosogi H, Ofuji T (1978) Effect of glucagon on growth hormone secretion in rats. *Horm Metab Res* **10**: 227-236

Tapia-Arancibia L, Arancibia S, Astier H (1984) K$^+$-induced thyrotropin-releasing hormone release from superfused mediobasal hypothalami in rats, inhibition by somatostatin. *Neurosci Lett* **45**: 47-52

Vigh S, Schally AV (1984) Interaction between hypothalamic peptides in a superfused pituitary cell system. *Peptides* **5**: [Suppl 1]: 241-247

Vijayan E, Samson WK, Said JJ, McCann SM (1979) Vasoactive intestinal peptide: evidence for a hypothalamic site of action to release growth hormone, luteinizing hormone and prolactin in conscious ovariectomized rats. *Endocrinology* **104**: 53-57

Waelbroeck M, Robberecht P, Coy DH, Camus JC, De Neef P, Christophe J (1985) Interaction of growth hormone-releasing factor (GRF) and 14 GRF analogs with vasoactive intestinal peptide (VIP) receptors of rat pancreas. Discovery of (N-AC-Tyr^1, D-Phe^2)-GRF(1-29)-NH$_2$ as a VIP antagonist. *Endocrinology* **116**: 2643-2649

Wardlaw SC, Wehrenberg WB, Férin M, Antunes JL, Frantz AG (1982) Effect of sex steroids on β-endorphin in hypophyseal portal blood. *J Clin Endocrinol Metab* **55**: 877-881

Wehrenberg WB, Baird A, Ying SY, Rivier C, Ling N, Guillemin R (1984) Multiple stimulation of the adenohypophysis by combinations of hypothalamic releasing factors. *Endocrinology* **114**: 1995-2001

Wehrenberg WB, Bloch B, Ling N (1985) Pituitary secretion of growth hormone in response to opioid peptides and opiates is mediated through growth hormone-releasing factor. *Neuroendocrinology* **41**: 13-16

Weiner RI, Findell PR, Kordon C (1988) Role of classic and peptide neuromediators in the neuroendocrine regulation of LH and prolactin. In: Knobil E, Neill J (eds) *The Physiology of Reproduction.* Raven, New York, pp 1235-1281

Willoughby JO, Terry LC, Brazeau P, Martin JB (1977) Pulsatile growth hormone, prolactin and thyrotropin secretion in rats with hypothalamic deafferentation. *Brain Res* **127**: 137-152

Willoughby JO, Beroukas D, Blessing WW (1987) Ultrastructural evidence for gamma aminobutyric acid-immunoreactive synapses on somatostatin-immunoreactive perikarya in the periventricular anterior hypothalamus. *Neuroendocrinology* **46**: 268-272

Yajima Y, Akita Y, Saito T (1986) Pertussis toxin blocks the inhibitory effects of somatostatin on cAMP-dependent vasoactive intestinal peptide and cAMP-independent thyrotropin-releasing hormone-stimulated prolactin secretion of GH3 cells. *J Biol Chem* **261**: 2684-2689

Advances in Growth Hormone and Growth Factor Research,
edited by E.E. Müller, D. Cocchi and V. Locatelli
Pythagora Press, Roma-Milano and Springer Verlag, Berlin-Heidelberg © 1989

Growth hormone releasing hormone and somatostatin in the physiology and pathophysiology of growth hormone secretion*

L. A. Frohman, P. Chomczynski, T. R. Downs, H. Katakami and J.-O. Jansson

Division of Endocrinology and Metabolism, Department of Internal Medicine, University of Cincinnati College of Medicine, Cincinnati, Ohio, U.S.A.

The regulation of growth hormone (GH) secretion involves a complex neuro-endocrine control system that includes the participation of neurotransmitters and feedback by hormonal and metabolic substrates. The final common integrative pathway for these signals consists of two neuropeptides, which are hypophysi-otropic hormones: GH-releasing hormone (GRH), a 40-44-residue peptide which exerts stimulatory effects on GH secretion, and somatostatin (SRIF), a tetrade-capeptide which exhibits an inhibitory influence. Although other hypothalamic neuropeptides and monoamines can be demonstrated to exert effects on GH secretion when injected or added to somatotrophs *in vitro*, there is as yet no convincing evidence for their direct effects on the pituitary under physiologic conditions. They may, however, exert important indirect effects on the regulation of GH secretion by modifying the secretion of GRH and SRIF. Similarly, GRH and SRIF participate in the regulation of GH secretion by neurotransmitter effects within the central nervous system (CNS) on one another or by autocrine effects on their own secretion via ultrashort loop feedback mechanisms.

Disturbances of the secretion of these two factors may have a profound influence on the secretion of GH, and an increasing number of experimental and spontaneous disorders of GH secretion in laboratory animals and clinical disorders of GH secretion in human beings may now be attributed to alterations in the

* The studies in the authors' laboratory were supported by USPHS grants DK 30667 and RR00068.

secretion or action of GRH and SRIF. This chapter reviews aspects of the normal physiology of GRH and SRIF on GH secretion and the consequences of their disordered secretion on somatotroph function.

THE PATTERN OF GH SECRETION

Growth hormone is secreted in an episodic manner in every species in which it has been examined. The secretory-pattern characteristics vary among species and have been most carefully examined in rats and human beings (Quabbe et al. 1966; Tannenbaum and Ling 1984). In the rat, the pattern of GH secretion is sexually dimorphic and greatly influenced by androgens during both the neonatal and the adult period (Jansson et al. 1985). In male rats, secretory bursts of GH occur at 3- to 3.3-h intervals, interspersed by trough periods during which levels are virtually undetectable. In contrast, GH pulses in female rats are more frequent, irregular in timing, and of lesser magnitude, and GH trough values are higher than in male rats. These differences result in a more continuous exposure of tissues to the hormone in females than in males. As a consequence, sexually dimorphic patterns of body growth, hepatic steroid metabolism, and hormone and growth factor receptor concentrations may, to a major extent, be attributable to the differences in patterns of GH secretion.

In monkeys, GH secretory bursts also occur at approximately 3-to 4-h intervals, though they appear less well synchronized than in rats, possibly because of the greater difficulty in eliminating stress during sampling in this species. In human beings pulsatile GH secretion exhibits greater variability, though patterns in individual subjects tend to be relatively consistent. While the majority of GH is secreted during the night, pulses of considerable magnitude, unrelated to sleep, activity, stress, or nutrient intake, occur throughout the day. GH secretory peaks are less pronounced in women than in men, but the differences are not as marked as in rodents.

ROLE OF GRH AND SRIF IN PULSATILE GH SECRETION IN THE RAT

The contributions of GRH and SRIF to the generation of GH secretory pulses have been examined in several species, but they have been most extensively studied in the rat, where two techniques have been employed to exclude selectively the effect of one of the hormones.

Passive immunization with anti-GRH or anti-SRIF serum has been used to bind (and presumably neutralize) circulating GRH and/or SRIF. Most studies utilizing anti-SRIF serum injection have demonstrated a preservation of pulsatile GH secretion despite an increase in basal GH levels (Ferland et al. 1976). Administration of anti-GRH serum, in contrast, has little effect on the already low basal GH levels but inhibits the secretory bursts (Wehrenberg et al. 1982). These findings have been interpreted to indicate that SRIF is important for maintaining low basal levels but that pulsatile GH secretion is dependent on GRH. Such conclusions are dependent on the presumption that elimination of GRH effect on the pituitary does not impair stimuli that are mediated through inhibition of SRIF. This may not be entirely correct, however, since very few, if any, stimuli for GH secretion are effective in the rat pretreated with anti-GRH serum or whose somatotrophs cannot respond to GRH.

A limitation of passive immunization experiments relates to the kinetics of antigen-antibody binding *in vivo*, which can be estimated only from *in vivo* measurements. Although the binding capacity of the injected antiserum is generally not an issue, the rapid dissociation that often occurs at 37°C (Frohman et al. 1979) and the extremely brief period of time required for transport of peptides from the median eminence to the pituitary can give unpredictable results. For example, continuous infusion rather than a single injection of anti-SRIF serum produces an inhibition of pulsatile GH secretion as well as an elevation of basal GH levels (Fig. 1).

Morphine-stimulated GH secretion, believed to be GRH mediated, is also inhibited. In contrast, the GH response to a maximal dose of exogenous GRH is enhanced. These results imply an inhibition of GRH secretion as a consequence of the elevated GH levels, either directly or through an action of SRIF within the CNS that is unaffected by the high circulating levels of anti-SRIF serum. The enhancement of the GH response to a near maximal dose of exogenous GRH in anti-SRIF-treated animals indicates persistence of the effects of the antiserum throughout the period of observation.

A second method used to eliminate the effects of hypophysiotropic hormones is ablation of the hypothalamic loci containing the perikarya in which they are synthesized or transection of the axonal fibers involved in their transport to the median eminence. Confirmation of such lesions is generally by demonstration of decreased content of the hormone in the median eminence. The limitation of this technique is that it is far from selective. The anatomic organization of the CNS is such that destruction of only one population of neurons is virtually impossible. Thus, other neural influences on the system under observation are necessarily modified and may complicate the interpretation of the observed results. Lesions of the medial preoptic (MPO) area of the hypothalamus or anterolateral hypothalamic deafferentation, which interrupt the pathways from the MPO to the median eminence, deplete median eminence SRIF, elevate basal GH levels for at least 8 days, and abolish the expected pattern of pulsatile GH secretion (Fig. 2). The elevated basal levels of GH secretion can be attributed in part to increased GRH secretion, which can be demonstrated *in vitro* (Fig. 3).

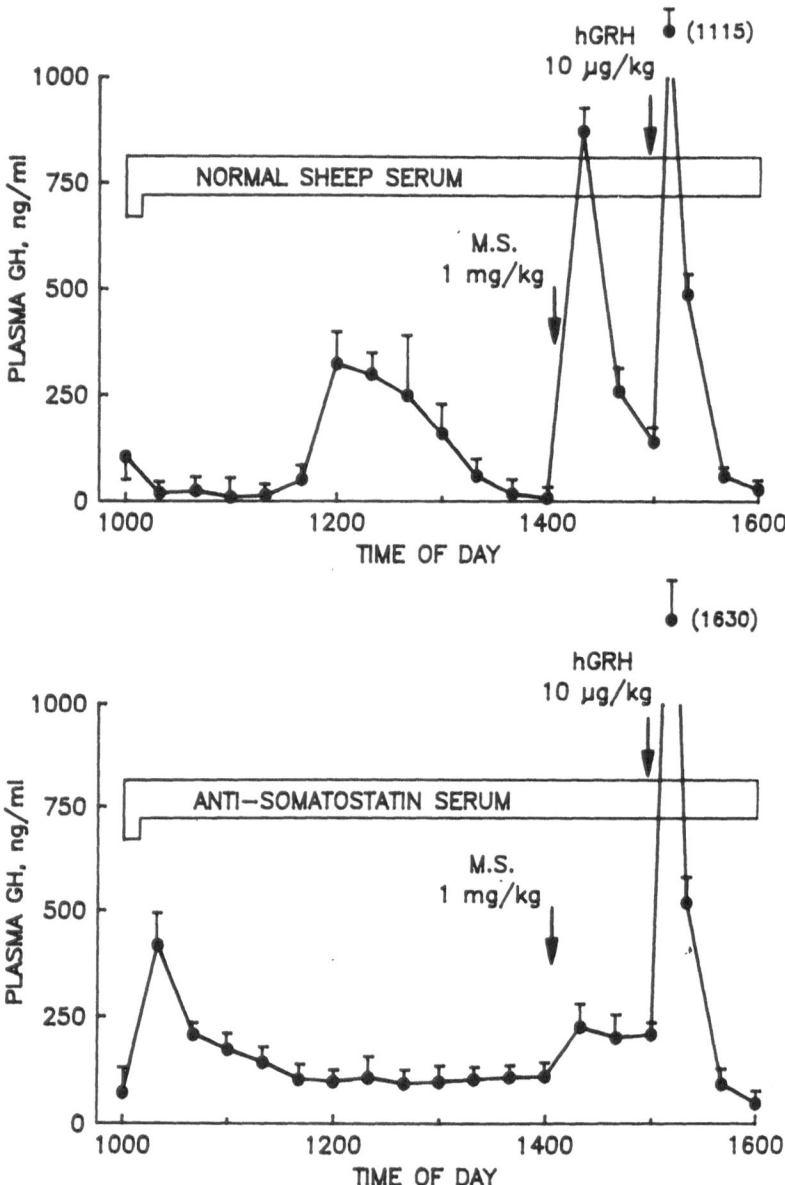

Fig. 1. Effect of infusion of anti-SRIF serum in unanesthetized rats on spontaneous GH secretion and on the responses to morphine sulfate (*M.S.*) and hGRH. Shown are the mean and SEM. Each group contained five animals. Anti-SRIF serum produced a sustained elevation of GH levels, an inhibition of pulsatile GH secretion, a near complete suppression of morphine-induced GH secretion, and an enhancement of GRH-stimulated GH secretion.

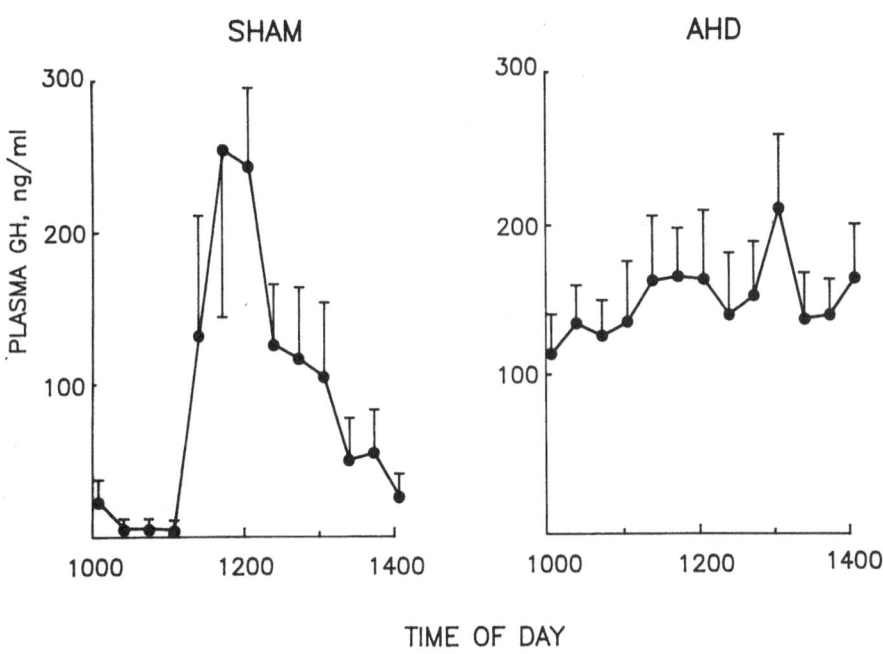

Fig. 2. Effect of bilateral anterolateral hypothalamic deafferentation (*AHD*) on spontaneous GH secretion in unanesthetized rats. Shown are the mean and SEM. Deafferented rats exhibited elevated basal levels and a loss of pulsatile GH secretion.

Thus, reduction or elimination of circulating SRIF tone, produced either by anti-SRIF serum or by hypothalamic destruction, not only elevates basal GH levels but also interferes with the normal periodicity of GH secretion. Together with the results of anti-GRH serum administration experiments, in which pulsatile GH secretion is also abolished, these findings are best explained by a bihormonal control of GH secretion with asynchronous periodic release of GRH and SRIF. Pulses of GH secretion would therefore be expected at times of maximal GRH and minimal SRIF secretion. Evidence in support of this model has been provided by (a) the demonstration that GH responses to exogenous GRH are greater at the time of GH peak (when SRIF tone is presumed low) than during trough periods (when SRIF tone should be high) (Tannenbaum and Ling 1984), and (b) a report of asynchronous GRH and SRIF peak in rat portal plasma (Plotsky and Vale 1985). Studies in sheep involving repeated hypothalamic-pituitary portal blood sampling are currently in progress to determine whether a similar pattern exists in the unanesthetized animal.

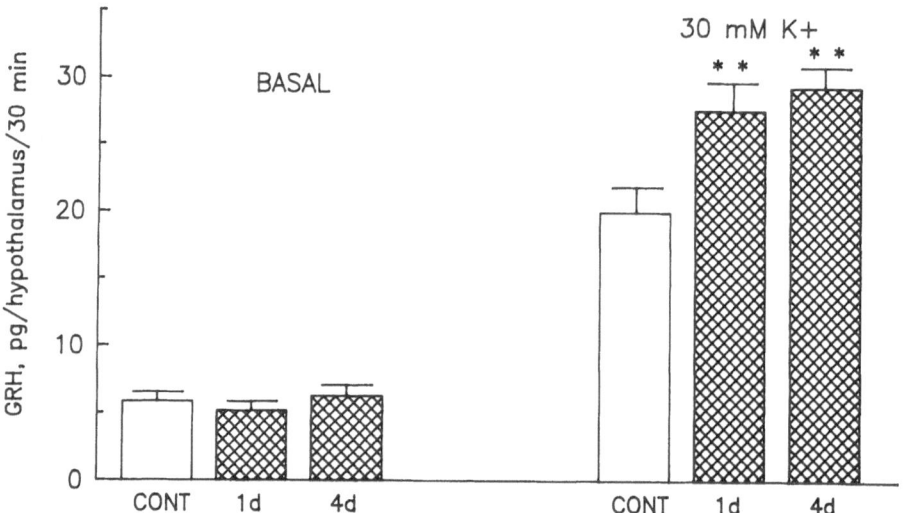

Fig. 3. Effect of anterolateral hypothalamic deafferentation on spontaneous and K⁺-induced GRF secretion from hypothalamic fragments in short-term incubations. Shown are the mean and SEM of quadruplicate incubations, each containing three hypothalami. Stimulated GRH secretion was significantly increased ($p < 0.01$) at 1 and 4 days.

ROLE OF SOMATOTROPH RECEPTORS IN MODULATING THE EFFECTS OF GRH AND SRIF

As with many other hormone-receptor systems, down-regulation of the GRH receptor has been demonstrated (Bilezikjian and Vale 1984) after prolonged exposure to GRH. The importance of this process in the physiologic regulation of GRH action is uncertain, since only about 30% receptor occupancy is required for full biologic activity. Repeated administration of GRH at short intervals (i.e., 2 h) or continuous infusions of GRH lead to decreased somatotroph responsiveness (Vance et al. 1984). However, several factors other than GRH receptors could account for this response attenuation. They include: (a) stimulation of GH, which in turn, stimulates SFIR release; (b) depletion of a finite GRH-releasable pool of GH within the somatotroph; and (c) desensitization or down-regulation at a post-receptor level. The response attenuation must necessarily be only partial, since continuous excessive stimulation by GRH is sufficient to deplete the majority of pituitary GH content (Wehrenberg et al. 1984) and to cause somatotroph hyperplasia and acromegaly (Frohman and Jansson 1986).

Somatostatin effectiveness is also dependent on the status of SRIF receptors, and receptor down-regulation has been demonstrated in response to GH-stimulated

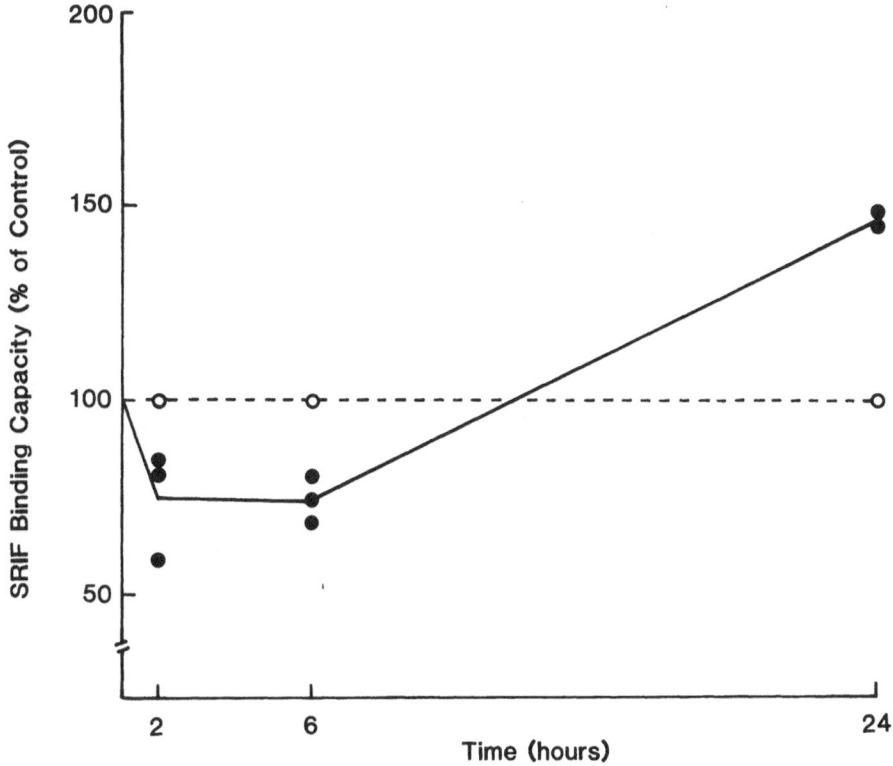

Fig. 4. Effect of GH treatment on SRIF binding capacity of pituitary membranes. Data are expressed as a percentage of the binding in anterior pituitary membranes from rGH-treated animals (*closed circles*) in relation to that in vehicle-treated controls (*open circles*), the mean binding of which was arbitrarily set at 100%. (Modified from Katakami et al. 1985).

release of endogenous SRIF (Katakami et al. 1985). Within 2 h after the injection of GH in intact rats, pituitary SRIF binding capacity was significantly decreased without alteration of SRIF binding affinity (Fig. 4). The effect persisted for at least 6 h but for less than 24 h. Continuous GH administration, however, led to persistent suppression of SRIF binding capacity. Preliminary data on normal human beings suggests a similar process (M. Kelijman and L.A. Frohman, unpublished observations).

ROLE OF GRH AND SRIF IN GH FEEDBACK REGULATION

The feedback regulation of GH has been evaluated by either increasing or decreasing the concentration of GH or GH-dependent factors (e.g., IGF-I). There

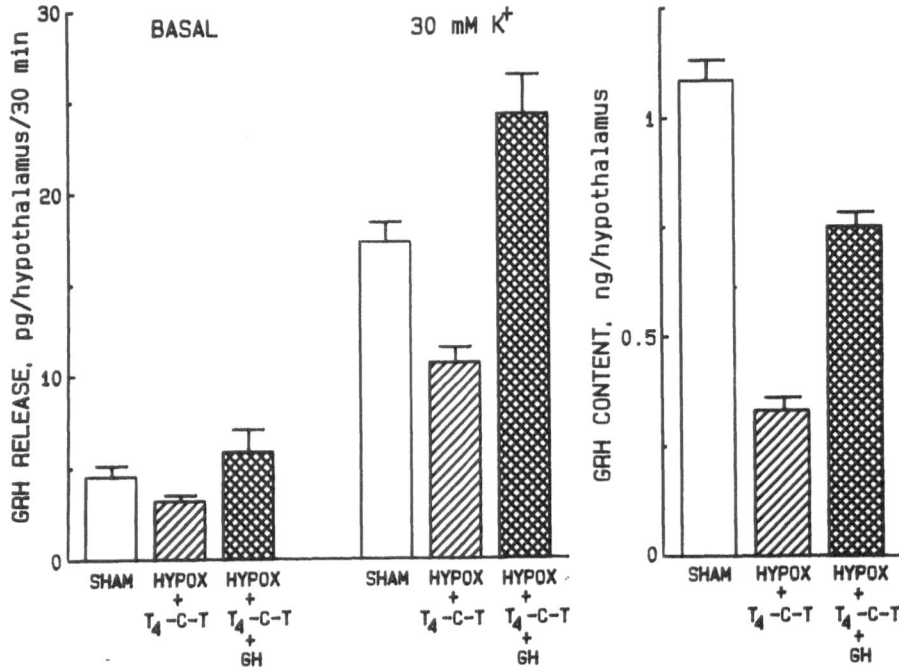

Fig. 5. Effect of hypophysectomy (*HYPOX*) and hormone-replacement treatment (T_4, thyroxine; C, cortisone; T, testosterone) on hypothalamic GRH release and content in the rat. Shown are the mean and SEM ($n = 4$ for each experimental group). (Reprinted from Katakami et al. 1987).

is now abundant evidence from laboratory animals (Abe et al. 1983) and man (Abrams et al. 1971; Rosenthal et al. 1986) that increasing the concentration of GH either in the peripheral circulation or within the central nervous system (by intracerebroventricular administration) results in a decrease in endogenous GH secretion or in the GH secretory response to GRH. The mechanism by which this effect occurs is complex, and present evidence implicates the mediation of both SRIF and GRH.

GH administration to normal rats increases SRIF levels in rat hypothalamus and also increases SRIF release from hypothalami *in vitro* (Berelowitz et al. 1981a). The exposure of hypothalami to GH or to IGF-I *in vitro* at concentrations comparable to those achieved *in vivo* results in a dose-dependent increase in SRIF release (Berelowitz et al. 1981a, b). Thus, increased GH levels activate a classical negative feedback system to inhibit GH secretion. Systemic administration of GRH leading to an increase in GH secretion is also followed by an inhibition of spontaneous GH secretion, presumably through a similar mechanism (Katakami et al. 1986). Intraventricular administration of GRH, however, results in a paradoxical suppression of spontaneous GH secretion. This effect is blocked by prior injection of anti-SRIF serum, indicating a SRIF-mediated mechanism. The pos-

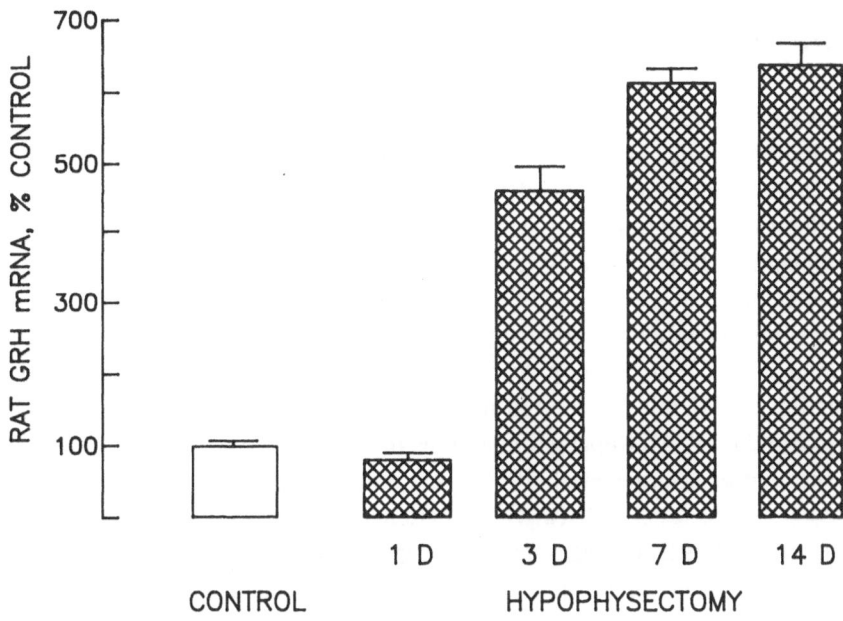

rGRH GENE EXPRESSION: ROLE OF HYPOPHYSECTOMY

Fig. 6. Effect of hypophysectomy on GRH mRNA levels in rat hypothalamus ($n = 4$ for each experimental group).

sible inhibition of endogenous GRH secretion by elevated circulating GH levels is suggested by studies already described (Fig. 1).

The absence of negative feedback is characteristically evaluated by target organ (pituitary) removal or by administration of antihormone (GH) serum. GH deficiency produced by hypophysectomy or anti-GH serum results in a decrease in hypothalamic SRIF levels and a decrease in SRIF release *in vitro* (Berelowitz et al. 1981b). Effects of GH deficiency on GRH secretion appear more complex. Hypophysectomy results in a decrease in hypothalamic GRH content and a decrease in GRH release *in vitro* (Katakami et al. 1987) (Fig. 5). This seeming paradox was further studied by determining the effect of hypophysectomy on hypothalamic GRH gene expression using both Northern blotting and a dot-blot technique (Chomczynski et al. 1987). Within 3 days of hypophysectomy, two- to threefold increase in GRH mRNA levels can be detected (Fig. 6), concomitant with a 30%–40% decrease in hypothalamic GRH content and only a minimal increase in GRH release. This finding indicates that GH deficiency stimulates GRH gene expression at the transcriptional level but that there is impairment in efficient translation of the GRH message. This impairment is unaffected by treatment with thyroid, adrenal, or gonadal hormones but is partially restored by treatment with GH. The mechanism responsible for the impairment in translation is not currently known.

PATHOPHYSIOLOGY OF GRH AND SRIF SECRETION

GH Deficiency

Growth hormone deficiency can be produced experimentally by destruction of the hypothalamic arcuate nucleus-ventromedial nucleus complex, the location of GRH-secreting neuronal perikarya. Both electrolytic and chemical (monosodium glutamate and gold- thioglucose) techniques have been used for this purpose. Hypothalamic GRH content is reduced in such animals, and pituitary sensitivity to GRH is intact.

The clinical counterpart to this model is represented by the majority of children with idiopathic GH deficiency. Such children respond to exogenous GRH with a stimulation of GH release, though in some the response is quantitatively reduced (Schriock et al. 1984). Repeated injections of GRH may result in a small increase in GH responses (Borges et al. 1981; Gelato et al. 1985), though the effect is not analogous to that seen with GnRH in patients with hypothalamic hypogonadism (Snyder et al. 1979). An occasional GH-deficient child does not respond to GRH (Schriock et al. 1984; Gelato et al. 1985) even with repeated stimulation, suggesting a problem with GRH action rather than with GRH secretion.

A similar disorder has been identified in the "little" strain of dwarf mice. The homozygous (lit/lit) mouse is growth retarded, with identifiable somatotrophs containing reduced levels of GH mRNA and of GH, while the heterozygotes are indistinguishable from control mice (Cheng et al. 1983; Jansson et al. 1986a). GH release is not increased by GRH either *in vivo* or *in vitro*, even after pretreatment for periods of up to 1 week. GRH fails to stimulate GH synthesis, cyclic AMP accumulation, or GH release *in vitro* (Fig. 7). In contrast, GH secretion from lit/lit pituitary cultures does respond to stimulation by cyclic AMP, forskolin, and cholera toxin, indicating that the defect is related to either the GRH receptor or to the G_s subunit of adenylate cyclase. When we probed other hormonal responses mediated by adenylate cyclase (glucagon, parathyroid hormone), we found fully intact responses, as reflected by measurement of plasma and urinary cyclic AMP levels respectively (Jansson et al. 1986b). Thus, the most likely explanation for the absent response to GRH is an absence of or abnormality in the GRH receptor or in its coupling to the G_s subunit. Receptor-binding experiments also remain to be performed to confirm these conclusions.

GH deficiency, as a consequence of excessive SRIF secretion, has not been recognized. It is possible, however, that reversible disorders of impaired GH secretion, e.g., psychosocial dwarfism, could be caused by excessive SRIF secretion as a consequence of altered CNS neurotransmitter activity rather than by impaired GRH secretion.

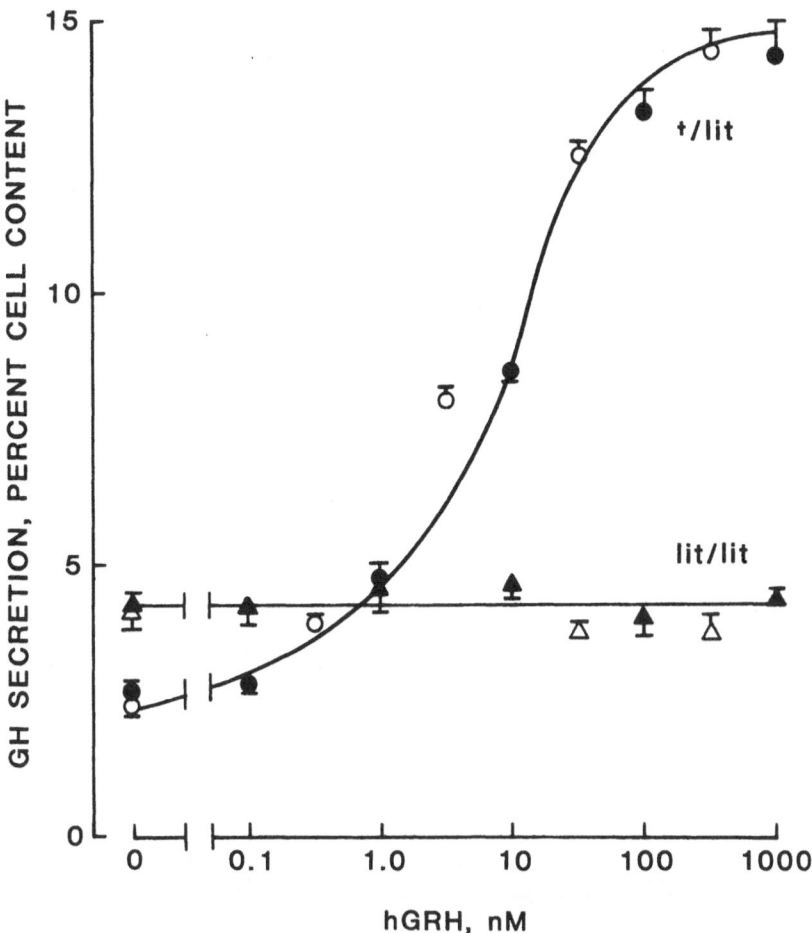

Fig. 7. Effect of hGRH on GH release from dwarf lit/lit and normal heterozygous +/lit mice from pituitary monolayer cultures during a 4-h incubation. Each point represents the mean and SEM of four to eight observations. (Reprinted from Jansson et al. 1986a).

GH Hypersecretion

Decreased SRIF function, whether produced by chemical (cysteamine), surgical (MPO lesions or anterolateral hypothalamic deafferentation), or immunologic (anti-SRIF serum) means, results in increased basal GH levels during short-term observation periods. Enhanced growth, a reflection of long-term GH over-production, has been reported after anterolateral hypothalamic deafferentation (Mitchell et al. 1973).

Decreased SRIF action appears to be an important contributing factor in the GH hypersecretion of patiens with acromegaly. Somatostatin sensitivity, as re-

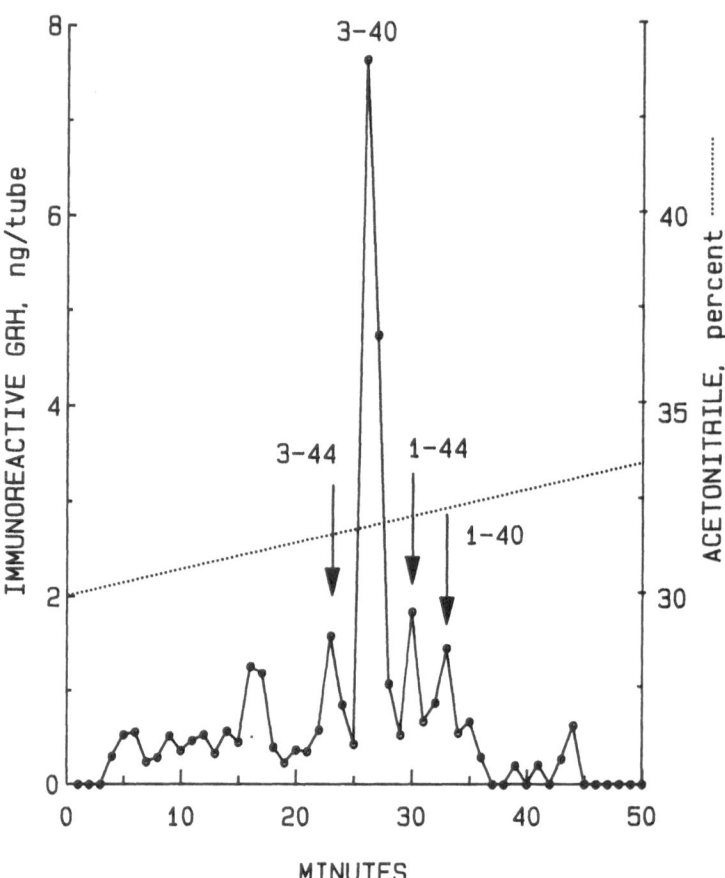

ECTOPIC GRH SECRETION: PLASMA HPLC

Fig. 8. HPLC profile of GRH immunoreactivity in plasma from a patient with ectopic GRH production. The retention times of GRH forms and metabolites are indicated by the *arrows*.

flected by suppression of basal GH levels and inhibition of GH responses to GRH, is significantly reduced in most acromegalics (Kelijman et al. 1987). Adenomatous somatotroph sensitivity to SRIF *in vitro* is highly correlated with that observed *in vivo*. In addition, SRIF receptor binding is diminished (Ikuyama et al. 1986; Reubi et al. 1987). While the role of SRIF in sustained GH hypersecretion may be secondary, there is convincing evidence for a primary role of GRH. Excessive GRH secretion, as occurs in patients with ectopic GRH-producing tumors, leads to acromegaly, somatotroph hyperplasia, and occasionally somatotroph adenoma formation. Plasma immunoreactive GRH levels are markedly elevated, with reported levels ranging from 0.3 to > 50 ng/ml (Frohman and Downs 1987). Recent studies have shown that much of the plasma immunoreactivity in patients with ectopic GRH secretion is not GRH (1-44) or GRH (1-40) but rather a biologically inactive metabolite lacking an aminoterminal dipeptide (Frohman et al. 1986, 1987) which has a markedly prolonged plasma half-life (Fig. 8). Similar observations have now also been made in hGRH transgenic mice, where the majority of GRH immunoreactivity in plasma is represented by GRH (3–44) NH_2 (L.A. Frohman et al., unpublished observations). These findings suggest that the long-term effects of GRH occur at plasma concentrations in the pg/ml range.

SUMMARY

The integrative role of GRH and SRIF in the regulation of GH secretion has been presented. Alterations in each have profound effects on GH secretion. Based on current information, it appears that disturbances in GRH secretion provide the most convincing argument for a pathophysiologic role of hypothalamic hormone secretion in clinically recognized GH disorders.

REFERENCES

Abe H, Molitch ME, van Wyk J, Underwood LE (1983) Human growth hormone and somatomedin-C suppress the spontaneous release of growth hormone in unanesthetized rats. *Endocrinology* **113**: 1319-1324

Abrams RL, Grumbach MM, Kaplan SL (1971) The effect of administration of human growth hormone on the plasma growth hormone, cortisol, glucose, and free fatty acid response to insulin: evidence for growth hormone autoregulation in man. *J Clin Invest* **50**: 940-950

Berelowitz M, Firestone SL, Frohman LA (1981a) Effects of growth hormone excess and deficiency on hypothalamus somatostatin content and release and on tissue somatostatin distribution. *Endocrinology* **109**: 714-719

Berelowitz M, Szabo M, Frohman LA, Firestone S, Chu L, Hintz RL (1981b) Somatomedin-C mediates growth hormone negative feedback by effects on both the hypothalamus and the pituitary. *Science* **212**: 1279-1281

Bilezikjian LM, Vale WW (1984) Chronic exposure of cultured rat anterior pituitary cells to GRF causes partial loss of responsivess to GRF. *Endocrinology* **115**: 2032-2034

Borges JLC, Blizzard RM, Gelato MC, Furlanetto R, Rogol AD, Evans WS, Vance ML, Kaiser DL, MacLeod RM, Merriam GR, Loriaux DL, Spiess J, Rivier J, Vale W, Thorner MO (1981) Effects of human pancreatic tumour growth hormone-releasing factor on growth hormone and somatomedin-C levels in patiens with idiopathic growth hormone deficiency. *Lancet* **ii**: 119-123

Cheng TC, Beamer WG, Phillips JA III, Bartke A, Mallone RL, Dowling AC (1983) Etiology of growth hormone deficiency in little, Ames and Snell dwarf mice. *Endocrinology* **113**: 1669-1678

Chomczynski P, Downs TR, Frohman LA (1987) Feedback regulation of growth hormone (GH)-releasing hormone (GRH) gene expression by growth hormone in rat hypothalamus. 69th Annual Meeting Endocrine Society (abstract 617)

Ferland L, Labrie F, Jobin M, Arimura A, Schally AV (1976) Physiological role of somatostatin in the control of growth hormone and thyrotropin secretion. *Biochem Biophys Res Commun* **68**: 149-156

Frohman LA, Downs TR (1987) Ectopic GRH syndrome. In: Robbins R, Melmed S (eds) *Acromegaly: A Century of Scientific and Clinical Progress* . Plenum, New York, pp 115-126

Frohman LA, Jansson J-O (1986) Growth hormone-relasing hormone. *Endocr Rev* **7**: 223-253

Frohman MA, Frohman LA, Goldman MB, Goldman JN (1979) Use of protein-A-containing staphylococcus aureus as an immuno-adsorbant in radioimmunoassays to separate antibody-bound from free antigen. *J Lab Clin Med* **93**: 614-621

Frohman LA, Downs TR, Williams TC, Heimer EP, Pan Y-CE, Felix AM (1986) Rapid enzymatic degradation of growth hormone-releasing hormone by plasma *in vitro* and *in vivo* to a biologically inactive, N-terminally cleaved product. *J Clin Invest* **78**: 906-913

Frohman LA, Downs TR, Heimer EP, Felix AM (1987) Characterization of the primary human growth hormone-releasing hormone-degrading enzyme in plasma as dipeptidyl-aminopeptidase, type IV. *Clin Res* **35**: 884A

Gelato MC, Ross JL, Malozowski S, Pescovitz OH, Skerda M, Cassorla F, Loriaux DL, Merriam GR (1985) Effects of pulsatile administration of growth hormone (GH)-releasing hormone on short-term linear growth in children with GH deficiency. *J. Clin Endocrinol Metab* **61**: 444-450

Ikuyama S, Nawata H, Kato K-I, Ibayashi H, Nakagaki H (1986) Plasma growth hormone responses to somatostatin (SRIH) and SRIH receptors in pituitary adenomas in acromegalic patients. *J Clin Endocrinol Metab* **62**: 729-733

Jansson J-O, Downs TR, Beamer WG, Frohman LA (1986a) Receptor-associated resistance to growth hormone-releasing hormone in growth hormone-deficient dwarf "little" mice. *Science* **232**: 511-512

Jansson J-O, Downs TR, Beamer WG, Frohman LA (1986b) The dwarf "little" (lit/lit) mouse is resistant to growth hormone-releasing peptide (GHRP-6) as well as to growth hormone-releasing hormone. 68th Annual Meeting Endocrine Society (abstract 397)

Katakami H, Berelowitz M, Marbach M, Frohman LA (1985) Modulation of somatostatin binding to rat pituitary membranes by exogenously administered growth hormone. *Endocrinology* **117**: 557-560

Katakami H, Arimura A, Frohman LA (1986) Growth hormone-releasing hormone stimulates hypothalamic somatostatin release: a inhibitory feedback effect on growth hormone secretion. *Endocrinology* **118**: 1872-1877

Katakami H, Downs TR, Frohman LA (1987) Effect of hypophysectomy on hypothalamic growth hormone-releasing hormone content and release in the rat. *Endocrinology* **120**: 1079-1082

Kelijman M, Williams TC, Downs TR, Frohman LA (1987) Comparison of the sensitivity of growth hormone secretion to somatostatin *in vivo* and *in vitro* in acromegaly. *Clin Res* **35**: 585A

Mitchell JA, Hutchins M, Schindler WJ, Critchlow V (1973) Increases in plasma growth hormone and naso-anal length in rats following isolation of the medial basal hypothalamus. *Neuroendocrinology* **12**: 161-173

Plotsky PM, Vale W (1985) Patterns of growth hormone- releasing factor and somatostatin secretion into the hypophyseal-portal circulation of the rat. *Science* **230**: 461-463

Quabbe H-J, Schilling E, Helge H (1966) Pattern of growth hormone secretion during a 24-hour fast in normal adults. *J Clin Endocrinol Metab* **26**: 1173-1177

Reubi JC, Heitz PU, Landolt AM (1987) Visualization of somatostatin receptors and correlation with immunoreactive growth hormone and prolactin in human pituitary adenomas: evidence for different tumor subclasses. *J Clin Endocrinol Metab* **65**: 65-73

Rosenthal SM, Hulse JA, Kaplan SL, Grumbach MM (1986) Exogenous growth hormone inhibits growth hormone-releasing factor-induced growth hormone secretion in normal men. *J Clin Invest* **77**: 176-180

Schriock EA, Lustig RH, Rosenthal SM, Kaplan SL, Grumbach MM (1984) Effect of growth hormone (GH)-releasing hormone (GRF) on plasma GH in relation to magnitude and duration of GH deficiency in 26 children and adults with isolated GH deficiency or multiple pituitary hormone deficiencies: evidence for hypothalamic GRH deficiency. *J Clin Endocrinol Metab* **58**: 1043-1049

Snyder PJ, Rudenstein RS, Gardner DF, Rothman JG (1979) Repetitive infusion of gonadotropin-releasing hormone distinguishes hypothalamic from pituitary hypogonadism. *J Clin Endocrinol Metab* **48**: 864-868

Tannenbaum GS, Ling N (1984) The interrelationship of growth hormone (GH)-releasing factor and somatostatin in generation of the ultradian rhythm of GH secretion. *Endocrinology* **115**: 1952-1957

Vance ML, Borges JLC, Kaiser DL, Evans WS, Furlanetto R, Thominet JL, Frohman LA, Rogol AD, MacLeod RM, Bloom S, Rivier J, Vale W, Thorner MO (1984) Human pancreatic tumor growth hormone-releasing factor (hpGRF-40): dose-response relationships in normal man. *J Clin Endocrinol Metab* **58**: 838-844

Wehrenberg WB, Brazeau P, Luben R, Böhlen P, Guillemin R (1982) Inhibition of the pulsatile secretion of growth hormone by monoclonal antibodies to the hypothalamic growth hormone-releasing factor (GRF). *Endocrinology* **111**: 2147-2148

Wehrenberg WB, Brazeau P, Ling N, Textor G, Guillemin R (1984) Pituitary growth hormone response in rats during a 24-hour infusion of growth hormone- releasing factor. *Endocrinology* **114**: 1613-1616

Advances in Growth Hormone and Growth Factor Research,
edited by E.E. Müller, D. Cocchi and V. Locatelli
Pythagora Press, Roma-Milano and Springer Verlag, Berlin-Heidelberg © 1989

Aspects of neurotransmitter control of GH secretion: basic and clinical studies*

F. Camanni[1], E. Ghigo[1], E. Mazza[1], E. Imperiale[1], S. Goffi[1], V. Martina[1], V. De Gennaro Colonna[2], S. G. Cella[2], D. Cocchi[2], V. Locatelli[2], F. Massara[1]† and E.E. Müller[2]

[1] *Dipartimento di Biomedicina Endocrino-Metabolica e Gastroenterologica, Divisione di Endocrinologia, Università di Torino; and* [2] *Dipartimento di Farmacologia, Università di Milano, Italy.*

† Deceased on February 11, 1987.

In all mammalian species studied so far, growth hormone (GH) is released in a pulsatile manner which would be crucial for proper somatomedin secretion and normal growth rhythm (Martin 1978; Müller 1987). This is true also in humans beings, in whom a deficiency in spontaneous GH pulsatility is associated with short stature (Howse et al. 1977; Albertsson-Wikland et al. 1983; Thorner et al. 1986). However, an shown by Spiliotis et al. (1984), a reduced spontaneous GH secretion during the day is not always due to a pituitary secretory inability. In fact, there are many children in whom, in spite of low 24-h GH secretion, a marked somatotroph responsiveness to provocative tests is present. These children have been defined as having GH neurosecretory dysfunction, i.e., an altered GH neuroregulation that leads to a GH deficiency. Therefore, from a clinical viewpoint, the importance of studying the neural control of GH secretion is evident.

It is now well established that the neural control of GH secretion is exerted mainly by two hypothalamic neuropeptides, one stimulatory, i.e., growth hormone-releasing hormone (GHRH) (Frohman and Jansson 1986) and one inhibitory, somatostatin (SS) (Martin 1978; Müller 1987). GHRH is essential for the appearance of GH pulses. In fact, in rats, GH pulsatility is abolished by GHRH antiserum, while SS antiserum only shifts plasma GH to higher levels during trough periods.

* Some of these studies were supported by a grant from the Ministro della Pubblica Istruzione and by a grant from the Italian National Research Council, Special Project Oncology, contract 2602266.44.

However, there is evidence that the characteristic pulsatile GH secretion results from a complex interaction between GHRH and SS (Wehrenberg 1982; Tannenbaum and Ling 1984; Plotsky and Vale 1985).

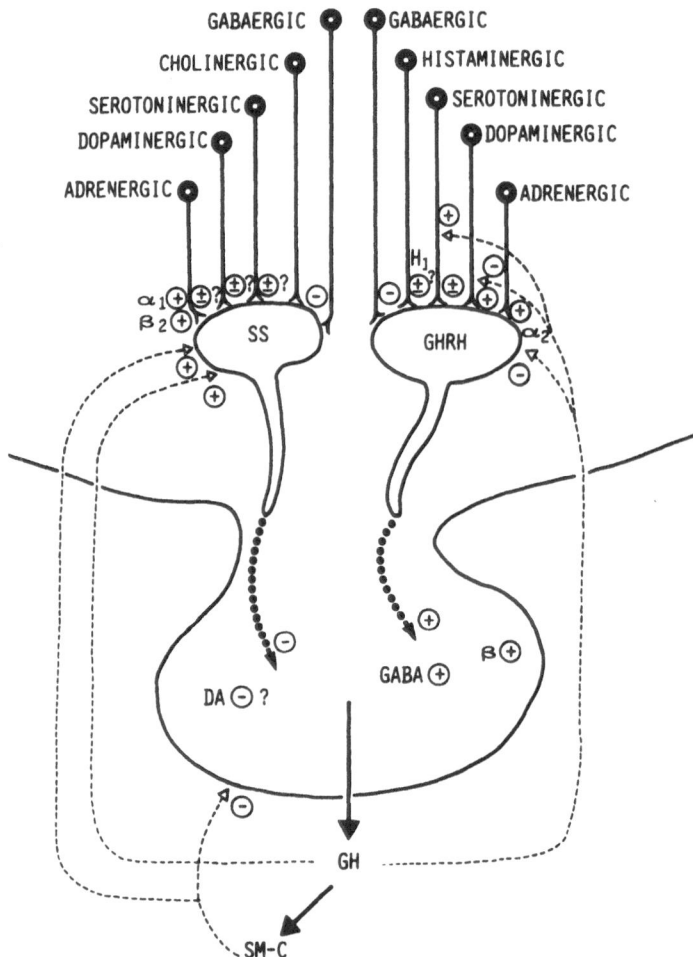

Fig. 1. Main neurotransmitter influences for the control of GH secretion, as derived from animal and human studies. Note the possible dual actions and the level at which they take place, a notable exception being cholinergic influences. Some elements of the mechanisms responsible for GH autoregulation are also depicted. Whenever possible, the proven or alleged action of neutransmitters at GHRH- or SS- secreting structures or anterior pituitary level is indicated. See text for details. +, stimulation; −, inhibition; ?, action questionable.

In addition to the specific neurohormones, a cohort of brain neurotransmitters is involved in the neural control of GH secretion (Müller 1987), with both stimulatory and inhibitory effects on GH release. These effects are exerted mainly via GHRH and/or SS. The ability of a given neurotransmitter to act on either GHRH or SS results in a dual mechanism of control and in the existence of receptor subtypes mediating the opposing influences. Alternatively, neurotransmitters may act directly at the pituitary level (Müller 1987). Another reason for complexity is added by the phenomenon of coexistence, which ultimately results in that of cotransmission, i.e., occurrence and then release of two neurotransmitters (or a neurotransmitter together with a neurohormone) present in the same nerve endings (Lundberg and Hökfelt 1983; Meister et al. 1986). As derived from animal and human studies, Figure 1 presents the main neurotransmitter influences for the control of GH release, showing the possible dual actions and the level at which they may take place.

In addition to classical neurotransmitters, other neuropeptides, including endogenous opioid peptides, vasoactive intestinal polypeptide, peptide histidine isoleucine amide, neuropeptide Y, motilin, galanin, cholecystokinin, and neurotensin, are known to stimulate GH secretion (McCann 1982; Müller 1987). Thyrotropin-releasing hormone (TRH) and gonadotropin-releasing hormone (GnRH) are also GH-releasers, though they are active especially in pathological conditions (Cocchi et al. 1984).

It can be said that, in general, these compounds do not act directly at the pituitary level but rather via the central nervous system (CNS), notable exceptions being motilin (Samson et al. 1984a), neuropeptide Y (McDonald et al. 1985), and secretin (Samson et al. 1984b). Regarding the underlying mechanism of action of these neuropeptides, it would seem that *in vivo* they do not act directly on the hypophysiotropic neurosecretory neurons but rather presynaptically, via mediation of "classical" neurotransmitters. This has been ascertained for opioid peptides (Katakami et al. 1981) and, more recently, for galanin (see below).

Coming back to classical neurotransmitters, in human beings catecholamines and acetylcholine seem to play a major role in the neuroregulation of GH secretion, and in this contribution these neurotransmitters are our main focus of interest.

ROLE OF BRAIN CATECHOLAMINES IN THE NEUROREGULATION OF GH SECRETION

With regard to the catecholaminergic system, it has been shown that clonidine, an α_2-receptor agonist, induces a clear-cut rise in plasma GH, both in animals (see Müller 1987, for references) and in man (Lal et al. 1975). As already reported in animals (Krulich et al. 1982), we have recently shown that yohimbine, a specific α_2-receptor antagonist, blunts the clonidine-induced GH rise in man (30 mg oral yohimbine 50 min before 0.15 mg clonidine infused intravenously over 10 min)

(Fig. 2). This finding shows that the GH-releasing effect of clonidine is specifically mediated by α_2-receptor activation, which probably acts via increased endogenous GHRH release. In fact, the ability of clonidine to increase GH release is preserved in reserpinized rats pretreated with SS antiserum (Edèn et al. 1981), while it is abolished in adult rats pretreated with GHRH antiserum (Miki et al. 1984) and in rats with monosodium glutamate-induced destruction of GHRH-secreting neurons (Katakami et al. 1984). In addition, clonidine is able to release GHRH from rat hypothalami perifused *in vitro* (Kabayama et al. 1986) and, like GHRH, acute or short-term administration of clonidine to 5-day-old rats induces changes in GH secretion overlapping with those elicited by GHRH, thus indicating that α_2-receptor activation may ultimately stimulate GH release and synthesis (Cella et al. 1986; Cozzi et al. 1986).

The important role played by α_2-adrenoceptors in the neural control of GH secretion is also testified to by the ability of chronic clonidine treatment to stimulate

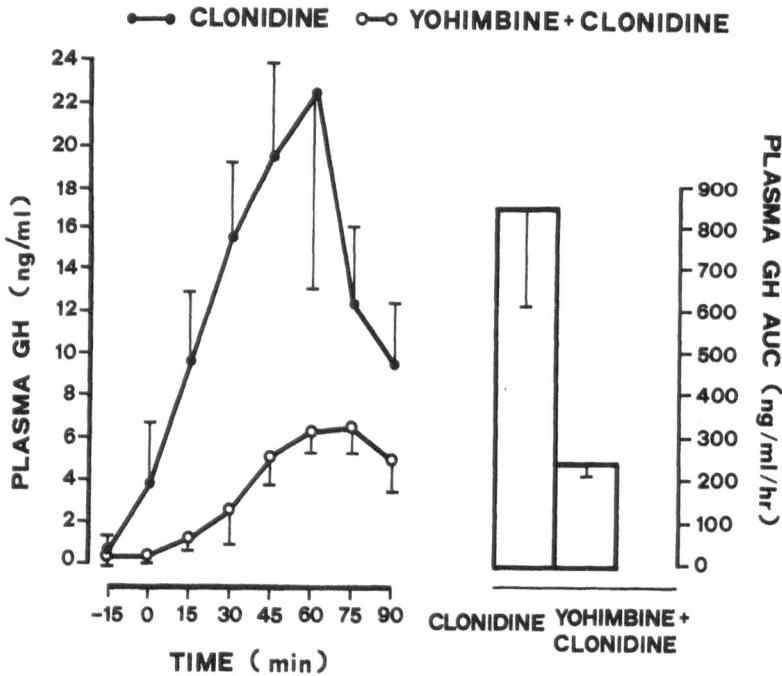

Fig. 2. Plasma GH response to clonidine alone (0.15 mg infused from 0 to +10 min) and preceded by yohimbine (30 mg orally at -50 min) in four normal subjects. On the right side, the integrated areas (*AUC*, ng/ml/h) are indicated. Values are expressed as mean and SEM.

linear growth in children with idiopathic GH deficiency or constitutional growth delay (Pintor et al. 1985; Pintor et al., this volume).

On the other hand, α-adrenoreceptors are also involved in the mechanism by which some neuropeptides, particularly galanin, elicit GH release. Galanin is a 29-amino acid peptide isolated and characterized from extracts of porcine intestine (Tatemoto et al. 1983) and subsequently localized to cell bodies of several hypothalamic nuclei and to nerve terminals in the external layer of the median eminence (Rokaeus et al. 1984). Galanin induces a clear-cut dose-related GH increase in rats when injected intracerebroventricularly, while it fails to alter somatotroph secretion from dispersed perifused pituitary cells (Ottlecz et al. 1986).

In a study by our group (Fig. 3, panel A) in 10-day-old rats, both an inhibitor

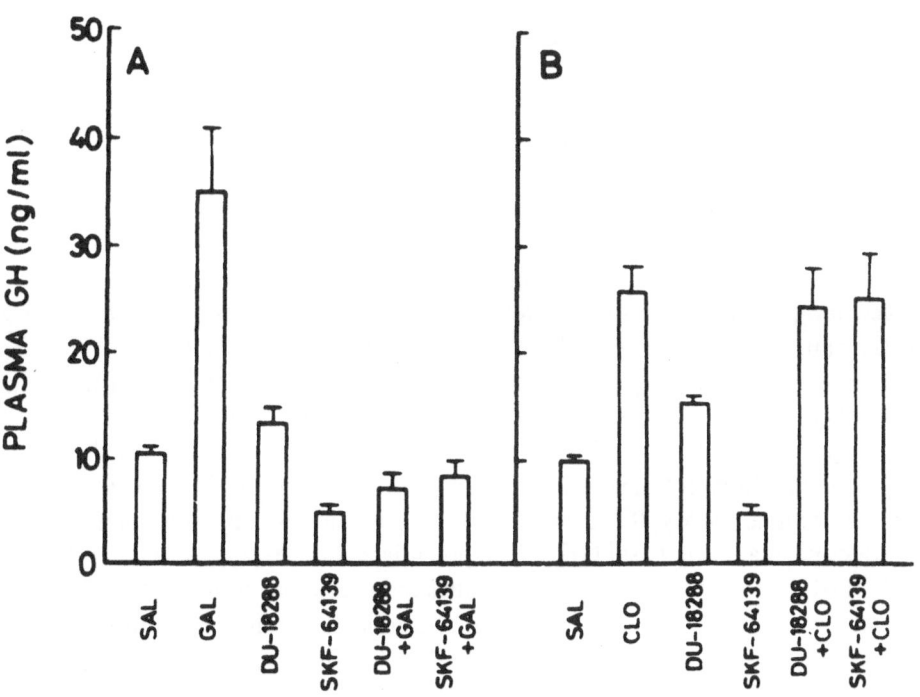

Fig. 3. Effect of synthetic porcine galanin (*GAL*, 25 μg/kg sc; *panel A*) or clonidine (*CLO*, 150 μg/kg sc; *panel B*) on plasma GH levels in 10-day-old rats pretreated or not with inhibitors of norepinephrine (DU-18288, 6 mg/kg ip) or epinephrine (SKF 64139, 50 mg/kg ip) synthesis. Each bar and vertical line represents the mean and SEM of 9–20 determinations made in duplicate).

of dopamine -β-hydroxylase (DU-18288, 6 mg/kg ip), the enzyme converting dopamine to norepinephrine, and a selective inhibitor of phenylethanolamine methyl-transferase (SKF 64139, 50 mg/kg ip), the enzyme converting norepinephrine to epinephrine, were effective in suppressing galanin (25 μg/kg sc)-induced GH release. Interestingly, inhibition of epinephrine biosynthesis induced per se a clear-cut reduction of basal plasma GH levels. Therefore, the GH-releasing effect of galanin, which is totally prevented by GHRH antiserum (data not shown), seems to occur via activation of catecholamines, namely epinephrine. As shown in Figure 3, panel B, the same inhibitors of norepinephrine and/or epinephrine biosynthesis did not modify the GH-releasing effect of clonidine (150 μg/kg sc), thus indicating that, also in 10-day-old rodents, this effect is a postsynaptic event.

In all, it could be speculated that galanin-like peptides may be profitably exploited to increase our understanding of the role of epinephrine on GH release and on interactions between epinephrinergic neurons and GHRH- secreting neurons. In addition, testing with clonidine and galanin, both potent GH secretagogues in human beings (Gil-Ad et al. 1978; Bauer et al. 1986), could unravel subtle alterations of catecholamine neurotransmission in subjects with growth disorders.

Concerning α_1-adrenergic receptors, there is evidence from animal studies for an inhibitory effect on GH release (Krulich et al. 1982). Recently (Cella et al. 1984), it was shown that in freely moving dogs methoxamine, an α_1-receptor agonist, suppressed the clonidine-induced GH rise, and the suppressive effect of methoxamine was abolished by pretreatment with prazosin, a specific α_1-receptor antagonist. The mechanism by which, at least in animals, α_1-receptor-mediated inhibition takes place is apparently via somatostatin. In fact, in infant rats pretreated with SS antiserum, methoxamine failed to suppress GH levels (Cella et al. 1987).

In man the role of α_1-adrenoceptors in GH control is yet to be clarified. In our own experience, pretreatment with methoxamine blunted the GH-releasing effect of clonidine in only two of four healthy men investigated (data not shown).

The inhibitory role exerted by β-receptor activation on GH release has been known for many years. It is a fact that β-receptor blockade by propranolol, though ineffective in altering basal GH levels, potentiates GH responses to several stimuli (Müller 1987) and promotes a GH response to intravenous infusion of epinephrine, which is ineffective per se (Massara and Strumia 1970). More recently, a marked enhancing effect of propranolol on the GHRH-induced GH rise has been shown in children (Chihara et al. 1985). Conversely, salbutamol, a β_2-receptor agonist (10 mg/min infused from -5 to +15 min), strongly inhibits the GH response to GHRH (1 μg/kg iv at 0 min) (Fig. 4), thus suggesting that β_2-receptor activation mediates the β-inhibitory effect on GH release. These data seem to rule out the possibility that the β-adrenergic effect is mediated via endogenous GHRH inhibition and suggest that it probably occurs by an increase in endogenous somatostatin release. In agreement with this hypothesis are data obtained *in vitro* showing the SS-releasing effect of isoproterenol on perifused rat pancreas (Samols et al. 1978) and the increased SS-like immunoreactivity in hypophyseal portal blood after administration of isoproterenol (Epelbaum et al. 1981). Irrespective of the mechanism underlying β-receptor mediated influences on GH secretion, from a clinical viewpoint β-receptor antagonists, namely propranolol, could be of diagnostic value in unmasking "false negative" GH responses to various stimuli, including GHRH. The usefulness of these compounds in the treatment of short stature does not seem to rest on a firm foundation, as propranolol fails to increase basal GH levels in Caucasian subjects (Massara and Strumia 1970). Concerning β$_2$-agonists, a drawback for linear growth may be envisaged in treating asthmatic children with these drugs.

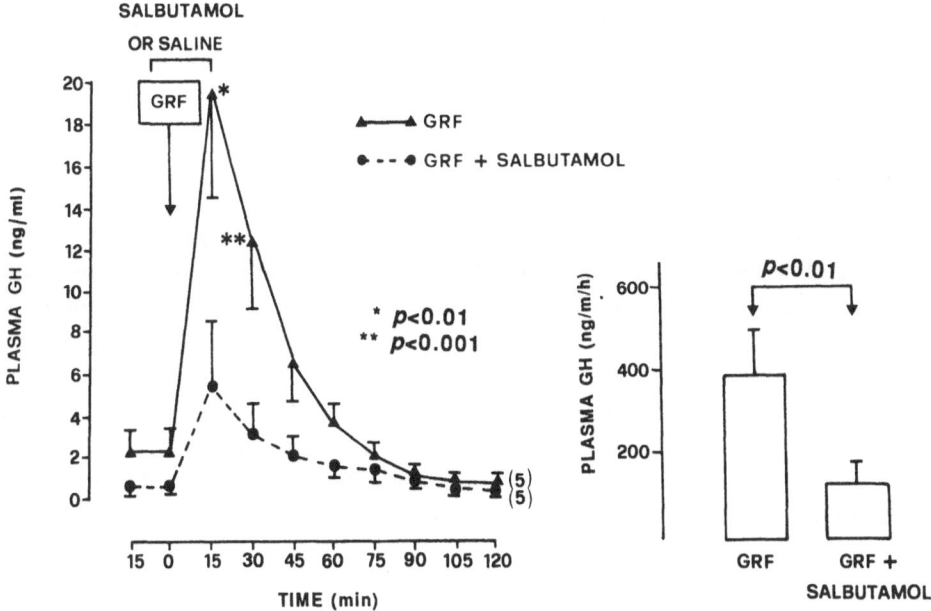

Fig. 4. Plasma GH response to GHRH (GRF) alone (1 μg/kg iv at 0 min) and preceded by salbutamol (10 mg/min infused from -5 to +15 min) in five normal subjects. On the *right*, the integrated areas (AUC, ng/ml/h) are indicated. Values are expressed as mean and SEM.

ROLE OF THE CHOLINERGIC SYSTEM IN THE NEUROREGULATION OF GH SECRETION

A great deal of evidence has been presented in recent years for a stimulatory role of the cholinergic system on GH secretion both in animals (Bruni and Meites 1978; Casanueva et al. 1983) and in man. In the latter, antagonists of cholinergic muscarinic receptors, such as methscopolamine (Mendelson et al. 1978), atropine (Casanueva et al. 1984), and pirenzepine (Delitala et al. 1982, 1983a,b; Massara et al. 1984b; Taylor et al. 1985; Peters et al. 1986) have been shown to abolish the rise in plasma GH elicited by several stimuli such as sleep, exercise, arginine, the enkephalin analogue FK33-824, glucagon, and the catecholaminergic drugs apomorphine, levodopa, and clonidine.

More recently, it has been shown in man that muscarinic antagonism, e.g., pirenzepine (Massara et al. 1984a) and atropine (Massara et al. 1986a; Casanueva et al. 1986), abolishes the GH response to GHRH. Conversely, the enhancement of the cholinergic tone by pyridostigmine, a cholinesterase inhibitor, induces a significant increase in basal plasma GH levels and, furthermore, clearly potentiates the somatotroph responsiveness to GHRH (Massara et al. 1986a). It is known that repeated GHRH bolus injections result in somatotroph refractoriness to GHRH itself (Losa et al. 1984). Pretreatment with pyridostigmine reinstates and even potentiates the blunted responsiveness to a second GHRH bolus in man (Massara et al. 1986b).

Concerning the mechanism by which the GH stimulatory effect of cholinergic activation takes place, there is evidence that it is exerted via somatostatin. In fact, acetylcholine and neostigmine inhibit SS release from rat hypothalami in culture and muscarinic cholinergic antagonism by atropine blocks this inhibition (Richardson et al. 1980). More recently, it has been shown in rats that maneuvers aimed at inducing somatostatin depletion, e.g., anterolateral deafferentation of the mediobasal hypothalamus or treatment with cysteamine, an SS depleting agent, blunt the potentiating effect of pilocarpine, a direct cholinergic agonist, and the inhibitory effect of both pirenzepine and atropine on GHRH-induced GH response (Locatelli et al. 1986). On the other hand, the observation that cholinergic antagonists abolish the GHRH-induced GH response would rule out the possibility that cholinergic inputs act via GHRH. Another potential site for cholinergic-GHRH interaction is the pituitary, where specific muscarinic cholinergic receptors have been found (Casanueva et al. 1986). However, failure of muscarinic cholinergic agonists and antagonists to alter the GHRH- induced GH release both *in vitro* and *in vivo* makes this possibility unlikely (Locatelli et al. 1986).

As will be alluded to below, these results have opened new avenues to the understanding of the pathophysiology, diagnosis, and therapy of GH hypersecretory and hyposecretory states.

Hypersecretory states

Pirenzepine fails to inhibit both basal and TRH-induced GH release in acromegaly, only inconstantly reducing somatotroph responsiveness to GHRH (Massara et al. 1986a; Jordan et al. 1986).

Reportedly, in type 1 diabetes mellitus basal GH levels may be elevated (Hansen and Johansen 1970), and an involvement of GH hypersecretion has been postulated for the occurrence of the dawn phenomenon (Campbell et al. 1985) and of diabetic microangiopathy (Lundbaek et al. 1970).

Recent data (Page et al. 1987; Martina et al. 1987) demonstrate that in type-1 diabetic patients pirenzepine abolishes the enhanced nocturnal GH secretion. In eight patients previously made euglycemic by biostator, pirenzepine was effective in doing so when administered both intravenously (20 mg bolus injection followed by 30 mg pirenzepine infused over 9 h from 11 p.m. to 8 a.m.) and orally (100 mg oral pirenzepine at 11 p.m.) during constant infusion of 0.15 mU/kg/min of regular insulin (Fig. 5). Interestingly, of three patients exhibiting a typical dawn phenomenon, pirenzepine abolished it in two. Therefore, it emerges from these data that the oral use of a safe and well-tolerated drug such as pirenzepine to inhibit nocturnal GH surges may be a useful adjunct for the treatment of type-1 diabetes mellitus.

Hyposecretory states

The notion alluded to in the introduction, that children with short stature may have a primary neurosecretory dysfunction, prompted us to study the functional activity of the cholinergic system in a group of these subjects (Ghigo et al. 1987a,b). We studied 28 children, who, according to usual growth criteria, were divided into four groups: normal children (n=5), children with familial short stature (FSS, n=7), children with constitutional growth delay (CGD, n=10) and children with GH deficiency (GHD, n=6). In the last group, GH deficiency was confirmed by a limited GH release after insulin-induced hypoglycemia and/or other provocative tests and low plasma Sm-C levels (i.e., <0.4 U/ml). All children underwent, in random order, the following four tests: (a) insulin hypoglycemia (IH) (0.15 IU/ kg iv), (b) GHRH (1 µg/Kg iv bolus), (c) pyridostigmine (PD) (60 mg orally), (d) PD+GHRH. As shown in Figure 6, pyridostigmine alone induced a GH increase which was not statistically different from those induced by insulin- hypoglycemia and GHRH alone in normal, FSS and CGD groups. Moreover, in all groups PD increased the GH responses to GHRH, inducing the highest plasma GH rise. This clear-cut potentiating effect of PD was present in all children with FSS and CGD and in at least two children with GHD (Fig. 7).

In all, these data demonstrate that PD plus GHRH is the most powerful single

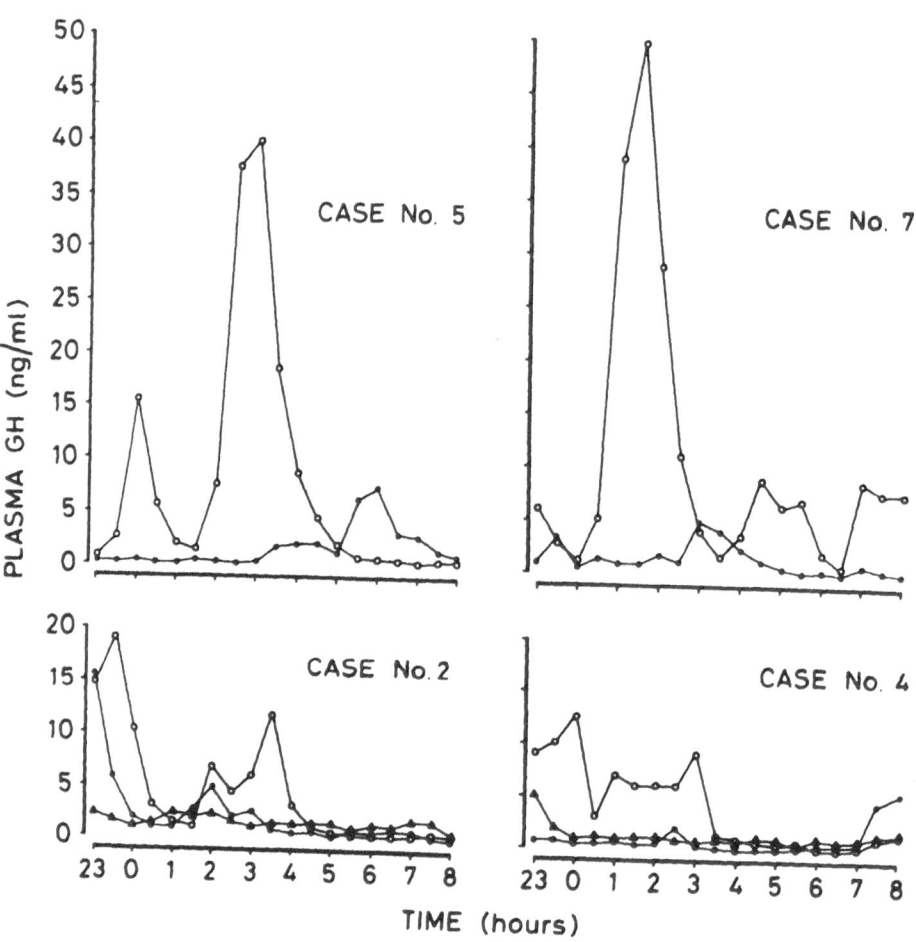

Fig. 5. Nocturnal plasma GH levels (ng/ml) after saline, (○—○), intravenous pirenzepine (20 mg bolus injection, followed by 30 mg infused from 11.00 to 8.00 a.m.), ●—●, and after oral pirenzepine (100 mg at 11.00 p.m.), ▲—▲, in four of eight type-1 diabetic patients. (From Martina et al. 1987, with permission).

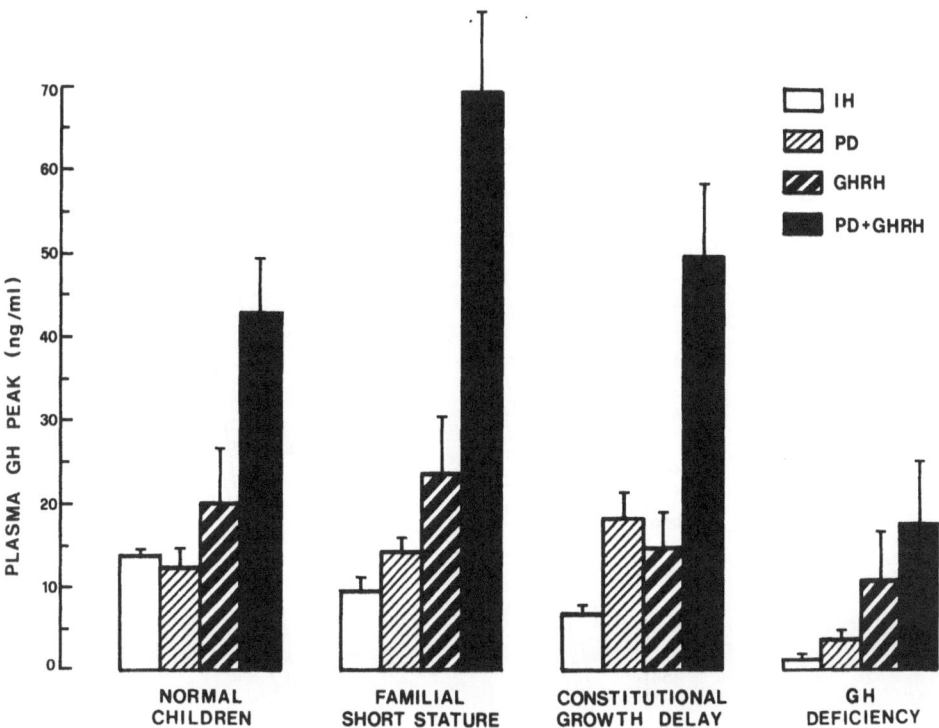

Fig. 6. Mean and SEM peak plasma GH levels (ng/ml) after insulin hypoglycemia (*IH*), pyridostigmine (*PD*), GHRH, and PD+GHRH in children with normal stature, familial short stature, constitutional growth delay, and GH deficiency.

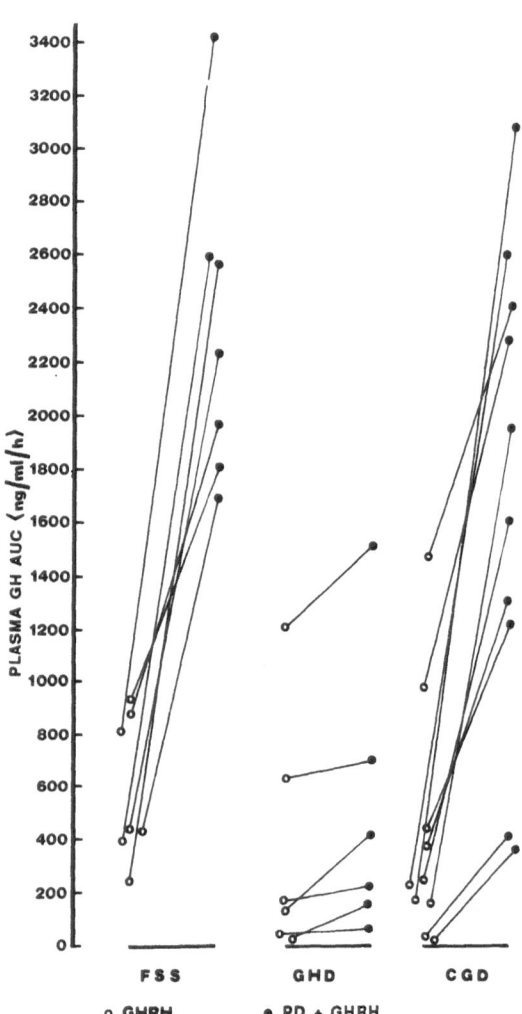

Fig. 7. Individual GH responses (AUC,ng/ml/h) to GHRH alone (○) and preceded by pyridostigmine (*PD*; ●)in 23 children with familial short stature (*FSS*), GH deficiency (*GHD*), or constitutional growth delay (*CGD*).

test for assessing the secretory integrity of somatotrophs. It differentiates GH deficiency due to the secretory inability of the pituitary from that due to hypothalamic dysfunction.

Attention has to be paid, however, to the significance of GH responsiveness to PD alone. Our data show that PD induces a GH increase higher than that observed after insulin hypoglycemia in a large group of both normal and short children (Ghigo et al. 1987a,b). Considering that cholinergic drugs stimulate GH secretion probably by inhibiting SS release (see above), a positive GH response to this drug alone would indicate the functional integrity of both hypothalamic GHRH-secreting structures and somatotrophs.

The diagnostic significance of cholinergic agonists in clinical practice is best exemplified by Figure 8, in which we show the GH responses to various provoca-

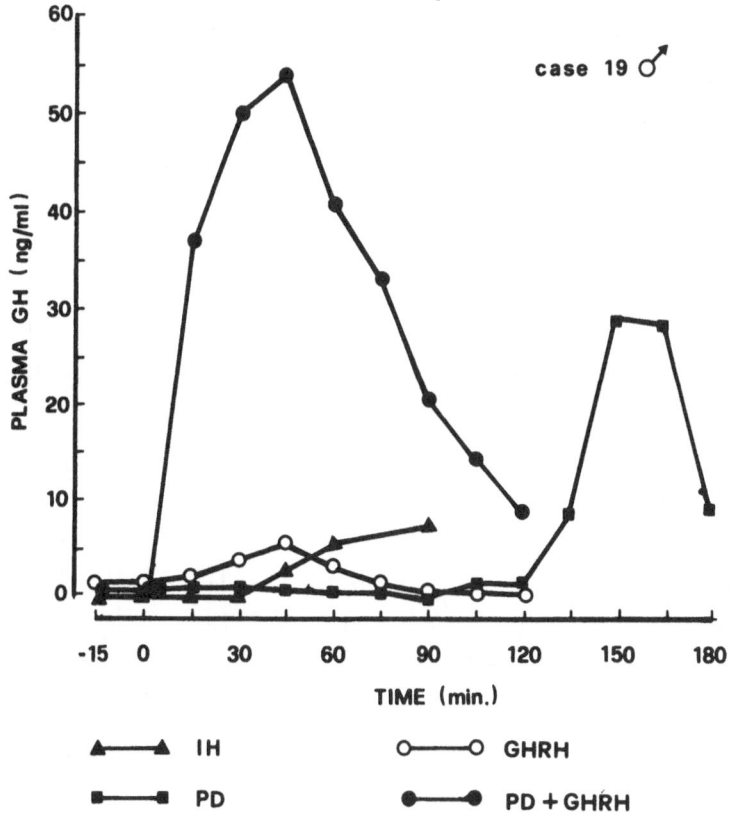

Fig. 8. Plasma GH levels after insulin hypoglycemia (*IH*), pyridostigmine (*PD*), GHRH, and PD + GHRH in a child classified as having a constitutional growth delay. (From Ghigo et al. 1987a, with permission).

tive tests in a child classified as having a CGD. In this child, in whom defective GH responses to IH and GHRH were repeatedly present, PD induced a clear-cut GH increase (peak 30 ng/ml) and promoted a marked GH response to GHRH. These data therefore allowed us to demonstrate that this subject did not have a pituitary secretory inability, as might be suspected on the basis of a deficient GH response to both IH and GHRH. As this patient also had an insufficient 24-h spontaneous GH secretion (Fig. 9, panel A) he could be considered to have a neurosecretory dysfunction, possibly involving a defective cholinergic tone.

On these premises, we made an attempt to enhance the hypothalamic cholinergic tone of this child, treating him cronically with oral PD. First, we administered 30 mg PD three times daily for 2 months, a dose which did not modify the spontaneous GH secretion (Fig. 9, panel B). In contrast, doubling of the dose clearly enhanced it (Fig. 9, panel C). After 6 months of treatment at this dose increases in growth rate and in plasma Sm-C levels were also observed (Fig. 9, panel C). No side effects were recorded in this child or in another four children so treated. Therefore, pyridostigmine could represent a new approach to therapy for short

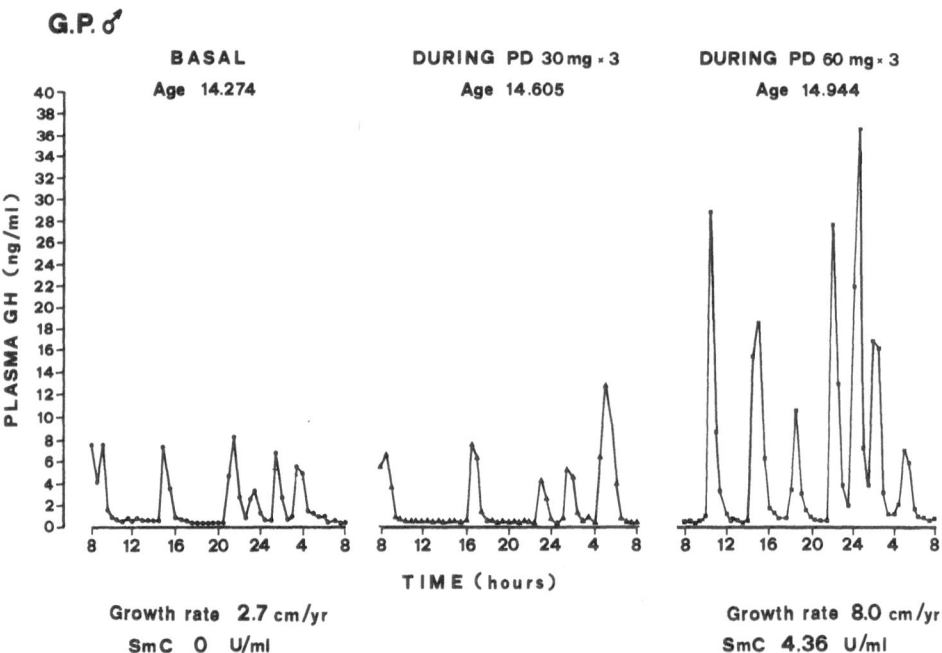

Fig. 9. Twenty-four- hour GH levels before (*panel A*) and during both 30 mg (*panel B*) and 60 mg (*panel C*) oral pyridostigmine (*PD*) three times daily in a child classified as having a constitutional growth delay (see also Fig. 8). Plasma Sm-C levels and growth rate before and after 6 months' PD treatment with 60 mg three times daily are also shown.

stature. Obviously, this is not always the case. In fact, different from what was shown in this child, who was a GHRH hyporesponder and had a low 24-h GH secretion, in a group of seven short children in whom GHRH responsiveness and spontaneous rises in nocturnal GH levels were present, 60 mg PD orally increased basal GH levels in the morning but failed to potentiate the enhanced nocturnal GH secretion (Ghigo et al. 1988). Among the possible interpretations for the inability of PD to potentiate nocturnal GHRH surges thought may be given to a spontaneous reduction of the somatostatinergic tone occurring nightly. Supporting this view, though inferentially, is the observation that repeated GHRH pulses induce somatotroph refractoriness during the morning but not at night (Hulse et al. 1986). Since in the morning PD reinstates GH responsiveness to repeated GHRH administration (Massara et al. 1986b), probably by inhibiting a somatostatinergic tone, the persistent GH response to repeated GHRH boli in the night is suggestive of the presence of a spontaneously reduced somatostatinergic tone. In this vein, the alleged SS hypofunction present at night may underlie the existence of an already maximally stimulated cholinergic tone at night. This view is inferentially supported by the observation that the cholinergic system plays a key role in generating sleep phases in the mammalian sleep cycle (Sitaram et al. 1980; Baghdoyan et al. 1984), which are related to the initial nocturnal surge of GH secretion (Takahashi et al. 1968).

More direct evidence for a strict functional relationship existing between cholinergic function and the nocturnal GH surge stems from the ability of cholinergic antagonists to completely abolish it (Mendelson et al. 1978; Taylor et al. 1985; Peters et al. 1986; Martina et al. 1987). All together, it would seem that further work is needed to ascertain whether probing of the cholinergic function by PD in subjects with different forms of growth disorders may be helpful to unravel interindividual alterations in the cholinergically mediated GH secretion.

CONCLUSIONS

The notion that in hypothalamic and extrahypothalamic structures neurotransmitters and hypophysiotropic hormones interact functionally to ensure the physiological secretion of GH implies that a primary neurotransmitter/neuropeptide dysfunction may underlie specific neuroendocrine disorders. It is evident from the data here presented that the increased knowledge of the neurotransmitter/neuropeptide mechanism(s) of action and the awareness that some growth disorders may result from a CNS alteration and not from a pituitary defect add a new dimension to the pathophysiology of these illnesses and to therapy with CNS-acting drugs.

REFERENCES

Albertsson-Wikland K, Isaksson O, Rosberg S, Westphal O (1983) Secretory pattern of growth hormone in children of different growth rates. *Acta Endocrinol* (Copenh) [Suppl. 103]: 72

Baghdoyan HA, Monaco AP, Rodrigo-Angulo ML, Assens F, McCarley RW, Hobson JA (1984) Microinjection of neostigmine into the pontine reticular formation of cats enhances desynchronized sleep signs. *J Pharmacol Exp Ther* **231**: 173-180

Bauer FE, Venetikou M, Burrin JM, Ginsberg L, MacKay DJ, Bloom SR (1986) Growth hormone release in man induced by galanin, a new hypothalamic peptide. *Lancet* ii: 193-194

Bruni JF, Meites J (1978) Effects of cholinergic drugs on growth hormone release. *Life Sci* **23**: 1351-1357

Campbell PJ, Bolli GB, Cryer PE, Gerich JE (1985) Sequence of events during development of the dawn phenomenon in insulin-dependent diabetes mellitus. *Metabolism* **34**: 1100-1104

Casanueva FF, Betti R, Cella SG, Müller EE, Mantegazza P (1983) Effects of agonists and antagonists of cholinergic neurotransmission on growth hormone release in the dog. *Acta Endocrinol* (Copenh) **103**: 15-20

Casanueva FF, Villanueva L, Cabranes JA, Cabezas-Cerrato J, Fernandez-Cruz A (1984) Cholinergic mediation of growth hormone secretion elicited by arginine, clonidine and physical exercise in man. *J Clin Endocrinol Metab* **59**: 526-530

Casanueva FF, Villanueva L, Dieguez C, Cabranes JA, Diaz Y, Szoke B, Scanlon MF, Schally AV, Fernandez-Cruz A (1986) Atropine blockade of growth hormone (GH)-releasing hormone-induced GH secretion in man is not exerted at pituitary level. *J Clin Endocrinol Metab* **62**: 186-191

Cella SG, Morgese M, Mantegazza P, Müller EE (1984) Inhibitory action of α_1-adrenergic receptor on growth hormone secretion in the dog. *Endocrinology* **114**: 2406-2408

Cella SG, Locatelli V, De Gennaro V, Pellini C, Pintor C, Müller EE (1986) *In vivo* studies with GH-releasing factor and clonidine in rat pups: ontogenetic development of their effect on GH release and synthesis. *Endocrinology* **119**: 1164-1170

Cella SG, Locatelli V, De Gennaro V, Wehrenberg WB, Müller EE (1987) Pharmacological manipulations of α-adrenoceptors in the infant rat and effects on GH secretion. Study of the underlying mechanisms of action. *Endocrinology* **120**: 1639-1643

Chihara K, Kodama H, Kaji H, Kita T, Kashio Y, Okimura Y, Abe H, Fujita T (1985) Augmentation by propranolol of growth hormone-releasing hormone (1-44) NH_2-induced growth hormone release in normal and short children. *J Clin Endocrinol Metab* **61**: 229-233

Cocchi D, Locatelli V, Müller EE (1984) Nonspecific responses to hypothalamic hormones in basic and clinical research. In: Shah NS, Donald AG (eds) *Psychoneuroendocrine Dysfunction*. Plenum, New York, pp 173-208

Cozzi MG, Zanini A, Locatelli V, Cella SG, Müller EE (1986) Growth hormone-releasing hormone and clonidine stimulate biosynthesis of growth hormone in neonatal pituitaries. *Biochem Biophys Res Commun* **138**: 1223-1230

Delitala G, Frulio T, Pacifico A, Maioli M (1982) Participation of cholinergic muscarinic receptor in glucagon- and arginine- mediated growth hormone secretion in man. *J Clin Endocrinol Metab* **55**: 1231-1233

Delitala G, Grossman A, Besser GM (1983a) Opiate peptides control growth hormone through a cholinergic mechanism in man. *Clin Endocrinol* (Oxf) **18**: 401-406

Delitala G, Maioli M, Pacifico A, Brianda S, Palermo M, Mannelli M (1983b) Cholinergic receptor control mechanism for L-dopa-, apomorphine-, and clonidine-induced growth hormone secretion in man. *J Clin Endocrinol Metab* **57**: 1145-1149

Edèn S, Eriksson E, Martin JB, Modigh K (1981) Evidence for a growth-hormone releasing factor mediating alpha-adrenergic influence on growth hormone secretion in the rat. *Neuroendocrinology* **33**: 24-27

Epelbaum J, Tapia-Arancibia L, Kordon C (1981) Noradrenaline stimulates somatostatin release from incubated slices of the amygdala and the hypothalamic preoptic area. *Brain Res* **215**: 393-397

Frohman LA, Jansson J-O (1986) Growth hormone-releasing hormone. *Endocr Rev* **3**: 223-253

Ghigo E, Mazza E, Imperiale E, Rizzi G, Benso L, Müller EE, Camanni F, Massara F (1987a) The enhancement of the cholinergic tone by pyridostigmine promotes both basal and growth hormone (GH)-releasing hormone-induced GH secretion in children of short stature. *J Clin Endocrinol Metab* **65**: 452-456

Ghigo E, Mazza E, Imperiale E, Molinatti P, Bertagna A, Camanni F, Massara F (1987b) Growth hormone responses to pyridostigmine in normal adults and in both normal and short children. *Clin Endocrinol* (Oxf) **27**: 669-673

Ghigo E, Imperiale E, Mazza E, Goffi S, Procopio M, Müller EE, Camanni F (1988) Cholinergic enhancement by pyridostigmine potentiates spontaneous diurnal but not nocturnal growth hormone secretion in short children. *Neuroendocrinology*, in press

Gil-Ad I, Topper E, Laron Z (1978) Oral clonidine as a growth hormone stimulation test. *Lancet* **ii**: 278-280

Hansen AP, Johansen K (1970) Diurnal patterns of blood glucose, serum free fatty acids, insulin, glucagon and growth hormone in normals and juvenile diabetics. *Diabetologia* **6**: 27-33

Howse PM, Rainer PHW, Williams JW, Rudd BT, Bertrande PV, Thompson CRS, Jones LA (1977) Nyctohemeral secretion of growth hormone in normal children of short stature and in children with hypopituitarism and intrauterine growth retardation. *Clin Endocrinol* **6**: 347-359

Hulse JA, Rosenthal SM, Cuttler L, Kaplan SL, Grumbach MM (1986) The effect of pulsatile administration, continuous infusion, and diurnal variation on the growth hormone (GH) response to GH-releasing hormone in normal men. *J Clin Endocrinol Metab* **63**: 872-878

Jordan V, Dieguez C, Valcavi R, Artioli C, Portioli I, Rodriguez-Arnao MD, Gomez-Pan A, Hall R, Scanlon MF (1986) Influence of dopaminergic, adrenergic and cholinergic blockade and TRH administration on GH responses to GRF 1-29. *Clin Endocrinol* (Oxf) **24**: 291-294

Kabayama Y, Kato Y, Murakami Y, Tanaka H, Imura H (1986) Stimulation by alpha-adrenergic mechanism of the secretion of growth hormone-releasing factor (GRF) from perifused rat hypothalamus. *Endocrinology* **119**: 432-434

Katakami H, Kato Y, Matsushita N, Hiroto S, Shimatsu A, Imura H (1981) Involvement of alpha adrenergic mechanisms in growth hormone release induced by opioid peptides in conscious rats. *Neuroendocrinology* **33**: 129-135

Katakami H, Kato Y, Matsushita N, Imura H (1984) Effects of neonatal treatment with monosodium glutamate on growth hormone release induced by clonidine and prostaglandin E_1 in conscious male rat. *Neuroendocrinology* **38**: 1-5

Krulich L, Mayfield MA, Steel MK, McMillen BA, Mc Cann SM, Koening JI (1982) Differential effects of pharmacological manipulations of central α_1- and α_2- adrenergic receptors on the secretion of thyrotropin and growth hormone in male rats. *Endocrinology* **110**: 796-804

Lal S, Tolis G, Martin JB, Brown GM, Guyda H (1975) Effect of clonidine on growth hormone, luteinizing hormone, follicle stimulating hormone, and thyroid stimulating hormone in the serum of normal man. *J Clin Endocrinol Metab* **41**: 703- 708

Locatelli V, Torsello A, Redaelli M, Ghigo E, Massara F, Müller EE (1986) Cholinergic agonist and antagonist drugs modulate in the rat the growth hormone-response to growth hormone releasing hormone. Evidence for mediation by somatostatin. *J Endocr* **111**: 271-278

Losa M, Bock L, Schopohl J, Stalla GK, Müller OA, von Werder K (1984) Growth hormone releasing factor infusion does not sustain elevated growth hormone levels in normal subjects. *Acta Endocrinol* (Copenh) **107**: 462-470

Lundbaek K, Christensen NJ, Jensen VA, Johansen K, Hamsen AP, Orskov M, Osterby R (1970) Diabetes, diabetic angiopathy and growth hormone. *Lancet* **i**: 131-133

Lundberg JM, Hökfelt T (1983) Coexistence of peptides and classical neurotransmitters. *Trends Neurosci* **6**: 325-331

Martin JB (1978) Brain regulation of growth hormone secretion. In: Martini L, Ganong WF (eds) *Frontiers in Neuroendocrinology*, vol 4. Raven, New York, pp 128-168

Martina V, Tagliabue M, Maccario M, Bertagna A, Ghigo E, Massara F, Camanni F (1987) Pirenzepine blunts the nocturnal growth hormone release in insulin-dependent diabetes. *Horm Metab Res* **19**: 449-450

Massara F, Strumia E (1970) Increase in plasma growth hormone concentration in man after infusion of adrenaline- propranolol. *J Endocr* **47**: 95-100

Massara F, Ghigo E, Goffi S, Molinatti GM, Müller EE, Camanni F (1984a) Blockade of hp-GRF-40-induced GH release in normal man by a cholinergic muscarinic antagonist. *J Clin Endocrinol Metab* **59**: 1025-1026

Massara F, Tangolo D, Goffi S, Ghigo E, Molinatti GM (1984b) Cholinergic muscarinic receptors mediate apomorphine- induced growth hormone secretion in man. In: *Proc 7th Int Congress of Endocrinology*. Excerpta Medica, Amsterdam, p 1174

Massara F, Ghigo E, Demislis K, Tangolo D, Mazza E, Locatelli V, Müller EE, Molinatti GM, Camanni F (1986a) Cholinergic involvement in the growth hormone releasing hormone-induced growth hormone release. Studies in normal and acromegalic subjects. *Neuroendocrinology* **43**: 670-675

Massara F, Ghigo E, Molinatti P, Mazza E, Locatelli V, Müller EE, Camanni F (1986b) Potentiation of cholinergic tone by pyridostigmine bromide reinstates and potentiates the growth hormone responsiveness to intermittent administration of growth hormone-releasing factor in man. *Acta Endocrinol* (Copenh) **113**: 12-16

McCann SM (1982) The role of brain peptides in the control of anterior pituitary hormone secretion. In: Müller EE, MacLeod RM (eds) *Neuroendocrine Perspectives*, vol 1. Elsevier, Amsterdam, pp 1-22

McDonald JK, Lumpkin MD, Samson WK, McCann SM (1985) Neuropeptide Y affects luteinizing hormone and growth hormone secretion in rats. *Proc Natl Acad Sci USA* **82**: 561-564

Meister B, Hökfelt T, Vale W, Sawchenko PE, Swanson L, Goldstein M (1986) Coexistence of tyrosine hydroxylase and growth hormone-releasing factor in a subpopulation of tuberoinfundibular neurons of the rat. *Neuroendocrinology* **42**: 237-247

Mendelson WB, Sitaram N, Wyatt RJ, Gilling JC, Jacobs LS (1978) Methscopolamine inhibition of sleep-related growth hormone secretion. Evidence of cholinergic secretory mechanism. *J Clin Invest* **61**: 1683-1690

Miki N, Ono M, Shizume K (1984) Evidence that opiatergic and alpha-adrenergic mechanisms stimulate rat growth hormone release via growth hormone-releasing factor. *Endocrinology* **114**: 1950-1952

Müller EE (1987) Neural control of somatotropic function. *Physiol Rev* **67** (3): 962-1053

Ottlecz A, Samson WK, McCann SM (1986) Galanin: evidence for a hypothalamic site of action to release growth hormone. *Peptides* (Fayetteville) **7**: 51-53

Page MD, Koppeschaar HPF, Dieguez C, Gibbs JT, Hall R, Peters JR, Scanlon MF (1987) Additive effects of growth hormone releasing factor and insulin hypoglycemia on growth hormone release in man. *Clin Endocrinol* (Oxf) **26**: 355-359

Peters JR, Evans PJ, Page MD, Hall R, Gibbs JT, Dieguez C, Scanlon MF (1986) Cholinergic muscarinic receptor blockade with pirenzepine abolishes slow-wave sleep-related growth hormone release in normal adult males. *Clin Endocrinol* (Oxf) **25**: 213-218

Plotsky PM, Vale W (1985) Patterns of growth hormone releasing factor and somatostain secretion into the hypophysial-portal circulation of the rat. *Science* **230**: 461-463

Pintor C, Cella SG, Corda R, Locatelli V, Puggioni R, Loche S, Müller EE (1985) Clonidine accelerates growth in children with impaired growth hormone secretion. *Lancet* **i**: 1482-1484

Richardson SB, Hollander GS, D'Eletto RD, Greenleaf PW, Than C (1980) Acetylcholine inhibits the release of somatostatin from rat hypothalamus *in vitro*. *Endocrinology* **107**: 1837-1842

Rokaeus A, Melander T, Hökfelt T, Landberg MM, Tatemoto K, Carlquist M, Mutt V (1984) A galanin- like peptide in the central nervous system and intestine of the rat. *Neurosci Lett* **47**: 161-166

Samols E, Weir GC, Ramseur R, Day JA, Patel YC (1978) Modulation of pancreatic somatostatin by adrenergic and cholinergic agonism and by hyper- and hypoglycemic sulfonamides. *Metabolism* **9** [Suppl 1]: 1219-1221

Samson WK, Lumpkin MD, McCann SM (1984a) Presence and possible site of action of secretin in the rat pituitary and hypothalamus. *Life Sci* **34**: 155-163

Samson WK, Lumpkin MD, Nilaver G, McCann SM (1984b) Motilin; a novel growth hormone-releasing agent. *Brain Res Bull* **12**: 57-62

Sitaram N, Gillin JC (1980) Development and use of pharmacological probes of the CNS in man: evidence of cholinergic abnormality in primary affective illness. *Biol Psychiatry* **15**: 925-931

Spiliotis BE, August GP, Hung W, Sonis W, Mendelson W, Bercu B (1984) Growth hormone neurosecretory dysfunction. A treatable cause of short stature. *JAMA* **251**: 2223-2230

Takahashi Y, Kipnis D, Daughaday W (1968) Growth hormone secretion during sleep. *J Clin Invest* **47**: 2079-2085

Tannenbaum GS, Ling N (1984) The interrelationship of growth hormone (GH)-releasing factor and somatostatin in generation of the ultradian rhythm of GH secretion. *Endocrinology* **115**: 1952-1957

Tatemoto K, Rokaeus A, Jornvall H (1983) Galanin - a novel biologically active peptide from porcine intestine. *FEBS Lett* **164**: 124-128

Taylor BJ, Smith PJ, Brook CGD (1985) Inhibition of physiological growth hormone secretion by atropine. *Clin Endocrinol* (Oxf) **22**: 497-504

Thorner MO, Vance NL, Blizzard RM, Rogol AD, Evans WS, Klingensmith G, Brook C, Smith P, Rivier J, Vale W (1986) Clinical studies with growth hormone releasing hormone. In: Müller EE, Mac Leod RM (eds) *Neuroendocrine Perspectives*, vol 5. Elsevier, Amsterdam, pp 87-99

Wehrenberg WB, Brazeau P, Luben R, Böhlen P, Guillemin R (1982) Inhibition of the pulsatile secretion of growth hormone by monoclonal antibodies to the hypothalamic growth hormone-releasing factor (GRF). *Endocrinology* **111**: 2147-2151

GROWTH HORMONE DEFICIENCY STATES

Advances in Growth Hormone and Growth Factor Research,
edited by E.E. Müller, D. Cocchi and V. Locatelli
Pythagora Press, Roma-Milano and Springer Verlag, Berlin-Heidelberg © 1989

Growth hormone deficiency (GHD) and GHD-like syndromes

R.M. Blizzard

Department of Pediatrics, Children's Medical Center University of Virginia, Charlottesville, Virginia, U.S.A.

The purpose of this manuscript is to provide a foundation of knowledge which will permit the best interpretation possible of the manuscripts which follow. To accomplish this, some salient data which are known will be briefly reviewed, and the challenges of unanswered questions concerning growth hormone deficiency (GHD) and GHD-like syndromes will be presented.

Classical GHD, nonclassical GHD, and GHD-like syndromes will be considered. These can be described as follows:

1. Classical GHD is characterized by unequivocal GHD and by the frequently seen, characteristic chubby and infantile appearance of the GHD patient.

2. Nonclassical GHD is characterized by partial or questionable GHD, and the GHD phenotype is usually absent.

3. GHD-like syndromes are characterized by the clinical appearance of the GHD phenotype in most instances, but GH usually is present in normal or increased quantities.

CLASSICAL GROWTH HORMONE DEFICIENT STATES

Although the usual characteristics of GHD, i.e., the chubby infantile appearance during childhood and the delayed skeletal maturation, are usually present in patients with this syndrome, there may be great variability in the clinical appearance. For example, patients with apparent failure to thrive may have GHD. Such patients often have a tumor of the hypothalamus, with resultant anorexia as a cause for the weight being markedly reduced in relation to the height. Sometimes such appearances are spontaneous and without explanation.

The classifications of classical GHD may be based on etiology, i.e., organic or idiopathic, or on the basis of when the lesion or symptoms occurred, i.e., congenital or acquired. The causes of organic GHD include tumors, trauma, autoimmune disease, and absent structures such as the hypothalamic defects observed in septo-optic dysplasia. With the exception of craniopharyngiomas or congenital cysts, the same causes, i.e., trauma, autoimmunity, and late appearing pituitary tumors, along with infiltrative lesions such as those occurring with Hand-Schüller Christian disease, account for the instances of acquired GHD. Idiopathic GHD occurs for unexplained reasons, including unrecognizable structural anatomic abnormalities and biochemical defects or aberrations, such as GHRH deficiency or somatostatin (SRIF) excess, which are theoretical, but still unproven, causes for GHD.

A minimum of one in 100 cases of idiopathic GHD are genetic in origin. Probably many more cases of GHD are genetic than those currently recognized, since three of the six types of currently recognized, genetically inherited GHD are inherited as autosomal dominant diseases (Table 1). If only one instance of GHD occurs in a family, there is no means of determining that the patient has genetic GHD unless a gene abnormality such as that which occurs in type IA is present. Since the possibility of a gene abnormality is infrequent, gene abnormalities are not sought in isolated cases of GHD, and it often is only when there are two or more cases in a family that genetic GHD is suspected.

Of the six types of genetic GHD, four have isolated GHD without other tropic hormone deficiencies and two have multiple tropic hormone deficiencies. One of each of these two major types is genetically inherited as an X-linked recessive disease. Three of these six are inherited as autosomal recessive diseases, and one is inherited as an autosomal dominant disease.

Four living generations of this latter type in one family were observed in our clinic. The great- grandmother was 84 years of age when last seen, and the great-grandfather died at the age of 77. GHD certainly did not interfere with longevity.

No abnormality of the GH or GHRH gene, except for an absent hGH-N gene in type IA, has been found in any of these entities by restricted fragment length polymorphism.

Type IA is much more significant in its occurrence than the other five types because this type of GHD is relatively less likely to respond to long-term growth hormone therapy, due to the marked predisposition of these patients to develop GH antibodies of the growth inhibiting type. A few patients with type-IA GHD have grown successfully in spite of development of antibodies against growth hormone. Patients with type-IA deficiency may have only an absence of the hGH-N gene or an absence also of the hGH-V gene and of one or more of the three human chorionic somatotropin genes (Fig. 1). None of these genes seem necessary for life.

Table 1. Genetic types of growth hormone deficiency

Isolated GHD		Inheritance	GH gene defect	GHRH gene defect	Comment
Type	IA	AR	GH-N Absent	Probably N1	Often develop GH antibodies
	IB	AR	None	None	
	II	AD	None	None	
	III	X-linked	?	?	Hypoglobulinemia
Multiple hormone defects:					
I		AR	None	None	
	III	X-linked	?	?	

I = AR, autosomal recessive; II = AD, autosomal dominant; III = X-linked (I, II, III to equal AR, AD, and X-linked are accepted designations for genetic inheritance. Unfortunately no type II for multiple hormone defects has been described and the use of III is misleading. However, it is advantageous to be consistent and follow acceptable nomenclature.)

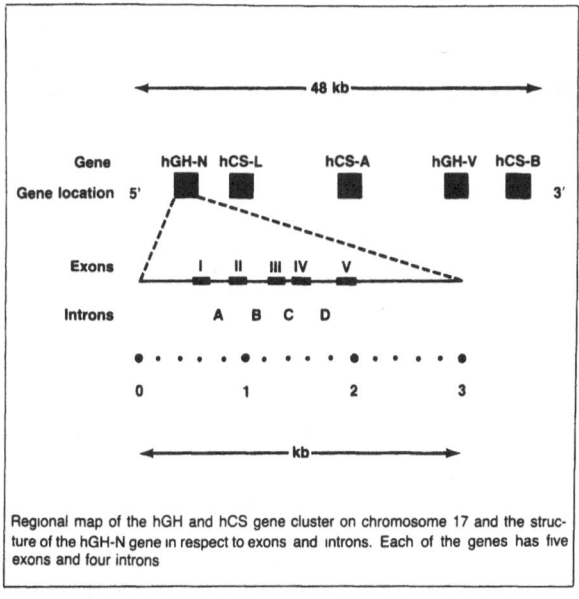

Regional map of the hGH and hCS gene cluster on chromosome 17 and the structure of the hGH-N gene in respect to exons and introns. Each of the genes has five exons and four introns

Fig. 1. Location of the two GH and three HCS genes on chromosome 17. Individuals with type 1A GHD have absence of the GH-N gene.

The other congenital forms of GHD, such as septo-optic dysplasia, congenital absence of the pituitary, and GHD secondary to a craniopharyngioma, are genetic, as no repeat occurrence has been reported among siblings.

NONCLASSICAL GROWTH HORMONE DEFICIENCY

Several entities may be included in this category including (a) partial GHD, (b) possibly constitutional delayed growth and adolescence (CDGA), (c) neurosecretory GHD (NSGHD), and (d) psychosocial short stature (PSS).

Partial GHD frequently is not characterized by the usual overt signs and symptoms of GHD. This entity must be suspected in children who grow less than 5.0 cm per year during the chronological ages of 3 through 12 years. The release of growth hormone to pharmacological stimuli is usually marginal, and the insulin-like-growth factor-I (IGF-I) or somatomedin-C (Sm-C) determination is usually in the low normal or slightly low range. There may be difficulty in differentiating partial GHD from CDGA and on some occasions from PSS. The aberrant signs and symptoms of PSS may assist in making the differential diagnosis.

Unequivocally, sex steroids increase the production of GH in patients with CDGA and even in some cases of GHD. Bierich (1983) reported that the growth hormone secreted in stages 1 and 3 of puberty is less in patients with CDGA than in controls of the same ages and stages of sexual development (Table 2).

Argente et al. (1987) presented data that support the hypothesis of Bierich. Growth hormone releasing hormone concentrations were measured in a group of

Table 2. GH secretion in CDGA (From Bierich 1983)

| | Puberty stages | | | |
| | P1 | | P3 | |
	Patients	Controls	Patients	Controls
Number	86	18	7	7
GH secretion ng/ml x 330 min	2469* ±1068	4349 ±1134	3580* ±1633	9794 ±1711
Peak [GH] ng/ml	+22.5	+38.8	+21.1	+66.2

* Patients vs Controls P<0.01

patients with short stature and compared with those in patients who had CDGA. Following L-dopa stimulation the latter group secreted significantly less GHRH than the former. The mean GH concentration was also less in the latter group, but it was not statistically significantly different from that of the first group.

Growth hormone was administered to 13 boys with CDGA by Bierich (1983) for periods of 0.5–3.5 years. Twelve of these had accelerated growth velocity (mean 4.1 cm per year before treatment and 8.3 cm per year with treatment). These data were interpreted to indicate that some patients with CDGA are relatively growth hormone deficient.

However, not all investigators agree that patients with CDGA secrete less GH than normals. Lanes et al. (1986) reported no differences in the mean overnight outputs of GH, total basal GH outputs, total nocturnal GH pulses, mean peak nocturnal GH levels and IGF-I values in prepubertal children with CDGA and controls.

Similar results were previously reported by Stubbs et al. (1985). Moreover, Richter et al. (1987) studied a group of children with CDGA and a group with genetic short stature utilizing the ^{15}N tracer technique, and found no difference before and after GH treatment in the CDGA group. They concluded that their results yielded no evidence of partial GHD in children with CDGA or with genetic short stature.

Neurosecretory GHD was described by Blatt et al. (1984) as reduced pulsatile GH secretion in children following therapy for acute lymphoblastic leukemia. This term was extended by Spilliotis et al. (1984) to children who had not been irradiated. The hypothesis of Spilliotis et al. (1984) was that some short children secrete inadequate GH during each 24-h period, although these patients are not GHD when evaluated by pharmacological testing. In addition, they believe that these children respond to GH treatment with increased growth. The data obtained for GH concentrations over 24-h when serum was tested every 20 min were comparable for patients with GHD and for those believed to have neurosecretory dysfunction of GH secretion. Zadik et al. (1985) reported similar findings. The integrated concentrations of GH were 6.6 ± 1.9 ng/ml in 46 children with normal stature, 3.8 ± 2.3 ng/ml in 71 who were below the third percentile but who had normal GH secretion by pharmacological testing, and 1.6 ± 0.6 ng/ml in 19 patients who were below the third percentile and who did not respond significantly to pharmacological testing for GH release (GHD patients). Zadik et al. (1985) did not use the term "NSGHD" for their patients who had low growth hormone production over 24-h and who released normal amounts of GH with pharmacological testing, but they described these findings as representative of a variation of GH secretion within the population, according to whether short stature was present. The short children with normal GH responses to pharmacological testing could not be distinguished by bone age from the GHD patients. Albertsson-Wikland et al. (1983) also measured serial and integrated concentrations of GH in normal prepubertal children of various ages over a period of 24 h. They found that short prepubertal children with a growth rate below -2 SD of the mean growth rate had GH levels just at the detection limit, or irregular broad bursts of growth hormone release of low amplitude, or one high peak of GH release but otherwise low levels. These data prompt many investigators to avoid the term "NSGHD", although neurosecretory dysfunction of GH secretion does occur unequivocally in some patients receiving cranial irradiation, as reported in this volume by Rappaport et al.

Psychosocial short stature results from psychological abuse and is often associated with decreased GH release, as compared with normal, when the patient is administered pharmacological agents known to stimulate GH release (Powell et al. 1967). These patients may appear malnourished, with protruding abdomens, or may be overweight for height. Their appearance often simulates that of the

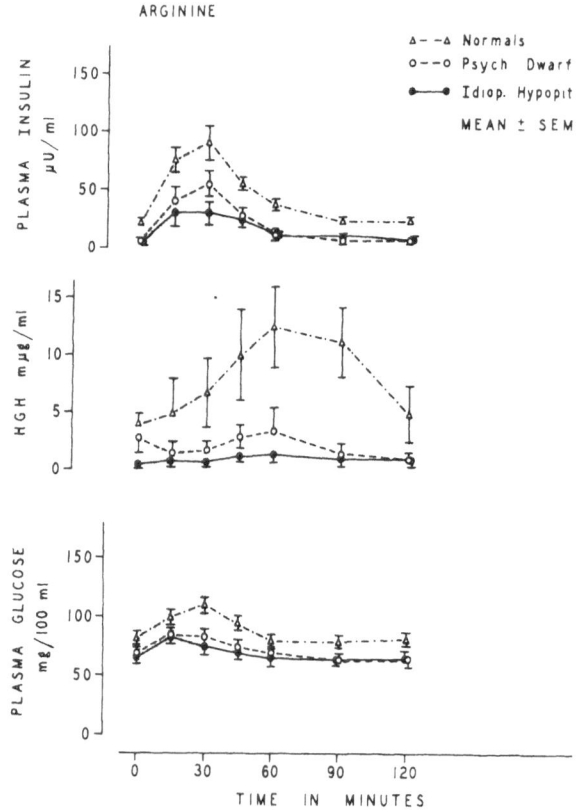

Fig. 2. The response of 11 patients in each of 3 groups (normal children, GHD patients, and PSS patients) to arginine infusion. The GHD and PSS children respond similarly with regard to release of GH, insulin, and glucose.

patient with GHD. The differentiation is made on the basis of history and/or a significant increase in growth velocity when the hostile environment is altered. Few serial determinations or integrated concentrations of growth hormone over 24 h have been measured. One such patient (CA = 6.1 yr, HA = 4.0 yr, BA = 4.0 yr, with peak GH at 3.4 ng/ml to exercise and 6.8 ng/ml to L-dopa, and a growth velocity of 2.5 cm per year) released essentially no GH over 24 h when serum was tested every 20 min. The integrated concentration was 0.5 ng/ml and the IGF-I concentration 0.17 units/ml. When given hGH at a dose of 0.1 units/kg body weight over 12 months his growth velocity increased but only to 5.0 cm per year and his IGF-I increased to 0.41 unit/ml. Holmes et al. (1984) reported similar findings in three patients whose IGF-I increased modestly and whose growth rate with GH increased over the pretreatment rate, but not to the extent expected for patients with GHD. Others have reported an apparent and partial peripheral resistance to GH administration.

While peripheral resistance to GH and nutritional factors may play a role in the growth retardation of these patients, they are not the major factors (Frasier and Rallison 1972). A major factor is diminished GH production, as demonstrated in the middle panel of Figure 2. Growth hormone, insulin, and glucose levels of 11 GHD patients and 11 patients with PSS are compared with those of 11 normals. The diminished GH production is similar to that of patients with GHD. The GH decrease in PSS differs from the decreased production in other patients with GHD in that it is transient. GH production returns to significant levels when the hostile environment is remedied, and the growth velocity accelerates markedly, which produces catch-up growth. Malnutrition as a major factor in the growth retardation in this entity can be dismissed, as GH concentrations increase with starvation and malnutrition; patients with PSS usually have decreased GH.

GROWTH HORMONE DEFICIENCY-LIKE SYNDROMES

By definition, patients falling into this broad category should by all appearances have GHD, but they do not. Their GH concentrations are normal or increased. The IGF-I concentrations may be normal, increased, or low, depending upon the specific syndrome present, and their responses to GH with IGF-I generation and increased growth velocity are variable, also according to the syndrome present. These data and a listing of the syndromes to be discussed are presented in Table 3.

The patient shown in Figure 3 was 3.5 years of age when I studied her case in 1955, during which time she received purified bovine GH prepared by Dr. C. H. Li for 2 weeks of a 6- week balance study. Her height age was 6 months but, surprisingly, her bone age was 3.5 yr. The latter finding is not expected in Laron's dwarfism (Laron et al. 1966), as a delay in skeletal maturation is usual. Severe hypoglycemia eventually required removal of 90% of her pancreas. In 1986 she was restudied. The GH production to arginine and insulin stimulation was excessive for a 33 yr-old woman, as was the integrated concentration obtained over a 24-h period (6.9 ng/ml). The IGF-I determinations were 0.20, 0.23, 0.21 unit/ml and did not increase over 5 days while she received GH at a dose of 0.1 unit/kg body weight each day. With the exception of no delay in bone age throughout childhood and no mental retardation, she is representative of the most common type of GHD-like syndrome, Laron's dwarfism.

In 1984 Laron reviewed the data of 72 patients he had seen or followed up since 1966 (Table 4). By evaluating GH receptors on liver cells obtained by biopsy, he and his collaborators demonstrated that either the number of receptors or their capacity for binding GH is diminished in this entity.

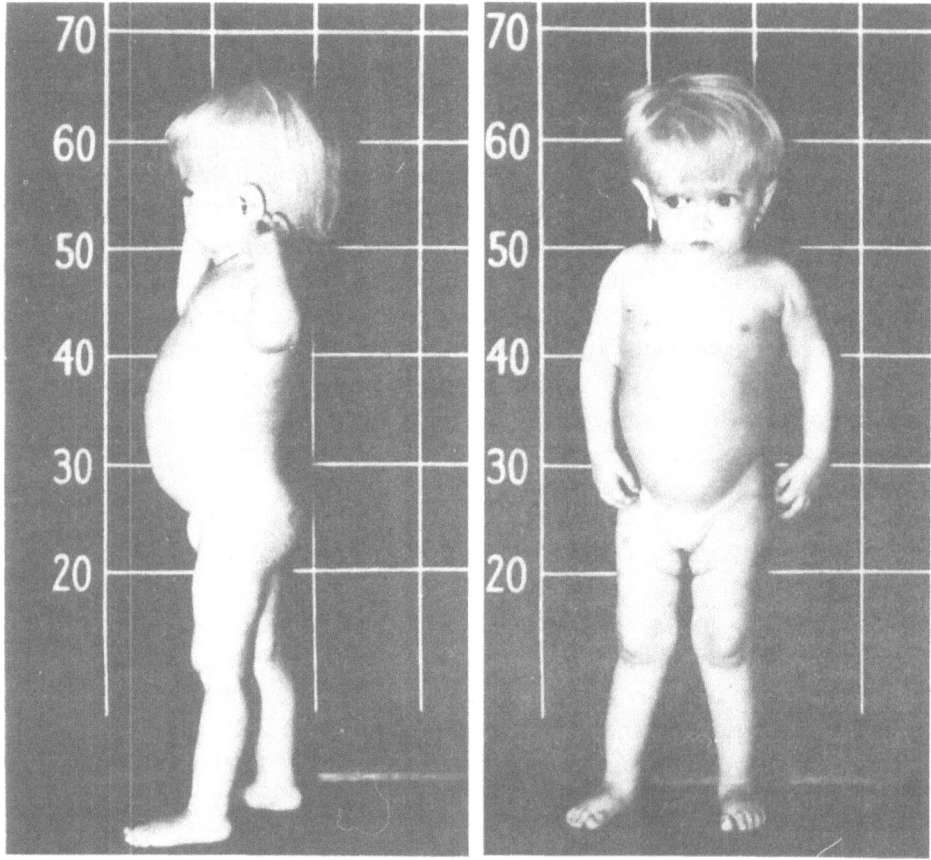

Fig. 3. The chronological and bone ages of this child with Laron's dwarfism were 3.5 years. The height age was 6 months. GH was released in large quantities but the IGF-I levels remained in the GHD range.

Table 3. GHD-like syndromes

Dwarfism type	Basal GH	Basal Sm-C	Response to GH Sm-C	Response to GH GV*
Laron	N or ↑	↓	↓	No
Sm resistance	N or ↑	↑	→	No
** Imm. react.-bio. inact. ** ↓ GH responsiveness	N or ↑	↓	↑	↑
Psychosocial	N or ↓	↓	↗	↗
Pygmy	N	↓	→	?

* GV, growth velocity.
** May be one or two similar syndromes.

Table 4. Characteristics of Laron dwarfism

Total known cases (many non-Jewish)	72
Birth weight >2500 g	18/21
Birth length <-2 SD of mean	10/16
Skeletal and/or mesenchymal dysmorphology	36/72
Fontanel closure	3-7 yr
Ultimate heights: males	119-142 cm
females	108-136 cm
Motor and language development	Slow
Hypoglycemia	Frequent
Squeaky voice	Frequent
Menarche and puberty	Delayed
Response to GH	Negative

Since these patients do not respond to GH, the treatment of choice should be IGF-I. With the availability of significant amounts of IGF-I in the future, made by recombinant DNA techniques, we will soon learn the effectiveness of this therapy for this entity (see Zapf et al., this volume).

Somatomedin-C-or IGF-I resistant dwarfism has been described by Lanes et al. (1980), Carmina et al. (1980), and Bierich et al. (1984). A fourth case was studied at the University of Virginia. Bierich et al. (1984) reported that the fibroblasts from their patient had decreased affinity for IGF-I as compared with fibroblasts from normal individuals. This rare syndrome deserves further investigation.

Immunoreactive-bioinactive GHD was postulated by Kowarsky et al. (1978). They reported on a dwarfed patient who produced significant amounts of GH, but who had decreased IGF-I determinations and presumably a decreased affinity of his GH for its receptors. This presumption was based on the fact that this patient's GH had an increased ratio of radioimmunoassayable GH to radioreceptor measurable growth hormone compared with normals. The radio-receptor assayable GH was postulated to be "bioactive" GH. This patient and those reported by Rudman et al. (1981) had similar ratios and were responsive to GH treatment.

Bright et al. (1981), from our institution, subsequently reported two similar patients but did not compare GH measurements by radioimmunoassay and radioreceptor assay. At a chronological age of 13.5 yr, the first of these boys had a height and bone age of 6 yr. These latter parameters were comparably delayed in the second. Growth hormone was released at peak values of 28 and 29 ng/ml in these boys in response to arginine stimulation. The serial concentrations obtained over 24 h were characterized by multiple pulses of GH with significant amplitudes. The integrated concentrations were in the normal range for chronological age. With growth hormone administration of 0.1 unit/kg body weight three times weekly the growth velocities accelerated markedly (8 and 12 cm per year) and IGF-I concentrations increased appropriately to 0.92 and 0.84 unit/ml from values of 0.36 and 0.24 unit/ml respectively. The significant contribution by Bright et al. (1983) was to measure the integrated concentrations of GH while these patients were receiving the doses of GH usually given to patients with GHD (0.1 unit/kg three times weekly). The integrated concentrations of 13.0 and 12.7 ng/ml were more than two times the pretreatment values of 4.8 and 5.6 ng/ml. These authors postulated that patients with this syndrome treated with the usual amounts of GH might have, for an unexplained reason, a relative GH resistance which was overcome when GH concentrations increased significantly. This is a possible explanation for some or all patients suspected of having this syndrome.

Valenta et al. (1981) reported studies on the GH in the serum of a 14- yr old boy with a height age of 10 yr, who was thought to have this or a similar syndrome. His GH responses to pharmacologic stimuli were normal and the IGF-I concentration was 1.7 units/ml. There was marked acceleration of growth during exogenous GH therapy. The RIA:RRA was 2.0 using IM 9 cells in the RRA technique. On column chromatography the usual three peaks of GH species were noted — i.e., big-big, big, and little (85,000, 45,000 and 20,000 daltons) — but were present in unusual proportions — 60% –90% as GH polymers (big-big or big) rather than the more usual 14% –40%. Further physical analysis revealed that the units of the polymers were joined by disulfide (covalent) linkage rather than by the usual noncovalent forces. Although the chemical analysis of the circulating GH species was very thorough, and a convincing case for a distinct abnormality in the physical and chemical properties of GH in this patient was presented, it is uncertain whether these abnormalities were the cause of the patient's short stature. The patient did not evidence growth failure before 8 or 9 years of age, which would be distinctly unusual for a congenital growth problem. In addition, the baseline IGF-I determination was at the upper limit of normal, 1.7 unit/ml, rather than subnormal, which should be the case if these molecular species were unable to cause the liver to produce IGF-I.

In spite of the reports of all these investigators, in my opinion, there has been no convincing case reported to date of unequivocally identified immunoreactive-bioinactive growth hormone. Studies done subsequent to those reported above, using the radioreceptor assay, have cast doubt on the meaning and/or interpretation of this assay. The use of monoclonal antibodies to study circulating GH in such patients may offer the opportunity to identify such patients.

Psychosocial dwarfism might be classified as either a GHD-like syndrome or a nonclassical GHD syndrome. In my opinion, it is more appropriately classified as a nonclassical GHD. Additional studies of GH production in this group of patients are needed to permit better classification of the syndrome.

In certain respects the pygmy can be classified as having a GHD-like syndrome. Several clinical and biochemical findings resemble those seen in GHD. These include prolonged hypoglycemia after administration of insulin and insulinopenia after glucose and arginine infusion although no fall in BUN occurs with GH administration. There is no increase in radioimmunoassayable or bioassayable IGF-I such as that observed in GHD with GH administration (Table 5) (Rimoin et al. 1969).

In 1981, Merimee et al. reported that pygmies had decreased IGF-I by radioimmunoassay but normal IGF-II determinations, and they attributed the normal sulfation activity to normal IGF-II concentrations. Recently, Merimee et al. (1987) restudied IGF-I and IGF-II concentrations and related these determinations to growth velocities at various ages. Allegedly, IGF-I concentrations do not increase in the pygmies at adolescence. The values were one third those of

Table 5. Pygmies - comparison with GHD patients

	[GH] to ARG	[IRI] to GLU	[GLU] to insulin	[BUN] to GH	[IGF-I] to GH
Normals	↑	↑	↘↗	↓	↑
GHD patients	→	→	↘↗	↓	↑
Pygmies	↑	→	↘↗	→	→

GH, growth hormone; IRI, immunoreactive insulin.
GLU, glucose; BUN, blood urea N$_2$; ARG, arginine.
IGF-I, insulin-like growth factor-I.

adolescent controls, and no adolescent growth spurt was observed, although testosterone and IGF-II values were not significantly different in the various groups. These authors suggested that the short stature of adult pygmies is due primarily to a failure of growth at puberty and postulated that IGF-I is the principal factor responsible for normal pubertal growth. Certain caution must be applied in accepting these interpretations, as expressed by Tanner (1988) in *Growth, Genetics and Hormones*. He is justifiably critical of the auxological approach used by the authors and concludes that while the authors may be correct, they have not proven their hypothesis.

SUMMARY AND CONCLUSIONS

There are certain facts which were known about GH and GHD-like states at the time this manuscript was written: (a) The etiologies of GHD are multiple and the phenotypes are varied. (b) Complete and extensive testing for GHD is essential in children with severe retardation of growth velocity when this is unexplained. (c) Most cases of GHD are not genetic in origin. (d) The true incidence of GHD is unknown because GHD of genetic origin is usually inherited as a mendelian recessive trait and two affected siblings with GHD are necessary to verify that inherited GHD is present. (e) Only one genetic type (IA) is known to involve an abnormality of the hGH gene. (f) Most patients with idiopathic GHD have a hypothalamic defect and not a pituitary defect.

Similarly, there are certain questions about GHD and GHD-like states. For example, is there a syndrome in which the defect lies in the production of a radioimmunoassayable but biologically inactive growth hormone? Is there an entity of NSGHD in patients other than those who receive cranial irradiation? What is the role, if any, of somatostatin and/or GHRH in PSS? What is the defect in GHD of genetic origin when there is no apparent abnormality of the known genes for production of GH or GHRH? Will the patient with Laron's dwarfism grow when given IGF-I? Will the pygmy grow at adolescence if given IGF-I? Some of these questions are addressed in other manuscripts published in this volume.

REFERENCES

Albertsson-Wikland K, Rosberg S, Isaakson O (1983) Secretory pattern of growth hormone in children of different growth rates. *Acta Endocrinol* (Copenh) [Supplement 256]: 72
Argente J, Evain-Brion D, Donnadieu M, Garnier P, Vaudry H, Job JC (1987) Impaired response of growth hormone releasing hormone measured in plasma after L-dopa

stimulation in patients with idiopathic delayed puberty. *Acta Paediatr Scand* **76**: 266-270

Bierich J (1983) Treatment of constitutional delay of growth and adolescence with human growth hormone. *Klin Padiatr* **195**: 309-316

Bierich JR, Moeller H, Ranke MB, Rosenfeld RG (1984) Pseudo- pituitary dwarfism due to resistance to somatomedin: a new syndrome. *Eur J Pediatr* **142**: 187-188

Blatt J, Bercu BB, Gillan JC, Mendelson WB, Poplack DG (1984) Reduced pulsatile growth hormone secretion in children following therapy for acute lymphoblastic leukemia. *J Pediatr* **104**: 182-186

Bright GM, Rogol AD, Johanson AJ, Blizzard RM (1983) Short stature associated with normal growth hormone and decreased somatomedin-C responses: response to exogenous growth hormone. *Pediatrics* **71**: 576-580

Carmina E, Lo Coco R, Porcelli P, Lanzara P, Jannì A (1980) Dwarfism with high somatomedin activity and delayed bone. A syndrome of receptor insensitivity to somatomedin? In: La Cauza C, Root AW (eds) *Problems in Pediatric Endocrinology.* Academic, London, pp 147-151

Frasier SD, Rallison ML (1972) Growth retardation and emotional deprivation: relative resistance to treatment with growth hormone. *J Pediatr* **80**: 603-609

Holmes NE, Blethen SL, Weldon VV (1984) Case report: Somatomedin-C response to growth hormone in psychosocial growth retardation. *Am J Med Sci* **288**: 86-88

Kowarski AA, Schneider J, Ben-Gaum E, Weldon VV, Daughaday WH (1978) Growth failure with normal serum RIA GH and low somatomedin activity: somatomedin restoration and growth acceleration after exogenous growth hormone. *J Clin Endocrinol Metab* **47**: 461-464

Lanes R, Plotnik LP, Spencer EM, Daughaday WH, Kowarski AA (1980) Dwarfism associated with normal serum growth hormone and increased bioassayable, receptor assayable, and immunoassayable somatomedin. *J Clin Endocrinol Metab* **50**: 485-488

Laron Z, Pertzelan A, Mannheimer S (1966) Genetic pituitary dwarfism with high serum concentrations of growth hormone. A new inborn error of metabolism? *Isr J Med Sci* **2**: 152-155

Laron Z (1984) Laron type dwarfism (hereditary somatomedin deficiency): a review. In: *Advances in Internal Medicine and Pediatrics.* Springer, Berlin Heidelberg New York, vol 51, pp 118-122

Merimee TJ, Zapf J, Froesch ER (1981) Dwarfism and the pygmy: an isolated deficiency of insulin-like growth factor 1. *N Engl J Med* **305**: 965-968

Merimee TJ, Zapf J, Hewlett B, Cavalli-Sforza LL (1987) Insulin like growth factors in pygmies. The role of puberty in determining final stature. *N Engl J Med* **316**: 906-911

Powell GF, Brasel JA, Blizzard RM (1967a) Emotional deprivation and growth retardation simulating idiopathic hypopituitarism. I. Clinical evaluation of the syndrome. *N Engl J Med* **276**: 1271-1278

Powell Gf, Brasel JA, Raiti S, Blizzard RM (1967b) Emotional deprivation and growth retardation simulating idiopathic hypopituitarism. II. Endocrinological evaluation of the syndrome. *N Engl J Med* **276**: 1279-1283

Rimoin DL, Merimee TJ, Rabinowitz D, Cavalli-Sforza LL, McKusick VA (1969) Peripheral subresponsiveness to human growth hormone in the African pygmies. *N Engl J Med* **281**: 1383-1388

Rogol AD, Blizzard RM, Foley TP, Furlanetto R, Selden R, Mayo K, Goodman HM, Evans

RM, Thorner MO (1985) Growth hormone releasing hormone and growth hormone:
    genetic studies in familial growth hormone deficiency. *Pediatr Res* **19**: 489-492
Rudman D, Kutner MH, Blackston RD, Cushman RA, Bain RP, Patterson JH (1981)
    Children with normal variant short stature: treatment with growth hormone for six
    months. *N Engl J Med* **305**: 123-130
Spilliotis BC, August GS, Hung W, Sonis W, Mendelson W, Bercu BB (1984) Growth
    hormone neurosecretory dysfunction. A treatable cause of short stature. *JAMA* **251**:
    2223-2230
Tanner JM (1973) Resistance to exogenous human growth hormone in psychosocial short
    stature (emotional deprivation). *J Pediatr* **82**: 171-172
Tanner JM (1988) Editorial comment. *Growth, Genetics, and Hormones* Vol 4 (1), in press
Valenta LJ, Siegel MD, Lesniak M, Elias AN, Lewis UJ, Friesen HG, Kershnar AK (1985)
    Pituitary dwarfism in a patient with circulating abnormal growth hormone polymers.
    N Engl J Med **312**: 214-216
Zadik Z, Chalew SA, Raiti S, Kowarski AA (1985) Do short children secrete insufficient
    growth hormone? *Pediatrics* **76**: 355-360

Advances in Growth Hormone and Growth Factor Research,
edited by E.E. Müller, D. Cocchi and V. Locatelli
Pythagora Press, Roma-Milan and Springer Verlag, Berlin-Heidelberg © 1989

Use of provocative tests and measurement of spontaneous GH-secretion to diagnose GH deficiency

J.R. BIERICH, G. BRÜGMANN, and E. KIESSLING

Department of Pediatrics University of Tübingen, Tübingen, Federal Republic of Germany

This contribution pursues two aims: the first is to critically examine the usual methods of investigating the somatotropic function of the pituitary, — on the one hand the provocation tests, on the other the assessment of the diurnal or nocturnal spontaneous GH secretion. The second aim is to give a condensed survey of the outcome of these investigations into the various forms of stunted growth. The test profiles obtained may then show which investigations are particularly useful in the various conditions.

METHODS OF INVESTIGATION

The provocation tests are usually classified into physiological and pharmacological tests. The physiological stimuli most frequently employed are physical exercise and nocturnal sleep onset. The *exercise* should represent a vigorous physical stress in standardized form, e.g., using a bicycle ergometer. If an increase of GH to 5 ng/ml is taken as the limit of the norm, such a value will be reached by 53% – 91% of normal controls (Buckler 1972; Keenan et al. 1972; Okada et al. 1972; Schönberg et al. 1975; Butenandt et al. 1978). The percentage of falsely positive, i.e., virtually pathological tests is with 9% – 47% extremely high. Apparently exercise is not a sufficient and reliable stimulus for a maximal activation of the pituitary. Therefore, the test is useful only as a screening procedure.

During *deep sleep*, i.e., in children with a slow- wave EEG, a marked rise of GH always ensues, which exceeds the peak following exercise. Several investigators have proposed drawing blood only once, 60 – 90 minutes after the registration of deep sleep (Parker et al. 1967; Eastman et al. 1971; Underwood et al. 1971; Mace et al. 1972; Van Gelderen and Heremans 1975). However, according to our own experience, one can by no means expect to catch the sleep-induced maximum with a single determination. If this is intended, serial determinations during the first 2 h are necessary. We well consider this point later on.

With the *pharmacological provocation tests*, the following are the main substances in use: arginine, insulin, L-dopa, glucagon and clonidine. The agent most frequently employed is arginine. Most authors are in agreement about the outcome of the arginine test (AT) in prepubertal children. The mean value for the maximal peak calculated from six large series is 15.2 ng/ml (Frohman et al. 1967; Parker et al. 1967; Sperling et al. 1970; Weldon et al. 1973; Job 1981; Bierich 1983). In the course of puberty the maxima gradually increase (Root et al. 1969; Sperling et al. 1970). Adults attain levels approximalety twice as high as those in children; the mean value from five large series of adults is 28.9 ng/ml (Knopf et al. 1965; Rimoin et al. 1966; Parker et al. 1967; Nakagawa 1971; Franchimont and Burger 1975). Women exhibit higher values than men. The variation of all control series published is rather high. For a correct interpretation of a single value it is therefore of the utmost importance to compare it with the normal values of the corresponding pubertal stage. The reliability of the published control series is additionally called into question by the fact that these groups consist mainly of children with small stature. Many of them suffer from constitutional delay which is connected with a partial GH deficency (GHD), so these values are all more or less biased.

Nearly as often as the AT, the *intravenous insulin tolerance test* (ITT) is employed. In order to gain reliable results standardized conditions are necessary. Blood glucose has to decrease by at least 50 % after insulin. If this is attained, the test delivers results which are approximately equal to those of the AT. The mean value of 5 large series of healthy prepubertal children was 16.6 ng/ml (Parker et al 1967; Weldon et al. 1973; Job 1981; Bercu et al. 1986; Lanes 1986). Not included in these series are the patients of Minuto et al. (1981) who had extraordinarily low values. Roughly 80 % of all healthy children attain GH levels > 7 ng/ml. Thus, the number of false-positive tests or "nonreponders" also corresponds to that with the AT. Figure 1 demonstrates the relationship between the AT and the ITT in 78 healthy children investigated by Job (1981). The frequency of the false-positive results was the same in both tests. Incidentally it should be mentioned that today it is common procedure to always perform two provocative tests to ascertain GHD. In many centers the AT and the ITT are performed consecutively on the same day.

The GH maxima in the *L-dopa test* are not different in principle from those ensuing from the above-mentioned two tests (Root et al. 1969; Boyd et al. 1970; Hayek and Crawford 1972; Porter and Rosenfield 1972; Chakmakjian et al. 1973;

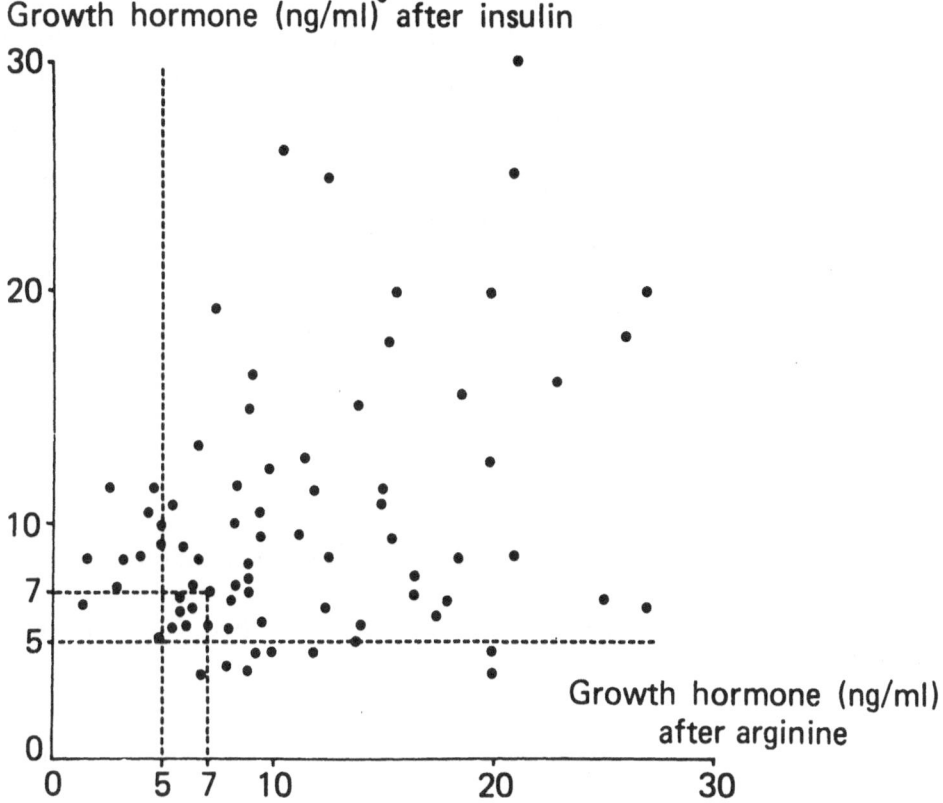

Fig. 1. Results of arginine and insulin tolerance tests in 78 healthy children (From Job 1981).

Weldon et al. 1973; Collu et al. 1978). Roughly 88 % of healthy prepubertal children attain GH levels of 5 ng/ml; 73 % attain levels > 7 ng/ml.

The same is true for the *glucagon test,* which also reliably permits the diagnosis of a GHD. False-positive results are as frequent as with the other tests mentioned. Parks et al. (1973) and Ruangwit et al. (1972) have recommended the additional administration of propranolol in order to reduce the false-positive results.

In the past few years the *clonidine test* has gained increased popularity. Gil-Ad et al. (1979), Lanes and Hurtado (1982) and Bercu et al. (1987) claim to have obtained higher maxima with clonidine than with insulin. This is in contrast to the experience gained by the British GH Committee (Health Services 1981), which performed both tests in 64 children and obtained similar results. The authors consider the two tests equally reliable but emphasize that the clonidine test is the safer investigation.

With *growth hormone releasing hormone* (GHRH) it is reliably possible to stimulate the GH output from the anterior pituitary of healthy subjects. The peak

GH levels obtained during childhood are higher than those in adults. The mean peak level reached in four series of healthy children was 36.1 ng/ml (Schriock et al. 1984; Takano et al. 1984; Pintor et al. 1985; Ranke et al. 1986). The quantitative response to GHRH and the chronologic age were inversely correlated (Schriock et al. 1984). Generally, the lower limit of a normal response is considered to be 10 ng/ml (Rogol et al. 1984; Schriock et al. 1984; Takano et al. 1984; Van Vliet et al. 1984; Butenandt et al. 1986; Ranke et al. 1986).

Figure 2 depicts the results of a large series of studies performed at our clinic (Ranke et al. 1986). Many patients with pituitary dwarfism responded with peak GH rises within the normal limits. When the injections were repeated over several days more than 50 % of the patients with GHD reacted with plasma GH levels > 10 ng/ml (Takano et al. 1984; Butenandt et al. 1986). Hence, it follows that the primary defect in the majority of patients with pituirary dwarfism is located in the hypothalamus and not in the pituitary gland.

Diagnostically, the GHRH test chiefly permits precise localization of the primary lesion in cases of GHD, a goal that is not only of scientific but also of therapeutic interest. Only when a primary hypothalamic alteration is present is replacement therapy by administration of GHRH feasible (see Ross et al., this volume).

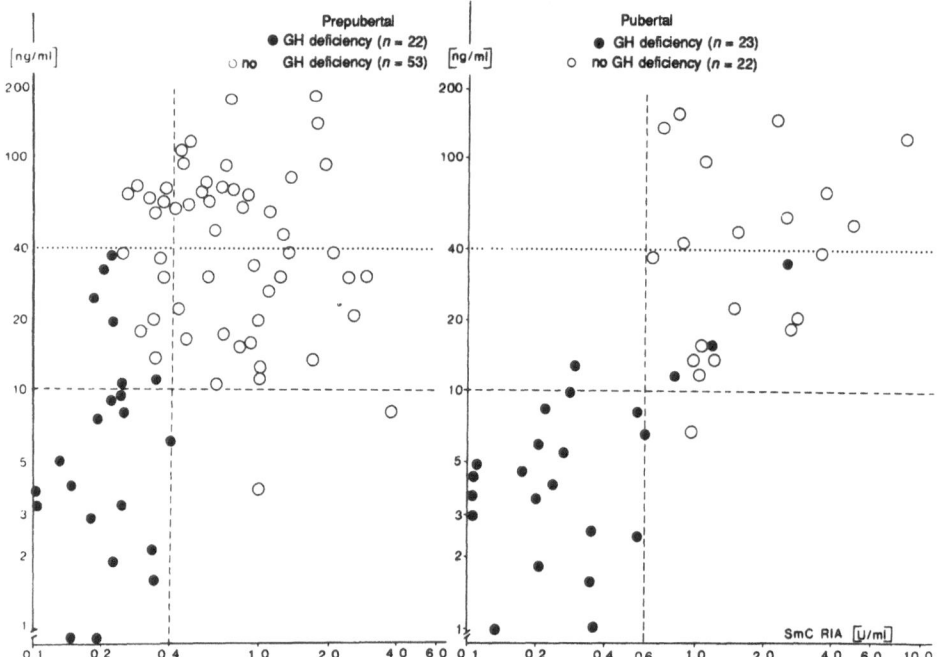

Fig. 2. Maximal GH levels to GHRH and IGF-I levels in subjects with (●) and without (○) GHD. Lines parallel to the axes indicate limits of normality (-----). The *dotted line* (......) indicates upper limits of maximal GH levels in patients with GHD.

Assessment of spontaneous GH secretion. From the 1960s on, investigations of spontaneous GH secretion have been carried out in increasing frequency, mostly over 24 h. Hunter et al. (1966a,b) and Quabbe et al. (1966) showed that the hormone is always secreted in single short bursts and that the most frequent pulses occur at night. As it turned out, the preponderance at night is not due to a superimposed circadian rhythm but is associated with sleep. Deep sleep stimulates, REM-sleep impedes the secretion. Prior to puberty, GH is produced almost exclusively at night. An important phenomenon is that puberty coincides with a vigorous increase in the secretion of GH, which is responsible for the pubertal growth spurt. Figure 3 shows examples of sleep- associated GH secretory profiles determined in the pubertal stages 1, 2, and 3. In order to correctly interpret patients' data one should always refer to comparative values of the same pubertal stage.

Initially, serial hormone determinations were generally performed over 24 h. Later on, the awareness of the preponderance of the nocturnal secretion led to the assessment of spontaneous secretion only at night. Table 1 reports the various modalities of such investigations. The first five research groups performed blood sampling only during the first 5–6 h after sleep onset, the other seven groups through 8–12 h. In our studies we have limited the length of observation time to 5.5 h and have standardized the protocol with 50 control children. If the child is sleeping deeply, a 5- to 6-h test time always suffices to detect two GH spontaneous bursts. We record the following items: (a) the maximal GH peak attained; (b) the

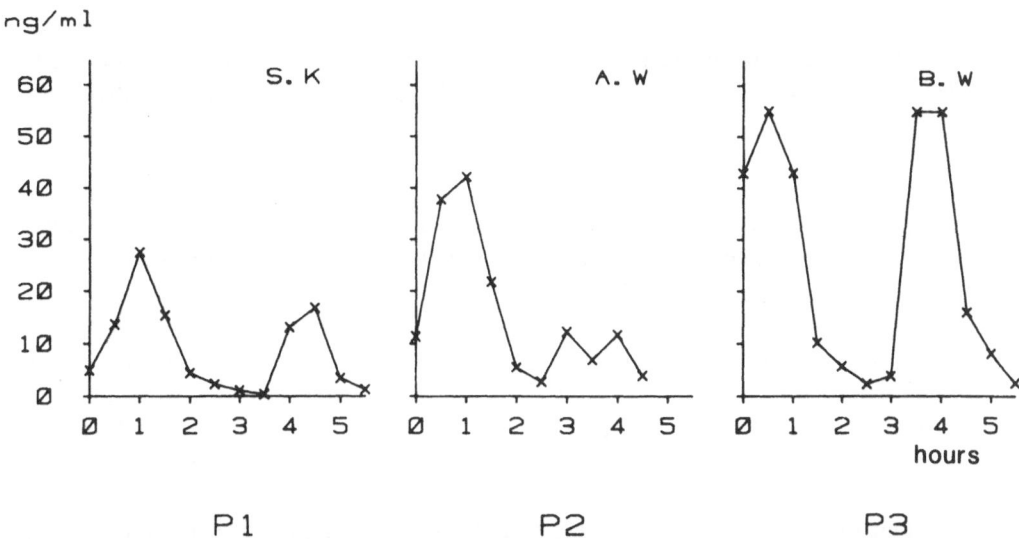

Fig. 3. Sleep-associated GH secretion in three healthy children in pubertal stages P_1, P_2 and P_3. The same symbols for pubertal development are used in the following figures.

Table 1. Sleep-induced secretion of growth hormone

Studies	Duration of observations (hours)
Howse et al. (1977)	5
van Gelderen and Heremans (1975)	5.5
Bierich and Potthoff (1979)	5.5
Zabransky and Kenkel (1987)	6
Stube et al. (1985)	6
Siegel et al. (1984)	8
Takahashi et al. (1968)	8.5
Lanes (1986)	9
Rochiccioli et al. (1985)	11
Greene et al. (1985)	11
Hopwood et al. (1985)	12
Holl et al. (1986)	12

integrated total GH secretion; (c) the number of GH peaks above 5 ng/ml. Since the last figure does not change among the different experimental groups, we shall disregard this index in the following discussion.

RESULTS

We will turn in the second part of this contribution to the results of the tests. In over 300 investigations we have combined one to three of the provocative tests alluded to previously with the assessment of sleep-associated GH secretion. As provocative tests, arginine was employed always and insulin was employed usually too; in addition, in some studies clonidine was also used. In the following only the highest GH peak attained in the various provocative tests is reported. Apart from the above results, we shall also report the data of other groups.

Figures 4–6 give the results obtained in healthy controls, subdivided according to pubertal stage. Figure 4 shows the integrated total secretion during a period of 5.5 h. Adolescents in P_3 secrete more than twice the amount of GH secreted by prepubertal children. Figure 5 shows the maximal GH peaks which were attained. It is evident that considerably higher GH peaks are reached durig sleep than with any of the provocative tests; in fact, the peaks are nearly twice as high as those of the same control children (Fig. 6).

Our findings in 64 *pituitary dwarfs* are depicted in Figure 7; from left to right, the total GH secretion, the highest GH peaks, the peak GH response to the arginine test; data are always compared with those of control children.

In a group of 93 pituitary dwarfs we compared the outcomes of the AT and

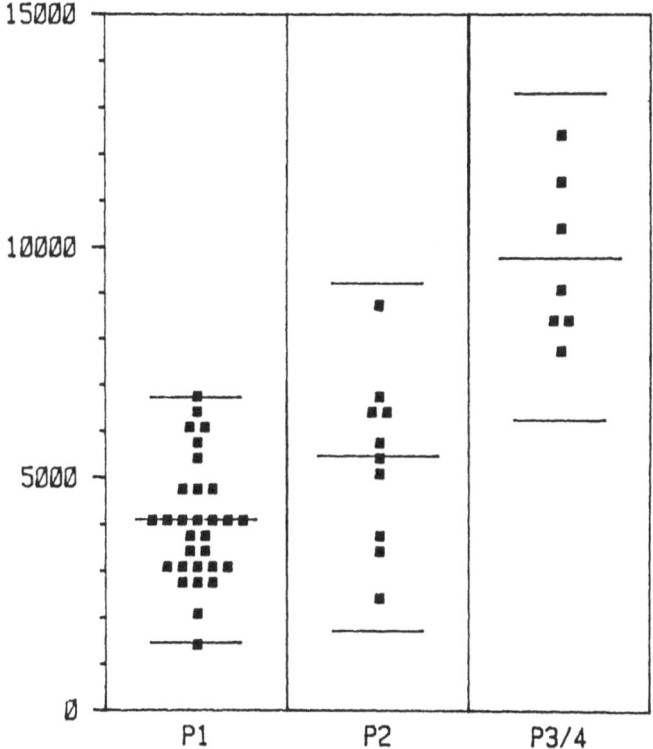

Fig. 4. Sleep-associated GH secretion in healthy controls; integrated total secretion.

the ITT. The results obtained were in full accordance, the mean peak GH values of the AT and ITT being 2.8 ng/ml and 2.4 ng/ml respectively.

The largest single group which we have investigated were 133 children with *constitutional delay of growth and adolescence* (CDGA). Figure 8 is derived from our first publication on this disorder 8 years ago (Bierich and Potthoff 1979). The results of three typical children are depicted in the left panel, those of matched pairs in the right panel. In two cases the sleep-associated GH secretion was examined twice in order to be positive on the extraordinarily low GH titers recorded, which, however, were confirmed. Figure 9 shows the three parameters alluded to previously for all patients in pubertal stage 1. It is evident that the patients secreted much smaller amounts of GH and exhibited significantly lower GH peak values. In contrast, the provocative tests yielded absolutely comparable values. The same is true for the patients in pubertal stage 2 (Fig. 10) and those in P_3-P_4 (Fig. 11). In the latter stage no overlapping was seen between patients and controls;

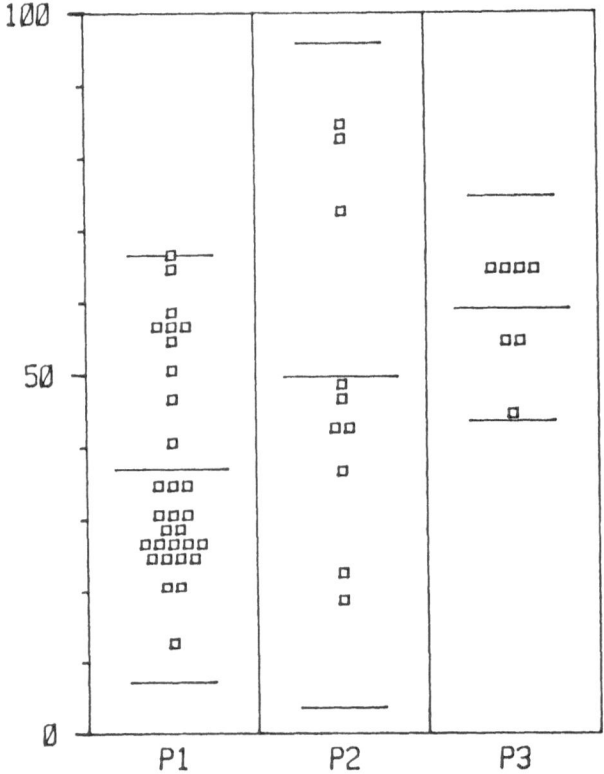

Fig. 5. Same children as in Fig. 4; highest GH peaks attained.

again, the provocative tests gave normal results. This important observation is once again corroborated by the findings given in Table 2, where the arginine tests of the CDGA patients are compared with those of the controls. In fact, there are no statistical differences between peak GH levels in patients and controls. This discrepancy is typical for CDGA and represents the main difference from partial pituitary GHD, which is characterized by a defective GH response to the provocative tests as well.

Some years ago, Bercu and his group, in particular Spiliotis, described a series of patients with similar clinical and laboratory characteristics including the discrepancy just mentioned. In 1984 Spiliotis et al. wrote: "We feel that the children with GH neurosecretory dysfunction may represent a substantial number of short

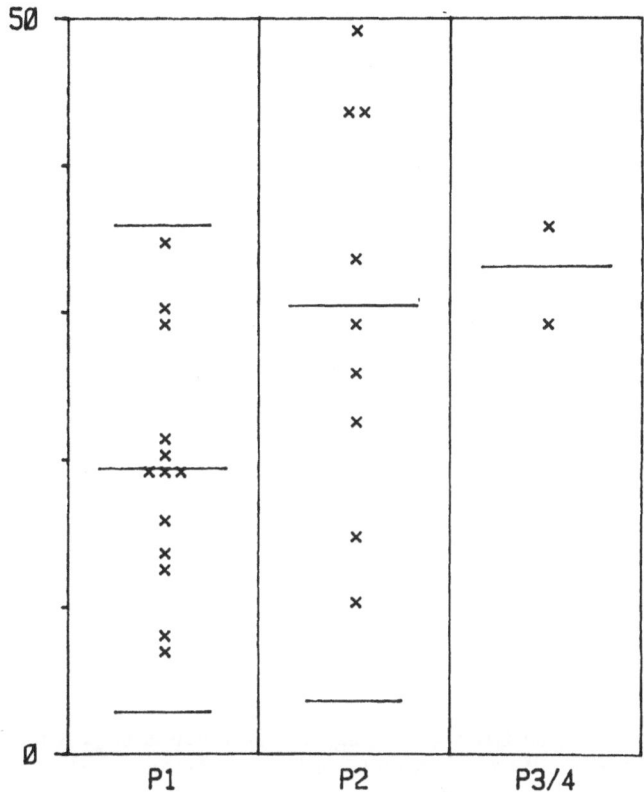

PROV. HGH SECRETION
IN HEALTHY CHILDREN, MAX. ng/ml

Fig. 6. Same children as in Figure 4; results of the arginine test.

children previously diagnosed as having constitutional delay of growth and puberty." Thus, we think that this new designation is, in most cases, simply another name for an old, well-known disorder. Confirmations of our findings, particulary of the discrepancy between stimulated and spontaneous GH secretion, have been published by Hopwood et al. (1985) and by Costin and Kaufman (1987).

Constitutional growth delay is accompanied not only by slow growth but also by retarded bone age and delayed puberty. As we have previously reported in several papers, the biological counterpart of CDGA, so-called early normal puberty, is related to and probably caused by increased spontaneous GH secretion (Bierich 1979; Bierich and Potthoff 1979). In 1983 Albertsson-Wikland et al. showed that the GH output over a 24-h period is significantly higher in fast growing than in slowly growing children. Corresponding results, also based on 24-h. GH

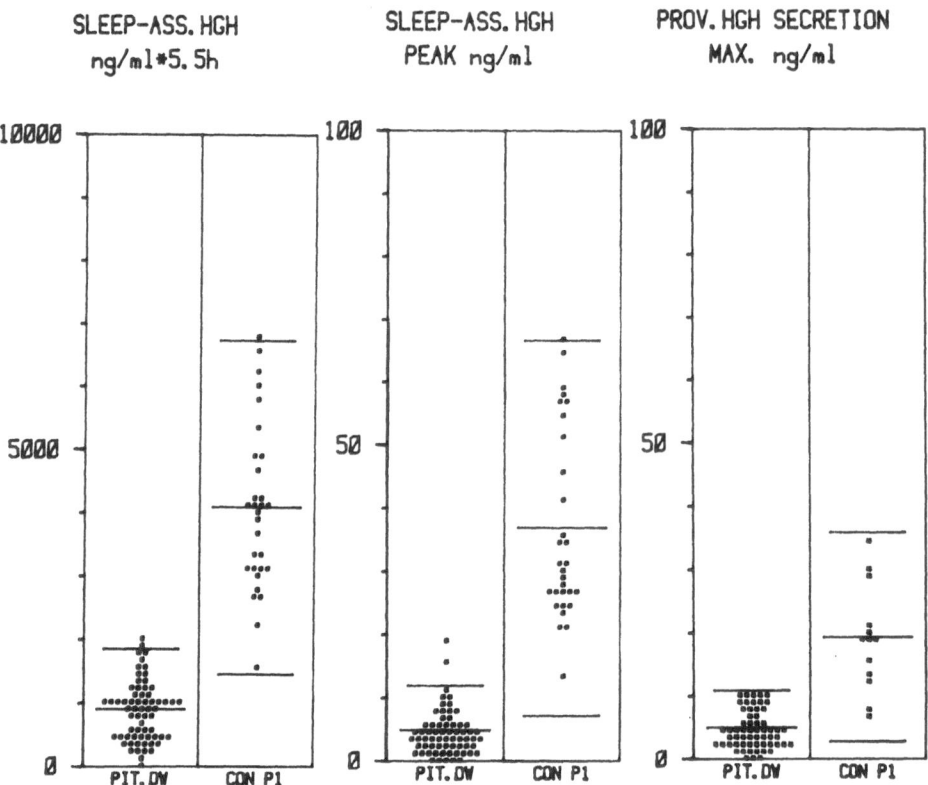

Fig. 7. Sleep-associated GH secretion and GH secretion after provocative tests in 64 pituitary dwarfs (*PIT.DW*) and in a group of healthy controls (*CON*).

assessment, have been presented by Hindmarsh et al. (1986). Thus, there is general agreement that GH secretion rate and growth velocity are interrelated processes.

An important growth disorder which can be clearly differentiated from CDGA, although the symptomatology is the same, was described in 1978 by Kowarski et al. and Hayek et al. The measurement of plasma GH gave normal quantitative values but a qualitatively abnormal, biologically less active GH level and consequently low plasma somatomedin levels. Figure 12 illustrates the three typical parameters of seven such patients. All children were young and belonged to pubertal stage 1. Mean total GH output during the 5.5–h blood sampling was 5.37 ng/ml, a value significantly above that of controls. The same was true for the maximal GH peaks attained, as well as for the GH peaks in the arginine tests. Apparently, lack of a physiological counter-regulatory feedback mechanism was involved. In two of these cases, an abnormal chromatographic pattern of the isolated GH was evident (courtesy of Drs. Heize and Schleyer, Ulm.)

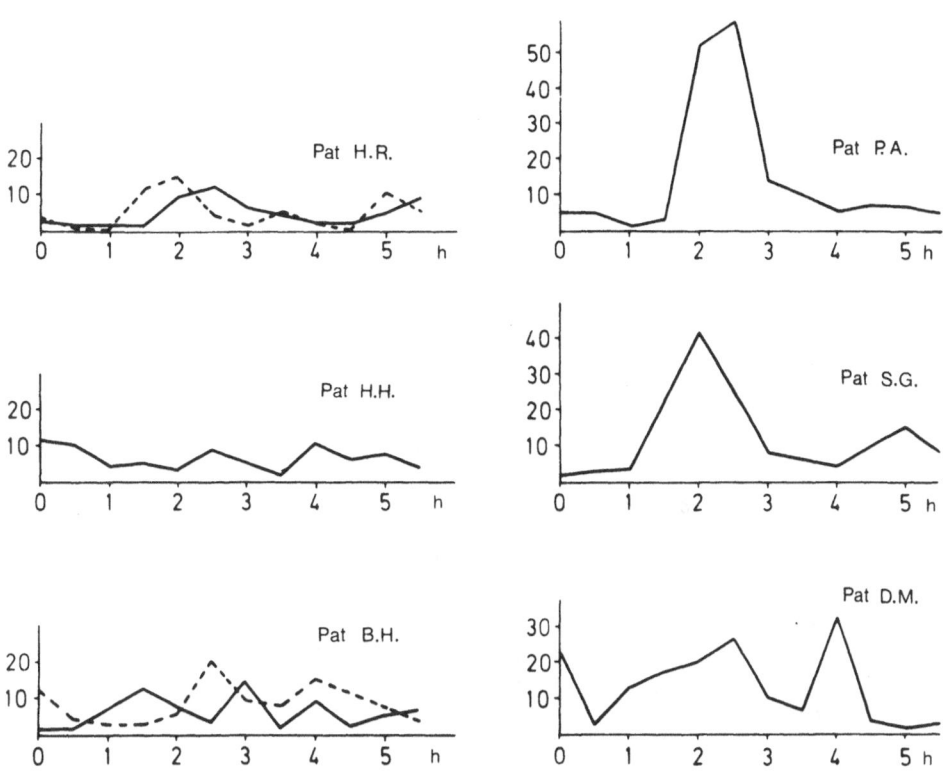

Fig. 8. Sleep-associated GH secretion in 3 children with CDGA (*left panel*), compared with control children of similar bone age (*right panel*).

With regard to *Turner's syndrome,* a crucial issue for the clinician is whether or not there is a substantial GHD. At present the answer is equivocal. The stunted growth in childhood is very likely due to the dysostosis of the patients, who radiologically show osteopenia and histochemically abnormal collagen.

However, during adolescence an additional factor comes into play, i.e., gonadal dysgenesis, which impedes the pubertal growth spurt. Taking this into account, we subdvided 40 patients with Turner's syndrome into two groups, i.e., those under (TS-I) and those over (TS-II) 9 years of age (Table 3).

In fact, the younger children are endocrinologically normal, whereas the adolescents exhibit very low GH secretion rates at night, although the arginine tests do not show values typical for GHD. Our results correspond fully to those of Ross et al. (1985), who performed their GH determinations during a 24-h. period, although our values are still lower than the average for prepubertal healthy children. These findings may provide the rationale for replacement therapy with hGH. An additional possibility is to perform therapy with sex hormones which are lacking.

Finally, we would like to briefly mention three further growth disorders. They

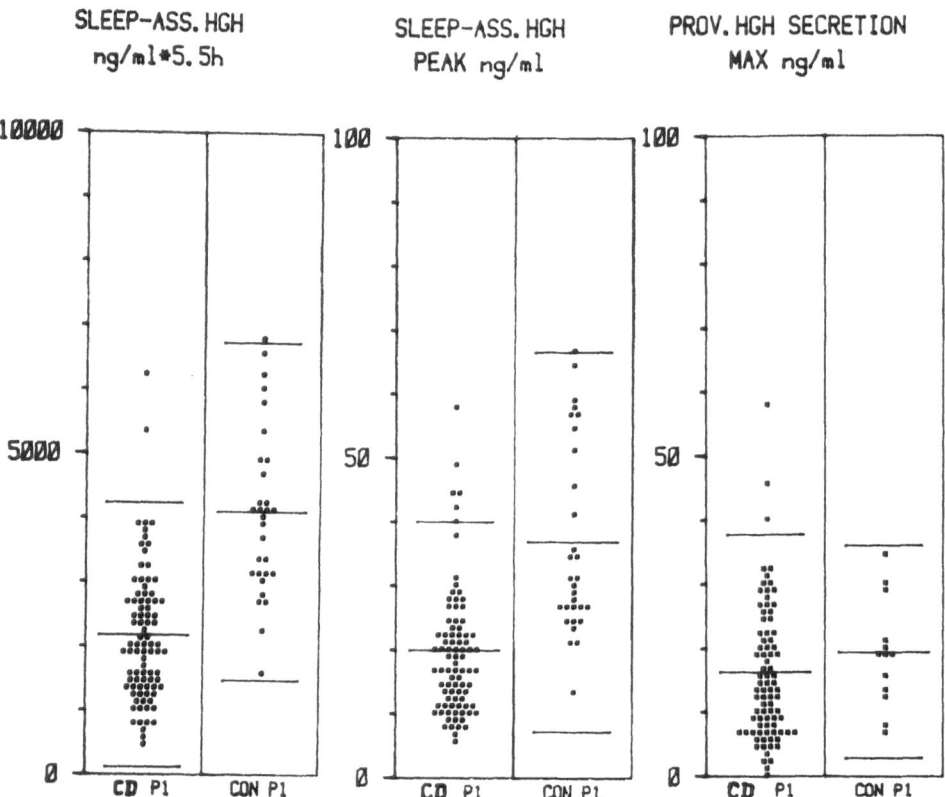

Fig. 9. Sleep-associated GH secretion and GH secretion after provocative tests in P_1-children with constitutional delay of growth (*CD*) compared with healthy controls (*CON*).

are not as problematic as the ones previously mentioned since stimulated and spontaneous GH secretion follow the same trend.

The first is *intrauterine growth retardation*, exemplified by ten patients who were born small for date. Four had the Silver- Russell-syndrome. Figure 13 depicts the striking variability of the results, which can be explained by the heterogeneity of the group. Nevertheless, most patients had results within the normal range (Table 4).

In Figure 14 the data from 11 children with *severe obesity* are depicted. The integrated GH secretion was maximally depressed, even more than in pituitary dwarfs, and the highest peaks attained and the peaks after provocative tests were lowered (Table 5). The explanation lies in the counter-regulating feedback mechanisms exerted by the increased levels of insulin and somatomedin-C.

The last group are children with *increased plasma levels of corticosteroids*. Six children who suffered from chronic inflammatory conditions like Crohn's disease were given 2 mg/kg prednisone per day. Three other children had Cushing's

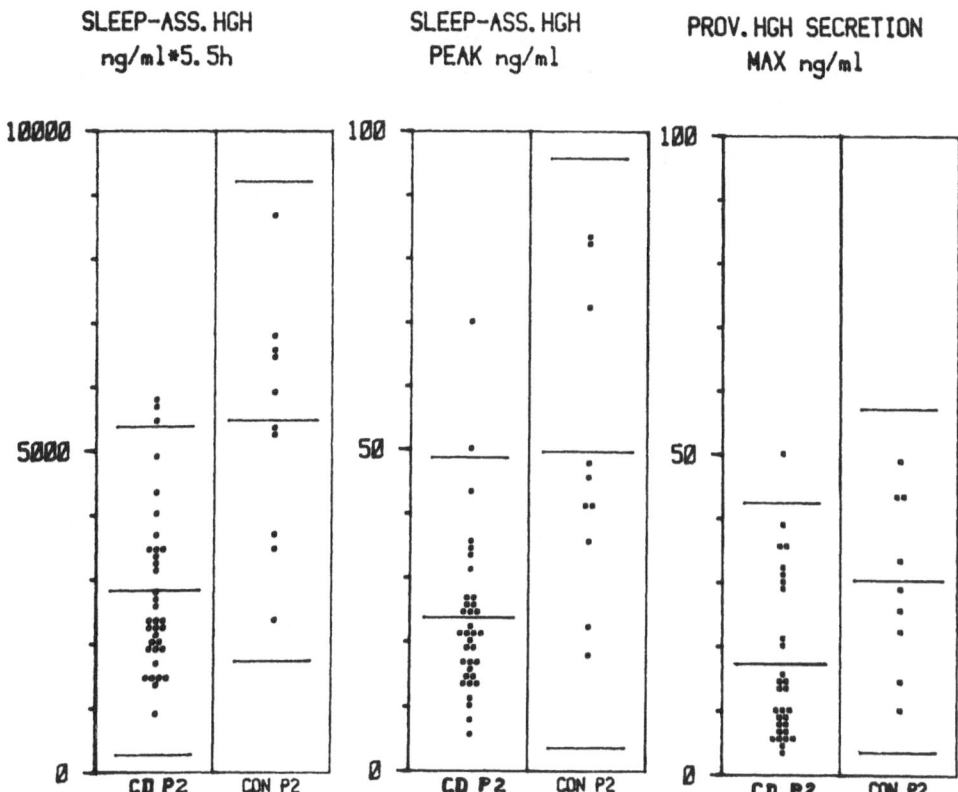

Fig. 10. Sleep-associated GH secretion and GH secretion after provocative tests in P$_2$-children with constitutional delay of growth (*CD*) compared with healthy controls (*CON*).

syndrome (Fig. 15). The patients' GH secretion rate was extremely low. It recovered gradually in those cases where either the corticosteroids were discontinued or the adrenals removed.

Clinically, we learn from these findings that the severe catabolic state which is induced by corticosteroids results not only from the direct catabolic actions of the steroids themselves but also from the total abolition of GH, the most active anabolic counterpart of the corticosteroids. From these considerations a new indication for the additional treatment of such patients with GH may follow, which we have commenced in several cases since the last year. Summing up, we would like to state the following:

1. The classical forms of GHD can be reliably diagnosed by the usual provocative tests for GH secretion. Two tests should always be performed.
2. Among the nonclassical forms of GHD, particulary in CDGA and in some of the postirradiation syndromes, discrepancies frequently appear between the

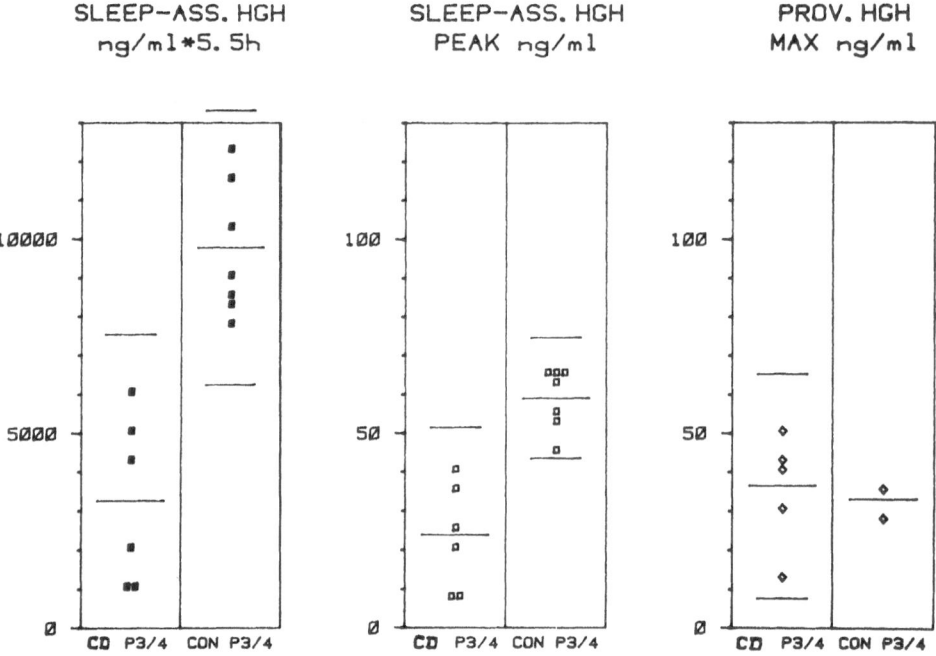

Fig. 11. Sleep-associated GH secretion and GH secretion after provocative tests in adolescents (P_3/P_4) with constitutional delay of growth (*CD*) compared with healthy controls (*CON*).

Table 2. Effect of the arginine test on plasma GH levels (ng/ml) of children with constitutional delay of growth and adolescence (CDGA) and healthy children

| Group | Pubertal stage | | |
	P_1	P_2	P_3
CDGA	16.3 ± 10.7	17.3 ± 12.6	36.5 ± 14.4
Controls	19.4 ± 8.3	30.4 ± 13.4	31.1 ± 12.3

Values are expressed as mean ± SEM.

outcome of the provocative tests, which provide normal results, and the assessment of spontaneous GH secretion, either measured during a 24-h period or sleep associated, which is often considerably diminished. This discrepancy points to a cybernetic distubance of the hypothalamic control mechanisms (neuroendocrine dysfunction). In these instances, the spontaneous GH secretion must be determined, for diagnostic as well as for therapeutic purposes.

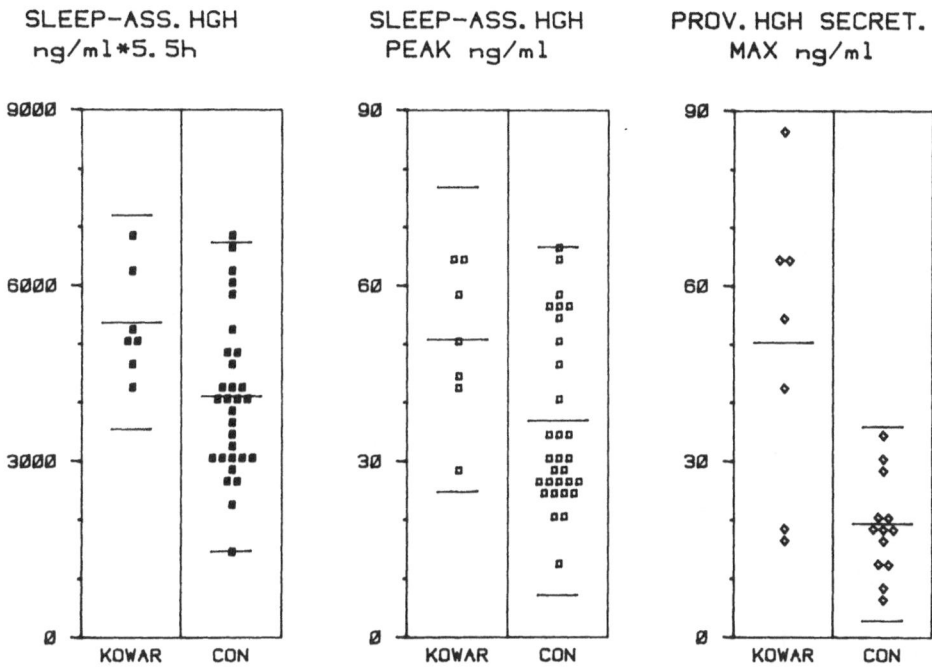

Fig. 12. Sleep-associated GH secretion and GH secretion after provocative tests in three patients with Kowarski-Hayek syndrome (*Kowar*) compared with healthy controls (*CON*).

Table 3. Evaluation of spontaneous versus stimulated growth hormone secretion in subjects with Turner's syndrome (TS)

Group	Sleep-associated (ng/ml/5.5 h)	Sleep-associated peak (ng/ml)	Arginine test peak (ng/ml)
TS-I	3087 ± 863 (6)	28 ± 6 (7)	23 ± 19 (5)
TS-II	1669 ± 880 (33)	18 ± 11 (32)	21 ± 13 (25)
Controls	4100 ± 1320 (30)	37 ± 15 (30)	19 ± 8 (13)

Values are expressed as mean ± SEM; number of subjects in parentheses.

3. In Turner's syndrome, the investigations show normal results in the prepubertal age-group. Thereafter, the spontaneous nocturnal GH secretion of the majority of patients is markedly diminished, in accordance with the missing pubertal growth spurt of these girls.

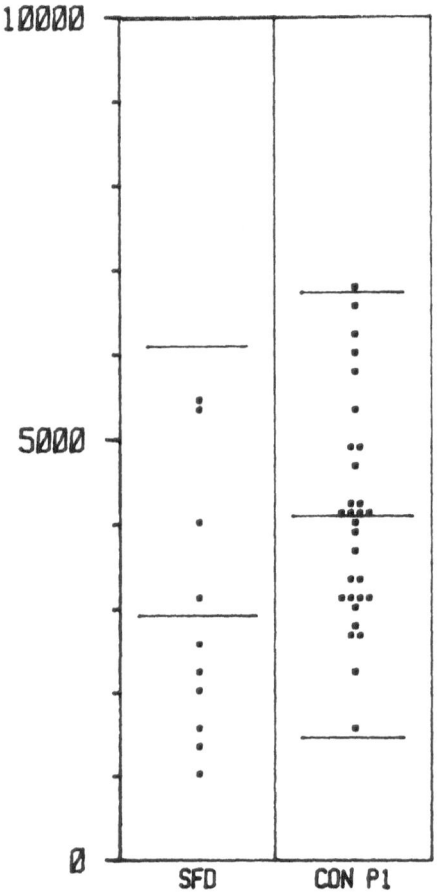

SLEEP-ASS. HGH
ng/ml*5.5h

Fig. 13. Sleep-associated GH secretion in patients with intrauterine growth retardation (small for date) (*SFD*) compared with healthy controls (*CON*).

Table 4. Evaluation of spontaneous versus stimulated growth hormone secretion in subjects with intrauterine growth retardation

Group	Sleep-associated (ng/ml/5.5 h)	Sleep-associated peak (ng/ml)	Arginine test peak (ng/ml)
SFD	2924 ± 1590 (10)	24 ± 17 (10)	21 ± 9
Controls	4100 ± 1320 (30)	37 ± 15 (30)	19 ± 8

SFD, Small for date

Values are expressed as mean ± SEM; number of subjects in parentheses

SLEEP-ASS. HGH
ng/ml

Fig. 14. Sleep-associated GH secretion in children with severe obesity (*FS*) compared with healthy controls (*CON*).

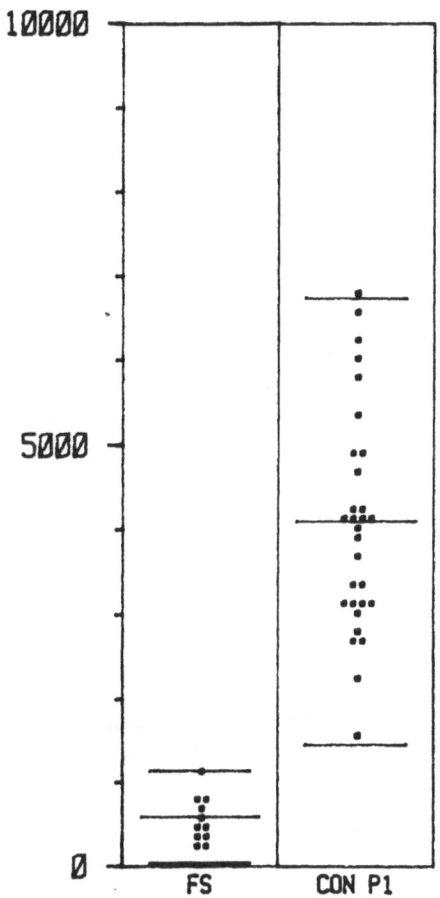

Table 5. Evaluation of spontaneous versus stimulated growth hormone secretion in children with severe obesity

Group	Sleep-associated (ng/ml/5.5 h)	Sleep-associated peak (ng/ml)	Arginine test peak (ng/ml)
Severely obese	588 ± 275 (11)	4.5 ± 2 (11)	11 ± 13 (10)
Controls	4100 ± 1320 (30)	37 ± 15 (30)	19 ± 8 (13)

Values are expressed as mean ± SEM; number of subjects in parentheses.

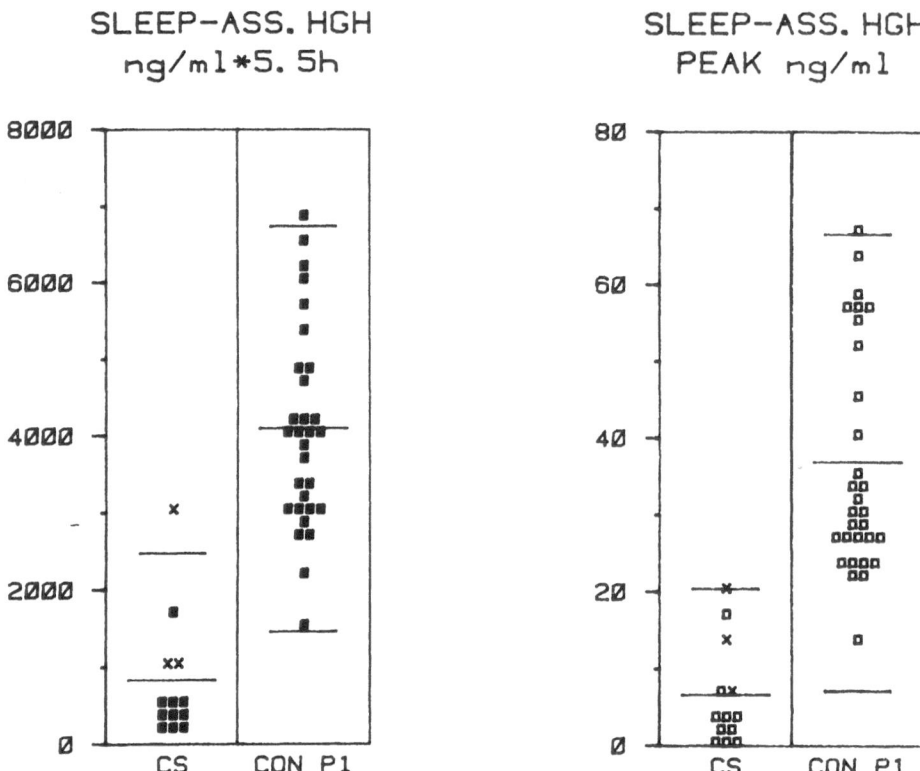

Fig. 15. Sleep-associated GH secretion in children on corticosteroid treatment (*CS*) (*n* = 6) or Cushing's syndrome (*n* = 3). (x) indicates cessation of steroid therapy (*n* = 2) or adrenalectomy (*n* = 1). Values are compared with those of healthy controls (*CON*).

REFERENCES

Albertsson-Wikland K, Rosberg S, Isaksson O (1983) Secretory pattern of growth hormone in children of different growth rates. (Abstr.) *Acta Endocrinol* [Suppl] **256**: 72

Bercu BB, Shulman D, Root AW, Spiliotis B (1986) Growth hormone provocative testing frequently does not reflect endogenous GH secretion. *J Clin Endocrinol Metab* **63**: 709-716

Bierich JR (1979) Growth velocity and spontaneous secretion of hGH in healthy children. Int. Paed. Conf. Europ. Countries, *The Healthy Child*, UNEPSA, Moscow, p 23

Bierich JR (1983) Minderwuchs. *Monatsschr Kinderheilkd* **131**: 180-192

Bierich JR (1987) Serum growth hormone levels in provocation tests and during nocturnal spontaneous secretion: a comparative study. Int. Symp. Diagnosis and Treatment Impaired GH Secretion, Vienna 1987. *Acta Pediatr Scand* [Suppl] **337**: 48-59

Bierich JR, Potthoff K (1977) Cause of constitutional delay of growth and adolescence: diminished secretion of growth hormone. Recent Results. 9th Meet Eur Soc Pediatr Res, Venice, Sept 7-10, 1977

Bierich JR, Potthoff K (1979) Die Spontansekretion des Wachstumshormons bei der konstitutionellen Entwicklungsverzögerung und der frühnormalen Pubertät. *Monatsschr Kinderheilkd* **127**: 561-565

Boyd AE, Lebowitz HE, Pfeiffer JB (1970) Stimulation of human growth hormone secretion by L-dopa. *N Engl J Med* **283**: 1425-1430

Buckler JMH (1972) Exercise as a screening test for growth hormone release. *Acta Endocrinol* (Copenh) **69**: 219-229

Butenandt O, Dörr H, Rüdisser A, Joswig R (1984) Wachstumshormonsekretion und Plasma-Renin-Aktivität nach Gabe von Clonidin. *Monatsschr Kinderheilkd* **132**: 776-779

Butenandt O, Eder R, Wohlfahrt K, Bidlingmaier F, Knorr D (1976) Mean 24-hour growth hormone and testosterone concentrations in relation to pubertal growth spurts in boys with normal or delayed puberty. *Eur J Pediatr* **122**: 85-92

Butenandt O, Grisar Th, Hager C (1978) Ausschluaβdiagnostik einer defizienten Wachstumshormonsekretion in der kinderärztlichen Praxis. *Klin Padiatr* **190**: 460-464

Chakmakjian ZH, Marks JF, Fink CW (1973) Effect of levodopa on serum growth hormone in children with short stature. *Pediatr Res* **7**: 71-74

Chalew SA, Armour KM, Raiti S, Kowarski A (1985) Response to growth hormone therapy in children with normal GH by provocative stimulation but deficient integrated concentration of GH (Abstr. 506). *Pediatr Res* **19**: 195A

Collu R, Brun G, Milsant F, Leboeuf G, Letarte J, Ducharme JR (1978) Reevaluation of levodopa-propranolol as a test of growth hormone reserve in children. *Pediatrics* **61**: 242-244

Costin G, Kaufman FR (1987) Growth hormone secretory pattern in children with short stature. *J Pediatr* **110**: 362-368

Eastman CJ, Lazarus L (1973) Growth hormone release during sleep in growth-retarded children. *Arch Dis Child* **48**: 502-507

Finkelstein JW, Rottwarg HP, Boyar RM, Kream J, Hellman L (1972) Age- related change in the 24-hour spontaneous secretion of growth hormone. *J Clin Endocrinol Metab* **36**: 665-670

Franchimont P, Burger H (eds) (1975) *Human Growth Hormone and Gonadotrophins in Health and Disease*. North Holland, American Elsevier, Amsterdam

Frohman LA, Aceto T, MacGillivray MH (1967) Studies on growth hormone secretion in children: normal, hypopituitary and constitutionally delayed. *J Clin Endocrinol Metab* **27**: 1409-1417

Gil-Ad I, Topper E, Laron Z (1979) Oral clonidine as a growth hormone stimulation test. *Lancet* **ii**: 278-280

Greene S, Rees L, Adlard P, Jones J, Chantler C, Preece M (1986) Abnormal overnight hormone profiles in adolescents with renal disease receiving long- term steroid therapy. (Abst 158) *Pediatr Res* **20**: 1204

Hayek A, Crawford DD (1972) L-dopa and pituitary hormone secretion. *J Clin Endocrinol Metab* **34**: 764-766

Hayek A, Peake GT, Greenberg RE (1978) A new syndrome of short stature due to biologically inactive but immunoreactive growth hormone. (Abstr.). *Pediatr Res* **12**: 415

Health Services Human Growth Hormone Committee (1981) Comparison of the intravenous insulin and oral clonidine tolerance tests for growth hormone secretion. *Arch Dis Child* **56**: 852-854

Hindmarsh P, Smith PJ, Pringle PJ, Brook CGD (1986) The relationship between height velocity and 24-hour growth hormone secretion. (Abstr. 10). *Pediatr Res* **20**: 1179

Holl R, Hartmann P, Heinze E, Sorgo W (1986) GH secretion during sleep and GRF-stimulation in GH-deficient and short normal children. (Abstr. 66). *Pediatr Res* **20**: 1189

Hopwood NJ, Bacon GE, Beirins IZ, Hale PM, Mendes TM, Kelch RP (1985) Growth hormone secretory patterns — aid to diagnosis of growth problems. (Abstr. 31). *Pediatr Res* **19**: 608

Howse PM, Rayner PHW, Williams JW, Rudd BT, de Bertram PV, Thompson CRS, Jones LA (1977) Nyctohemeral secretion of growth hormone in normal children of short stature and in children with hypopituitarism and intrauterine growth retardation. *Clin Endocrinol* **6**: 347-359

Hunter WM, Rigal WM (1966) The diurnal pattern of growth hormone concentration in children and adolescents. *J Endocr* **34**: 147-153

Hunter WM, Friend JAR, Strong JA (1966) The diurnal pattern of plasma growth hormone concentration in adults. *J Endocr* **34**: 139-146

Illig R, Sthal M, Henrichs J, Hecker A (1971) Growth hormone release during slow-wave sleep. Comparison with insulin and arginine provocation in children with small stature. *Helv Paediatr Acta* **26**: 655-672

Job JC (1981) The pituitary. In: Job JC, Pierson M (eds) *Pediatric Endocrinology.* Wiley, New York

Keenan BS, Killmer LB, Sode J (1972) Growth hormone response to exercise: a test of pituitary function in children. *Pediatrics* **50**: 760-764

Knopf RF, Conn JW, Fajans SS, Floyd JC, Gunthsche EM, Rull JA (1965) Plasma growth hormone response to intravenous administration of aminoacids. *J Clin Endocrinol Metab* **25**: 1140-1144

Kowarski AA, Schneider J, Ben Galim E, Weldon VV, Daughaday WH (1978) Growth failure with normal serum RIA-GH and low somatomedin activity: somatomedin restoration and growth acceleration after exogenous GH. *J Clin Endocrinol Metab* **47**: 461-464

Lanes R (1986) Growth hormone secretion in patients with constitutional delay of growth and pubertal development. *J Pediatr* **109**: 781-783

Lanes R, Hurtado E (1982) Oral clonidine — an effective GH-releasing agent in prepubertal subjects. *J Pediatr* **100**: 710-714

Mace JW, Gotlin RW, Beck P (1972) Sleep-related human growth hormone (GH) release: a test of physiologic growth hormone secretion in children. *J Clin Endocrinol Metab* **34**: 339-341

Minuto F, Barreca A, Ferrini S Mazzocchi G, Del Monte P, Giordano G (1982) Growth hormone secretion in pubertal and adult subjects. *Acta Endocrinol* (Copenh) **99**: 161-165

Nakagawa K (1971) Probable sleep dependency of diurnal rhythm of plasma growth hormone response to hypoglycemia. *J Clin Endocrinol Metab* **33**: 854-856

Okada Y, Hikita T, Ishitobi K, Wada M, Santo Y, Harada Y (1972) Human growth hormone secretion after exercise and oral glucose administration in patients with short stature. *J Clin Endocrinol Metab* **34**: 1055-1058

Parker ML, Hammond JM, Daughaday WH (1967) The arginine provocative test: an aid in the diagnosis of hyposomatotropism. *J Clin Endocrinol* **27**: 1129-1136

Parks JS, Amrhein JA, Vaidya V, Moshang T Jr, Bongiovanni AM (1973) Growth hormone responses to propranolol-glucagon stimulation: a comparison with other tests of growth hormone reserve. *J Clin Endocrinol Metab* **37**: 85-92

Pertzelan A, Keret R, Bauman B, Josefsberg Z, Laron Z (1985) hGH response to GH-RH 1-44 in obesity of various etiologies. (Abstr. 40). *Pediatr Res* **19**: 610

Pintor C. Corda R, Puggioni R, Cella SG, Locatelli V, Loche S, Müller EE (1985) Clonidine accelerates growth in children with impaired growth hormone secretion. *Lancet* i: 1482-1484

Plotnick LP, Lee PA, Migeon CJ, Kowarski AA (1979) Comparison of physiological and pharmacological tests of growth hormone function in children with short stature. *J Clin Endocrinol Metab* **48**: 811-815

Porter BA, Rosenfield RL, Lawrence AM (1972) L-dopa stimulation of growth hormone release in children and its significance. (Abstr. 90). *Pediatr Res* **6**: 350

Quabbe HJ, Schilling E, Helge H (1966) Pattern of growth hormone secretion during a 24-hour test in normal children. *J Clin Endocrinol* **26**: 1173-1177

Ranke MB, Gruhler M, Blum WFP, Bierich JR (1986) Testing with growth hormone-releasing factor and somatomedin-C measurement for the evaluation of growth hormone deficiency. *Eur J Pediatr* **145**: 485-492

Rimoin DL, Merimee TJ, McKusick VA (1966) Growth hormone deficiency in man: an isolated recessively inherited defect. *Science* **152**: 1635-1637

Rochiccioli P, Sanz MT, Calvet U. Arbus L, Chabelain P, Bernard MT, Dutau G, Sabbayrolles B, Enjaume C (1985) Etude de la sécrétion somatotrope du sommeil dans 60 cas de retards staturaux de l'enfant. *Arch Fr Pediatr* **42**: 665-670

Rogol AD, Blizzard RM, Johanson AJ, Furlanetto RW, Evans WS, Rivier J, Vale WW, Thorner MO (1984) Growth hormone release in response to human pancreatic tumor growth hormone releasing hormone 1- 40 in children with short stature. *J Clin Endocrinol Metab* **59**: 580-586

Root AW, Saenz-Rodriguez C, Bongiovanni AM, Eberlein WR (1969) The effect of arginine infusion on plasma growth hormone and insulin in children. *J Pediatr* **2**: 187-197

Ross JL, Meyerson Long L, Loriaux DL, Cutler GB Jr (1985) Growth hormone secretory dynamics in Turner syndrome. *J Pediatr* **106**: 202-206

Ruangwit A, Cavallo A, Fisher FN, Grigler JF (1972) Propranolol enhancement of immunoreactive growth hormone response to glucagon in children. (Abstr.). *Pediatr Res* **6**: 350

Schönberg DK, Puttkamer C von, Klemm W, Gupta D, Bierich JR (1975) Kurz- und Kombinationstests zur Prüfung hormonaler Insuffizienz bei minderwüchsigen Kindern. *Monatsschr Kinderheilkd* **123**: 331-334

Schriock E, Lustig RH, Rosenthal SM, Kaplan SL, Grumbach MM (1984) Effect of growth hormone (GH)-releasing hormone (GRH) on plasma GH in relation to magnitude and duration of growth hormone deficiency in 26 children and adults with isolated GHD or multiple pituitary hormone deficiency. *J Clin Endocrinol Metab* **58**: 1043-1049

Shibasaki T, Shizume K, Nakahara M, Masuda A, Jibiki K, Demura H, Wakabayashi I, Ling N (1984) Age-related changes in plasma growth hormone response to growth hormone-releasing factor in man. *J Clin Endocrinol Metab* **58**: 212-214

Siegel SF, Becker DJ, Lee PA, Gutai JP, Foley TP, Rash AL (1984) Comparison of physiologic and pharmacologic assessment of growth hormone secretion. *Am J Dis Child* **138**: 540-543

Sperling MA, Kenny FM, Drash AL (1970) Arginine-induced growth hormone responses in children: effect of age and puberty. *J Pediatr* **77**: 462-465

Spiliotis BE, August GP, Wellington H, Sonis W, Mendelson W, Bercu BB (1984) Growth hormone neurosecretory dysfunction. A treatable cause of short stature. *JAMA* **251**: 2223-2230

Stubbe P, Jakat K, Heidemann P (1985) Growth hormone secretion in constitutional delay of growth and development. (Abstr.). *Ped Res* **19**: 197

Takahashi Y, Kipnis DM, Daughaday WH (1968) Growth hormone secretion during sleep. *J Clin Invest* **47**: 2079-2090

Takano K, Hizuka N, Shizume K, Asakasa K, Miyakawa M, Hirose N, Shibasaki T, Ling NC (1984) Plasma GH response to GHRF in children with short stature and patients with pituitary dwarfism. *J Clin Endocrinol Metab* **58**: 236-241

Underwood LE, Azumi K, Voina SJ, Van Wyk JJ (1971) Growth hormone levels during sleep in normal and growth hormone-deficient children. *Pediatrics* **48**: 946-954

Van Gelderen HH, Hermans G (1975) Growth hormone levels during the first hours of sleep in children (Abstr.). *Pediatr Res* **9**: 870

Van Vliet G, Styne DM, Kaplan SL, Grumbach MM (1984) Growth hormone treatment for short stature. *N Engl J Med* **309**: 1016-1022

Weldon VV, Gupta SK, Haymond W (1973) The use of L-dopa in the diagnosis of hyposomatotropism in children. *J Clin Endocrinol Metab* **36**: 42-46

Zabransky S, Kenkel B (1987) Diagnostik des hypophysären Minderwuchses. *Excerpta Paediatr* **11**: 123-127

Zadik Z, Chalew SA, Raiti S, Kowarski AA (1985) Do short children secrete insufficient growth hormone? *Pediatrics* **76**: 355-360

Advances in Growth Hormone and Growth Factor Research,
edited by E.E. Müller, D. Cocchi and V. Locatelli
Pythagora Press, Roma-Milano and Springer Verlag, Berlin-Heidelberg © 1989

Growth hormone secretory dysfunction after hypothalamic and pituitary irradiation*

R. Rappaport, M. Fontoura, and R. Brauner

Clinical Unit of Pediatric Endocrinology and Diabetes, Research Laboratory INSERM U-30, Paris, France

External cranial radiation is a major therapeutic tool in the treatment of various malignant diseases in children. As growth retardation has become a significant complication of this therapy, there have been quite a number of studies designed to elucidate its mechanism, especially its effects on hypothalamic and pituitary function. Ever since the earliest studies (Tan and Kunaratnam 1966; Larkins and Martin 1973; Samaan et al. 1975; Shalet et al. 1975; Perry-Keene et al. 1976; Richards et al. 1976; Czernichow et al. 1977) it has been very clear that growth hormone (GH) secretion is the most frequently altered function, with obvious consequences for growth. These studies demonstrated that the pituitary and hypothalamic injury was dose dependent, which explains why long-term consequences of cranial radiation differ according to the different therapeutic protocols used for each disease (Table 1).

The natural course of endocrine disease is now clear, as a result of a few prospective studies (Shalet et al. 1978; Brauner et al. 1985a; Lam et al. 1987) which have provided a general view of the progressive alteration of GH secretion and its impact on statural growth. More generally, it can be stated that cranial radiation has provided a model of induced GH deficiency. It has stimulated a number of studies aimed at a more critical evaluation of GH secretion. Even more interesting are the data comparing the GH secretion of these children with their pattern of growth. If there is any quantitative relationship between biological data related to GH, somatomedin-C (Sm-C), and growth, these patients should provide an opportunity for a more critical description of the so-called complete or partial GH

* This work was supported by a grant from the Association pour la Recherche contre le Cancer (ARC no. 415-1984) and the INSERM.

Table 1. Frequency of growth hormone deficiency after cranial radiation according to radiation dose (Brauner R, Rappaport R., unpublished data)

Radiation dose (Gy)	Etiology	No. of cases	Cases of GHD (%)
24	ALL	86	52
20-45	Head/neck tumors	56	68
30-45	Medulloblastoma	56	76
50-55	Optic glioma	48	100

GHD = GH (AITT) peak < 8 ng/ml.

deficiency. In addition, a number of unexpected situations indicate that GH secretion is not the sole factor regulating growth and final stature in these patients. These studies have proven to be of great interest to the pediatric endocrinologist, but they should also help to improve the therapeutic protocols used for the treatment of a great variety of malignant diseases.

GROWTH HORMONE SECRETION

Evidence from studies in head-irradiated primates suggests that the disturbance of GH secretion is predominantly the result of disturbed hypothalamic control (Chrousos et al. 1982). A few studies using GRF stimulation in man tend to support the hypothesis of hypothalamic GRF deficiency. These data are based on the finding that GH-deficient children respond to the administration of hGRF (Ahmed and Shalet 1984; Lustig et al. 1985; Crosnier et al. 1986) and on the relative radioresistance of the pituitary gland (Lawrence et al. 1937) compared with the hypothalamus (Arnold 1954). However, the distinction between a hypothalamic and a pituitary site of damage is difficult to assess. Some children were low responders to GRF, and radiation damage to the pituitary somatotrophs may also have occurred. This distinction is of some consequence if a treatment with synthetic hGRF is considered.

Experimental Studies

Chrousos et al. (1982) showed that, in subhuman primates, there were distinct discrepancies in the response to various types of pharmacological stimulation: the GH response to insulin-induced hypoglycemia was impaired; the response to arginine and to L-dopa stimulation remained normal. Despite this normal response they found that 1 year after radiation there was a marked decrease in the frequency and magnitude of spontaneous GH pulses. These data suggest that radiation could principally affect the hypothalamic centers generating pulsatile GH secretion and/ or the secretion of GRF. These data also led to the concept of neurosecretory dysfunction (Spiliotis et al. 1984; Bercu et al. 1986), based upon the discrepancies between the GH response to various pharmacological tests and the more physiological spontaneous profile of GH secretion, as will be discussed further in this report. Studies by Mosier et al. (1985) showed that irradiated rats had normal periodicity of bursts of GH secretion. However, the amounts of GH secreted, as indicated by integrated concentrations, were significantly reduced. Unfortunately, these experimental studies did not provide more data as to the nature of the hypothalamic and pituitary lesions induced by radiation, their respective roles, and their changes over time after radiation.

Clinical Studies

It has become apparent from previous studies that GH secretory dysfunction has multiple presentations according to the methods used to quantitate GH secretion. A general pattern of progressive pituitary failure is now rather clear. GH deficiency is the first to appear and is the most frequent postradiation pituitary defect in the cranial radiation dosage of 24-45 Gy or more (Table 2).

Table 2. Frequency of pituitary deficiencies in children after cranial radiation (Brauner R, Rappaport R., unpublished data)

Radiation dose Gy		GH (%)	TSH (%)	ACTH (%)	Abnormal puberty (%)[a]
24	(n = 86)	52	2	0	18
30-45	(n = 56)[b]	68	35	7	13
30-45	(n = 56)[c]	76	47	8	21

[a] Evaluated among children with pubertal bone age.
[b] Cranial irradiation for head and neck tumors.
[c] Cranial and spinal irradiation for medulloblastoma.

Evaluation of GH secretion

In an effort to better correlate growth with the level of GH secretion different methods have been used, assuming that physiological GH secretion should better reflect the hypothalamic pituitary status. Because of the experimental data of Chrousos et al. (1982), several studies have compared the information obtained from pharmacological testing and from the sleep-related or 24-h GH profiles. A discrepancy between the responses obtained by these two distinct procedures has been well documented. In children who had previously received cranial radiation for acute lymphoblastic leukemia (ALL) a reduced total spontaneous GH output with growth retardation was observed, in contrast to a normal GH response to hypoglycemia (Blatt et al. 1984). Romshe et al. (1984) reported discrepant results between the daytime pulsatile GH secretion and the GH response to arginine, L-dopa, and insulin, concluding that there was no reliable combination of stimulation tests to indicate which child was GH deficient and should be treated with hGH. The frequency and the significance of these discrepancies remain a matter of debate because of the small number of patients appropriately evaluated. In fact, the main issue is the correlation with growth and the usefulness of these investigations to diagnose GH deficiency.

There is some evidence that hypoglycemia-induced GH secretion is correlated with the decrease of pulsatile GH secretion in children receiving doses of 24 Gy or more (Dickinson et al. 1978; Romshe et al. 1984; Ahmed et al. 1986). However, in some cases GH peak values after arginine or L-dopa remained normal or

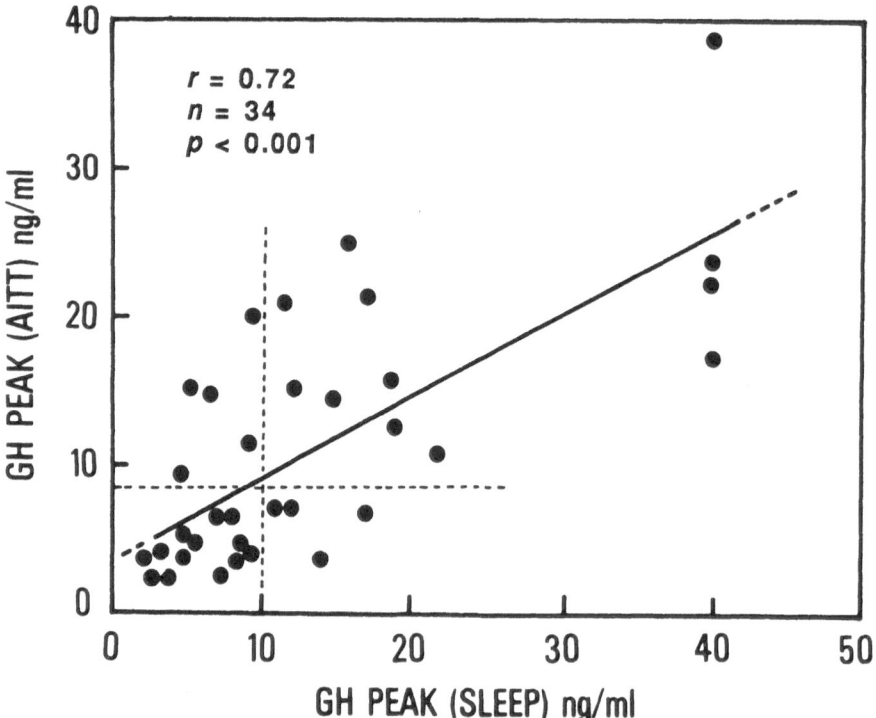

Fig. 1. Comparison between peak GH response to AITT and peak GH during sleep obtained in prepubertal children who received cranial radiation at doses of 32 - 40 Gy (Rappaport R, Brauner R., unpublished data).

subnormal, in spite of the lack of response to hypoglycemia (Dickinson et al. 1978). In a group of 14 children who had received doses of 24 Gy or more, all showed a blunted response to hypoglycemia with a low 24-h GH profile and 6-h sleep GH output. Based on these data, it was suggested that the insulin tolerance test was the test of choice because of marked radiation sensitivity (Ahmed et al. 1986).

We performed a similar evaluation as part of a prospective study in a group of children who received doses ranging from 32-40 Gy for medulloblastoma (Fig. 1). We compared the peak GH values after arginine-insulin tolerance test (AITT) with the night-sleep GH peak. These values were significantly correlated (r = 0.72, $P < 0.001$). Only one of the nine children with complete GH deficiency (AITT peak < 5 ng/ml) had a normal sleep-related GH peak. Conversely of 13 children with an AITT response > 8 ng/ml, five had a low sleep-related GH peak. A similar distribution was observed in partial responders to AITT (5-8 ng/ml peak). It may be concluded that a response to AITT below 5 ng/ml remains a valid indicator of GH deficiency. Discrepant results are more likely to be observed in partial and normal responders to AITT. Although there are not many studies taking into account the role of radiation dosage and the length of time since therapy, previous results and our own data suggest that discrepancies are more likely to occur at an early postradiation period and when radiation doses were in the low range, as in ALL affected children. For practical purposes, we suggest that pharmacological testing of these patients remains a valuable tool. Assuming that physiological studies are preferable, one can obtain sleep-related GH profiles only if the GH response to these tests is normal in contrast to retarded growth. Repeated studies of the dynamics of spontaneous GH secretion might provide additional information on possible time-related changes in the pulse frequency and amplitude after radiation. In fact, the more critical issue is the correlation of these GH secretion data with the growth velocity and the overall growth pattern.

Influence of age at the time of radiation and the radiation dose

Early data on children suggested that younger children were more vulnerable (Shalet et al. 1976). More recently, this age-related susceptibility was clearly demonstrated in a well-defined group of children receiving 24 Gy (Brauner et al. 1986a) (Fig. 2). This susceptibility is probably lower in the adult, as observed during the early follow-up of adult patients irradiated for cranial tumors distant from the pituitary region (Lam et al. 1987). Comparisons between pediatric and adult studies are often difficult because of differences in dose, time interval, and etiology, which add more variables interfering with the evaluation of the dose-related effect. However, the radiation dose is also an important risk factor. A rather clear pattern emerges from a few retrospective studies and from our own results as presented in Table 3 and Figure 3. Based upon the GH response to AITT, the

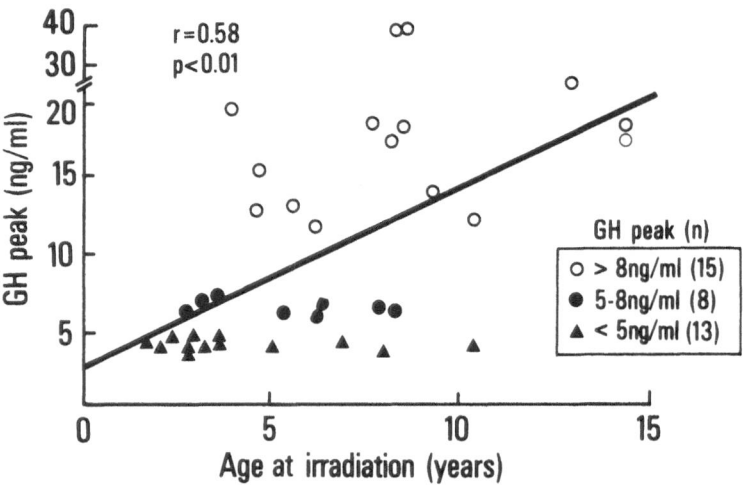

Fig. 2. Relationship between age at cranial irradiation and occurrence of GH deficiency about 5 yr later in prepubertal children. (From Brauner et al. 1986a, published with permission of Journal of Pediatrics).

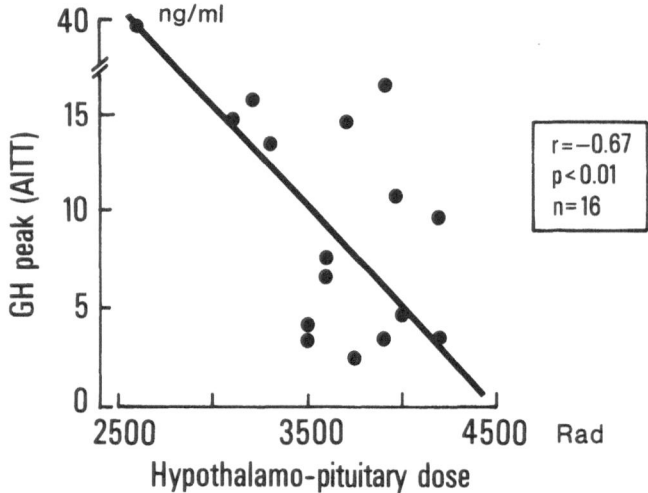

Fig. 3. Correlation between the calculated radiation dose delivered to the hypothalamic and pituitary regions and the GH response to AITT (Rappaport R, Brauner R., unpublished data).

Table 3. Distribution of the GH peak response to arginine-insulin stimulation in relation to the radiation dose (from Rappaport et al. 1988)

Pituitary estimated radiation dose (Gy)	No. of cases	Percentage of cases according to AITT GH peak (ng/ml)		
		<5	5-8	>8
<24	8	0	0	100
24	86	30	22	48
25-30	11	27	0	73
30-45	40	45	25	30
>45	12	75	8	17

frequency of GH deficiency largely depends on the estimated radiation dose delivered to the hypothalamic and pituitary zone. Shalet et al. (1976) showed that not only the total dose but also the fractionation schedule influenced the biological effectiveness of radiation in children receiving 24 Gy. In our experience, covering a range between 24 and 45 Gy (or more) with interval times since irradiation of 4-7 years, we found that complete GH deficiency (peak < 5 ng/ml) occurred in 30% of the patients who received 24 Gy and in 75% of those given more than 30 Gy. Doses below 18 Gy are probably not harmful, the critical zone being around

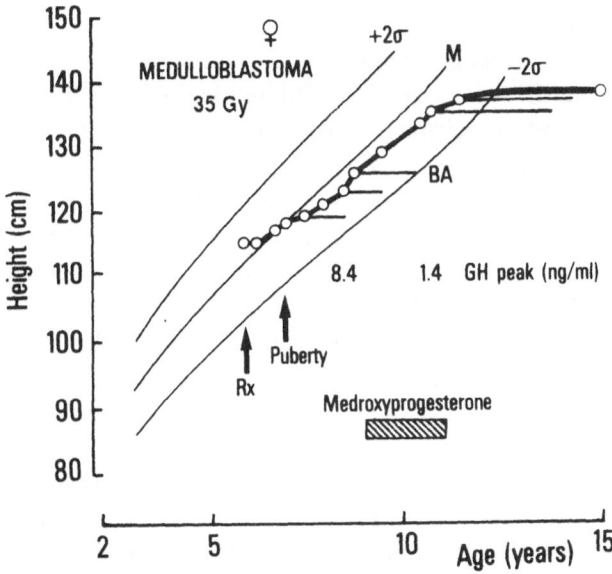

Fig. 4. Growth after cranial irradiation in a child with GH deficiency and precocious puberty.

20 Gy. Doses above 45 Gy induced GH deficiency in almost all patients studied at least 2 years after their irradiation (Rappaport et al. 1987). Furthermore, high radiation doses led to multiple hormonal deficiencies: these are, in order of frequency, TSH, gonadotropin, and ACTH secretion (Table 2). There was no diabetes insipidus. Paradoxical precocious puberty has also been reported, and may interfere with GH deficiency to modify growth and reduce final stature (Fig. 4).

The early course of pituitary dysfunction after radiation

We performed a prospective study of children who had received 30-42 Gy cranial doses (Brauner et al. 1985a). Nine children were studied 1 and 2 yr after radiation. At the 1-yr follow-up, two children were GH deficient, and at 2 yr five children had a peak GH response to AITT below 8 ng/ml. The simultaneous study of night GH profiles showed, as expected, a partial correlation with the GH response to provocative testing. Because of the large range of values in the normal control population for both the GH response to provocative tests and the GH sleep-related profile (Costin et al. 1987). the occurrence of occasional discrepant results seems unavoidable. We propose that the first testing be performed 2 yr after radiation, and that testing be repeated periodically according to the growth pattern. Because of its potential impact on growth, it should be mentioned that TSH deficiency also occurs soon after radiation. By comparison with the data obtained during retro-spective studies one can summarize the course of GH secretory dysfunction in children receiving the most frequently used radiation doses of 30–45 Gy: 2 yr after radiation we found that nine out of 16 children had GH peak responses below 8 ng/ml, and six of these had complete GH deficiency (peak < 5 ng/ml). This indicates frequencies of 56% and 37% respectively. After an average follow-up duration of 5 yr in a larger group of patients, as shown in Table 3, these frequencies were 70% and 45% respectively. It can therefore be concluded that GH deficiency occurred within a few years after radiation. None of our patients presented with spontaneous improvement of GH secretion. According to Dacou Voutetakis et al. (1975, 1977) spontaneous night secretion of GH may be low immediately after radiation and recover later. This acute effect of radiation may be quite different from the progressive and irreversible deterioration of GH secretion described on more prolonged follow-up of these patients. In some patients, in fact, repeated testing clearly showed a progressive deterioration of the pituitary function. More long-term data in these patients will be needed to assess their GH secretion during and after puberty. It is likely that most of them will not be able to adequately respond to the pubertal sex-steroid stimulation. This critical period might disclose further abnormalities of GH secretion, principally in the group of patients who received lower radiation doses.

GROWTH AND GROWTH HORMONE DEFICIENCY

The relationship between these abnormalities of GH secretion and the statural growth of the affected children is most important. Studies have aimed at a better understanding of the GH requirement for normal growth, as these children appear to have a wide distribution of postradiation GH secretion values, ranging from normal to severely impaired or even totally suppressed secretion. How these data relate to growth, and possibly to circulating Sm-C/IGF-I values, is of importance for a decision on hGH therapy.

Different patterns of prepubertal growth were observed in cranially irradiated children, as shown in Figures 5 and 6, which depict growth according to the cranial radiation dose. A peculiar pattern of growth is observed in children who have also received spinal radiation at high doses, as performed in conventional medulloblastoma therapy (Shalet et al. 1987). At the onset of puberty children who had received 24 Gy reached a mean height score at 0.7 standard deviation (SD) below

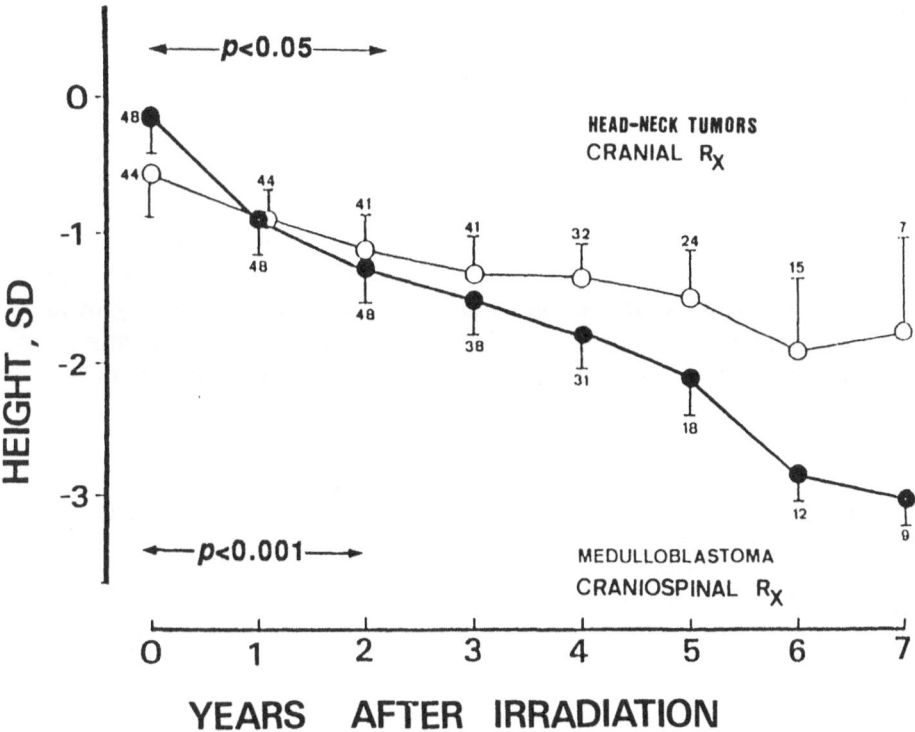

Fig. 5. Growth before puberty of children who received cranial radiation (30-45 Gy) for head and neck tumors (*open circles*) or medulloblastoma (*closed circles*). The medulloblastoma group received additional spinal radiation (Brauner et al., unpublished data). Height is expressed by the SD score.

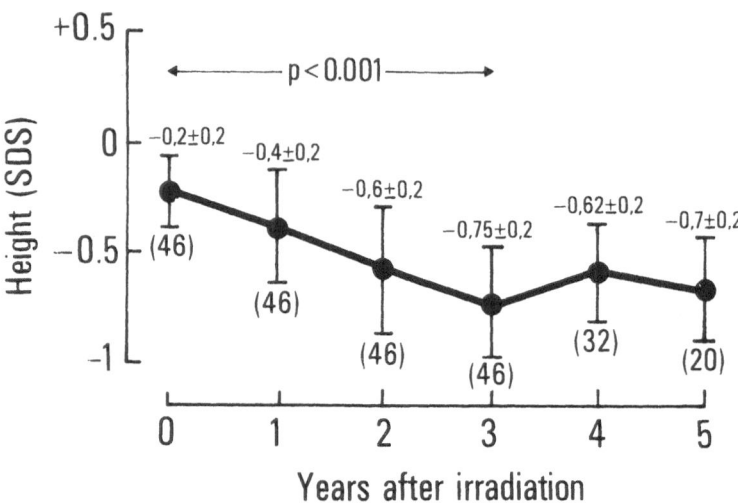

Fig. 6. Growth before puberty of children who received 24 Gy prophylactic cranial radiation for acute lymphoblastic leukemia. Height is expressed in the SD score. (From Brauner et al. 1986b, published with permission of Acta Endocrinologica).

normal. When radiation doses were in the range of 30-45 Gy they reached a mean height score of –1.7 SD before puberty, with an average prepubertal height of -1.1 SD (Rappaport et al. 1987). These growth data should again be compared with the GH data indicating that complete GH deficiency was present in 30% and 45% of the cases in the first and the second group respectively. Table 4 indicates how prepubertal growth is related to the GH response to AITT. When patients received more than 24 Gy, a lack of GH responses to AITT was accompanied by growth retardation. In contrast, in the group irradiated with 24 Gy, normal growth rates were observed in spite of the GH responses to AITT being below 6 ng/ml. Therefore, and for practical purposes, it should be stressed that growth rate cannot

Table 4. Comparison between growth and GH secretion according to the cranial radiation dose in children evaluated at least 2 yr after therapy (Brauner et al., unpublished data)

Cranial radiation dose	GH peak (AITT)	Growth Normal (no. of patients)	Decreased*
24 Gy	> 8 ng/ml	14/16	2/16
	< 8 ng/ml	17/34	17/34
> 25 Gy	> 8 ng/ml	7/7	-
	< 8 ng/ml	-	18/18

* Height loss of >1 SD since time of radiation

be anticipated from the biological data in children who received radiation doses in the range of 24 Gy. Some important groups of patients will be discussed separately in this report because of specific features related to their primary diseases and, more likely, to the radiation dose delivered to the hypothalamic and pituitary regions.

In normal children, puberty is accompanied by an increase in GH secretion and circulating Sm-C/IGF-I values (Blizzard et al. 1974; Miller et al. 1982). These changes, at least during the first part of puberty, are correlated with increased growth rates (Rosenfield et al. 1983). Cranial irradiation leads to complex situations, such as gonadotropin deficiency secondary to high cranial radiation doses (Rappaport et al. 1982) or true central precocious puberty (Brauner et al. 1984). In both cases, all the patients with abnormal puberty had a simultaneous GH deficiency. We reported earlier on a group of children treated for medulloblastoma, astrocytoma, facial tumor, or ALL with radiation doses of 24-45 Gy: they developed precocious puberty with, as the main consequence, an accelerated bone maturation which was not compensated for by proportional growth because of the combined GH deficiency (Brauner et al. 1984).

SPECIFIC FEATURES RELATED TO DISEASES

There is presently a fairly precise description of the consequences for GH secretion and growth in patients treated with cranial radiation doses above 25 Gy (Duffner et al. 1985; Shalet 1986; Rappaport et al. 1987). This report will present some more recent data on two groups of patients at risk for pituitary dysfunction as well as growth retardation and for whom the therapeutic indication for using hGH is not yet clearly defined. Children treated for medulloblastoma have a severe and specific growth retardation due to their spinal radiation at doses superior to 20 Gy given in addition to the cranial radiation (Fig. 5). Spinal radiation reduces growth as early as during the first year of treatment (Brauner et al. 1985b; Shalet et al. 1987). Therefore, the relationship between pituitary dysfunction and growth cannot be analyzed as easily in these patients as in the other groups.

Acute Lymphoblastic Leukemia

Any study of growth velocity in survivors of ALL should be interpreted in relation to sites and doses of radiation. It is likely that cranial radiation doses inferior to 20 Gy do not modify GH secretion (Shalet et al. 1976; Pomarede et al. 1984). At the total dose of 24-25 Gy the effect on GH secretion depends upon the fractionation schedule of the dose (Shalet 1986). Although these patients do not receive spinal irradiation, it was recently reported that some children had received

8.5–10 Gy on the spine with no reduction in their sitting height (Sainsbury et al. 1985). In a retrospective analysis of medical records of children with prolonged remission from ALL, the average rate of growth decreased during initial treatment and the potential for final adult height was not regained: the mean final height loss was about 4% of the adult stature (Starceski et al. 1987). However, in this study the authors did not account for the GH secretion level. In fact, we found a mean final height at -1 SD in a small group of children irradiated with 24 Gy. This moderate growth retardation was in contrast to the presence of GH deficiency in most of the patients (Fig. 6). A more comprehensive description of the growth pattern of these patients and of its correlation with the biological data is therefore necessary.

As shown in Table 4, normal GH secretion was accompanied by normal growth before puberty. But the remaining patients with GH deficiency (most of whom had a GH peak response to AITT < 6 ng/ml) could be divided into two groups. Some patients had a concomitant growth retardation and were obvious candidates for hGH therapy; others presented with the paradoxical situation of a combined GH deficiency (GHD) and normal growth rate (Shalet et al. 1979). In a separate study of 86 children (data presented at this meeting) we found 25 cases (29%) with GH peaks <5 ng/ml. Eleven of them had a normal growth velocity, with plasma Sm-C values ranging from low to normal values for age (0.08 – 1.4 unit/ml). One patient who had repeated testing, which confirmed the persistence of GHD and low SM-C values (<0.15 unit/ml), presented a normal pubertal growth spurt and had a height loss of only 1 SD when he reached his final height. The persistence of normal growth in these children in spite of GHD could not be explained by any hormonal or weight changes as reported in similar conditions (Bucher et al. 1983).

Plasma Sm-C values of 0.25 ± 0.15 (SD) unit/ml were in the hypopituitary range in the prepubertal subjects, who presented with GH peak values <5 ng/ml, and were significantly lower than the mean values obtained in the group with partial GHD or normal GH secretion, with values of 0.68 ± 0.50 and 0.49 ± 0.30 unit/ml respectively. Preliminary data from these patients with normal GH secretion indicate an unexplained trend towards low RIA Sm-C values (using the nonequilibrium technique of Furlanetto et al. [1977]).

In conclusion, the ALL group of children showed unusual features. Although one group of patients could be considered GH deficient by all criteria, others were able to maintain a close-to-normal growth rate in spite of GHD and possibly low circulating Sm-C values. This growth rate was not related to simultaneous precocious puberty, which we reported as occurring in some other cases. Therefore, one could speculate that, in these cases, in spite of low or borderline biological data, the requirements of GH secretion were met. We did not find significant changes in blood prolactin or any cases of overweight among these patients (Bucher et al. 1983). This group deserves further evaluation. At least it is presently clear that hGH treatment should not be decided only on the basis of the hormone data but should depend on the growth pattern of each child.

Total Body Irradiation

Total body irradiation (TBI) has become part of the treatment of hematologic malignancies, as a conditioning step prior to bone marrow transplantation. Because of the increasing number of long-term survivors, growth and GH secretion have become an important issue. In a recent study by Sanders et al. (1986), GHD was present in 17 of 25 children who received TBI in addition to previous cranial irradiation (8-24 Gy) and in six of 18 patients who had not received cranial irradiation. The respective roles of GHD- and TBI-induced lesions on growth remained unclear, as growth velocity was decreased in all patients who had not received cranial irradiation. The authors also stated that 3 yr after transplant the patients who had received fractionated TBI grew significantly taller than those who had received a single TBI exposure. We therefore selected patients who received TBI exclusively, at doses of 8–12 Gy. Our data (Fontoura et al. 1988) for a group of 25 patients evaluated after a mean interval of 3.5 yr add the following information: (a) GHD does not appear as a complication of TBI at the radiation dosages used in our conditioning protocol. (b) Growth retardation was seen only after single-exposure TBI and is probably due to diffuse lesions of the metaphyseal growth cartilage. A more careful study of GH secretion and Sm-C changes during prolonged follow-up and a comparison with height gains are in progress and must be completed before appropriate therapeutic recommendations can be made. It is important to decide which TBI protocol carries the smallest risk of growth retardation and remains fully efficient as conditioning for bone marrow transplantation.

GROWTH HORMONE TREATMENT

Treatment with growth hormone is another important issue. Its efficacy has been evaluated in several studies (Shalet et al. 1981; Romshe et al. 1984; Winter and Green 1985; Brauner et al. 1985b). Our results dealing with children receiving more than 30 Gy can be divided into two groups. Before puberty the mean annual height gain increased from 3.5 cm to 6.5 cm and remained superior to the control value until the fourth year, allowing a mean catch up of 1 SD. These values were slightly lower than the height gain obtained in idiopathic GHD with similar hGH dosage. When hGH was initiated during early puberty the mean height gain during the first year was 6.5 cm. Similar results were obtained in patients treated for ALL. We suggest that hGH therapy deserves further evaluation by prospective studies, as many children appeared to be rather poor responders. One cause of failure could be accelerated bone maturation due to early or precocious puberty (Winter and Green 1985). It is likely that the response to hGH also depends on hGH dosage and mode of administration. Another important issue is the selection of the patients to be treated. It is easy with children whose growth retardation is clearly due to GHD in the group receiving high doses of cranial radiation. When spinal radiation is also performed, as for medulloblastoma, one should consider that growth will be poorly stimulated by hGH (Brauner et al. 1985b), and prolonged treatment may even become questionable when the child enters puberty. At that age the height gain is dependent mainly on spinal growth, and children with craniospinal irradiation are unable to develop a satisfactory pubertal growth spurt. It is less easy to decide in the group of children treated with low doses of radiation for ALL, as they may present unexpected dissociations between GH secretion values and growth rates. Therefore, we suggest that hGH therapy be considered for these patients only if growth retardation is well documented beyond the first year following radiation. It is clear for all irradiated children that hGH therapy should be proposed only if a significant benefit can be expected. Fortunately, there is no evidence that GH therapy might precipitate a recurrence of the original malignant disease (Arslanian et al. 1985; Shalet 1986; our personal experience). As GHD may not be the only pituitary defect, one should also carefully evaluate thyroid and gonadal functions which contribute to the control of growth.

Acknowledgments

We acknowledge the collaboration of the following oncologists, hematologists and neurosurgeons: Drs. C. Griscelli and J.F. Hirsch (Hôpital des Enfants Malades), J.M. Zucker and P. Bataini (Institut Curie), J. Lemerle and D. Sarrazin (Institut Gustave Roussy, Villejuif), G. Schaison (Hôpital St-Louis), and D. Machover (Hôpital Tenon). The manuscript was prepared with the skillful assistance of Mrs. C. Chamot, secretary.

REFERENCES

Ahmed SR, Shalet SM (1984) Hypothalamic growth hormone-releasing factor deficiency following cranial irradiation. *Clin Endocrinol* **212**: 483-488

Ahmed SR, Shalet SM, Beardwell CG (1986) The effects of cranial irradiation on growth hormone secretion. *Acta Paediatr Scand* **75**: 255-260

Arnold A (1954) Effects of X-irradiation on the hypothalamus: a possible explanation for the therapeutic benefits following X-irradiation of the hypophyseal region for pituitary dysfunction. *J Clin Endocrinol Metab* **14**: 859-868

Arslanian SA, Becker DJ, Lee PA, Drash AL, Foley TP (1985) Growth hormone therapy and tumor recurrence. *Am J Dis Child* **139**: 347-350

Bercu BB, Shulman D, Root AW, Spiliotis BE (1986) Growth hormone (GH) provocative testing frequently does not reflect endogenous GH secretion. *J Clin Endocrinol Metab* **63**: 709-716

Blatt J, Bercu BB, Gillin JC, Mendelson WB, Poplack DG (1984) Reduced pulsatile growth hormone secretion in children after therapy for acute lymphoblastic leukemia. *J Pediatr* **104**: 182-186

Blizzard RM, Thompson RG, Baghdassarian A, Kowarski A, Migeon CJ, Rodriguez A (1974) The interrelationship of steroids, growth hormone and other hormones in pubertal growth. In: Grumbach MM, Grave GD, Mayer EE (eds) *The Control of the Onset of Puberty*. Wiley, New York, pp 223-239

Brauner R, Czernichow P, Rappaport R (1984) Precocious puberty after hypothalamic and pituitary irradiation in young children. *N Engl J Med* **311**: 920

Brauner R, Czernichow P, Prévot C, Guyda HJ, Rappaport R (1985a) Longitudinal study of GH secretion, somatomedin and growth in the two years following cranial irradiation. *Pediatr Res* **19**: 613

Brauner R, Czernichow P, Rappaport R (1985b) Croissance staturale après irradiation du système nerveux central pour médulloblastome de la fosse postérieure. Analyse rétrospective de 45 observations. *Arch Fr Pediatr* **42**: 219-223

Brauner R, Czernichow P, Rappaport R (1986a) Greater susceptibility to hypothalamopituitary irradiation in younger children with acute lymphoblastic leukemia. *J Pediatr* **108**: 332

Brauner R, Prévot C, Roy MP, Rappaport R (1986b) Growth, growth hormone secretion and somatomedin C after cranial irradiation for acute lymphoblastic leukemia. *Acta Endocrinol* (Copenh) [Suppl 279]: 178-182

Bucher H, Zapf J, Torresani T, Prader A, Froesch ER, Illig R (1983) Insulin-like growth factors I and II, prolactin, and insulin in 19 growth hormone-deficient children with excessive, normal, or decreased longitudinal growth after operation for craniopharyngioma. *N Engl J Med* **309**: 1142-1146

Chrousos GP, Poplack D, Brown T, O'Neill D, Schwade J, Bercu BB (1982) Effects of cranial radiation on hypothalamic-adenohypophyseal function: abnormal growth hormone secretory dynamics. *J Clin Endocrinol Metab* **54**: 1135-1139

Costin G, Kaufman F, Brasel JA (1987) Growth hormone secretory patterns in normal-statured subjects. (Abstr). *Pediatr Res* **21**: 246 A

Crosnier H, Brauner R, Prévot C, Rappaport R (1986) GH response to hGRF as a sensitive index of GH neurosecretory dysfunction after cranial irradiation. (Abstr). *Pediatr Res* **20**: 1189

Czernichow P, Cachin O, Rappaport R, Flamant F, Sarrazin D, Schweisguth O (1977) Séquelles endocriniennes des irradiations de la tête et du cou pour tumeurs extracraniennes. *Arch Fr Pediatr* **34**: 154-164

Dacou Voutetakis C, Haidas St., Zannos-Mariolea L (1975) Radiation and pituitary function in children. *Lancet* **ii**: 1206-1207

Dacou Voutetakis C, Xypolyta A, Haidas St, Constantinidis M, Papavasiliou C, Zannos-Mariolea L (1977) Irradiation of the head. Immediate effect on growth hormone secretion in children. *J Clin Endocrinol Metab* **44**: 791-794

Dickinson WP, Berry DH, Dickinson L, Irvin M, Schedewie H, Fiser RH, Elders MJ (1978) Differential effects of cranial radiation on growth hormone response to arginine and insulin infusion. *J Pediatr* **92**: 754-757

Duffner PK, Cohen ME, Voorhess ML, MacGillivray MH, Brecher ML, Panahon A, Gilani BB (1985) Long-term effects of cranial irradiation on endocrine function in children with brain tumors. *Cancer* **56**: 2189-2193

Fontoura M, Brauner R, Rappaport R, Fischer A, Quintana E, Zucker JM, Bernaudin F, Vilmer E, Devergie A (1988) Growth and endocrine function after bone marrow transplantation (BMT) with or without total body irradiation. (Abstr). *Pediatr Res* **23**: 111

Furlanetto RW, Underwood LE, Van Wyk JJ, D'Ercole AJ (1977) Estimation of somatomedin-C levels in normals and patients with pituitary disease by radioimmunoassay. *J Clin Invest* **60**: 648-657

Lam KSL, Tse VKC, Wang C, Yeung RTT, Ma JTC, Ho JHC (1987) Early effects of cranial irradiation on hypothalamic-pituitary function. *J Clin Endocrinol Metab* **64**: 418-428

Larkins RG, Martin FIR (1973) Hypopituitarism after extracranial irradiation: evidence for hypothalamic origin. *Br Med J* **1**: 152-153

Lawrence JH, Nelson WO, Wilson H (1973) Roentgen irradiation of the hypophysis. *Radiology* **29**: 446-454

Lustig RH, Schriok CA, Kaplan SL, Grumbach MM (1985) Effect of growth hormone-releasing factor on growth hormone release in children with radiation-induced growth hormone deficiency. *Pediatrics* **76**: 274-279

Miller JD, Tannenbaum GS, Colle E, Guyda HJ (1982) Daytime pulsatile growth hormone secretion during childhood and adolescence. *J Clin Endocrinol Metab* **55**: 989-994

Mosier HD, Jansons RA, Swingle KF, Sondhaus CA, Dearden LC, Halsall LC (1985) Growth hormone secretion in the stunted head-irradiated rat. *Pediatr Res* **19**: 543-548

Perry-Keene DA, Connelly JF, Young RA, Wettenhall HNB, Martin FIR (1976) Hypothalamic hypopituitarism following external radiotherapy for tumours distant from the adenohypophysis. *Clin Endocrinol (Oxf)* **5**: 373-380

Pomarede R, Czernichow P, Zucker JM, Schlienger P, Haye C, Rosenwald JC, Labib A, Rappaport R (1984) Incidence of anterior pituitary deficiency after radiotherapy at an early age: study in retinoblastoma. *Acta Paediatr Scand* **73**: 115-119

Rappaport R, Brauner R, Czernichow P, Thibaud E, Renier D, Zucker JM, Lemerle J (1982) Effect of hypothalamic and pituitary irradiation on pubertal development in children with cranial tumors. *J Clin Endocrinol Metab* **54**: 1164-1168

Rappaport R, Fontoura M, Brauner R (1988) Growth hormone secretion and growth in central nervous system irradiated children. In: Bercu BB (ed) *Basic and Clinical Aspects of Growth Hormone*. Plenum, New York, pp 143-155

Richards GE, Wara WM, Grumbach MM, Kaplan SL, Sheline GE, Conte FA (1976) Delayed onset of hypopituitarism: sequelae of therapeutic irradiation of central nervous system, eye, and middle ear tumors. *J Pediatr* **89**: 553-559

Romshe CA, Zipf WB, Miser A, Miser J, Sotos JF, Newton WA (1984) Evaluation of growth hormone release and human growth hormone treatment in children with cranial irradiation-associated short stature. *J Pediatr* **104**: 177-181

Rosenfield RL, Furlanetto R, Bock D (1983) Relationship of somatomedin-C concentrations to pubertal changes. *J Pediatr* **103**: 723-728

Sainsbury CPQ, Newcombe RG, Hugues IA (1985) Weight gain and height velocity during prolonged first remission from acute lymphoblastic leukaemia. *Arch Dis Child* **60**: 832-836

Samaan NA, Bakdash MM, Caderao JB, Cangir A, Jesse RH, Ballantyne AJ (1975) Hypopituitarism after external irradiation. Evidence for both hypothalamic and pituitary origin. *Ann Int Med* **83**: 771-777

Sanders JE, Pritchard S, Mahoney P, Amos D, Buckner CD, Witherspoon RP, Deeg HJ, Doney KC, Sullivan KM, Appelbaum FR, Storb R, Thomas ED (1986) Growth and development following marrow transplantation for leukemia. *Blood* **68**: 1129-1135

Shalet SM (1986) Irradiation-induced growth failure. *Clin Endocrinol* **15**: 591-606

Shalet SM, Beardwell CG, Morris-Jones PH, Pearson D (1975) Pituitary function after treatment of intracranial tumours in children. *Lancet* ii: 104-107

Shalet SM, Beardwell CG, Pearson D, Morris-Jones PH (1976) The effect of varying doses of cerebral irradiation on growth hormone production in childhood *Clin Endocrinol* (Oxf) 5: 287-290

Shalet SM, Price DA, Beardwell CG, Morris-Jones PH, Pearson D (1979) Normal growth despite abnormalities of growth hormone secretion in children treated for acute leukemia. *J Pediatr* 94: 719-722

Shalet SM, Whitehead E, Chapman AJ, Beardwell CG (1981) The effects of growth hormone therapy in children with radiation-induced growth hormone deficiency. *Acta Paediatr Scand* 70: 81-86

Shalet SM, Gibson B, Swindell R, Pearson D (1987) Effect of spinal irradiation on growth. *Arch Dis Child* 62: 461-464

Spiliotis BE, August GP, Hung W, Sonis W, Mendelson W, Bercu BB (1984) Growth hormone neurosecretory dysfunction. A treatable cause of short stature. *JAMA* 251: 2223-2230

Starceski PJ, Lee PA, Blatt J, Finegold D, Brown D (1987) Comparable effects of 1800 and 2400 rad (18 and 24 Gy) cranial irradiation on height and weight in children treated for acute lymphocytic leukemia. *Am J Dis Child* 141: 550-552

Tan BC, Kunaratnam N (1966) Hypopituitary dwarfism following radiotherapy for nasopharyngeal carcinoma. *Clin Radiol* 17: 302-304

Winter RJ, Green OC (1985) Irradiation-induced growth hormone deficiency: blunted growth response and accelerated skeletal maturation to growth hormone therapy. *J Pediatr* 106: 609-612

ROUND TABLE

APPROACH TO THERAPY

.

Advances in Growth Hormone and Growth Factor Research,
edited by E.E. Müller, D. Cocchi and V. Locatelli
Pythagora Press, Roma-Milano and Springer Verlag, Berlin-Heidelberg © 1989

A critical overview of the state of the art

R.D.G. MILNER

Department of Paediatrics, University of Sheffield; Children's Hospital, Sheffield, United Kingdom

This «overview» attempts to give a précis of the current literature; it is deliberately subjective and larded in places with personal anecdotes.

Growth hormone (GH) therapy is in a state of flux because the availability of potentially unlimited supplies of GH has revealed a diagnostic jungle; previously, when treatment was possible only with scarce pituitary GH, the selection of unequivocally GH-deficient children was relatively straightforward. In addition, our better understanding of the physiology and pathophysiology of endogenous GH secretion has led to other approaches, in which endogenous GH release is stimulated using growth hormone-releasing hormone (GHRH) or certain neuro-transmitters. Happily, the jungle is not so impenetrable as it might first seem if the clinician takes care to distinguish problems of organic pathology from those focused on social issues.

Traditionally, GH deficiency has been diagnosed on the basis of measurements of GH in the blood following pharmacological or physiological stimulation (Milner and Burns 1982), and more recently the description of diurnal profiles has been employed to try to identify children who, while not overtly GH deficient, have a day-by-day GH secretion insufficient for optimal growth (Hindmarsh et al. 1987). The other aim of diagnosis is the careful measurement of growth in height before and during therapy. Originally, and even currently, auxology has been perceived by many to be supplementary to a biochemical diagnosis, but in fact it is auxology and not biochemistry which is the more important aspect of both diagnosis and therapy. This is because the final arbiter of GH therapy is whether the child grows faster and achieves a greater adult height, and the only way that outcome can be monitored is by careful documentation of velocity before and during treatment, coupled with repeated estimations of osseous maturation. In a recent national study of the efficacy of Somatrem, the biochemical diagnosis of GH deficiency was found to be slightly less predictive of a response to therapy than a pretreatment whole-year height velocity of less than 5 cm/yr, irrespective of age (Milner et al. 1987).

The weakness of biochemical tests of GH deficiency lies in their lack of precision. Pharmacological tests such as the insulin tolerance test or clonidine stimulation have a combined false-positive and false-negative rate on the order of 25% (Health Services Human Growth Hormone Committee 1981). Much effort has been devoted to the analysis of physiological GH profiles, with attention being paid to pulse frequency, height and area under the curve (Shulman and Bercu 1987), and this has helped us to understand that there is a continuum of GH release, from the unequivocally normal through a grey zone to the unequivocally subnormal. Profiles have also clarified the effect of CNS irradiation on GH secretion (Blatt et al. 1984; Romshe et al. 1984) and how GH secretion in Turner's syndrome deteriorates in later childhood (Ross et al. 1985). Unfortunately, there is little information on the reproducibility of profiles, so that this biochemical approach is unlikely, in my view, to be more helpful than a pharmacological test of GH secretion in diagnosing a need for GH therapy in an individual patient. This prejudice is reinforced when we reflect on the inconvenience of a profile to the patient and the considerable extra workload on both the clinician and the laboratory arising from a profile study. A word on semantics: the adjectives *deficient* and *insufficient* are now being used to discriminate between patients conventionally described as GH deficient and those who manifest the auxological characteristics of GH deficiency but may have apparently normal peak plasma GH levels when tested biochemically (Milner 1986). This is an illustration of the shift in the conceptual balance that is taking place among clinicians, from diagnosing patients with GH lack to recognising patients who are GH responsive; the only criterion by which response to GH therapy can be reliably judged is auxology. Thus, there is a pragmatism creeping into clinical practice: if the patient is growing slowly, will he or she grow faster if given GH therapy? Such an attitude of mind is both laudable and potentially dangerous: praiseworthy, because it indicates a more open-minded approach to the spectrum of patients who may benefit from GH therapy; dangerous, because it could lead to a superficial and incomplete assessment of what is wrong.

Failure to grow is the ultimate paediatric bioassay of chronic ill health, and it may arise from a wide spectrum of causes, both organic and psychological. Some of the misdiagnoses which may masquerade as GH deficiency have been reviewed recently (Milner and King 1986), and they include pathology of every organ system in the body. Possibly the most important single aspect in the assessment of the short, slowly growing child is psychosocial. Twenty years have passed since the discovery that psychosocial deprivation could result in slow growth, stunting, and biochemistry that was characteristic of GH deficiency (Powell et al. 1967). It is now clear that the aetiology of this functional GH deficiency is neuroendocrine and that the condition is reversible by correction of the environment in which the child lives. This is sufficiently accepted to be useful for medicolegal purposes; improvement in growth on changing place of residence is acceptable as evidence of psychosocial deprivation (Taitz and King 1986). Children suffering from chronic psychosocial stress may be difficult to recognise unless the clinician maintains a very enquiring and open-minded attitude in consultation.

Another aspect of the psychosocial side of GH therapy is the consultation where the physician is pressured, usually by one or both parents, to offer therapy for a child who is short but healthy and growing normally. Delicacy is needed to disinguish the short, normal child from one who may grow appreciably more if given GH therapy, and repeated predictions of final adult height (Tanner et al. 1983) are often helpful in buying time before a final recommendation is made on whether or not to treat.

The expectations of families with short children will vary nationally, depending on the publicity given to the subject of short stature and GH deficiency by pressure groups, whose ability to manipulate the media may be considerable. The medical and related professions may heighten expectations by actively screening for short stature at clinics and schools.One of the most important socioeconomic influences on GH therapy is who pays for it. In countries where treatment is available there is a wide spectrum of financing the expensive drug bill, varying from direct payment by the state, to state-subsidized or private insurance, to payment by the individual. This is an influential factor for the progress of diagnosis and treatment of short, slowly growing children.

CHILDREN WHO MAY BENEFIT FROM GH

Current evidence indicates that there are three categories of patient who may benefit from GH therapy. Children with either isolated *GH deficiency* or disease involving more that one pituitary hormone are more common than was previously accepted (Vimpani et al. 1977) but they are still being diagnosed late (Herber and Milner 1986). Despite a surprising assertion to the contrary (Karlberg et al. 1987), GH deficiency may present in the neonatal period when hypoglycaemia is a common feature (Herber and Milner 1984). Microgenitalia in male children should alert the clinician to the possibility of panhypopituitarism (Salisbury et al. 1984), and in one such patient a first-year growth velocity of 32.7 cm/yr was achieved (Fig. 1); to the best of our knowledge, this is the greatest therapeutic response to GH documented.

The term *GH insufficiency* is used to categorise children who are growing abnormally slowly, are probably shorter than predicted from parental height, and who may or may not have a retarded bone age. By definition, when tested with a conventional stimulus to GH secretion they have a normal peak plasma response, which has led to appellations such as «normal variant short stature, NVSS» (Rudman et al. 1981), «constitutional growth delay, CGD» (Pintor et al. 1987),

G.H. THERAPY IN FIRST YEAR

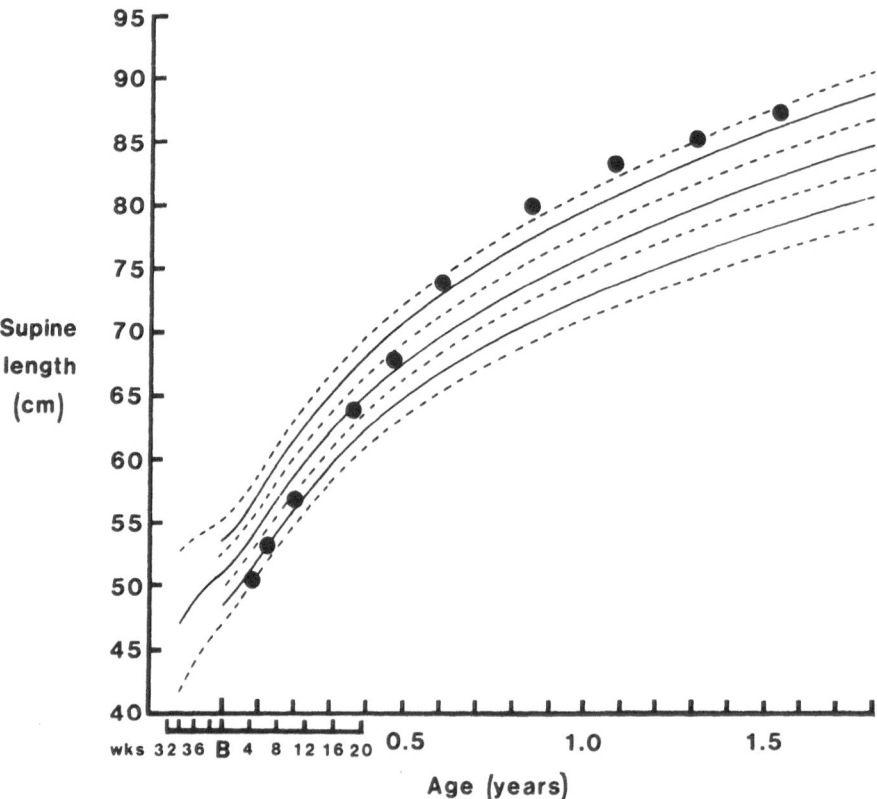

Fig. 1. Length chart of a male infant in whom multiple pituitary hormone deficiency was diagnosed at birth and who was treated with thyroxine, cortisone and GH by daily sc injection. His growth velocity was 32.7 cm/yr in the first year of treatment, 45.6 cm/yr in the first 6 months and 50.0 cm/yr in the first 3 months.

and «short slowly growing cildren» (Buchanan et al. 1987). More recently, it has been asserted that these children fall into a grey zone in physiological GH profile tests, with the inference that although they are capable of responding to a pharmacological challenge they secrete insufficient GH day by day to achieve optimal growth (Zakid et al. 1985). Previously, when pituitary GH supplies were scarce, these children were disadvantaged by being deemed normal. More recently, a number of studies have reported sustained improvement in height velocity when they were given GH therapy in conventional dosage (Van Vliet et al. 1983; Gertner et al. 1984; Hindmarsh and Brook 1987).

This group of patients presents the physician with the greatest diagnostic and therapeutic dilemma. Figure 2 shows the height velocity chart of a toddler who was the healthiest, most mischievous extrovert I had met for a long time. His mother would not go away, despite our showing that the boy had a peak plasma GH of 36 mU/l in response to oral clonidine and by telling both parents that she was neurotic. Two years of persistent nagging led to my capitulation, and the boy's response, first to pituitary GH and then to Somatrem, remains a profoundly educative experience for me. This example is recounted to illustrate the essentiality of high-quality auxology. It was only by documenting height velocity that we were able to conclude with confidence that the boy was GH responsive.

The obverse illustration comes from the only existing placebo-controlled trial of GH therapy (Buchanan et al. 1987). The majority of slowly growing short children given GH showed an impressive improvement in height velocity standard deviation score, but two of 11 grew more slowly on GH. In the placebo-treated group, three of 7 grew appreciably faster on placebo therapy. Trials of placebo therapy are not appropriate for everyday clinical practice, and auxology must be

G.H. THERAPY IN SHORT, SLOWLY GROWING BOY

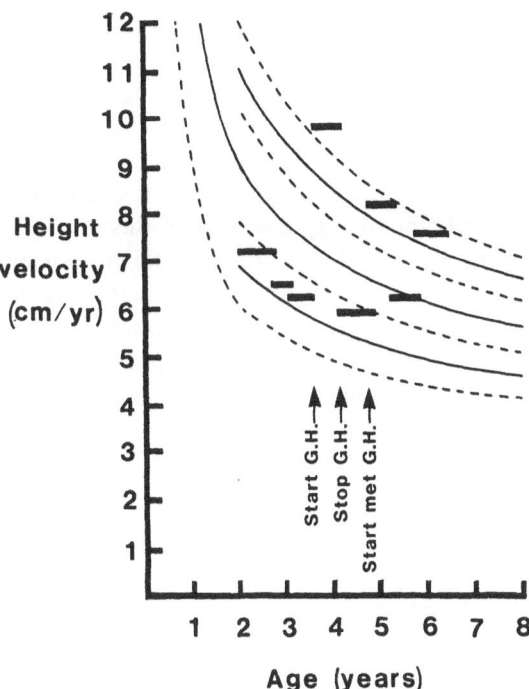

Fig. 2. Height velocity chart of a boy who was endocrinologically normal to formal tests of GH secretion, and who was treated first with pituitary GH and then with Somatrem.

used to judge if patients diagnosed as GH deficient or insufficient do not show a whole-year improvement in height velocity of more than 2 cm/yr. In such cases GH therapy should be discontinued, and the physician must search assiduously for alternative pathology which may be organic or psychological.

Of all her problems, the girl with *Turner's syndrome* is likely to find short stature the most crippling (Brook 1986). Despite early reports that GH therapy was ineffective in Turner's syndrome (Tanner and Whitehouse 1967), further clinical trials were undertaken using both pituitary GH and Somatrem with or without oestrogens. No patient with Turner's syndrome responded to placebo injections, but six of nine had an impressive increase of height velocity when treated with pituitary GH for 6 months (Buchanan et al. 1987). In a large multicentre study in the United States, both Somatrem and oxandrolone improved growth velocity over 1 yr, and a significantly better result was obtained when both were used simultaneously (Rosenfeld et al. 1986). However, the simultaneous use of oxandrolone and Somatrem caused the bone age to advance faster than the chronological age, and the authors were commendably cautious about whether combined therapy would produce the greatest improvement in final height prognosis.

Turner's syndrome has a wide spectrum of phenotypes, and not every girl with the condition responds to GH therapy (Milner 1987). There is an especial need to monitor auxology carefully so that treatment is not prolonged inappropriately when no benefit is accruing. This leads to the hoary question of what is the minimum interval in which a stable estimate of long-term height velocity can be made. Wales and Milner (1987a) demostrated that measurements at monthly intervals for 5 months were necessary to predict a 6-month height velocity within 10% and that monthly measurements for shorter periods were unacceptable. Despite this caveat, Wales and Milner (1987b) attempted to use knemometry (short-term growth of the lower leg) as a predictor of response to GH therapy in girls with Turner's syndrome. In a small trial, none of five girls who failed to respond to GH therapy as judged by knemometry had an increase in height velocity as judged by conventional auxology over a 6-month period, whereas four of eight who responded knemometrically in the short term also responded in the longer term. Thus, although knemometry may be a useful tool to sieve nonresponders out of a Turner cohort, conventional auxology over 6-month or longer intervals remains the best way to identify long-term responders.

MODALITIES OF GH THERAPY

Exogenous GH

Pituitary GH, Somatrem and authentic recombinant somatotropin all have similar growth-promoting potency. Pituitary GH is unlikely to be used therapeutically now because of the hazard of slow virus contamination and the risk of Creutzfeldt-Jakob disease (Milner 1985). Market forces will dictate which of the biosynthetic products will be chosen for therapy, but GH in one form or another will remain the backbone of therapy because it is the final common messenger, whatever the aetiology of GH deficiency.

Questions that remain topical and pertinent to GH therapy are dose, frequency and route of administration. The comprehensive and authoritative review by Frasier (1983) clearly demonstrated that 0.1 units/kg thrice weekly produced optimal results in GH-deficient patients. In some countries, GH is prescribed on a body weight-related basis; in others, all older children receive 12 or 14 units/week as three doses of 4 units or daily doses of 2 units for 6 or 7 days. The weight-invariate

THRICE WEEKLY OR DAILY G.H.

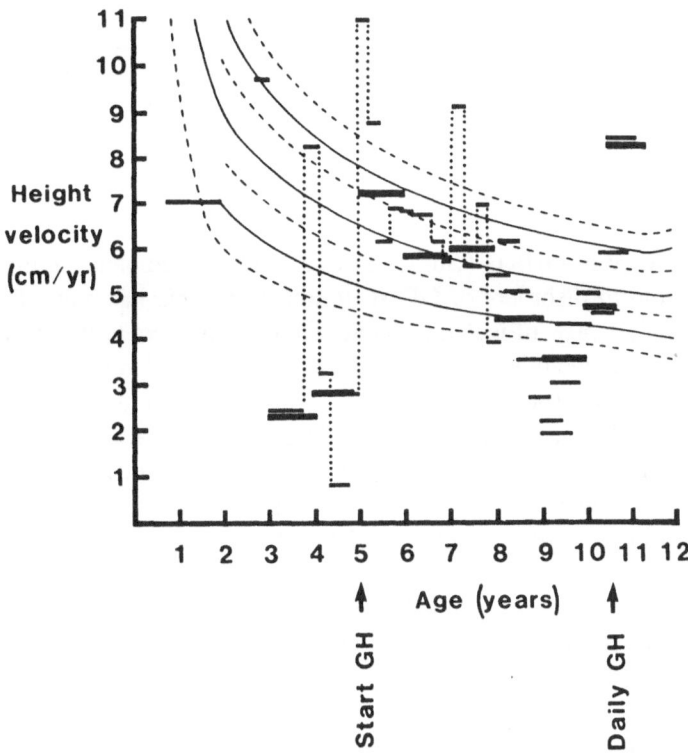

Fig. 3. Height velocity chart of a boy with GH deficiency treated first with GH by thrice-weekly injection and then by daily injection.

regimen is generous, giving 0.1 or more units/kg to the majority of patients, who are children weighing 30 kg or less, and a satisfactory dose to patients weighing up to 60 kg, since the dose-response curve is linear. Of more importance than dose is frequency of injection; a number of reports now exist showing that children given GH daily grow faster than those given thrice-weekly injections (Kastrup et al. 1983; Albertsson-Wikland et al. 1986). The evidence is convincing, but in clinical practice one personal anecdote is worth a mountain of other people's papers. Figure 3 shows the growth chart of a patient who had been first on pituitary GH and then Somatrem since the age of 5 years, following the conventional practice of three injections per week. He clearly grew faster on treatment than off it, but his response had never been as gratifying as the best. In light of the mounting evidence he changed to daily injections with the same weekly total dose and experienced a very satisfactory increase of height velocity which has been sustained for 6 months.

Historically, GH injections were given intramuscularly. A number of studies have compared subcutaneous with intramuscular injections (Sandahl-Christiansen et al. 1983; Wilson et al. 1985). There is no significant difference in response to therapy or in side effects such as the development of antibodies between the two routes. Since subcutaneous injection is less painful this is now the preferred route, and syringes with very fine needles manufactured for insulin injection are often employed to reduce the discomfort further.

Enhancement of Endogenous GH Secretion

The discovery of GHRH (Guillemin et al. 1984) prompted vigorous study of its possible role in disorders of GH secretion. From this has come the assertion that approximately 50% of isolated GH deficiency is due not to pituitary pathology but rather to disordered release of GHRH from the hypothalamus (Rogol et al. 1984; Shriock et al. 1984; Takano et al. 1984). The logical sequel: treatment of GH deficiency by GHRH soon followed. After the first report that GHRH therapy could produce a sustained increase in height velocity (Thorner et al. 1985), a number of reports have shown that a therapeutic result is achieved in only two thirds to three quarters of a cohort (Smith et al. 1986; Ross et al. 1987; Rochiccioli et al. 1987) and that a better height velocity may be obtained in the same patient when he or she is given GH (Smith and Brook 1987). This is just what might be expected, since only a fraction of isolated GH deficiency is hypothalamic; the present experience with GHRH therapy reinforces the view that GH remains the reference standard treatment for GH deficiency.

The arguments for seeking alternative approaches to GH centre on economic grounds and patient convenience. If GHRH were considerably cheaper than GH or could be administered enterally as opposed to parenterally it might attract serious interest. At present there is no stable pricing relationship between GH and GHRH, so the first possibility remains unresolved. There has been speculation that it might be possible to package GHRH in liposomes which would release the drug for absorption across the colonic mucosa. This'raises the intriguing possibility of taking a pill at bedtime to promote growth, and if and when this becomes reality, society and the medical profession will be faced with a number of vexing ethical questions.

The idea of promoting GH release by altering the neuroendocrine tone of the hypothalamo-pituitary axis has also advanced in another direction. A recent comprehensive review on the neural control of somatotropic function (Müller 1987) forms an admirable background for recent clinical experiments in which drugs such as clonidine have been used chronically to increase GH secretion and linear growth. Pintor and co-workers have claimed success with clonidine in accelerating the growth of children with impaired growth hormone secretion (Pintor et al. 1985) and, more recently, in the treatment of children with short stature (Pintor et al. 1987). Careful inspection of their results shows that the effect is heterogeneous; clonidine appears to work on only a fraction of the cohort, and in those for whom success is claimed the effect is not uniformly sustained. Others have failed to observe a growth-promoting effect of clonidine (Castro-Magana et al. 1987; Cassio et al. 1987; Bernasconi et al. 1987). Even if clonidine worked it would be unlikely to have a specific and selective effect on GH secretion, leaving the physician with the dilemma of contemplating long-term, subtle side effects.

REFERENCES

Albertsson-Wikland K, Westphal O, Westgren U (1986) Daily subcutaneous administration of human growth hormone in growth hormone-deficient children. *Acta Paediatr Scand* **75**: 89-97

Bernasconi S, Chizzoni L, Romanini F, Volta C, Virdis R, Giovanelli G (1987) Clonidine treatment in children with consitutional growth delay (CGD). (Abstr. 122). Eur Soc Paed Endocr 26th Meeting, Toulouse

Blatt J, Bercu BB, Gillin JC, Mendelson WB, Poplack DG (1984) Reduced pulsatile growth hormone secretion in children after therapy for acute lymphoblastic leukaemia. *J Pediatr* **104**: 182-186

Brook CGD (1986) Turner syndrome. *Arch Dis Child* **61**: 305-309

Buchanan Cr, Law CM, Milner RDG (1987) Growth hormone in short slowly growing children and those with Turner's syndrome. *Arch Dis Child* **62**: 912-916

Cassio A, Stefanini C, Tassoni P, Tonini P, Tassinari D, Zuchini S, Cacciari E (1987) Treatment with clonidine in subjects with constitutional growth delay (CGD): a controlled double-blind study (Abstr. 121) Eur Soc Paed Endocr 26th Meeting, Toulouse

Castro-Magana M, Angulo M, Fuentes B, Canas A, Sharp A (1987) Failure of clonidine to improve the growth hormone respone to L-dopa administration. (Abstr). *Pediatr Res* **21**: 245A

Frasier SD (1983) Human pituitary growth hormone (hGH) therapy in growth hormone deficiency. *Endocr Rev* **4**: 155-170

Gertner JM, Genel M, Gianfredi SP, Hintz RL, Rosenfeld RG, Tamborlane WV (1984) Prospective clinical trial of human growth hormone in short children without growth hormone deficiency. *J Pediatr* **104**: 172-176

Guillemin R, Brazeau P, Böhlen P, Esch F, Ling N, Wehrenberg WB, Bloch B, Mougin C, Zeytin F, Baird A (1984) Somatocrinin, the growth hormone-releasing factor. *Recent Prog Horm Res* **40**: 233-299

Health Services Human Growth Hormone Committee (1981) Comparison of the intravenous insulin and oral clonidine tolerance tests for growth hormone secretion. *Arch Dis Child* **56**: 852-854

Herber SM, Milner RDG (1984) Growth hormone deficiency presenting under age 2 years. *Arch Dis Child* **59**: 557-560

Herber SM, Milner RDG (1986) When are we diagnosing growth hormone deficiency? *Arch Dis Child* **61**: 110-112

Hindmarsh PC, Brook CGD (1987) Effect of growth hormone on short normal children. *Br Med J* **295**: 573-577

Hindmarsh P, Smith PJ, Brook CGD, Mathews DR (1987) The relationship between height velocity and growth hormone secretion in short prepubertal children. *Clin Endocrinol* (Oxf) **27**: 581-591

Karlberg J, Engström I, Karlberg P, Fryer JG (1987) Analysis of linear growth using a mathematical model. 1. From birth to three years. *Acta Paediatr Scand* **76**: 478-488

Kastrup KW, Christiansen JS, Anderson JK, Ørskov H (1983) Increased growth rate following transfer to daily s.c. administration from three weekly i.m. injections of hGH in growth hormone-deficient children. *Acta Endocrinol* (Copenh) **104**: 148-152

Milner RDG (1985) Growth hormone 1985. *Br Med J* **291**: 1593-1594

Milner RDG (1986) Which children should have growth hormone therapy? *Lancet* i: 483-485

Milner RDG (1987) Current views on the treatment of Turner's syndrome. *Acta Paediatr Scand* [Suppl] **331**: 53-59

Milner RDG, Burns C (1982) Investigation of suspected growth hormone deficiency. *Arch Dis Child* **57**: 944-947

Milner RDG, King JM (1986) Growth hormone 1985. In: Meadow R (ed) *Recent Advances in Paediatrics*. Churchill Livingstone, Edinburgh, pp 33-34

Milner RDG, Barnes ND, Buckler JMH, Carson DJ, Hadden DR, Hughes IA, Johnston DI, Parkin JM, Price DA, Rayner PHW, Savage DCL, Savage MO, Smith CS, Swift PGF (1987) United Kingdom multicentre clinical trial of Somatrem. *Arch Dis Child* **62**: 776-779

Müller EE (1987) Neural control of somatotropic function. *Physiol Rev* **67**: 962-1053

Pintor C, Cella SG, Locatelli V, Puggioni R, Loche S, Müller EE (1985) Clonidine accelerates growth in children with impaired growth hormone secretion. *Lancet* i: 1482-1485

Pintor C, Cella SG, Corda R, Locatelli V, Puggioni R, Loche S, Müller EE (1987) Clonidine treatment for short stature. *Lancet* i: 1226-1230

Powell GF, Brasel A, Blizzard R (1967) Emotional deprivation and growth retardation simulating idiopathic hypopituitarism. *N Engl J Med* **276**: 1271-1283

Rochiccioli PE, Tauber M-T, Conde F-X, Amone M, Morre M, Uboldi F, Barbeau C (1987) Results of 1-year growth hormone (GH)-releasing hormone (1-44) treatment on growth, somatomedin-C, and 24-hour GH secretion in six children with partial GH deficiency. *J Clin Endocrinol Metab* **65**: 268-274

Rogol AD, Blizzard RM, Johanson AJ, Furlanetto RW, Evans WS, Rivier J, Vale WW, Thorner MO (1984) Growth hormone release in response to human pancreatic tumor growth hormone-releasing hormone-40 in children with short stature. *J Clin Endocrinol Metab* **59**: 580-586

Romshe CA, Zipf WB, Miser A, Miser J, Sotos JF, Newton WA (1984) Evaluation of growth hormone release and human growth hormone treatment in children with cranial irradiation-induced short stature. *J Pediatr* **104**: 177-182

Rosenfeld RG, Hintz RL, Johanson AJ, Brasel JA, Burstein S, Chernansek SD, Clabots T, Frane J, Gotlin RW, Kuntze J, Lippe BM, Mahoney PC, Moore WV, New MI, Saenger P, Stoner E, Sybert V (1986) Prospective randomized trial of methionyl human growth hormone and/or oxandrolone in Turner syndrome. *J Pediatr* **109**: 936-943

Ross JL, Meyerson Long L, Loriaux DL, Cutler DB jr (1985) Growth hormone secretory dynamics in Turner syndrome. *J Pediatr* **106**: 202-206

Ross RJM, Rodda C, Tsagarakis S, Davies PSW, Grossman A, Rees LH, Preece MA, Savage MO, Besser GM (1987) Treatment of growth-hormone deficiency with growth hormone-releasing hormone. *Lancet* **i**: 5-8

Rudman D, Kutner MH, Blackston RD, Cushman RA, Bain RP, Patterson JH (1981) Children with normal variant short stature: treatment with human growth hormone for six months. *N Engl J Med* **305**: 123-131

Salisbury DM, Leonard JV, Dezateux CA, Savage MO (1984) Micropenis: important early sign of congenital hypopituitarism. *Br Med J* **288**: 621-622

Sandahl-Christiansen J, Ørskov H, Binder C, Kastrup KW (1983) Imitation of normal plasma growth hormone profile by subcutaneous administration of human growth hormone to growth hormone-deficient children. *Acta Endocrinol* (Copenh) **102**: 6-10

Shriock EA, Lustig RH, Rosenthal SM, Kaplan SL, Grumbach MM (1984) Effect of growth hormone (GH)-releasing hormone (GRH) on plasma GH in relation to magnitude and duration of GH deficiency in 26 children and adults with isolated GH deficiency or multiple pituitary hormone deficiencies: evidence for hypothalamic GRH deficiency. *J Clin Endocrinol Metab* **58**: 1043-1049

Shulman DI, Bercu B (1987) Evaluation of growth hormone secretion: provocative testing vs endogenous 24-hour growth hormone profile. *Acta Paediatr Scand* [Suppl] **337**: 61-72

Smith PJ, Brook CGD (1988) Growth hormone-releasing hormone or growth hormone treatment in growth hormone insufficiency? *Arch Dis Child* **63**: 629-634

Smith PJ, Brook CGD, Rivier J, Vale W, Thorner MO (1986) Nocturnal pulsatile growth hormone-releasing hormone treatment in growth hormone deficiency. *Clin Endrocrinol* **25**: 35-44

Taitz LS, King JM (1986) Medical evidence in child abuse. *Arch Dis Child* **61**: 205-206

Takano K, Hizuka N, Shizume K, Asakawa K, Miyakawa M, Hirose N, Shibasaki T, Ling NC (1984) Plasma growth hormone (GH) response to GH-releasing factor in normal children with short stature and patients with pituitary dwarfism. *J Clin Endrocrinol Metab* **58**: 236-241

Tanner JM Whitehouse RH (1967) Growth response of 26 children with short stature given human growth hormone. *Brit Med J* **2**: 69-75

Tanner JM, Landt KW, Cameron N, Carter BS, Patel J (1983) Prediction of adult height from height and bone age in childhood. *Arch Dis Child* **58**: 767-776

Thorner MO, Reschke J, Chitwood J, Rogol AD, Furlanetto R, Rivier J, Vale W, Blizzard RM (1985) Acceleration of growth in two children with human growth hormone-releasing factor. *N Engl J Med* **312**: 4-9

Van Vliet G, Styne GM, Kaplan SL, Grumbach MM (1983) Growth hormone treatment for short stature. *N Engl J Med* **309**: 1016-1022

Vimpani GV, Vimpani AF, Lidgard GP, Cameron EHP, Farquhar JW (1977) Prevalence of severe growth hormone deficiency. *Br Med J* **2**: 427-430

Wales JKH, Milner RDG (1987a) Knemometry in assessment of linear growth: *Arch Dis Child* **6**: 166-171

Wales JKH, Milner RDG (1987b) Knemometry as a predictor of response to Somatrem in Turner's syndrome. *Acta Paediatr Scand* [Suppl] **337**: 37-40

Wilson DM, Baker B, Hintz RL, Rosenfeld RG (1985) Subcutaneous versus intramuscular growth hormone therapy: growth and acute somatomedin response. *Pediatrics* **76**: 361-364

Zadik Z, Chalew SA, Raiti S, Kowarski AA (1985) Do short children secrete insufficient growth hormone? *Pediatrics* **76**: 355-360

Advances in Growth Hormone and Growth Factor Research,
edited by E.E. Müller, D. Cocchi and V. Locatelli
Pythagora Press, Roma-Milano and Springer Verlag, Berlin-Heidelberg © 1989

Human growth hormone therapy in children with growth hormone deficiency*

K. Takano, K. Shizume, N. Hizuka, K. Asakawa, I. Sukegawa, R. Horikawa,
and the members of the Study Committee for hGH

*Department of Medicine, Institute of Clinical Endocrinology, Tokyo Women's
Medical College and Research Laboratory, the Foundation for Growth Science in
Japan, Tokyo, Japan*

The growth-promoting effect of human growth hormone (hGH) in children
with classical pituitary GH deficiency is widely accepted (Frasier 1983). However,
it is sometimes difficult to diagnose whether a child has enough endogenous GH
secretion or not. Once the safety and efficacy of recombinant hGH (r-hGH) were
established, it became possible to use r-hGH for the treatment of short statured
children other than just those with classical pituitary GH deficiency (Spiliotis et
al. 1984; Stahnke 1984).

Methionyl hGH (m-hGH) is the first preparation of r-hGH. Methionyl hGH
has a growth-promoting effect equal to that of pituitary-extracted hGH (p-hGH),
while its antigenicity is greater than that of p-hGH (Takano et al. 1986b; Kaplan
et al. 1986).

Recently, methionine-free hGH (mf-hGH) has become available due to the
advancements in gene technology. This preparation has not only a growth-
promoting effect but also very minimal antigenicity (Takano et al. 1987; Albertsson-Wikland 1987).

We have studied 24-h plasma GH profiles in normal but short children and
in patients with Turner's syndrome (Villadolid et al. 1988), who had normal GH
responses to GH-provocative stimuli. One third of these children had relatively
small amounts of GH secretion. These children grew with hGH treatment, indicating that they were lacking sufficient endogenous growth hormone.

* This work was partially supported by a Grant in Aid for Scientific Research from the Ministry of
Education, Science and Culture, Japan (No. 61440052, No. 61570566 and No. 61770872), by a
Research Grant from the Intractable Disease Division, Public Health Bureau, Ministry of Health and
by a Research Grant from the Foundation for Growth Science in Japan.

Recently, a highly sensitive enzyme immunoassay (EIA) for hGH was developed (Hashida et al. 1985). Using the EIA, we observed that the urinary GH value was significantly correlated with a mean 24-h plasma GH concentration (Sukegawa et al. 1988). Therefore, measurement of urinary GH seems to be helpful for screening children who would benefit from exogenous human GH therapy.

PATIENTS AND METHODS

Thirty-seven and 88 patients with GH deficiency were treated with m-hGH and mf-hGH respectively, at the dosage of 0.5 units/kg/week for 6–12 months. The diagnosis of GH deficiency was established based on the failure of plasma GH to respond to insulin-induced hypoglycemia, arginine loading, and/or the glucagon-propranolol test – so-called classical GH deficiency.

A 24-h endogenous GH secretion study (24-h GH study) was performed in four patients with GH deficiency, six patients with Turner's syndrome, and 35 normal but short children (NSC). NSC had normal responses to GH-provocation tests and had no physical or laboratory abnormalities other than short stature. An indwelling catheter was placed in a forearm vein and 2.5-ml samples were obtained every 20 min for 24 h using an ambulatory withdrawal pump (model ML-6-5H, Cormed, NY) to measure plasma GH concentration.

Plasma GH was measured by a double antibody RIA generously provided by The National Hormone and Pituitary Program of NIADDK, NIH (HGH:AFP-4793 B; antibody: AFP-97720133). Plasma somatomedin-C (Sm-C) levels were measured by RIA kit (Nichols Institute, San Juan Capistrano, CA). Urine was collected during 24 h with 50 μM sodium azide (NaN_3) at room temperature. The collected urine was dialyzed against 0.01 M phosphate-buffered saline (PBS) at 4°C for 16–24 h. Urinary GH values in dialyzed urine samples were directly determined using a highly sensitive enzyme immunoassay method (EIA).

RESULTS

Treatment of Pituitary Dwarfism with Recombinant Human Growth Hormone

Thirty-seven patients with pituitary dwarfism (naive patients) were treated with a highly purified m-hGH preparation (Somatonorm, ECP:3-4 ng/vial) obtained from KabiVitrum AB, Stockholm (Takano et al. 1986b). During 6 months of m-hGH treatment in 16 patients, height increased by between 2.6 and 6.1 cm, which calculated out to between 5.2 and 12.2 cm/yr, and during 12 months of m-hGH treatment in the other 21 patients, height increased by between 4.3 and 12.1 cm. The mean growth rate during m-hGH treatment (n=37) was 8.1 ± 0.3 cm/year, which was significantly greater than that for pretreatment (3.4 ± 0.2 cm/yr; P<0.001) (Fig. 1). There was no relationship between the growth rate on the one hand and chronological age and bone age on the other, taken at the start of treatment.

Eighty-eight patients with pituitary dwarfism were treated with three different kinds of mf-hGH (SM 9500 from KabiVitrum AB, Sweden; LY 137998 from Eli Lilly, USA; YM 17798 from Nordisk Ltd., Denmark). As shown in Table 1, the mean growth rates of naive patients ranged from 7.7 to 9.2 cm/yr and those of switched patients ranged from 5.8 to 7.4 cm/yr. These values were similar to those observed in previous treatment with p-hGH.

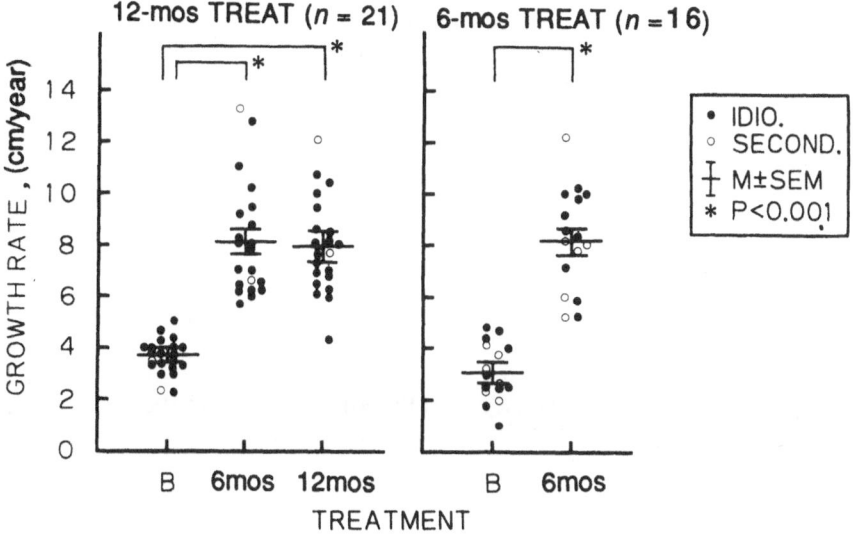

Fig. 1. Growth rate before and after m-hGH (Somatonorm III) treatment for 12 (*left panel*) and 6 (*right panel*) months. All naive patients.

Table 1. Growth rate (cm/yr, mean ± SD) in patients with GH deficiency being treated with recombinant mf-hGH

Type of mf-hGH	Naive patients		Switched patients	
	n	Growth rate (cm/yr)	n	Growth rate (cm/yr)
SM-9500	16	7.7 ± 1.8	9	5.8 ± 1.0
LY-137998	15	9.2 ± 1.6	12	7.4 ± 1.6
YM-17798	16	8.1 ± 1.5	20	6.7 ± 1.7

The appearance of hGH antibody (hGHAb) during m-hGH and mf-hGH treatment is shown in Table 2. The percentages of patients who had the antibody at the end of 6 months and 12 months of m-hGH treatment were 64.9% and 76.2% respectively. This incidence was greater than that observed in p-hGH treatment, which was 10%–15%. The incidence of anti-hGH antibody appearance was low in mf-hGH treatment, ranging from 0% to 12.5% at the end of 12 months of treatment. During the treatment with biosynthetic hGH, no special change in blood count, urinalysis, and/or routine chemistries was noted.

Table 2. hGH antibody appearance during 6 and 12 months treatment with m-hGH and mf-hGH

r-hGH preparation	Naive patients		Switched patients	
	6 months	12 months	6 months	12 months
m-hGH				
Somatonorm III	24/37	16/21	-	-
	(64.9%)	(76.2%)		
mf-hGH				
SM-9500	2/16	1/16	0/9	0/9
	(12.5%)	(6.3%)	(0%)	(0%)
LY-137998	1/15	0/15	0/12	012
	(6.7%)	(0%)	(0%)	(0%)
YM-17798	5/16	3/47	1/20	0/20
	(31.3%)	(12.5%)	(5.0%)	(0%)

24-h Endogenous GH Secretion Study

Four patients with hypopituitarism, six patients with Turner's syndrome, and 35 NSC underwent a 24-h endogenous GH secretion study. The 24-h plasma GH profiles in patients with GH deficiency and NSC are shown in Figure. 2.

According to the mean plasma 24-h GH concentration, we arbitrarily divided the NSC into two groups: a low endogenous GH group (L-GH: mean plasma 24-h GH concentrations below 3 ng/ml), and a normal endogenous GH group (N-GH: mean plasma 24-h GH concentrations above 3 ng/ml). The mean GH concentration in patients with pituitary dwarfism ranged from 1.1 to 1.5 ng/ml with a mean of 1.3 ± 0.2 ng/ml. In 12 of 35 NSC, the mean GH concentration was below 3 ng/ml, ranging from 1.3 to 2.9 ng/ml with a mean of 2.2. ± 0.5 ng/ml. It was above

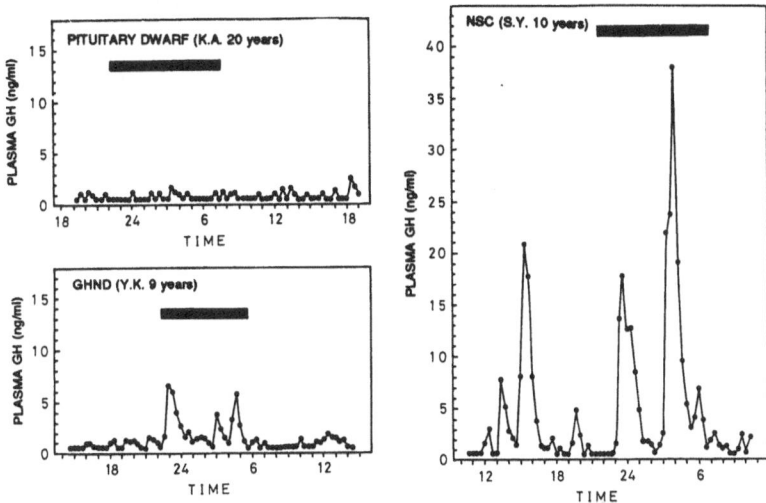

Fig. 2. Twenty-four-hour GH secretory profiles in a patient with pituitary dwarfism (*left upper panel*) and in two NSC (*left lower panel and right panel*). Blood samples were taken every 20 min.

3 ng/ml in 23 of 35 NSC, varying between 3.1 and 13.1 ng/ml with a mean of 6.7 ± 2.7 ng/ml. There was a positive correlation between the mean plasma GH and Sm-C, which was measured at the end of the 24-h GH study. The 24-h plasma GH profiles in the six patients with Turner's syndrome were also carried out and compared with those of normal but short girls (Figs. 3 and 4). The episodic secretion of plasma GH was observed in patients with Turner's syndrome; however, the mean plasma 24-h GH level was low, ranging between 1.2 and 4.6 ng/ml with a mean of 3.6 ± 0.6 ng/ml. This value was significantly lower than that of six normal short girls studied at the same time (7.1 ± 0.9 ng/ml; $P<0.01$).

GH Treatment in Normal but Short Children and in Patients with Turner's Syndrome

Five NSC with mean plasma 24-h values less than 3 ng/ml were treated with mf-hGH at the dosage of 0.1 units/kg/day for 6 months. The mean growth rate before treatment was 4.2 ± 0.5 cm/yr. During the treatment, height increased from 3.2 to 4.6 cm, which corresponded to from 6.4 to 9.2 cm with a mean of 7.8 ± 0.5 cm/yr.

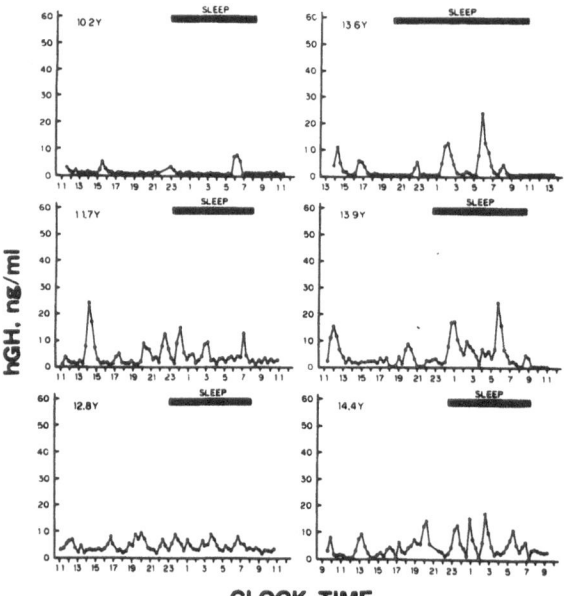

Fig. 3. Twenty-four-hour plasma GH secretory profiles in six patients with Turner's syndrome. Blood samples were taken every 20 min.

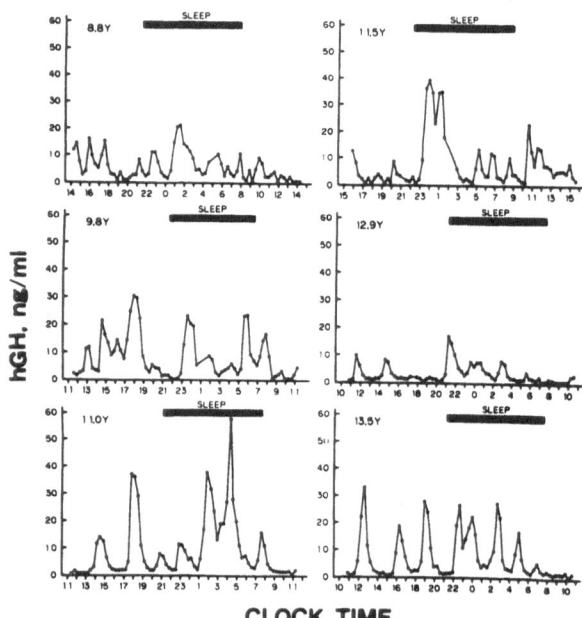

Fig. 4. Twenty-four-hour plasma GH secretory profiles in six normal but short girls, age-matched to six girls with Turner's syndrome, shown in Fig. 3. Blood samples were taken every 20 min.

The glucose metabolism during hGH treatment was studied. A glucose-loading test was performed before and 4–5 months after hGH treatment (Fig. 5). There was no glucose intolerance in these patients. The GH response to GH-provocation tests was also studied during hGH treatment. GH injection was stopped 2 days before the test, and the same kind of GH-provocation test as the patients had before was performed. Plasma GH response was calculated as the GH area under the curve (AUC; Fig. 6). There was no difference before and during hGH treatment in the GH response.

Twenty patients with Turner's syndrome were treated with m-hGH at the dosage of 0.5 units/kg/week for 6 months (Takano et al. 1986a). The height increase during the treatment was greater in 16 patients but smaller in four as compared with the pretreatment value. Height increased by between 1.4 and 3.7 cm, which corresponded to between 2.8 and 7.4 cm with a mean of 5.7 ± 0.4 cm/yr. This value was significantly greater than the pretreatment value of 3.7 ± 0.2 cm/yr; ($P<0.001$). The growth curve in a patient treated with hGH for 3.5 yr is shown in Figure 7.

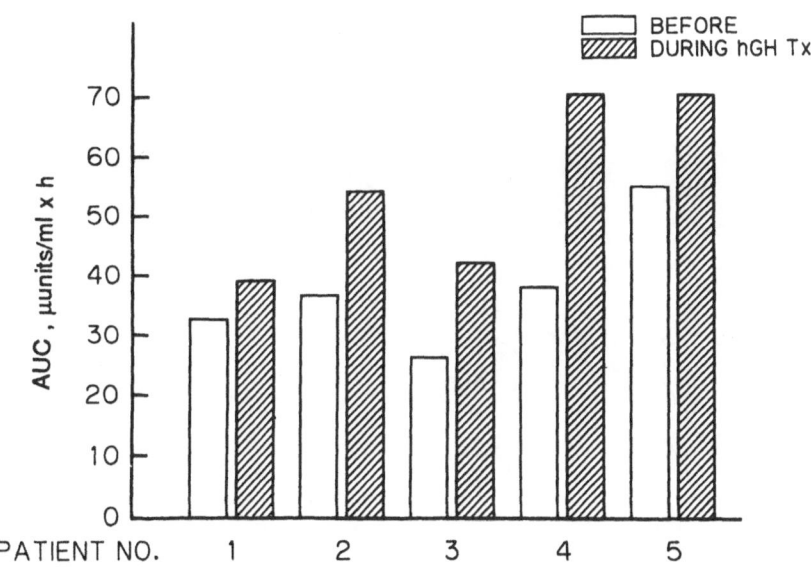

Fig. 5. Plasma insulin response to oral glucose test before and during hGH treatment. *AUC*, area under the curve.

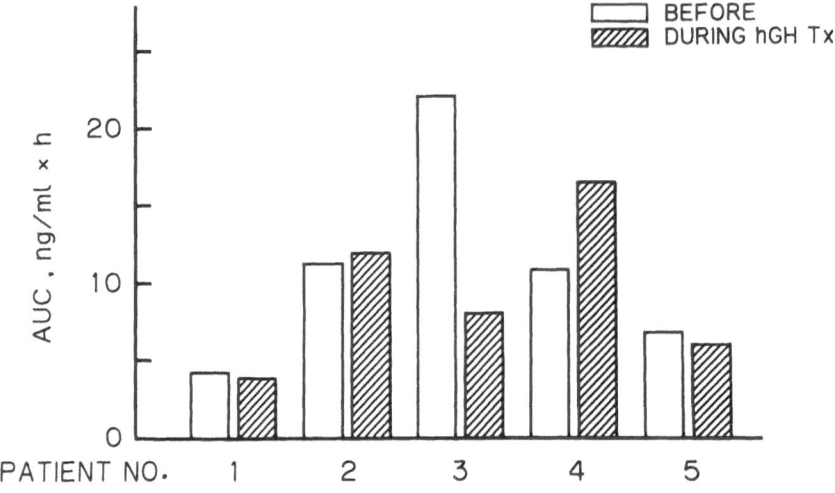

Fig. 6. GH responses to GH provocation test before and during hGH treatment. *AUC*, area under the curve.

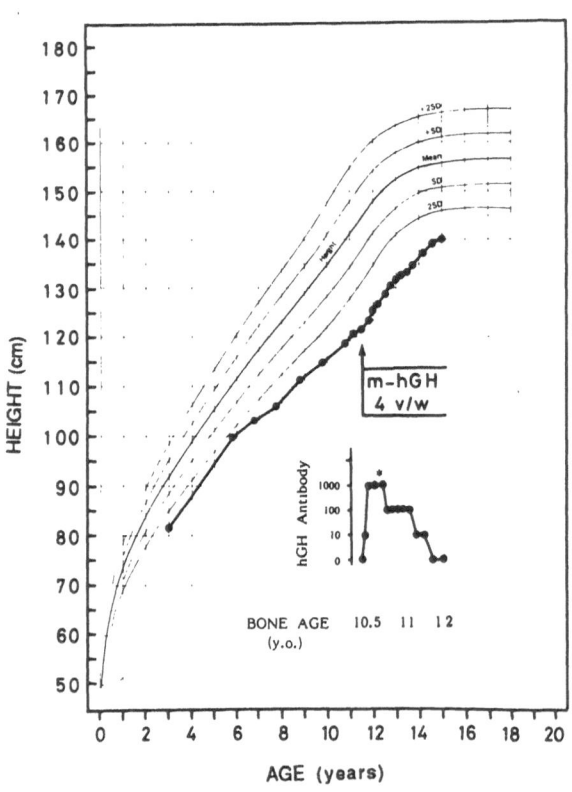

Fig. 7. Effects of m-hGH on linear growth and the production of hGH antibody in a patient with Turner's syndrome. At the age of 12.2 yr (*), the m-hGH preparation was changed from 82412 to 81000. ECP contents in these batches were 220 and 3 ng per vial, respectively.

How Can We Screen the Children Who Might Benefit from Exogenous Human GH Therapy?

The diagnosis of GH deficiency has long been based on a blunted GH secretory response to two or more GH-provocation tests. However, there are short-statured children who have normal GH secretion after provocation tests but markedly reduced endogenous GH secretion (Zadik et al. 1985; Bercu et al. 1986). Until now, evaluation of endogenous GH secretion has been based on multiple plasma GH measurements over 24 h, but for clinical studies blood sampling is not practical, and it is inconvenient for the children. Therefore, we have studied whether the measurement of urinary GH values is useful in evaluating endogenous GH secretion (Sukegawa et al. 1988).

Urinary GH values in patients with hypopituitarism, NSC, and patients with acromegaly are shown in Fig. 8. The values in six patients with hypopituitarism were undetectable. In 25 NSC the urinary GH values ranged from being undetectable to 55.8 ng/g Cr (creatinine). The values of urinary GH in the L-GH group and those in the N-GH group ranged from undetectable to 11.7 ng/g Cr and from 6.8 to 55.8 ng/g Cr, and mean values in the two groups were 6.4 ± 1.4 and 23.1 ± 3.2 ng/g Cr respectively. There were overlaps in individual values among the three groups; however, the mean values were significantly different from one

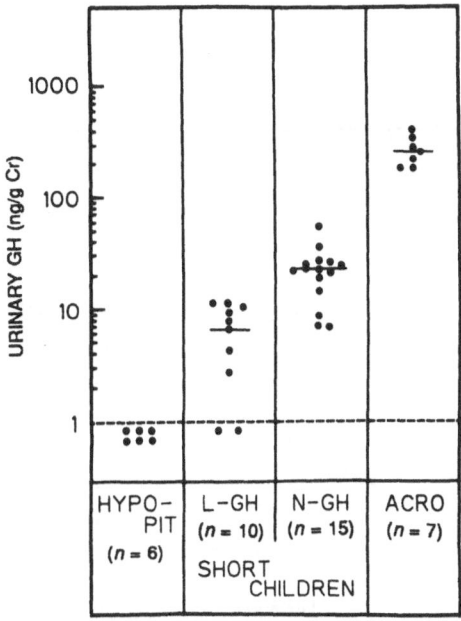

Fig. 8. Urinary GH values in hypopituitary, normal but short children, and acromegalic subjects.

another (P<0.001). The urinary GH values in seven patients with acromegaly ranged from 188.2 to 414.4 ng/g Cr with a mean of 274 ± 31.5 ng/g Cr; these values were ten times higher than those in normal adults.

The urinary GH values showed positive correlation with mean plasma GH concentrations (r=0.81, P<0.001) and plasma Sm-C levels (r=0.67, P<0.001) (Figs. 9 and 10).

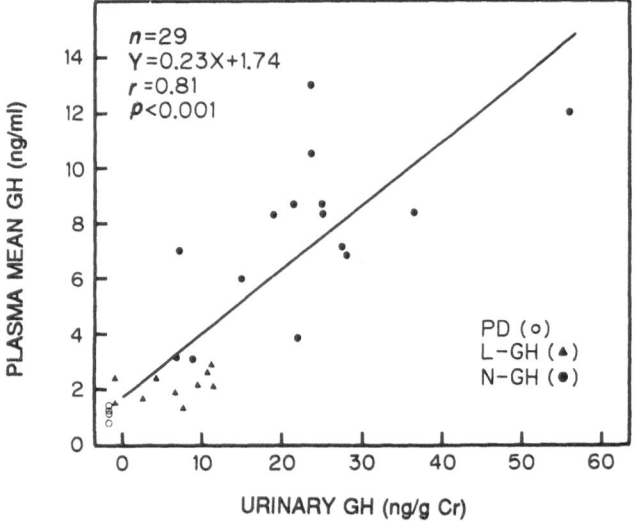

Fig. 9. Relationship between mean plasma GH concentration and urinary GH.

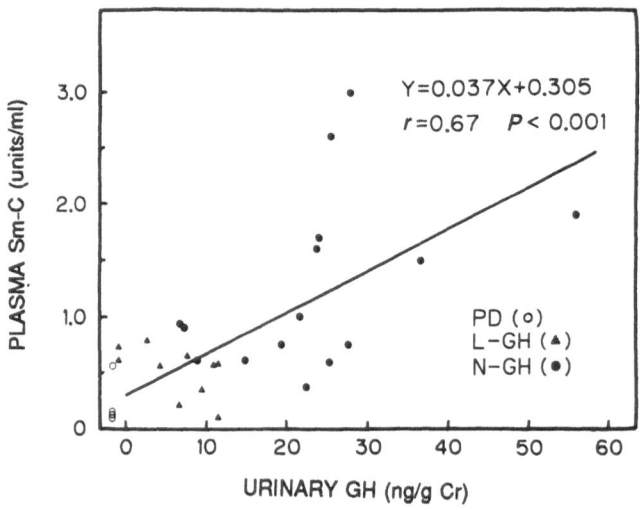

Fig. 10. Relationship between plasma SM-C and urinary GH.

DISCUSSION

It is clear that biosynthetic hGH has a growth-promoting effect in patients with GH deficiency. A remarkable difference between m-hGH and mf-hGH was the higher antigenicity of the former. With highly purified m-hGH, 76.2% of the patients had antibodies at the end of 1 year of treatment, while with mf-hGH only three of 47 patients had antibody with low titers. Recently we studied again the incidence of antibody production in naive or switched patients treated with m-hGH (Table 3). It is interesting to note that the incidence of hGHAb production is reduced in patients who have been treated with p-hGH previously (switched patients).

Knowledge of the safety and efficacy of recombinant hGH made it possible to extend treatment with r-hGH to short-statured patients other than those with classical pituitary GH deficiency. By means of the 24-h GH study, we observed that one third of NSC have low endogenous GH secretion, as do some patients with Turner's syndrome (Ross et al. 1985; Villadolid et al. 1988). Furthermore, these patients responded well to hGH treatment. At present a 24-h GH study is necessary to evaluate endogenous GH secretion. However, our data indicate that the measurement of urinary GH seems to be helpful for screening children who would benefit from exogenous human GH therapy.

Table 3. Number of patients who had hGH antibody at the end of 6 months of m-hGH (Somatonorm) treatment

Patients	Naive	Switched
Total no. of cases	160	51
Number of hGHAb (+) cases	74 (46.3%)	2 (3.9%)
Number with hGHAb titer		
10	14	1
10^2	36	1
10^3	22	0
10^4	2	0

REFERENCES

Albertsson-Wikland K (1987) Clinical trial with authentic recombinant somatropin in Sweden and Finland. *Acta Paediatr Scand* [Suppl] **331**: 28-37

Bercu BB, Shulman D, Root AW, Spiliotis BE (1986) Growth hormone (GH) provocative testing frequently does not reflect endogenous GH secretion. *J Clin Endocrinol Metab* **63**: 709-714

Frasier SD (1983) Human pituitary growth hormone (hGH) therapy in growth hormone deficiency. *Endocr Rev* **4**: 155-170

Hashida S, Ishikawa E, Nakagawa K, Ohtaki S, Ichioka T, Nakajima K (1985) Demonstration of human growth hormone in normal urine by a highly specific and sensitive sandwich enzyme immunoassay. *Anal Lett* **18**: 1623

Kaplan SL, Underwood LE, August GP, Bell JJ, Blethen SL, Blizzard RM, et al. (1986) Clinical studies with recombinant-DNA-derived methionyl human growth hormone in growth hormone-deficient children. *Lancet* **i**: 697-700

Ross JL, Long LM, Loriaux LD, Cutler GB Jr (1985) Growth hormone secretory dynamics in Turner's syndrome. *J Pediatr* **106**: 202-206

Stahnke N (1984) Human growth hormone treatment in short children without growth hormone deficiency. *New Engl J Med* **310**: 925-926

Spiliotis BE, August GP, Hung W, Sonis W, Mendelson W, Bercu BB (1984) Growth hormone neurosecretory dysfunction. A tretable cause of short stature. *JAMA* **251**: 2223-2230

Sukegawa I, Hizuka N, Takano K, Asakawa K, Horikawa R, Ishikawa E, Mohri Z, Murakami Y, Shizume K (1988) Measurement of urinary growth hormone (GH) values is useful for evaluating endogenous GH secretion. *J Clin Endocrinol Metab* **66**: 1119-1123

Takano K, Hizuka N, Shizume K (1986a) Treatment of Turner's syndrome with methionyl human growth hormone for six months. Acta Endocrinol *(Copenh)* **112**: 130-137

Takano K, Shizume K, Hizuka N, Okuno A, Umino T, Kobayashi Y, et al. (1986b) Treatment of pituitary dwarfism with methionyl human growth hormone in Japan. *Endocrinol Jpn* **33**: 589-596

Takano K, Shizume K, Hibi I, Okuno A, Hanyu K, Suwa S, et al. (1987) Treatment of pituitary dwarfism with authentic recombinant human growth hormone (SM-9500). *Endocrinol Jpn* **34**: 291-297

Villadolid MC, Takano K, Hizuka N, Asakawa K, Sukegawa I, Horikawa R, Shizume K (1988) Twenty-four-hour plasma GH, FSH and LH profiles in patients with Turner's syndrome. *Endocrinol Jpn* **35**: 71-81

Zadik Z, Chalew SA, Raiti S, Kowarski AA (1985) Do short children secrete insufficient quantities of growth hormone? *Pediatrics* **76**: 355-360

Advances in Growth Hormone and Growth Factor Research,
edited by E.E. Müller, D. Cocchi and V. Locatelli
Pythagora Press, Roma-Milano and Springer Verlag, Berlin-Heidelberg © 1989

Growth hormone-releasing hormone: a new treatment for growth hormone deficiency*

R.J.M. Ross[1], M.A. Preece[2], G.M. Besser[1], and M.O. Savage[1]

[1] *Department of Endocrinology, St Bartholomew's Hospital; and* [2] *Department of Growth and Development, Institute of Child Health, London, United Kingdom*

The identification and synthesis of growth hormone-releasing hormone (GHRH) coincided with the realisation that Jacob-Creutzfeldt disease could be transmitted by extracted cadaveric GH. Fortunately, the acceleration in supplies of biosynthetic GH enabled treatment regimens to be continued and even extended, and the field of growth and its therapy has since become a major focus of research. This review attempts to assess the current place of GHRH in the treatment of GH deficiency.

Although Green and Harris described the anatomical basis for the control of the pituitary by the hypothalamus in 1946, and Reichlin demonstrated that GH secretion was controlled by a hypothalamic factor in 1960, only in 1982 was GHRH finally extracted, sequenced and subsequently synthesised, Unlike the other hypothalamic regulatory peptides, it had not been possible to extract GHRH from animal hypothalami because of the small quantities present and because the overwhelming amount of somatostatin in the tissue interfered with the biossay for GHRH. In 1982, the two groups of Guillemin and Vale independently characterised peptides from pancreatic tumours of two patients who had acromegaly as a consequence of the ectopic production of GHRH. These peptides, which contained between 37 and 44 residues of which the first 40 were homologous, have subsequently proved to be identical to human hypothalamic GHRH. Both 40- and 44-residue peptides identical to the pancreatic peptides have been extracted from the human hypoythalamus. In addition, the entire human GHRH gene has been

* The work was supported by the Joint Research Board of St Bartholomew's Hospital, the Peel Medical Research Trust, and the Child Growth Foundation.

structurally characterised and mapped on chromosome 20. There has been considerable debate as to whether the 44- or the 40-residue peptide is the native human peptide, although, as outlined below, they have very similar biological activities.

GHRH ANALOGUES AND THEIR DOSE-RESPONSE RELATIONSHIPS

It appears that the biological activity of GHRH resides principally in the first 29 residues, as analogues of this length with deletions from the C-terminal end retain most of their GH-releasing activity, but the N-terminal amino acid needs to be preserved. Most human studies have been performed with the native peptides GHRH–NH$_2$ and GHRH–OH, or the shorter analogue GHRH–NH$_2$. All three, when given as an intravenous bolus, selectively promote GH release in normal male subjects at doses of 1–3 µg/kg (Gelato et al. 1983; Rosenthal et al. 1983; Grossman et al. 1983). The mean peak serum GH after a supramaximal iv bolus of GHRH, as reported from the above studies on normal men, ranged between 32 and 68 m units/l (1 m unit/l=0.5 ng/ml), and the time of the peak varied between 15 and 60 min after administration.

The only side effect of GHRH reported is facial flushing, which occurs in most subjects within 1 min of an intravenous injection of 100 µg and lasts no longer than 5 min. When compared in the same subjects, all three analogues elicit an identical pattern of GH release (Grossman et al. 1984b; Losa et al. 1984).

There is considerable individual variation in the response to GHRH, and rapid repetitive administration leads to a decreasing GH response. At a time when the pituitary is apparently refractory to GHRH it will still release GH in response to hypoglycaemia, suggesting that the lack of response to GHRH is not due to depleted pituitary stores of GH (Shibasaki et al. 1985). In addition, the prior administration of GH completely abolishes the GH response to GHRH (Ross et al. 1987a). It is likely that changes in somatostatin tone at the pituitary are responsible for the variable response to GHRH, with an increase in somatostatin being induced by GH secretion. Recent work suggests that GH feedback is mediated through somatostatin under cholinergic control, as pyridostigmine, an acetylcholinesterase inhibitor, blocks GH feedback (Ross et al. 1987b).

ROUTES OF ADMINISTRATION

GHRH may be given subcutaneously or intranasally, although 30- and 300-fold higher doses respectively are required to obtain a similar response to that seen after an iv bolus (Evans et al. 1985). From measurement of immunoreactive GHRH (ir-GHRH) after administration, it appears that GHRH is rapidly but very incompletely absorbed by both the intranasal and subcutaneous routes (Evans et al. 1985; Sassolas et al. 1985).

CLINICAL STUDIES IN GH DEFICIENT PATIENTS

The majority of patients with idiopathic GH deficiency, as defined by conventional tests of GH secretion (insulin-induced hypoglycaemia, clonidine, etc.) which act through the hypothalamus, will show a growth hormone response to an iv bolus of GHRH (Borges et al. 1983; Grossman et al. 1983; Sassolas et al. 1983) (Fig. 1). These patients thus have a defect in either the synthesis or the delivery of hypothalamic GHRH, rather than being truly "GH deficient". Patients who are GH deficient due to pituitary tumours, hypothalamic disease (including craniopharyngiomas and extrasellar germinomas), septo-optic dysplasia and following cranial irradiation have all been described as responding to GHRH, although the number of patients and the degree to which they respond varies. Patients with idiopathic

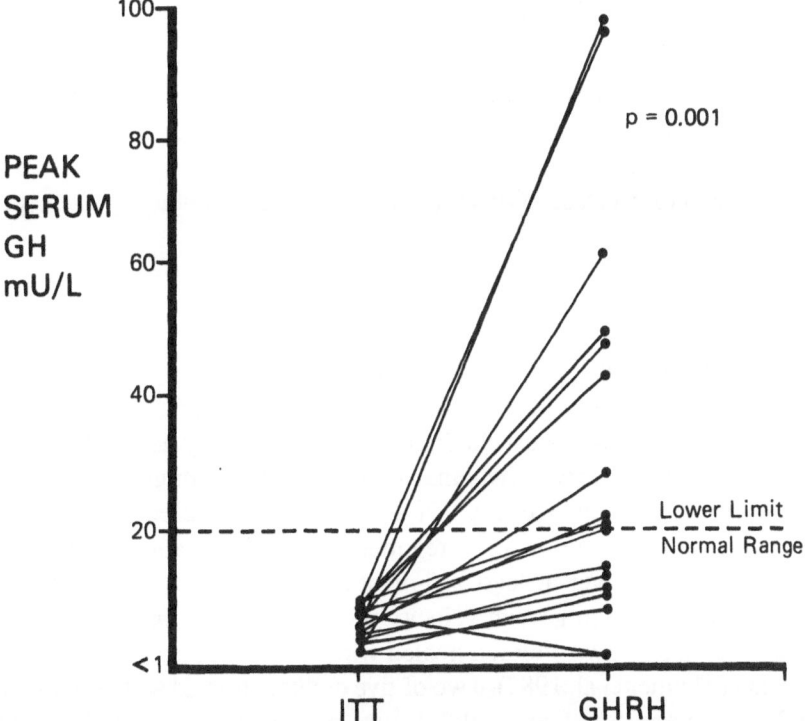

Fig. 1. Peak GH responses to insulin hypoglycaemia and GHRH in 25 GH-deficient children.

GH deficiency show a lesser response than normal children, although there is considerable overlap among all groups (Gelato et al. 1986). Hypothalamo-pituitary irradiation for both pituitary and nonpituitary disease results in GH deficiency, and again, the majority of these patients will respond to GHRH, although the longer the interval after irradiation, the poorer the response (Ahmed and Shalet 1984; Grossman et al. 1984a; Lusting et al. 1985). In patients with idiopathic GH deficiency, the more severe the deficiency and the longer its duration (i.e., the older the patient) the poorer the response to GHRH (Schriock et al. 1984).

IDIOPATHIC SHORT STATURE

We use the term 'idiopathic short stature' to include those groups of short children variously described as having constitutional short stature, constitutional growth delay, normal variant short stature, and GH neurosecretory dysfunction. These are children who are short and growing slowly, who have a normal GH response to conventional tests of GH secretion but subnormal physiological GH secretion as measured by nocturnal profiles (Spiliotis et al. 1984). In our experience, these children with idiopathic short stature (height below 3rd centile and a peak GH response to hypoglycaemia greater than 20 m units/l) usually show a greater response to GHRH, which is identical to that seen in young normal adults; this has also been found by other groups (Chalew et al. 1986; Gelato et al. 1986). The present evidence suggests that the poor growth in children with idiopathic short stature is related to decreased GHRH stimulation of the pituitary by the hypothalamus.

GHRH TREATMENT

The demonstration that most children with GH deficiency have a defect in either the synthesis or the delivery of hypothalamic GHRH to the pituitary suggested that GHRH, by promoting endogenous GH secretion, can be used to treat GH deficiency. Four studies have now been reported using GHRH administered subcutaneously as treatment over at least a 6-month period.

Taking an increase in growth velocity of greater than 2 cm/yr as evidence of an effect of treatment, two children treated with 3-hourly pulses of GHRH showed acceleration (Thorner et al. 1985). Two of five children treated with 3-hourly pulses overnight also grew (Smith et al. 1986). Five patients with partial GH deficiency also responded (Rochiccioli et al. 1987).

The largest study, by the authors, will be described in detail (Ross et al. 1987d).

This study was initiated in May 1985, following the withdrawal of pituitary hGH from use in the United Kingdom after the realisation that it might transmit Jacob-Creutzfeldt disease. Biosynthetic hGH was not widely available, leaving many GH-deficient children without treatment.

STUDY OF GHRH THERAPY IN 18 GH-DEFICIENT CHILDREN

Patients. Eighteen prepubertal patients, 16 boys and two girls aged 6.8–14.2 yr, were entered into the study. Ten had idiopathic GH deficiency, four multiple pituitary hormone deficiencies, three postirradiation GH deficiency, and one septo-optic dysplasia. Fifteen had previously received hGH treatment for between 4 months and 4.4 yr but stopped 3–6 months before the start of this trial. Patients defined as GH deficient showed no evidence of nonpituitary disease but had a subnormal growth velocity and peak GH concentrations of less than 7 m units/l during one of the following GH-stimulation tests; insulin-induced hypoglycaemia, glucagon or clonidine (one patient).

Treatment. The synthetic analogue $GHRH_{1-29}NH_2$ (KabiVitrum) was given twice daily sc in doses of 250 µg to eight children weighing less than 20 kg and in doses of 500 µg to ten children weighing more than 20 kg. Fourteen of the patients have been treated for 6–12 months and one for 18 months.

Tests. Standard intravenous GHRH tests using 100 µg bolus doses were performed before and after 3 months of treatment.

Auxology. Height was measured with a Holtain stadiometer (Holtain Ltd) by the same trained observer whose standard error for measurement of stature was 0.07 cm. An increment in height velocity of greater than 2 cm/yr on GHRH therapy was defined as a worthwhile response.

RESULTS

Growth. After 3–6 months of GHRH therapy, 12 of the 18 children showed an increase in height velocity, and in eight of these after 6 months this was greater than 2 cm/yr (Fig. 2). These eight "responders" have now been treated for 6–18 months, and their increase in height velocity has been maintained. The height velocity of all the patients at 3–6 months after GHRH treatment correlated well with that while the patients were on hGH (r=0.85, P<0.001), and in the GHRH responders height velocity with GHRH was similar to that with hGH.

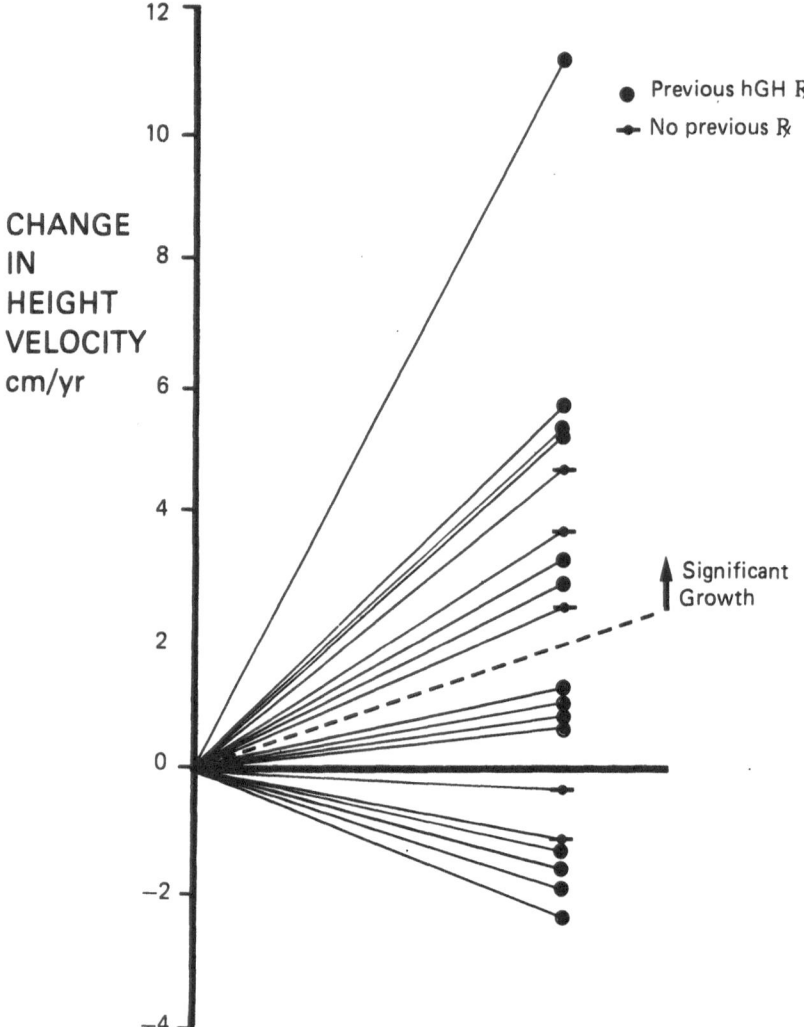

Fig. 2. Change in height velocity after 6 months' GHRH treatment in 19 GH-deficient children.

Growth hormone responses to GHRH. The peak GH responses to iv and sc GHRH were the same before treatment and after 1,2,3 and 6 months of GHRH therapy (Fig. 3).

The peak GH response to the first iv dose of GHRH was significantly higher in the group of growth responders than in the nonresponders (P=0.015). Only three responders had peak GH levels to the first dose of GHRH of under 30 m units/ l. There was no evidence of either a priming effect or a desensitisation effect of GHRH treatment on GH response to GHRH.

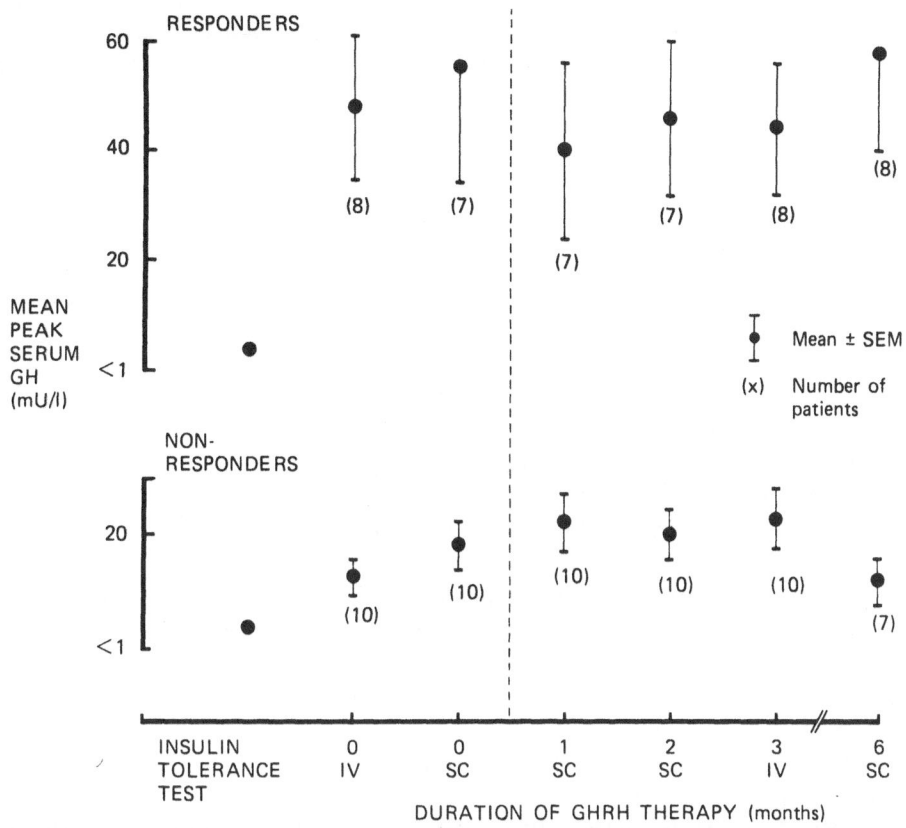

Fig. 3. Mean peak serum GH before and during treatment with 100 μg iv or 500 μg sc GHRH–NH$_{2}$.
$_{1-29}$

Somatomedin-C levels. These were subnormal in 16 of 18 patients off treatment — mean 0.22, range 0.12–0.43 units/ml (normal 0.4–1.5 units/ml). The Sm-C levels rose after 3 months of GHRH therapy in six of eight growth responders and in two of the nonresponders.

GHRH antibodies. These were found in 14 of 17 patients after 3 months of GHRH therapy. There was no significant difference in the GH response to GHRH in these patients. The circulating antibody showed a high affinity for GHRH K=10^{9}–10^{10} l/mol), but was present only in low titre (log titre 1.1–3.5) Seven of the eight growth responders had antibodies at 3 months of treatment and four were growing at a greater velocity at 6 months than at 3 months. Consequently, GHRH antibodies did not appear to inhibit the growth response to GHRH.

Side effect. There were no changes in routine tests of blood count, urea, electrolytes, and liver function or in a midstream specimen of urine after 3 and 12 months of therapy. The injections were well tolerated and 12 of the 18 children gave their own injections.

CURRENT STATUS OF GHRH THERAPY

It is clear from the above studies that GHRH will promote growth in at least 50% of children who are GH deficient based on conventional criteria.

However, these children are likely to be the most difficult children to treat with GHRH, as they have frequently had prolonged GHRH deficiency and therefore only a small pituitary pool of GH.

The group of patients who make up the largest number who require treatment for short stature are those that we have called patients with idiopathic short stature. These are children growing slowly who have low physiological release of GH but a normal GH response to conventional stimuli of GH secretion. These children, who have a structurally normal hypothalamo-pituitary axis and a good GH response to GHRH, are likely to grow well on GHRH and may be the group in whom this therapy would show the most benefit.

The recent finding that acetylcholinesterase inhibitors such as pyridostigmine enhance the GH response to GHRH has important therapeutic implications (Ross et al. 1987b). It may be that the addition of pyridostigmine to any GHRH treatment would greatly increase its effect.

Biosynthetic GH is now widely available and, given once daily, is very effective in promoting growth. The present regimens of GHRH treatment require at least twice-daily injections and therefore make it less practical than hGH therapy. However, GHRH has a number of important advantages over hGH. It is a smaller molecule and can therefore be given intranasally, and, as already mentioned, it preserves feedback at the pituitary and should therefore avoid any dangers of overtreatment. GHRH given as a continuous infusion augments the normal endogenous release of GH, thus preserving the pulsatile pattern. It has already been shown that GHRH infusions will augment GH release in partially GH-deficient children (Rochiccioli et al. 1986), suggesting that a depot preparation of GHRH may well promote growth. The development of a depot preparation of GHRH would provide an important alternative to hGH for the treatment of short stature, with the advantages that it could be given infrequently and would promote physiological GH release without the danger of overtreatment.

Acknowledgment

We are grateful to Alison Platts for her secretarial assistance.

REFERENCES

Ahmed SR, Shalet SM (1984) Hypothalamic growth hormone-releasing factor deficiency following cranial irradiation. *Clin Endocrinol* (Oxf) **21**: 483-488

Chalew SA, Armour KM, Levin PA, Thorner MO, Kowarski AA (1986) Growth hormone (GH) response to GH-releasing hormone in children with subnormal integrated concentrations of GH. *J Clin Endocrinol Metab* **62**: 1110-1115

Davies RR, Turner S, Johnston DG (1984) Oral glucose inhibits growth hormone secretion induced by human pancreatic growth hormone-releasing factor 1–44 in normal man. *Clin Endocrinol* (Oxf) **22**: 477-481

Evans WS, Vance ML, Kaiser DL, Sellers RP, Borges JLC, Downs TR, Frohman LA, Rivier J, Vale W, Thorner MO (1985) Effects of intravenous, subcutaneous and intranasal administration of growth hormone (GH)-releasing hormone-40 on serum GH concentrations in normal men. *J Clin Endocrinol Metab* **61**: 846-850

Frohman LA, Jansson J-O (1986) Growth hormone-releasing hormone. *Endocr Rev* **7**: 223-253

Gelato MC, Pescovitz O, Cassorla F, Loriaux L, Merriam GR (1983) Effects of a growth hormone-releasing factor in man. *J Clin Endocrinol Metab* **57**: 674-676

Gelato MC, Malozowski S, Caruso-Nicoletti M, Ross JL, Pescovitz OH, Rose S, Loriaux DL, Cassorla F, Merriam GR (1986) Growth hormone (GH) responses to GH-releasing hormone during pubertal development in normal boys and girls: comparison to idiopathic short stature and GH deficiency. *J Clin Endocrinol Metab* **63**: 174-179

Grossman A, Savage MO, Wass JAH, Lytras N, Sueiras-Diaz J, Coy D, Besser GM (1983) Growth hormone-releasing factor in growth hormone deficiency: demonstration of a hypothalamic defect in growth hormone release. *Lancet* **ii**: 137-138

Grossman A, Lytras N, Savage MO, Wass JAH, Coy DH, Rees LH, Jones AE, Besser GM (1984a) Growth hormone-releasing factor: comparison of two analogues and demonstration of hypothalamic defect in growth hormone release after radiotheraphy. *Br Med J* **288**: 1785-1787

Grossman A, Savage MO, Lytras N, Coy DH, Rees LH, Besser GM (1984b) Responses to analogues of growth hormone-releasing hormone in normal subjects, and in growth hormone-deficient children and young adults. *Clin Endocrinol* (Oxf) **21**: 321-30

Imaki T, Shibasaki T, Shizume K, Masuda A, Hotta M, Jibiki YK, Demura H, Tsushima T, Ling N (1985) The effect of free fatty acids on growth hormone (GH)-releasing hormone-mediated GH secretion in man. *J Clin Endocrinol Metab* **60**: 290-293

Losa M, Schopohl J, Müller OA, von Werder K (1984) Stimulation of growth hormone secretion with human growth hormone-releasing factors (GRF1-44, GRF1-40, GRF1-29) in normal subjects. *Klin Wochenschr* **62**: 1140-1143

Lustig RH, Shriock EA, Kaplan AL, Grumbach MM (1985) Effect of growth hormone-releasing factor on growth hormone release in children with radiation-induced growth hormone deficiency. *Pediatrics* **76**: 274-279

Press M, Tamborlane WV, Thorner MO, Vale W, Rivier J, Gertner JM (1984) Pituitary response to growth hormone-releasing factor in diabetes. *Diabetes* **33**: 804-806

Rochiccioli PE, Tauber M, Uboldi F, Coude F, Morre M (1986) Effect of overnight constant infusion of human growth hormone (GH)-releasing hormone (1-44) on 24-hour GH secretion in children with partial GH deficiency. *J Clin Endocrinol Metab* **63**: 1100-1105

Rochiccioli PE, Tauber M, Coude F, Arnone M, Morre M, Uboldi F, Barbeau C (1987) Results of 1-year growth hormone (GH)-releasing hormone (1-44) treatment on growth, somatomedin-C, and 24-hour GH secretion in six children with partial GH deficiency. *J Clin Endocrinol Metab* **65**: 268-274

Rosenthal SM, Schriock EA, Kaplan SL, Guillemin R, Grumbach MM (1983) Synthetic human pancreas growth hormone-releasing factor (hpGRF1-44-NH2) stimulates growth hormone secretion in normal men. *J Clin Endocrinol Metab* **57**: 677-679

Ross RJM, Borges F, Grossman A, Smith R, Nhagafoong L, Rees LH, Savage MO, Besser GM (1987a)Growth hormone pretreatment in man blocks the response to growth hormone-releasing hormone; evidence for a direct effect of growth hormone. *Clin Endocrinol (Oxf)* **26**: 117-123

Ross RJM, Tsagarakis S, Grossman A, Nhagafoong L, Touzel RJ, Rees LH, Besser GM (1987b) GH feedback occurs through modulation of hypothalamic somatostatin under cholinergic control; studies with pyridostigmine and GHRH. *Clin Endocrinol (Oxf)* **27**: 727-733

Ross RJM, Grossman A, Davies PSW, Savage MO, Besser GM (1987c) Stilboestrol pretreatment of children with short stature does not affect the GH response to growth hormone-releasing hormone. *Clin Endocrinol (Oxf)* **27**: 155-161

Ross RJM, Tsagarakis S, Grossman A, Preece MA, Rodda C, Davies PSW, Rees LH, Savage MO, Besser GM (1987d) Treatment of growth-hormone deficiency with growth hormone-releasing hormone. *Lancet* i: 5-7

Sassolas G, Biot-Laporte S, Cohen R, Charfi AE, Ferry S, Borson F (1983) Somatocrinin induces release of growth hormone in one case of growth hormone deficiency of hypothalamic origin. *CR Acad Sci* [III] **296**:527-529

Sharp PS, Foley K, Chahal P, Kohner EM (1984) The effect of plasma glucose on the growth hormone response to human pancreatic growth hormone-releasing factor in normal subjects. *Clin Endocrinol (Oxf)* **20**: 497-501

Shibasaki T, Hotta M, Masuda A, Toshihiro I, Obara N, Demura H, Ling N, Shizume K (1985) Plasma GH responses to GHRH and insulin-induced hypoglycemia in man. *J Clin Endocrinol Metab* **60**: 1265-1267

Shriock EA, Lustig RH, Rosenthal SM, Kaplan SL, Grumbach MM (1984) Effect of growth hormone (GH)-releasing hormone (GRH) on plasma GH in relation to magnitude and duration of growth hormone deficiency in 26 children and adults with isolated GHD or multiple pituitary hormone deficiencies: evidence for hypothalamic GRH deficiency. *J Clin Endocrinol Metab* **58**: 1043-1049

Smith PJ, Brook CGD, Rivier J, Vale W, Thorner MO (1986) Noctural pulsatile growth hormone-releasing hormone treatment in growth hormone deficiency. *Clin Endocrinol (Oxf)* **25**: 35-44

Spiliotis BE, August GP, Hung W, Sonis W, Mendelson W, Bercu BB (1984) Growth hormone neurosecretory dysfunction. A treatable cause of short stature. *JAMA* **251**: 2223-2230

Thorner MO, Reschke J, Chitwood J, Rogol AD, Furlanetto R, Rivier J, Vale W, Blizzard RM (1985) Acceleration of growth in two children treated with human growth hormone-releasing factor. *N Engl J Med* **312**: 4-9

Advances in Growth Hormone and Growth Factor Research,
edited by E.E. Müller, D. Cocchi and V. Locatelli
Pythagora Press, Roma-Milano and Springer Verlag, Berlin-Heidelberg © 1989

Growth hormone deficiency states: approach by CNS-acting compounds

C. Pintor[1], S. Loche[1], R. Puggioni[1], S. G. Cella[2], V. Locatelli[2], A. Lampis[1], and E.E. Müller[2]

[1] Department of Pediatrics, Chair of Pediatric Endocrinology, University of Cagliari, Cagliari, Italy; and [2] Department of Pharmacology, University of Milan, Milan, Italy

INTRODUCTION

Over the past few years, a number of new findings have led to a better understanding of the pathophysiology of growth hormone (GH) secretion. The isolation of GH-releasing hormone (GHRH) (Guillemin et al. 1982; Rivier et al. 1982) has led to the conclusion that most GH-deficient patients can produce normal amounts of GH when stimulated with GHRH (Grossman et al. 1983; Pintor et al. 1983; Thorner et al. 1983; Schriock et al. 1984). GH hyposecretory states may therefore result from a primitive abnormality of the somatotrophs or, more frequently, from reduced GHRH synthesis and/or release. The latter may be caused by hypofunction of stimulatory neurotransmitter pathways. More recently, it has been observed that a number of short children, with normal GH responses to standard pharmacological stimuli, have low 24-h GH secretion (IC-GH) and show a positive response to treatment with exogenous GH. This has strengthened the argument for the existence of a central abnormality of GH neuroregulation (Spiliotis et al. 1984; Zadik et al. 1985).

Based on these assumptions, in recent years the use of neuroactive drugs for the treatment of short children has been the focus of considerable interest. Huseman and coworkers have shown that administration of L-dopa or bromocriptine to children with intrauterine growth retardation or GH deficiency resulted in acceleration of growth (Huseman and Hassing 1984; Huseman 1985; Huseman et al. 1986). In addition, we (Pintor et al. 1985, 1987) and others (Castro Magana et al.

1986) have shown that chronic administration of the α_2-adrenergic agonist clonidine (Clo) also accelerates growth in children with constitutional growth delay (CGD). We report here our experience with the use of Clo in the chronic treatment of children with CGD.

MATERIALS AND METHODS

Study 1

Constitutional growth delay was studied in 34 children (20 boys and 14 girls). In all cases there was a strong familial pattern of normal childhood growth and adolescent development. There was no evidence of systemic diseases, malnutrition, dysmorphic syndrome or psychosocial disturbances, and none had received long-term medication, GH, or anabolic steroids prior to entering the study. The main clinical characteristics of the subjects are summarized in Table 1. Clo (Catapresan, Boehringer, Ingelheim) was given orally at a daily dose of 0.1 mg/m^2 in two doses. This treatment was continued in 34 children for 6 months and in 22 of them for up to 12 months.

The 22 children were then withdrawn from Clo for 6 months, during which time they were given a placebo. At the end of this period, Clo was reinstituted in six children, and height velocity (HV) was reevaluated 6 months later. Standard iv GHRH tests (hpGRF$_{1-40}$, 1 μg/kg), oral Clo tests (0.15 mg/m^2), and baseline GH and somatomedin C (Sm-C) determinations were performed before therapy, and after 2, 4, 6 and 12 months of therapy, 14 h after the last Clo dose. The 12-month evaluation was carried out 3-5 days after the last Clo dose. Height velocity was recorded at 2-month intervals using a Harpenden stadiometer.

Children were considered as responders (C-R) and nonresponders (C-NR) according to the increment in HV recorded at 6 and/or 12 months, the C-R group having an increase of at least 2 cm/yr.

Study 2

Eight children (six males and two females) with CGD aged 6 - 13 yr were also studied to evaluate the effect of chronic Clo administration on 24-h GH secretion. The main clinical and laboratory characteristics of this second group of children are shown in Table 2. All children were given Clo orally at the daily dose of 0.1 mg/m^2 at bed rest, and 24-h GH secretion studies were performed before and after 2 months of treatment which was continued for up to 6 months. Treatment was discontinued 36 h before the second 24-h GH study, and was reinstituted the day

Table 1. Clinical details of patients

Case No. Responders	Sex	Age (yr) Chron	Bone	Delay	Height (SDS)	Pubertal stage	HV (cm/yr) before Clo
1	F	11.1	8.6	-2.5	-3.5	P_0-P_2	2.0
2	F	11.3	9.1	-2.2	-3.5	P_2-P_4	7.5
3	F	7.2	3.2	-4.0	-4.8	P_0	4.2
4	M	11.6	10.1	-1.5	-2.0	P_0	3.5
5	M	9.1	7.3	-1.8	-2.8	P_0	5.0
6	F	10.5	9.0	-1.5	-4.6	P_0	3.7
7	M	14.5	10.2	-4.3	-3.0	P_2-P_4	2.5
8	M	8.5	7.5	-1.0	-3.9	P_0	0.8
9	M	15.2	11.6	-3.6	-5.2	P_2	2.4
10	F	13.2	11.7	-1.5	-3.6	P_0-P_1	4.0
11	F	11.3	10.2	-1.1	-2.0	P_2	4.2
12	F	6.2	5.2	-1.0	-2.8	P_0	3.2
13	M	9.7	9.1	-0.6	-2.0	P_0	3.6
14	M	13.1	9.5	-3.6	-2.0	P_1-P_3	3.7
15	M	6.2	3.8	-2.4	-3.8	P_0	3.8
16	F	13.5	11.2	-2.3	-4.4	P_2-P_4	5.2
17	M	5.2	2.6	-2.6	-5.0	P_0	2.5
18	M	5.8	3.5	-2.3	-5.0	P_0	2.2
19	F	9.4	6.1	-3.3	-2.4	P_0	2.7
20	M	7.0	—	—	-1.8	P_0	2.4
21	M	13.3	9.7	-3.6	-5.0	P_1-P_3	3.8
22	M	9.6	7.2	-2.4	-3.0	P_0-P_1	3.0
Mean ± SEM		10.1 (0.6)	7.9 (0.6)	-2.3 (0.2)	-3.4 (0.2)*		3.4 (0.2)†
Nonresponders							
23	F	9.8	8.1	-1.7	-3.8	P_1-P_2	6.0
24	F	11.2	8.6	-2.6	-2.9	P_0	3.1
25	M	8.2	6.6	-1.6	-4.1	P_0	4.5
26	M	5.1	—	—	-1.8	P_0	3.5
27	M	8.5	6.6	-1.9	-1.8	P_0-P_2	7.7
28	M	15.1	13.1	-1.8	-4.0	P_2	4.8
29	F	10.2	8.2	-2.0	-1.8	P_0	2.0
30	M	9.8	6.5	-3.3	-2.2	P_0	3.5
31	F	10.8	10.0	-0.8	-2.5	P_2-P_3	8.0
32	M	13.2	11.2	-2.8	-2.8	P_1-P_2	6.0
33	M	9.8	8.6	-1.2	-1.8	P_0	4.6
34	F	11.7	10.2	-1.5	-2.0	P_0	8.0
Mean ± SEM		10.2 (0.7)	8.9 (0.6)	-1.8 (0.2)	-2.6 (0.2)		5.5 (0.5)

Age, bone-age, and height SDS refer to data at the start of clonidine (Clo) therapy.
* Responders and non-responders significantly different ($p<0.05$) (unpaired Student's t-test).
† Responders and non-responders significantly different ($p<0.05$) (unpaired Student's t-test).

after; patient no. 2 received treatment for 2 months only. Basal Sm-C determinations and childrens' height measurements were carried out before and after 2 and 6 months of treatment. The IC-GH test was performed according to methods described elsewhere (Spiliotis et al. 1984; Zadik et al. 1985a,b). Blood was collected using a nonthrombogenic catheter and constant withdrawal pump, and the blood collection tubes were replaced every 20 min. Patient activity and food intake were not restricted during the 24-h period. After plasma separation, aliquots from each 20-min sample were combined producing a pool in which the IC-GH was determined. As described by Spiliotis and coworkers (Spiliotis et al. 1984), a GH pulse was identified as an increase of two or more consecutive GH values greater than 3 SD of the coefficient of variation of that assay. The mean pulse

Table 2. Main clinical and laboratory characteristics of the children studied in study 2

Case No.	Sex	Age (yr) Chron	Age (yr) Bone	Height (SDS)	Pubertal stage (Tanner)	GH peak[a] (ng/ml)	HV (cm/yr)
1	M	13.0	10.3	-2.0	P_1	27.5	2.0
2	M	12.3	10.6	-2.5	P_1	11.0	5.0
3	M	12.9	11.0	-2.2	P_1	15.9	2.5
4	F	12.3	9.6	-3.5	P_1	11.1	3.5
5	M	12.6	11.1	-2.3	P_1	15.1	4.0
6	M	6.0	4.0	-3.5	P_1	12.7	4.1
7	F	7.9	6.6	-2.0	P_1	10.0	1.0
8	M	9.5	8.0	-4.0	P_1	6.2[b]	3.0

Chron, chronological.
[a] GH peak after acute oral clonidine test (0.15 mg/m^2).
[b] GH peak of 17.0 ng/ml after insulin-hypoglycemia (0.1 U/kg iv).

amplitude was obtained by taking the mean of all GH peaks. For determining the number of pulses over 24 h, only GH peaks > 5.0 ng/ml were counted.

Assays. GH (Radim, Rome) and Sm-C (Nichols Institute Diagnostic, San Juan Capistrano, CA) were measured according to methods described elsewhere (Pintor et al. 1985, 1987; Loche et al. 1987).

Statistical Analysis. The statistical significance of the differences was evaluated using the paired or unpaired *t* test or the Wilcoxon matched-pair signed-rank test where appropriate. Linear regression analysis was used to correlate the clinical and the laboratory data. A discriminant analysis was done to obtain HV and SDS thresholds. All values are given as mean ± standard error of mean (SEM).

RESULTS

After 6 months of Clo therapy 25 of the 34 children (Table 1, nos. 1-5, 7-22, 25, 26, 30, 33) showed an increase in HV, and in 21 this was greater than 2 cm/yr (mean ΔHV 4.4 ± 0.5 cm/yr) (Fig. 1). Of the 22 children who were treated for 1 yr, the increment in HV was maintained in 13 (Table 1, nos. 1, 3, 4, 6-8, 11, 14-16, 19, 21, 22) (mean HV 3.4 ± 0.4 cm/yr; Fig. 1). In two children in the C-NR group (nos. 23 and 27) a clear-cut growth deceleration was evident with Clo. Withdrawal of Clo for 6 months in 14 C-R children resulted in unimpaired or only slightly reduced HV in 6 (nos. 1, 6, 7, 10, 14, 21), whereas in the remaining eight children (nos. 2, 3, 4, 8, 11, 15, 16, 22) HV returned close to or below pretherapy values. Overall, mean ΔHV during the 6 months off therapy in these children was 1.7 ± 0.8 cm/yr (Fig. 1). In patients nos. 4, 8, 11, 15, 16, and 22, reinstitution of Clo treatment caused a new increment in HV in four of them (mean ΔHV 4.6 ± 1.5 cm/yr).

Comparison of the pretreatment clinical indices in C-R and C-NR children revealed a significantly higher mean SDS and lower HV in the former that the latter (p<0.05 and p<0.01 respectively). The mean bone age was lower in C-R than in C-NR children. In the C-R children there was an inverse relationship between

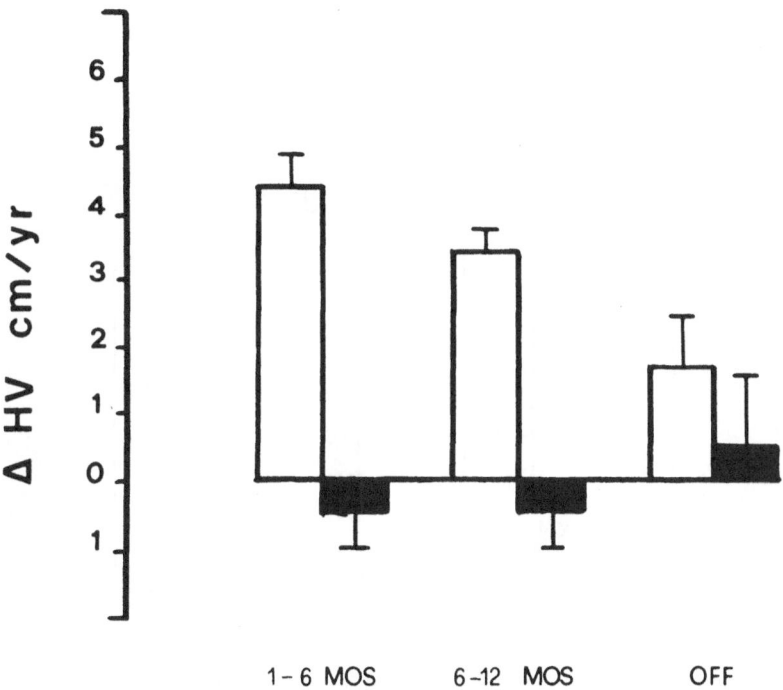

Fig. 1. Mean and SEM of ΔHV calculated over pretreatment HV in C-R (*open bars*) and C-NR (*closed bars*) children on and off clonidine treatment.

pretreatment HV and the increase in HV at 12 months ($r= -0.6$, $p< 0.01$). Discriminant analysis showed that a child could be classified with an 80% chance of being right as C-R when his/her HV and SDS were ≤ 4.6 cm/yr and ≤ -2.5 respectively. The probability of classifying as C-NR a child who may actually respond to Clo is then 20%.

In C-R children there was an initial rise in plasma GH (Fig. 2) and Sm-C levels during treatment (Fig. 3). At 12 months there was a further increase in plasma GH but not in Sm-C levels (Figs. 2 and 3). Peak GH responses to GHRH and Clo followed a similar trend (Fig. 2).

There was no significant difference in the frequency distribution of pubertal stages (Tanner) between C-R and C-NR.

In the second study Clo treatment for 2 months caused a significant ($p< 0.02$) augmentation of mean IC-GH which rose from 2.6 ± 0.4 to 4.6 ± 0.6 ng/ml (Table 3). The increase in IC-GH was mainly the result of increased GH-pulse amplitude which rose from 10.1 ± 1.1 to 15.9 ± 2.0 ng/ml ($p< 0.01$) (Table 3). The mean GH-pulse amplitude was significantly higher ($p< 0.02$) during sleep (17.6 ± 2.8 ng/ml) than during waking hours (10.1 ± 2.1 ng/ml) before treatment. After Clo

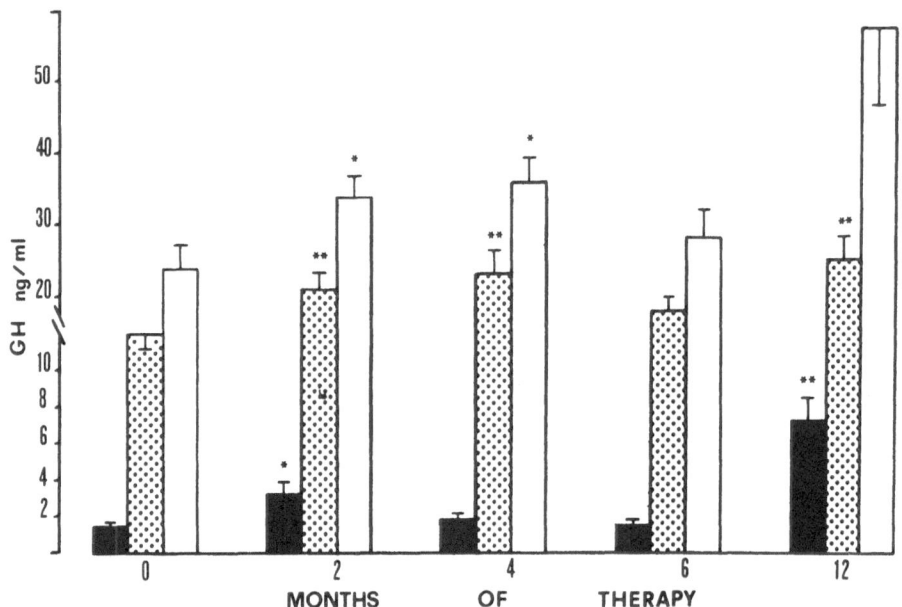

Fig. 2. Mean and SEM of baseline GH levels (*closed bars*), peak GH levels after acute clonidine (*shaded bars*), and GHRH (*open bars*) stimulation in C-R children before and during clonidine treatment. * $p< 0.05$, ** $p<0.01$ versus respective controls.

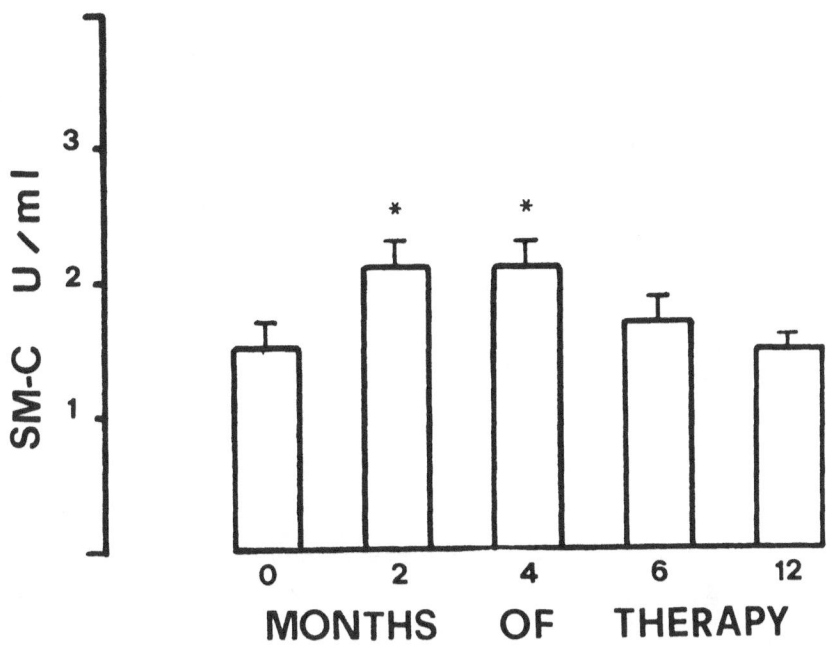

Fig. 3. Mean and SEM of plasma Sm-C levels in C-R children before and during clonidine treatment. * $p < 0.05$.

treatment, the mean GH-pulse amplitude during waking hours (15.6 ± 4.6 ng/ml) was not statistically different from than during hours of sleep (19.7 ± 1.7 ng/ml). Analysis of GH pulses during waking and sleep hours revealed that Clo did not affect GH-pulse amplitude in the same fashion in all subjects. In fact, GH-pulse amplitude increased mainly during waking hours in patients nos. 1, 4, 5, and 7, and mainly during sleep in patients nos. 2, 3, 6, and 8. In patient 3, GH-pulse amplitude during waking hours was reduced after treatment, and only the increase in nighttime GH secretion accounted for the slight augmentation of IC-GH. Individual plasma-GH profiles of patients nos. 1, 2, 6, and 7 during the 24-h period before and after Clo administration are shown in Figure 4. In the patients in this study Clo treatment caused a significant augmentation of HV which rose from 3.1 ± 0.5 to 10.2 ± 1.0 cm/yr after 2 months ($p < 0.001$) and to 7.0 ± 0.7 cm/yr after 6 months ($p < 0.02$). Only in three patients were Sm-C levels markedly increased after 2 months of treatment (Table 2, nos. 2, 4, and 8); however, a trend towards increased Sm-C levels was seen in all patiennts so that after 2 months mean Sm-C concentrations (1.4 ± 0.3 units/ml) were significantly higher ($p < 0.02$) than the corresponding pretreatment values (0.92 ± 0.24 units/ml) (Table 3) and remained so after 6 months (1.76 ± 0.43 units/ml; $p < 0.05$).

None of the children had noticeable side effects during treatment. Four children

Table 3. Effect of 2-month-Clo treatment on 24-h GH-pulse frequency, GH-pulse amplitude, IC-GH, and Sm-C levels

Case No.	GH-pulse frequency (n/24h)		GH-pulse amplitude (ng/ml)		IC-GH (ng/ml)		Sm-C (U/ml)		
	Before	After	Before	After	Before	After	Before	2 mos	6 mos
1	9	8	12.20	25.40	2.4	7.1	1.56	1.59	1.97
2	6	4	8.03	19.27	1.9	3.8	1.22	2.47	NA
3	6	4	11.80	13.70	4.2	4.7	0.39	0.55	0.62
4	7	5	7.80	9.90	2.3	2.8	0.60	1.60	1.90
5	7	8	10.60	16.90	3.6	6.3	0.60	0.90	3.30
6	6	9	4.90	9.00	1.1	3.6	0.20	0.30	0.28
7	7	7	13.90	19.90	4.0	5.4	1.51	1.70	1.65
8	4	6	12.10	15.30	2.6	3.4	1.39	2.10	2.60
Mean±	6.3	6.6	10.10	15.90*	2.6	4.6*	0.9	1.4**	1.76#
SEM	0.6	0.6	1.10	2.00	0.4	0.6	0.24	0.3	0.43

NA, not available; SEM, standard error of mean.
* $p < 0.01$; ** $p < 0.02$; # $p < 0.05$.

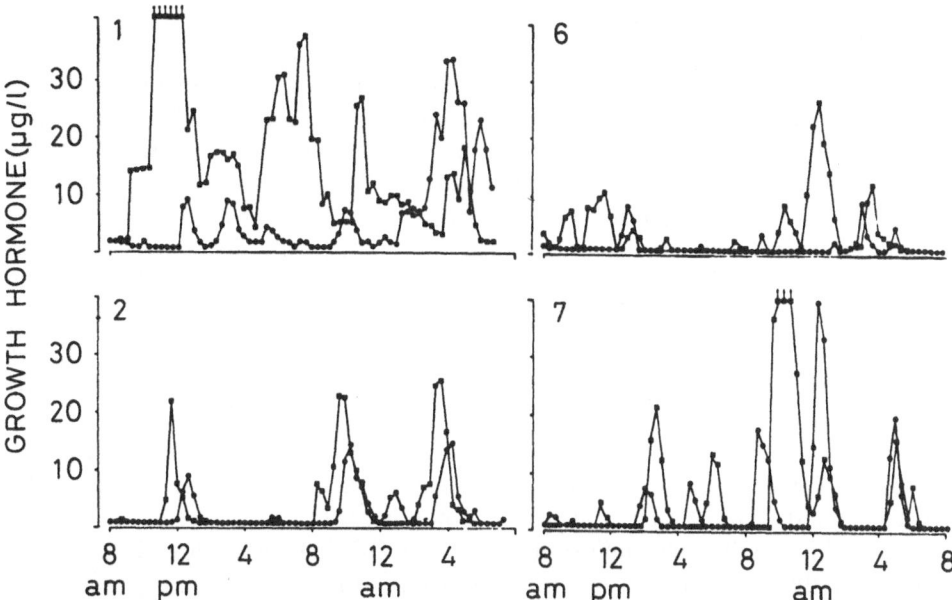

Fig. 4. Individual 24-h GH profiles of patients nos. 1, 2, 6, and 7 before (●) and after (▲) 2 months of clonidine treatment.

presented with drowsiness and sleepiness which diminished in three of them, despite continued use of Clo, and necessitated discontinuation of the drug in only one. Two children had a skin rash and urticaria and had to be excluded from the study. Acute Clo testing induced drowsiness or sleep and a slight reduction in systolic blood pressure both before and after chronic Clo treatment.

DISCUSSION

Clonidine is a potent GH secretagogue in both animals (Durand et al. 1977) and humans (Lal et al. 1975; Gil-Ad et al. 1979), and it acts via release of GHRH (Miki et al. 1984; Katakami et al. 1984). The positive response to Clo treatment seen in most of our patients indicates that these children may have a primary defect in the mechanism of GHRH release, but not in that of GHRH synthesis. The effect of Clo on linear growth in children with CGD is comparable to that of GH treatment in children with short stature (Van Vliet et al. 1983; Gertner et al. 1984).

Withdrawal of Clo after 12 months of treatment resulted in deceleration of linear growth to pretherapy or lower values in eight C-R children, but the enhancing effect of Clo on linear growth was maintained in the remaining six children. These findings are difficult to interpret; we suggest, however, that the children who continued to grow while not taking Clo had less severe impairment of hypothalamic adrenergic neurotransmission than the remaining C-R children, since pharmacological stimulation with Clo re-established normal neurotransmitter function. This response is similar to the maintainance of increased HV seen in CGD children after withdrawal of anabolic steroids (Bettmann et al. 1971), and fits in well with the transient impairment of GH secretion in some of these children (Gourmelen et al. 1979). Reinstitution of Clo for 6 months triggered a new growth increment in four of six children in whom Clo withdrawal had dramatically decreased HV.

The results of the 24-h GH secretion studies have confirmed the ability of Clo to accelerate linear growth in children with CGD. In our first study as well as in the study of Castro Magana et al. (1986), the increase in linear growth was accompanied by an increase of the plasma-GH response to acute Clo or GHRH stimulation.

A large body of evidence indicates that a normal GH response to a pharmacological stimulus does not prove per se that spontaneous GH secretion occurs (Chalew et al. 1986; Bercu et al. 1986). Spiliotis et al. (1984) recently found that some children with the clinical features of CGD may have low IC-GH, and suggested the term «neurosecretory dysfunction» to indicate that a defect in the central regulation of GH secretion is the underlying endocrine abnormality. Children connoted as having neurosecretory dysfunction or CGD respond well to treatment with exogenous GH (Spiliotis et al. 1984). Six subjects of the second study showed pretreatment IC-GH levels lower than 3 ng/ml, and, according to Spiliotis et al. (1984), they might fit into the category of children with «neurosecretory dysfunction». Consistent with our findings, a decreased mean GH secretion over a 5.5-h night study has been reported by Bierich (1986) in children with CGD. Whatever the nature of their growth retardation, Clo was capable of increasing IC-GH levels in all children studied, and increased mean baseline Sm-C levels, thus providing a basis for the understanding of the growth acceleration induced by the drug. When the pattern of GH-secretory episodes was analyzed, the increased mean IC-GH could be accounted for by an increased GH pulse amplitude, while GH pulse frequency was affected by treatment neither during waking nor during sleeping hours.

The pattern of GH secretion elicited by chronic Clo treatment in children with CGD strongly resembles that evoked by spontaneous puberty in normal adolescents or by sex-steroid-induced pubertal maturation in children with CGD. In both instances, the augmentation of GH secretion is a pulse amplitude-modulated phenomenon, independent of changes in pulse frequency (Link et al. 1986; Mauras et al. 1987). In all, these observations reinforce the view that the mechanisms triggered by spontaneous or induced sexual development and by α_2-adrenergic stimulation are the same or are remarkably similar. Sex steroids are known to affect brain catecholaminergic function (Engel et al. 1979). Interestingly, the GH response to α_2-adrenergic stimulation by Clo is drastically reduced in the male rat after castration, and restored to normal by testosterone replacement therapy (Jansson et al. 1982). It is therefore likely that sex steroids, as well as Clo, may interact with catecholaminergic pathways stimulatory to GHRH release resulting ultimately in an increased amplitude of GHRH secretory pulses. These observations are consistent with the view that a functional immaturity of neurotransmitter (catecholaminergic) pathways for the control of GH secretion (which is improved by Clo or remits spontaneously at puberty or pharmacologically under gonadal steroid treatment) may be responsible for the transient hypopituitarism of children with CGD.

Some clinical features appear to be useful in identifying the children with CGD who will respond to Clo; the responders have a clearly higher height SDS, a lower HV, and a greater delay in bone age than do nonresponders. Similar observations have been made in short stature children given hGH replacement therapy (Van Vliet et al. 1983). In our study, as already stated, discriminant analysis showed that values of HV and height SDS of ≤ 4.6 cm/yr and ≤ -2.5 cm/yr respectively, could predict responsiveness to Clo in 80% of the children.

Huseman and coworkers (Huseman 1985; Huseman and Hassing 1984; Huseman et al. 1986) who reported stimulation of linear growth in some hypopituitary children treated with L-dopa or bromocriptine, ascribed GH deficiency and short stature in these subjects to hypothalamic dopaminergic dysfunction. However, in their studies, the greater effectiveness of L-dopa (the precursor of both dopamine and noradrenaline) compared with bromocriptine (a specific dopamine agonist) makes impairment of noradrenergic function more likely. Our findings with Clo support this proposition. Whichever brain neurotransmitter is primarily involved, it seems very likely that many children with growth disorders have a primary neurosecretory dysfunction, and thus may benefit from a therapeutic approach with CNS-acting compounds.

Recently, two studies appearing in abstract form have reported the poor ability of Clo to stimulate linear growth in children with CGD (Bernasconi et al. 1987; Cassio et al. 1987; see also Milner, this volume). Due to the limited number of children studied and the considerable variability of the results, however, it is not possible to reach any firm conclusion.

Further studies will indicate whether prolonged therapy with Clo will induce only a temporary acceleration of longitudinal growth or allow a higher adult height, and whether this approach is suitable for children who are suffering psychologically from their small stature but in whom the growth defect should be only temporary (Stabler and Gilbert 1987). However, irrespective of these problems, our studies confirm that many of these children have a low level of GH secretion, and clearly show that Clo can restore to normal the secretory defect in GH secretion, thus resulting in acceleration of linear growth.

Acknowledgment

We are very grateful for the participation in these studies of Miss Anna Carla Muntoni and Miss Liliana Stabilini.

REFERENCES

Bercu BB, Shulman D, Root AW, Spiliotis BE (1986) Growth hormone (GH) provocative testing fequently does not reflect endogenous GH secretion. *J Clin Endocrinol Metab* **63**: 709-714

Bernasconi S, Ghizzoni L, Romanini F, Volta C, Virdis R, Giovannelli G (1987) Clonidine treatment in children with constitutional growth delay (CGD). Eur Soc Paed Endocr 26th Meeting, Toulouse (abstract 122)

Bettman HK, Goldman HS, Abramowicz K, Sobel EH (1971) Oxandrolone treatment of short stature: effect on predicted mature height. *J Pediatr* **79**: 1018-1023

Bierich JR (1986) Treatment by hGH of constitutional delay of growth and adolescence. In: Vicens-Calvet E, Flodh H (eds) *Recombinant Human Growth Hormone and Related Growth Factors. Acta Paediatr Scand* [Suppl] **325**: 71-75

Cassio A, Stefanini C, Tassoni P, Tonini P, Tassinari D, Zuchini S, Cacciari E (1987) Treatment with clonidine in subjects with constitutional growth delay (CGD): a controlled double blind study. Eur Soc Paed Endocr 26th Meeting, Toulouse (abstract 121)

Castro Magana M, Angulo M, Fuentes B, Castelar ME, Canas A, Espinoza B (1986) Effect of prolonged clonidine administration on growth hormone concentrations and rate of linear growth in children with constitutional growth delay. *J Pediatr* **109**: 784-787

Chalew SA, Armour KM, Levin PA, Thorner MO, Kowarski AA (1986) Growth hormone (GH) response to GH-releasing hormone in children with subnormal integrated concentrations of GH. *J Clin Endocrinol Metab* **62**: 1110-115

Engel J, Ahlenius S, Almgren O, Carlsson A, Larsson K, Sodersten P (1979) Effect of gonadectomy and hormone replacement on brain monoamine synthesis in the male rat. *Parmacol Biochem Behav* **10**: 149-154

Gertner JM, Genel M, Gianfredi SP, Hintz RL, Rosenfeld RG, Tamborlane WV (1984) Prospective clinical trial of human growth hormone in short children without growth hormone deficiency. *J Pediatr* **104**: 172-176

Gil-Ad I, Topper E, Laron Z (1979) Oral clonidine as a growth hormone stimulation test. *Lancet* **ii**: 278-280

Gourmelen R, Phaam-Hu-Trung MT, Girard F (1979) Transient partial hGH deficiency in prepubertal children with delay of growth. *Pediatr Res* **13**: 221-224

Grossman A, Savage MO, Wass JAH, Lytras N, Sueiras-Diaz J, Coy DH, Besser GM (1983) Growth hormone-releasing factor in growth hormone deficiency: demonstration of a hypothalamic defect in growth hormone release. *Lancet* **ii**: 137-138

Guillemin R, Brazeau P, Böhlen P, Esch P, Ling N, Wehrenberg WB (1982) Growth hormone releasing factor from a human pancreatic tumor that caused acromegaly. *Science* **218**: 585-587

Huseman CA (1985) Growth enhancement by dopaminergic therapy in children with intrauterine growth retardation. *J Clin Endocrinol Metab* **61**: 514-519

Huseman CA, Hassing JM (1984) Evidence for dopaminergic stimulation of growth velocity in some hypopituitary children. *J Clin Endocrinol Metab* **58**: 419-425

Huseman CA, Hassing JM, Sibilia MG (1986) Endogenous dopaminergic dysfunction: a novel form of human growth hormone deficiency and short stature. *J Clin Endocrinol Metab* **62**: 484-490

Jansson J-O, Eriksson E, Edèn S, Modigh K (1982) Effects of gonadectomy and testosterone replacement on growth hormone response to α_2 adrenergic stimulation in the male rat. *Psychoneuroendocrinology* **7**: 245-248

Katakami H, Kato Y, Matsushita M, Imura H (1984) Effects of neonatal treatment with monosodium glutamate on growth hormone release induced by clonidine and prostaglandin E₁ in consciuos male rats. *Neuroendocrinology* 38: 1-5

Lal S, Tolis G, Martin JB, Brown GM, Guyda J (1975) Effect of clonidine on growth hormone, prolactin, luteinizing hormone, follicle-stimulating hormone and thyroid-stimulating hormone in the serum of normal men. *J Clin Endocrinol Metab* 41: 827-832

Link K, Blizzard RM, Evans WS, Kaiser DL, Parker MW, Rogol AD (1986) The effect of androgens on the pulsatile release and the twenty-four-hour mean concentration of growth hormone in peripubertal males. *J Clin Endocrinol Metab* 62: 159-164

Loche S, Cappa M, Borrelli P, Faedda A, Crino A, Cella SG, Corda R, Müller EE, Pintor C (1987) Reduced growth hormone response to growth hormone releasing hormone in children with simple obesity: evidence for somatomedin C mediated inhibition. *Clin Endocrinol* (Oxf) 27: 145-153

Mauras N, Blizzard RM, Link K, Johnson ML, Rogol AD, Veldhuis JD (1987) Augmentation of growth hormone secretion during puberty: evidence for a pulse amplitude-modulated phenomenon. *J Clin Endocrinol Metab* 64: 596-601

Miki N, Ono M, Shizume K (1984) Evidence that opiatergic and alpha-adrenergic mechanisms stimulate rat growth hormone release via growth hormone releasing factor (GRF). *Endocrinology* 114: 1950-1952

Pintor C, Fanni V, Loche S, Locatelli V, Villa F, Cella SG, Minuto F, Corda R, Müller EE (1983) Synthetic hpGRF 1-40 stimulates growth hormone and inhibits prolactin secretion in normal children and in children with isolated growth hormone deficiency. *Peptides* 4: 929-933

Pintor C, Cella SG, Corda R, Locatelli V, Puggioni R, Loche S, Müller EE (1985) Clonidine accelerates growth in children with impaired growth hormone secretion. *Lancet* i: 1482-1485

Pintor C, Loche S, Corda R, Cella SG, Puggioni R, Locatelli V, Müller EE (1987) Clonidine treatment for short stature. *Lancet* i: 1226-1230

Rivier J, Spiess J, Thorner MO, Vale W (1982) Characterization of a growth hormone releasing factor from a human pancreatic tumor. *Nature* 300: 276-278

Schriock EA, Lusting RH, Rosenthal AL, Kaplan SL, Grumbach MM (1984) Effect of growth hormone (GH) releasing hormone (GRH) on plasma GH in relation to magnitude and duration of GH deficiency in 26 children and adults with isolated GH deficiency or multiple pituitary hormone deficiencies: evidence for hypothalamic GRH deficiency. *J Clin Endocrinol Metab* 58: 1043-1049

Spiliotis BE, August GP, Hung W, Sonis W, Mendelson W, Bercu BB (1984) Growth hormone neurosecretory dysfunction. A treatable cause of short stature. *JAMA* 251: 2223-2230

Stabler B, Gilbert MC (1987) Psychological effects of growth delay. In: Hintz RL, Rosenfeld RG (eds) *Growth Abnormalities.* Churchill Livingstone, New York, pp 225-274

Thorner MO, Rivier J, Spiess J, Borges JEC, Vance ML, Bloom SS, Rogol AD, Cronin MJ, Kaiser DL, Evans WS, Webster JD, MacLeod RM, Vale W (1983) Human pancreatic growth hormone releasing factor selectively stimulates growth hormone secretion in man. *Lancet* i: 24-28

Van Vliet G, Styne DM, Kaplan SL, Grumbach MM (1983) Growth hormone treatment for short stature. *N Engl J Med* 309: 1016-1022

Zadik Z, Chalew SA, McCarter MJ, Meistas M, Kowarski AA (1985a) The influence of age on the 24-hour integrated concentration of growth hormone in normal individuals. *J Clin Endocrinol Metab* **60**: 513-516

Zadik Z, Chalew SA, Raiti S, Kowarski AA (1985b) Do short children secrete insufficient growth hormone? *Pediatrics* **76**: 355-360

Discussion

J.R. Bierich: I will restrict myself to only one question, i.e., that dealing with who should be treated with growth hormone. Of course, classical GH deficiency includes the children undergoing irradiation discussed by Dr. Rappaport. To complement what has been said by Dr. Takano, I will describe only a few data from the Genotropin studies we are now running and on which we have recently reported in Vienna. They deal with 49 patients with hypopituitarism who were never treated before; it could be seen that during the first 3, 6, 9, and 12 months of treatment the height velocity increased from less than 5 cm/yr, namely 3.7 cm/yr to 15 cm/yr for the first 3 months and to 12 cm/yr for the whole year. This increase in height velocity is greater than that shown by Dr. Takano, but, reportedly, German children grow faster than Japanese children. We then treated 28 patients with GH and these children also had a good GH response, 5.6 cm/yr before treatment increasing to 10.7 cm/yr and 8.4 cm/yr for the whole year. We were surprised at these good results in these patients. In regard to the antibody formation, i.e., anti-ECP, it can be seen that the original plasma titers at the onset of therapy did not change during treatment; there was also no increment in the formation of antibodies against GH. Nonclassical GH deficiency consists of a variety of GH disturbances; in this vein I will consider only some aspects dealing with constitutional delay of growth. We have had two series published in the past few years, and I will discuss some of these cases and then summarize them. The first is a typical case; his growth velocity was 2-4 cm/yr before therapy and went up under treatment with GH (12 IU per week) from 2 to 8 cm/yr. This is, of course, an initial effect, but the next case shows it is maintained further; in this instance, the growth velocity rose from 3.5 to 14.5 cm/yr, before the pubertal spurt. The child has continued to grow for up to 2 years, and that indicates that perhaps his final height will be improved; shifted to a higher percentile growth. Among our cases were two children who are now adults; they gained 4.5 cm each, following the prognosis for final height made before treatment, and have now reached their final height. I am not saying that constitutional delay of growth has to be treated with GH–I am far from wanting to make this claim– but these children can be treated successfully, and it is a purely psychological question of whether to treat them or not. My last comment concerns a group of patients who have been previously mentioned in my lecture, those children who are treated over the long term with corticosteroids. They have no growth hormone; they have only the catabolic action of the corticosteroids but no anabolic hormones. To prevent osteoporosis, which many of them had in a very severe form, and to prevent dwarfism we started treating them with 3 IU GH per week.

R.D.G. Milner: Thank you. Now let me open the discussion. I propose that we start off by asking you to focus on whom should be treated. We have heard about classical GH deficiency and I think we can leave this group of patients out of the discussion, largely, unless there is anything specific; but certainly there is a group which has a variety of names, and then there is Turner's syndrome and there are specific groups of short children, and you may want to make comments on these types of children. Let us start now with: who should be treated?

R. Rappaport: I am not sure it is really time to ask such a question: Who should be treated? I mean, who should enter the trial for treatment? Because if you say, «Who should be treated?» already we have some clear view of who will benefit from treatment in the short or the long term, and we have a long way to go before we can really pose that type of question. I will make a few comments. First, I believe that we have to decide which group of patients are fit for therapeutic trials in the short term or over several years, and I think to start with I will leave out, and I am sure that Dr. Bierich will largely agree, the patients with constitutional delay of growth. It is clear they have been lacking GH for some time, I do not know how long, and it is clear that treating them with low doses of steroids, such as estradiol or equivalent ethynylestradiol or testosterone, does not impair final height and is quite efficient, probably inexpensive, and much easier to handle. Now if we come to short children, of course what we first want to see is if we can obtain results in very short children, and I think here we have to set strict criteria, whatever they would be, e.g., -2.5-3 SD, for selecting this population and go ahead and do trials. Another point we will face is what will we call a good result on average or in individual cases, and how many centimeters will be, in the short term, let's say for 2 or 3 years, significant for changing the image of the child; besides, this is something that will affect its future and eventually its behavior, through childhood and adolescence – sufficiently increased stature will help him psychologically. This is something that surely will come up at some time; at present we have no good tools for evaluating this, but we should not forget about it. For the meantime, I leave the discussion to others.

R.D.G. Milner: There may well be a body of opinion in this audience; that is to say, as for normal variant short stature, constitutional growth delay, etc, the trials have been done, the literature is extensive. The question is, do you accept the results of the literature in your own clinical practice, or do you really believe that truth is not truth until it is published in the language of your country, based on a child who comes from your country. And with respect to what is a good result, I was particularly impressed by Dr. Savage's point: he said that we need a height velocity change of better than 2 cm/yr, measured over the whole year, before we continue therapy. If it is less than that I am prepared to discontinue therapy after 1 year.

E.E. Müller: I would like to make a comment related to that of Dr. Rappaport. He was referring to the patients who should be treated and he mentioned constitutional delay of growth. But he soon switched to: What type of treatment for these patients? and he did mention GH and gonadal steroids. From the evidence existing in the literature, it would seem that children with constitutional delay of growth have a transient impairment in the ability to release GH after provocative stimuli (Gourmelen et al. 1979), and these stimuli, reportedly, act at the level of the CNS. Thus, why not consider the possibility, already alluded to by Dr. Pintor in his presentation, of the existence in these children of a defect in neurotransmitter, namely noradrenergic, function which, interestingly, is modified by gonadal steroids? (Engel et al. 1979). Thus, I think that from a more strictly etiological viewpoint, an approach with neuroactive drugs acting upstream from the GHRH-secreting neurons may be very appropriate, and I ask the pediatricians present in this audience to start thinking in practical terms of a neurotransmitter approach, when therapy is considered to be appropriate, more often than they have previously done. We have heard, in these past few days, repeated mention of «neurosecretory dysfunction»; what is the meaning of this term, which in my mind does not connote a clinical entity but rather a physiopathological mechanism? It means something that we are now capable of facing with the use of neuropharmacological tools. Incidentally, when we treat some of these children with GH it is possibile that we are not doing them any good. It is well known that there is GH feedback at the hypothalamus and possibly at the pituitary level, so it is conceivable that under sustained GH therapy their endogenous GHRH-GH function may be impaired.

R.D.G. Milner: There are many reasons why I want to stretch the discussion, but first I want to get back to the question of who should be treated, and this contribution from Dr. Müller is appropriate, since again it focuses the clinician's mind on whom we should treat. Here is an alternative modality of treatment for a particular group, i.e., constitutional delay of growth, normal variant short-stature, and I would like to hear what people have to say about it.

P.J. Smith: I just want to go back a little bit to who should be treated. I think we have to remember the work Tanner did some years ago: it has now been confirmed that the response depends on the pretreatment height velocity. So if you are beginning to treat children who are less insufficient or deficient, whatever word you want to use, you cannot expect as great a response as in the child who is classically GH deficient. Also it depends on how you give GH, the frequency of GH administration – which has been already mentioned – and the dose per week you are giving.

R.M. Blizzard: I would like to speak against the use of GH.

R.D.G. Milner: On economic grounds?

R.M. Blizzard: To a great extent, but also I think on a scientific basis. I am not totally against the use of GH, obviously, but I believe that it has been demanded by parents, as Dr. Milner has already said, and promoted by companies, as it has been observed that GH is a panacea that can do a lot of things for a lot of children. I do not happen to believe that this is the case. I believe that we should be exceedingly cautious as clinicians, and I am speaking as a clinician now, on the use of GH; we should rely upon control studies to determine what patient should and what patient should not be treated and these control studies should not cover 6 months or a year but longer periods of time. I say this with certain scientific observations in mind: First, we all know that in the treatment of Turner's syndrome, to use an analogy, Anovar works well in the first year and increases growth velocity, and it works reasonably well in the second year in some patients, increasing the growth velocity over that of pretreatment, but usually decreases it compared with the first year. We know that in the third year Anovar may as well not be used in most patients with Turner's syndrome, and that what we have accomplished is a modest increase in growth velocity; we do not even know yet if we have increased the ultimate height in these patients. So, on the basis of a 6-month or 12-month report, to state that normal variant short-stature children or short children of some or all categories will respond to GH is, I think, a mistake. Therefore, I think Dr. Rappaport and I concur in asking what kind of patients should be put into studies that will provide short-term and long-term data, which in turn will permit us to move ahead as scientists; I think all of us here regard ourselves as such, even if our time is spent primarily on clinical work.

J.J. Heinrich: I am afraid we are forgetting some factors, mainly some subtle effects of minor undernutrition and psychological aspects which are not usually mentioned in this kind of trial. Perhaps we should introduce some psychological measures to compare groups of children treated with a physiological approach with those who are just given a little more food. Those of us who live in developing countries know that some of those children who can improve if given clonidine may also improve if given a few more calories, and maybe we should introduce some control groups of this type.

R.D.G. Milner: I think a very important point has been made. I would not want the audience to feel that there were an incompatibility between what I said and what Dr. Rappaport was saying. While one accepts Dr. Blizzard's and Dr. Rappaport's concept about the way to advance knowledge, it cannot have any bearing on the consultation we may have tomorrow. In that respect, Dr. Heinrich, I think your point is well taken, because unless the pediatrician or the general practitioner sees the small child, looks at the whole person, he will not actually get to the solution of that particular person's problem, which could in some cases be occult celiac disease or inflammatory bowel disease or renal failure masquerading as growth delay.

M.O. Savage: I just want to complement what Dr. Blizzard said. We know very little about the long-term potential metabolic complications of growth hormone treatment. In a recent paper in the *British Medical Journal*, Drs. Hindmarsh and Brook showed that fasting insulin levels were significantly elevated in normal short children during GH treatment. I really think that this point should be addressed. We do not know anything about the long-term effects of fasting insulin, the acquisition of insulin resistance, or the reversibility of insulin resistance. I think that some careful studies should be performed on insulin dynamics, looking at insulin resistance, particularly reversibility, in children who are given GH.

M. Cattaneo: I want to know if there are any data on surfactant maturation in children with intrauterine growth retardation and on the possible effects of GH treatment.

R.D.G. Milner: I do not know of any information in the literature on that point. There is nobody else on that specific point? No, I do not see any hands. I'm sorry, I know this is not a satisfactory answer. I would like to return to a previous point. Dr. Rosenfeld probably has the most information on the use of GH therapy in Turner's syndrome, and Dr. Takano probably has the most experience in terms of numbers. I have two specific questions. One, would all Turner's patients respond to GH, or does the spectrum of phenotypes within Turner's syndrome suggest that there are some patients who will not respond? Two, would you just say a word or two about the GH-oxandrolone combination and the change in bone age based on the most up-to-date information?

Dr. Rosenfeld: We are now, in the U.S.A., in the fourth year of a multicenter study initially involving 70 girls with Turner's syndrome and now including more than 150 girls. The observations at this point are quite incomplete because all the girls are still growing and most of them have not yet reached their adult height, but I think that we can make several conclusions already on a few points. Number one, whatever the reality about spontaneuous GH secretion in girls with Turner's syndrome — and we have heard a lot in meetings about decreases and integrated GH secretion abnormality in response either to provocation testing or possibly to GHRH, whatever the reality of that situation is, these children do not behave like GH-deficient children, they do not respond to GH therapy the way classical GH hypopituitary children do. In the Japanese experience, the average growth rate for the first year of therapy was 5.7 cm/yr; in our experience it was 6.6 cm/yr during year 1, 5.5 cm/yr during year 2, and 4.7 cm/yr during year 3. Now, in all 3 years the growth rate was significantly greater than that for the control group during the 1-year control period, and this is true whether you express data as cm/yr or as standard deviation for either chronological age or bone age. There is no doubt in my mind that GH therapy alone, at least for a 3-year period, will cause some significant increase in growth velocity. I think it is important to agree that these children should not be expected to grow the way that simple hypopituitary children grow. They just do not grow that way, and my only bias is that Turner's syndrome does represent some form of skeletal dysplasia. The second question had to do with the combination of growth hormone and oxandrolone. The data now for the 3 years of this study is that during year 1 the children who received the combination GH-oxandrolone grew 9.6 cm/yr; during year 2, despite the fact that we lowered the oxandrolone dose by about 50%, they still had a very good growth rate of about 7.8 cm/yr, and during year 3 it was 6.8 cm/yr. So the combination of GH and oxandrolone does work better, at least in the short-term, than GH alone We do not have all the 3-year bone-age data back yet, but those data that we have so far indicate that — at least using the Bailey-Pinneau prediction of adult height, with all the limitations of that method — the combination therapy does seem to be superior to GH alone or to oxandrolone alone in terms of predictive adult height. But again, these are only 3-year data and they need to be taken with a grain of salt; I absolutely agree with the comments of both Dr. Blizzard and Dr. Rappaport that we need long-term data before we can make any conclusion. And this brings me to the last point I wanted to make, which was touched on just recently and which is a problem near and dear to the Food and Drug

Administration in the U.S.A.: i.e., What should be the criteria for determining efficacy of treatment? I think this is particularly important because this is in large part a pediatric audience and I think we need to come to the conclusion that therapy, i.e., growth-promoting therapy, whether it be clonidine or GHRH or GH, really has two potential benefits; one is an alteration in the final adult height but the other is short-term growth acceleration, and there is no question in my mind that there is a very real therapeutic benefit to short-term growth acceleration, which does not mean that final adult height is not important; the ideal therapy will give us both, but we need to critically evaluate the role of short-term growth acceleration and we do need psychological testing very much. I think it has been a glaring deficiency in the studies that have been done and, I am sorry to say, also in many of the other studies that people in the audience have been involved in, that we have neglected the opportunity to do very careful psychological assessments of children whether they be GH-deficient children, or children with normal short stature, or Turner's syndrome patients.

G. Costin: I would also like to talk about what Dr. Blizzard and Dr. Rosenfeld said regarding the control studies of so-called normal variant short children and Turner's children. To assess whether the treatment that we are now trying to promote is really good and whether we should give it, I want first to draw attention to the fact that making the diagnosis of neurosecretory growth deficiency based on a 24-h growth hormone profile or concentration of < 3 ng/ml, as suggested in previous work, is probably a mistake. We have looked at about 40-42 children of normal stature who had mean GH concentrations between 1.9 and 6.0 ng/ml for 24 h, and those children had normal growth velocity and no illness and had absolutely no problem; you could look at them and say that they were normal children. Also, it was alluded to previously that what we measure, what I measure, as 5 ng/ml of GH may be 7 ng/ml in someone else's laboratory. So we really cannot compare results from different laboratories to make a diagnosis on the basis of which we are going to treat those children, and I think that unless we all use the same radioimmunoassay to measure GH, we really cannot propose treatments based on someone else's data. I would be very careful in proposing treatments based on these 24-h profiles; they are very interesting in terms of the physiology of GH secretion, but certainly, they are far from being a good diagnostic tool. I also would like to make a comment on the point raised by Dr. Heinrich regarding calories. We really tend to forget how important nutrition is for the growth of children, and this applies not only to normal variant short stature but even to GH-deficient children. At least in my experience, children who were thin, who were picky eaters, responded less to treatment even if they were GH deficient than did the children who ate better, and if you check the insulin levels you'll see that there is a discrepancy between insulin levels in children growing better and those in children growing less; this means that we need to come back to nutrition, in addition to all the other criteria that we now propose.

R.D.G. Milner: Thank you. I think the last point is one that has been experienced by many of us; we have a patient with overt organic GH deficiency on treatment, and then he or she gets anorexia nervosa then stops growing. I had forgotten to mention that.

Can we switch now to the second half of the discussion? Let's concentrate not so much on who should be treated, but, accepting that the treatment is needed, on what the treatment should be. Let us focus on the less conventional modalities of treatment; perhaps you may want to put questions or make comments specifically to Dr. Pintor and Dr. Savage, related to neuroendocrine drugs and to GHRH.

P.J. Smith: Just a comment on GHRH to add a little bit more to the discussion. Over the past 2 years we have been treating 16 children with two regimens of GHRH, using both $GHRH_{1-40}$ and $GHRH_{1-29}$ peptides. Our results show that we have a 73% success rate in increasing growth velocity of over 2 cm/yr, if we use that definition. But the pulsatile regimen, either the 12- or 24-h regimen, is a more efficient use of the peptide than the only twice-daily regimen. So we could use half the daily dose in a pulsatile manner, giving 4–8 pulses a day; then we could give the same dose twice daily. And so, if we have to think of different peptides, we have to decide on the regimen we use. A final point is the intravenous studies. I think it is very difficult to use an intravenous GHRH study to determine which child you are going to use this new type of therapy for, particularly with assay problems, sensitivity and specificity, or whether you use a probability theory on how you are going to use predictors.

R.D.G. Milner: Do you have any difference in pump compliance from bid injection regimen compliance?

P.J. Smith: No, there seems to be good compliance with both regimens.

R.D.G. Milner: It is not a decision you would like to make in the context of treating a diabetic child; it is quite a serious therapeutic consideration. Any further comment about clonidine?
     May I ask you, Dr. Pintor, how you consider, or whether you are considering, within the Department the use of pyridostigmine plus or minus clonidine? This question is related to the communication we received, I think yesterday (Camanni et al., this volume)

C. Pintor: I think that the combined use of clonidine and pyridostigmine may represent a better association for the future treatment of children.

R.D.G. Milner: So we know what clonidine will do, we know what pyridostigmine will do; the possibility then arises of a clonidine-pyridostigmine combination. Have you started?

C. Pintor: Not yet.

E.E. Müller: Let me comment a little bit on this point. There are some very preliminary results that suggest that the combination of clonidine and pyridostigmine is not better than clonidine or pyridostigmine alone. To support these results there are some data from rats that show that clonidine is capable of decreasing acetylcholine function in many brain regions, including the hypothalamus (Buccafusco 1984); so from a strictly pharmacological viewpoint, this drug association would be meaningless. We think that the neurotransmitter defect present in many of these short-stature children is not confined to the noradrenergic system but may involve many neurotransmitters and possibly also neuropeptides. So, possibly in the future we will know more about these neurotransmitter defects and we will be able to develop more specific, discrete neuropharmacological approaches to the treatment of these subjects. For instance, the group in Turin (Camanni et al., this volume) have some children of short stature under treatment with pyridostigmine, and one or two of these children are actively growing because of the action of this compound. It means that in these cases there was no major alteration of the noradrenergic neurotransmission; likely, a disturbance related to cholinergic function was involved. Interestingly, in two of the clonidine-nonresponder children alluded to previously by Dr. Pintor there was an initial good response to administration of prydostigmine, which reinforces the proposition promulgated above.

R.D.G. Milner: An absolutely fascinating matter.

E. Ghigo: Concerning pyridostigmine and clonidine, in our experience with associating the drugs, we have been unable to show an increase in GH responsiveness, the reason probably being that alluded to previously by Dr. Müller, i.e., an inhibitory effect of clonidine on cholinergic neurons. For this reason, I think the therapeutic association of clonidine with pyridostigmine will be a difficult one[1].

J. Marek: We have some experience with clonidine. During the first 6 months, children with constitutional delay of growth did very well under clonidine, but during the second 6- month period the growth rate decreased and was less than that reported here during the lecture of Dr. Pintor. So I wonder if there will be a positive effect on the final height of these children. There was one substantial difference in our experience: the height response after 1 year was positive in only a few children.

L. Cavallo: A question to Dr. Pintor, about the bone maturation observed after 1 year of clonidine treatment, and a general comment about clonidine therapy and GH therapy in Turner's syndrome. It has been said that there was, in these instances, a fair improvement in height for a short period, and this is a psychologically positive effect, but each of these therapies shows a progressive decrease in improvement. So it may possibly be that the final height will not actually be better than the final height of an untreated subject. Now, in a person who has been injected and tested for some years with the aim of improving his or her final height and has not really grown as we had hoped, do you think that the psychological effect would really be a good one?

---

[1] Subsequent investigations have shown, however, that in adult volunteers combined clonidine-pyridostigmine administration results in an additive effect on GH release, though with disturbing side effects (Ghigo et al., unpublished results).

R.D.G. Milner: Your point is well taken, as one clinician listening to another clinician, I say your point is very well taken.

R.J.M. Ross: To return to the concept of using pyridostigmine as an adjuvant to therapy, we have been using this drug in combination with GHRH. Our feeling is that the use of GHRH as a twice-daily injection has a few advantages over the use of GH. We think that the development of a depot preparation could have considerable advantages, and, with a view to this problem, we perform subcutaneous infusions with GHRH and have seen augmentation of endogenous pulses of human GH release. The addition of pyridostigmine to subcutaneous infusion of GHRH greatly augments the endogenous pulses of GH release, and if you have a monthly depot preparation of peptide, the once-nightly administration of pyridostigmine will greatly augment its growth-promoting effect. It is something that we have to investigate for the future.

R.M. Blizzard: I want to follow up on what Dr. Ross said with respect to the use of atenolol, a β_1-adrenergic receptor antagonist, in patients who are treated with GHRH. Atenolol will enhance GH release as pyridostigmine apparently does and that may be an augmentative phenomenon which will have an advantage for the future.

May I ask Dr. Takano one question? It deals with the antibody results she was referring to concerning the different GH preparations. Did you measure the GH antibodies in your laboratory in all these instances?

K. Takano: We have measured all the antibodies in the same system.

R.M. Blizzard: So the percentages you are quoting are derived from your laboratory, not from the drug company laboratories. Regarding the met-GH data that you presented, was that met-GH which was used as an early preparation?

K. Takano: Yes, that was a highly purified one, with an ECP content of 3 ng per vial.

R.D.G. Milner: I have a very important, final comment to make; that is, that all of you who remain are very conscious of the other 50%, whose hypoglicemia conquered their intellectual curiosity. So I congratulate those of you who have had the self-discipline to stay with this interesting discussion, and I ask you to join me in thanking the other speakers and all the discussants before you, too, run to join the queues for whichever restaurants remain open.

REFERENCES

Buccafusco JJ (1984) Inhibition of regional brain acetylcholine biosynthesis by clonidine in spontaneously hypertensive rats. *Drug Devel Res* **4**: 627-633

Engel J, Ahlenius S, Almgren O, Carlsson A, Larsson K, Sodersten P (1979) *Pharmacol Biochem Behav* **10**: 149-154

Gourmelen M, Phan-Huu-Trung T, Girard F (1979) Transient partial hGH deficiency in prepubertal children with delay of growth. *Pediatr Res* **13**: 221-224

Hindmarsh P, Brook CG (1987) Effect of growth hormone on short children. *Brit Med J* **295**: 573-577

GROWTH HORMONE HYPERSECRETORY STATES

Advances in Growth Hormone and Growth Factor Research,
edited by E.E. Müller, D. Cocchi and V. Locatelli
Pythagora Press, Roma-Milano and Springer Verlag, Berlin-Heidelberg © 1989

Growth hormone-releasing factor in acromegaly*

T. Shibasaki, A. Masuda, M. Hotta, T. Imaki, N. Hizuka, K. Takano, and K. Shizume

Department of Medicine, Institute of Clinical Endocrinology, Tokyo Women's Medical College, Tokyo, Japan

The secretion of growth hormone (GH) from the pituitary is controlled by two antagonistic hypothalamic hormones, a GH–releasing factor (GRF) and a GH–release–inhibiting hormone, somatostatin. Since the structure of GRF was determined in 1982 (Guillemin et al. 1982; Rivier et al. 1982), synthetic GRF and its antiserum have been used to investigate the regulatory mechanisms of GH secretion (Wehrenberg et al. 1982 a,b,c). In acromegaly that shows chronic hypersecretion of GH, the normal interactive relationship between the hypothalamus and the pituitary may be disturbed. In this study, therefore, we tried to elucidate the pathophysiology of acromegaly in terms of GRF release from the hypothalamus.

MATERIALS AND METHODS

Materials

GRF–NH$_2$ was synthesized by solid–phase methodology as described previously (Ling et al. 1984). SMS 201–995 was generously supplied by Sandoz Ltd. (Basel, Switzerland). Thyrotropin–releasing hormone (TRH) and gonadotropin-releasing hormone (LHRH) were purchased from Tanabe Research Laboratory (Osaka, Japan) and Daiichi Pharmaceutical Co. (Tokyo, Japan) respectively.

* This work was supported in part by research grants from the Japanese Ministry of Education, Science and Culture, the Ministry of Health and Welfare.

GRF, TRH, and LHRH Tests

Forty–one active acromegalic patients, who were diagnosed as having acromegaly due to a pituitary GH–producing adenoma on the basis of clinical signs, endocrinological data, and roentgenographic studies, were chosen for this study. Patients who were being treated with bromocriptine stopped taking the drug at least 1 week before receiving GRF, TRH, or LHRH. One hundred micrograms of synthetic GRF were administered intravenously (iv) as a bolus dose, as described previously (Shibasaki et al. 1984a; Imaki et al. 1985; Masuda et al. 1985; Hotta et al. 1985). Five hundred micrograms of TRH or 200 µg LHRH were injected iv at intervals of 2–3 days. Blood samples were drawn at 15 min before and 0, 15, 30, 45, 60, 90, and 120 min after the injection or as indicated in Figure 1. GRF tests were performed before and after surgery in seven patients who had a pituitary microadenoma and underwent a total adenomectomy through a transsphenoidal approach. Plasma GH levels were measured by radioimmunoassay (RIA) using reagents generously provided by the NIADDK Pituitary Hormone Distribution Program (USA), as described previously (Shibasaki et al. 1985).

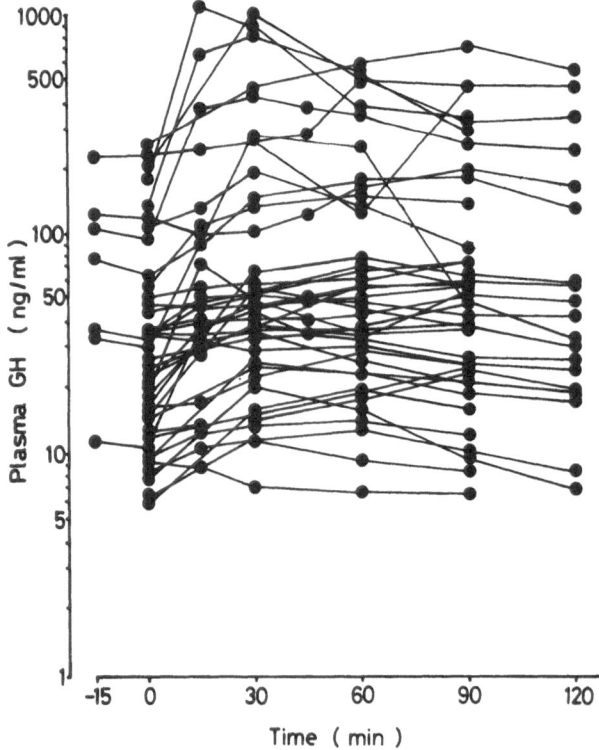

Fig. 1. Response of plasma GH secretion to GRF in 41 acromegalic patients.

Effect of SMS 201–995 on GRF Secretion

Fifty micrograms of SMS 201–995, a long–acting somatostatin analog, were subcutaneously injected in normal young male subjects and acromegalic patients. Blood samples were drawn to determine plasma GH and GRF levels. Plasma GRF concentrations were measured with an RIA using the anti–GRF serum (TR712) (Shibasaki et al. 1984b) after extraction with silicic acid. The responses of GRF to SMS 201–995 were compared between the normal subjects and the acromegalic patients.

Concentration of GRF in CSF

The concentrations of GRF in the cerebrospinal fluid (CSF) of the acromegalic patients and of patients with nonendocrine diseases were measured by RIA for GRF (Shibasaki et al. 1984b).

RESULTS

The responses of plasma GH to GRF in 41 active acromegalic patients are shown in Figure 1. Each patient showed a different change in plasma GH levels in response to GRF. In Figure 2, the responsiveness of GH secretion to GRF is expressed as the ratio of the peak GH to the basal GH in the GRF test. There were 35 patients with a ratio higher than 1.5; these were defined as GRF responders. There were six GRF nonresponders with a ratio lower than 1.5. Of the GRF responders, 75% had a ratio of between 1.5 and 4.0, although the ratio ranged from 1.5 to 9.2.

The mean age, estimated period of disease, and basal GH level of the patients were compared between the GRF responders and nonresponders (Table 1). The mean estimated period of disease of GRF non responders (17.5 ± 2.7 yr: mean ± SEM) was significantly longer than that of GRF responders (8.8 ± 0.7 yr), although there was no significant difference in the mean age or the basal GH level between the GRF responders and nonresponders.

The responsiveness of GH to GRF, TRH, and LHRH in the 41 patients is shown in Table 2. Plasma GH response to GRF was shown by 85% of the patients, and 78% and 17% of the patients showed paradoxical responses of plasma GH to TRH and LHRH respectively. The combination of the responsiveness of GH to each provocative test is shown in Figure 3. Only four patients showed responses of GH to all three: GRF, TRH, and LHRH. The six GRF nonresponders showed plasma GH response to TRH. There was no patient who showed plasma GH response only to LHRH. No correlation was found between the ratio of peak GH to basal GH

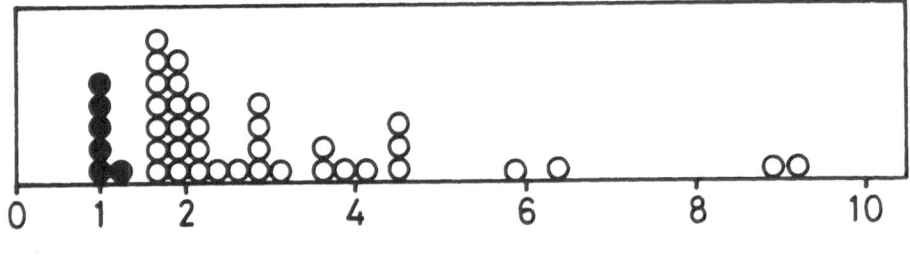

Peak GH / Basal GH

Fig. 2. Ratio of plasma peak GH to basal GH in GRF test of 41 acromegalic patients. ○, GRF responder; ●, GRF non responder.

Table 1. Comparison between GRF responder and GRF nonresponder acromegalic patients

	No. of patients	Age (yr)	Lenght of disease (yr)	Basal GH (ng/ml)
Responders	35	44.6 ± 2.0	8.8 ± 0.7	66.7 ± 13.4
Nonresponders	6	47.5 ± 6.1	17.5 ± 2.7*	44.3 ± 12.7

* $P < 0.005$ vs Responders.

Table 2. Responsiveness to GRF, TRH, and LHRH in 41 acromegalic patients

Test	Responders (%)	Nonresponders (%)
GRF	85.4	14.6
TRH	78.0	22.0
LHRH	17.1	82.9

in the GRF test and that in the TRH test in GRF–and TRH–responsive patients, as shown in Figure 4.

The GRF test was performed before and 2 or 3 weeks after surgery in ten acromegalic patients whose pituitary adenoma was completely removed through a transsphenoidal operation (Fig. 5). The mean basal GH level before surgery was 37.5 ± 11.4 (SEM) ng/ml with the mean peak GH level of 132.6 ± 85.6 ng/ml at 15 min after GRF administration. On the other hand, after total adenomectomy, the mean basal GH level was 2.5 ± 0.4 ng/ml and the mean peak GH level was 15.2 ± 4.7 ng/ml.

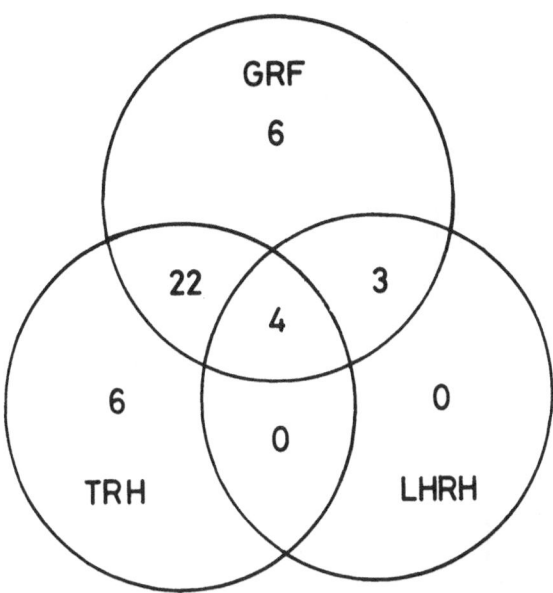

Fig. 3. Responsiveness of GH secretion to GRF, TRH, and LHRH in 41 acromegalic patients.

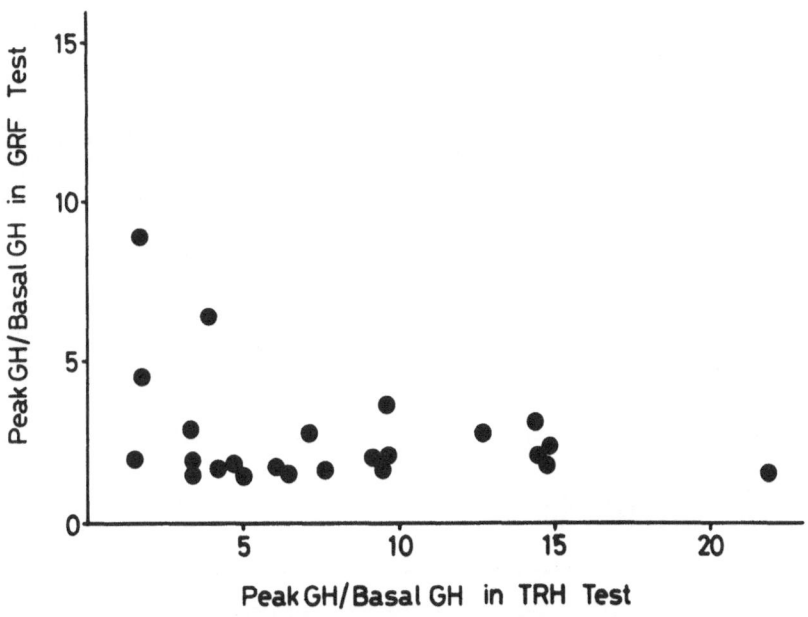

Fig. 4. Relationship between ratio of plasma peak GH to basal GH in GRF test and that in TRH test of GRF-and TRH-responsive patients with acromegaly.

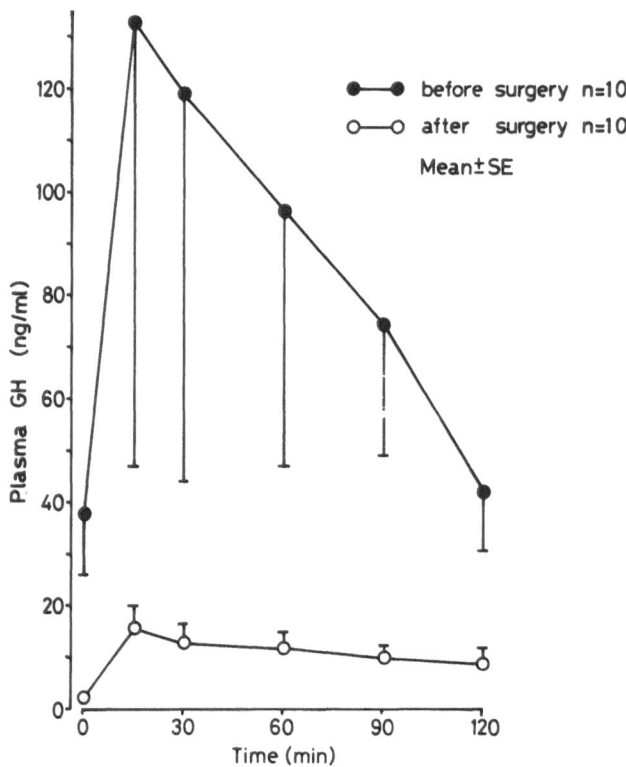

Fig. 5. Response of GH secretion to GRF before and after total adenomectomy in ten acromegalic patients.

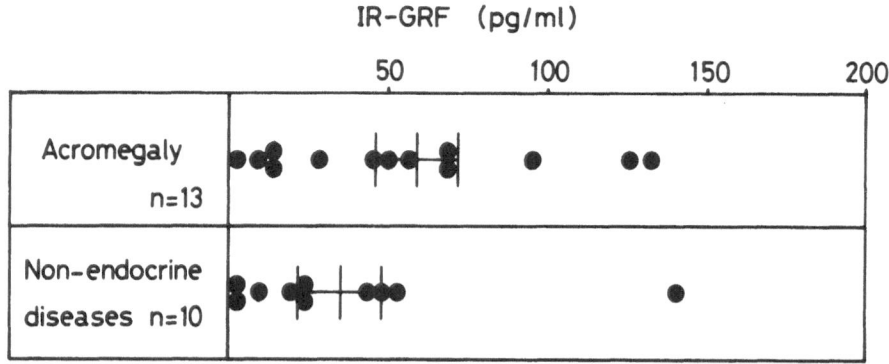

Fig. 6. Concentration of GRF in CSF of acromegalic patients and patients with nonendocrine diseases.

Figure 6 shows the concentrations of GRF in the CSF of patients with acromegaly and nonendocrine diseases. The mean concentration of GRF in the CSF of acromegalic patients, 59.3 ± 13.0 (SEM) pg/ml, was not significantly different from that in patients with nonendocrine diseases, 35.3 ± 13.1 pg/ml.

The administration of 50 μg SMS 201–995 induced no consistent tendency in the change of plasma GRF levels, whereas plasma GRF levels were significantly decreased in response to 50 μg SMS 201–995 in normal subjects.

DISCUSSION

We and others have reported that iv administration of synthetic GRF induces plasma GH elevation in acromegalic patients (Wood et al. 1983; Shibasaki et al. 1984 c; Chiodini et al. 1985; Gelato et al. 1985). The ratio of plasma peak GH level to basal GH level varied from 1.0 to 9.2 in the present study. We previously reported that four of 25 acromegalic patients showed no response of plasma GH to GRF (Shibasaki et al. 1986). The four GRF nonresponders included in the present study had significantly longer estimated periods of disease compared with GRF responders (Shibasaki et al. 1986). The present study again demonstrated that the estimated period of disease was significantly longer in GRF nonresponders than in GRF responders. It is therefore suggested that the characteristics of the GRF receptors on the adenoma cells change during the long period of growth and that they lose their responsiveness. The pituitary adenoma of two of the GRF non-responders was completely removed in transsphenoidal surgery. The GRF test was performed again 3 weeks after surgery in those patients. One patient showed a peak GH level of 3.1 ng/ml with a basal GH level of 1.4 ng/ml, and the other showed a peak GH level of 4.3 ng/ml with a basal level of 1.5 ng/ml in the GRF test after surgery. Since the mean GH peak level in the GRF test was 15.2 ± 4.7 ng/ml in ten acromegalic patients whose pituitary adenoma was completely removed, the responsiveness of the nontumorous somatotrophs to GRF was also extremely decreased in addition to the sensitivity of the adenoma cells to GRF in these two GRF nonresponders. The results suggest that there are some acromegalic patients whose GH secretion is independent of GRF, although it is not clear whether GRF–nonresponsive adenoma cells developed from GRF–nonresponsive somatotrophs or whether adenoma cells developed from somatotrophs and the adenoma cells and nonadenomatous somatotrophs became nonresponsive or extremely low in responsiveness to GRF during long periods of growth.

Each patient had a different response pattern to GRF, TRH, or LHRH. The variable combinations of responsiveness among patients imply that each adenoma contains different receptors. We previously suggested that TRH and LHRH stimulate GH secretion from pituitary adenoma cells through receptors different from those for GRF, based on the finding that TRH still stimulates GH secretion even when desensitization of adenoma cells to GRF occurs (Shibasaki et al. 1986). This hypothesis is also supported by the finding that there is no correlation between the ratios of peak GH to basal GH in the GRF test and those in the TRH test, as shown in Figure 4.

The mean concentration of GRF in the CSF of acromegalic patients was not different from that of patients with nonendocrine diseases. This result implies that chronic hypersecretion of GH does not influence GRF concentration in CSF.

We observed that administration of 50 μg SMS 201–995 lowered not only plasma GH levels but also plasma TSH and GRF levels in normal subjects. This result implies that TSH secretion is suppressed by somatostatin in man as well as in the rat. However, plasma GRF levels did not decrease in response to 50 μg SMS 201–995 in all the patients, although plasma TSH levels were reduced by 50 μg SMS 201–995. These results suggest that the inhibition of the response of GRF release to somatostatin is disturbed in some cases of acromegaly.

Acknowledgment
We thank Dr. Nicholas Ling for supplying us with synthetic GRF.

REFERENCES

Chiodini PG, Liuzzi A, Dallabonzana D, Oppizzi G, Verde GG (1985) Changes in growth hormone (GH) secretion induced by human pancreatic GH-releasing hormone-44 in acromegaly: a comparison with thyrotropin-releasing hormone and bromocriptine. *J Clin Endocrinol Metab* **60**: 48-52

Gelato MC, Merriam GR, Vance ML, Goldman JA, Webb C, Evans WS, Rock J, Oldfield EH, Molitch ME, Rivier J, Vale W, Reichlin S, Frohman LA, Loriaux DL, Thorner MO (1985) Effects of growth hormone-releasing factor on growth hormone secretion in acromegaly. *J Clin Endocrinol Metab* **60**: 251-257

Guillemin R, Brazeau P, Böhlen P, Esch F, Ling NC, Wehrenberg WB (1982) Growth hormone-releasing factor from a human pancreatic tumor that caused acromegaly. *Science* **218**: 585-587

Hotta M, Shibasaki T, Masuda A, Imaki T, Wakabayashi I, Demura H, Ling N, Shizume K (1985) The inter-and intra-subject variabilities of plasma GH response to human growth hormone-releasing hormone (1-44)NH₂ in men. *Endocrinol Jpn* **32**: 673-680

Imaki T, Shibasaki T, Shizume K, Masuda A, Hotta M, Kiyosawa Y, Jibiki K, Demura H, Tsushima T, Ling N (1985) The effect of free fatty acids on growth hormone (GH)-releasing hormone-mediated GH secretion in man. *J Clin Endocrinol Metab* **60**: 290-293

Ling N, Esch F, Böhlen P, Brazeau P, Wehrenberg WB, Guillemin R (1984) Isolation, primary structure and synthesis of human hypothalamic somatocrinin: growth hormone-releasing factor. *Proc Natl Acad Sci USA* **81**: 4302-4306

Masuda A, Shibasaki T, Nakahara M, Imaki T, Kiyosawa Y, Jibiki K, Demura H, Shizume K, Ling N (1985) The effect of glucose on growth hormone (GH)-releasing hormone-mediated GH secretion in man. *J Clin Endocrinol Metab* **60**: 523-526

Rivier J, Spiess J, Thorner M, Vale W (1982) Characterization of growth hormone-releasing factor from a human pancreatic islet tumor. *Nature* **300**: 276-278

Shibasaki T, Shizume K, Nakahara M, Masuda A, Jibiki K, Demura H, Wakabayashi I, Ling N (1984a) Age-related changes in plasma growth hormone response to growth hormone-releasing factor in man. *J Clin Endocrinol Metab* **58**: 212-214

Shibasaki T, Kiyosawa Y, Masuda A, Nakahara M, Imaki T, Wakabayashi I, Demura H, Shizume K, Ling N (1984b) Distribution of growth hormone-releasing hormone-like immunoreactivity in human tissue extracts. *J Clin Endocrinol Metab* **59**: 263-268

Shibasaki T, Shizume K, Masuda A, Nakahara M, Hizuka N, Miyakawa M, Takano K, Demura H, Wakabayashi I, Ling N (1984c) Plasma growth hormone response to growth hormone-releasing factor in acromegalic patients. *J Clin Endocrinol Metab* **58**: 215-217

Shibasaki T, Hotta M, Masuda A, Imaki T, Obara N, Demura H, Ling N, Shizume K (1985) Plasma GH responses to GHRH and insulin-induced hypoglycemia in man. *J Clin Endocrinol Metab* **60**: 1265-1267

Shibasaki T, Hotta M, Masuda A, Imaki T, Obara N, Hizuka N, Takano K, Wakabayashi I, Demura H, Ling N, Shizume K (1986) Studies on the response of growth hormone (GH) secretion to GH-releasing hormone, thyrotropin-releasing hormone, gonadotropin-releasing hormone, and somatostatin in acromegaly. *J Clin Endocrinol Metab* **63**: 167-173

Wehrenberg WB, Ling N, Brazeau P, Esch F, Böhlen P, Baird A, Ying S, Guillemin R (1982a) Somatocrinin, growth hormone-releasing factor, stimulates secretion of growth hormone in anesthetized rats. *Biochem Biophys Res Commun* **109**: 382-387

Wehrenberg WB, Ling N, Böhlen P, Esch F, Brazeau P, Guillemin R (1982b) Physiological roles of somatocrinin and somatostatin in the regulation of growth hormone secretion. *Biochem Biophys Res Commun* **109**: 562-567

Wehrenberg WB, Brazeau P, Luben R, Böhlen P, Guillemin R (1982c) Inhibition of the pulsatile secretion of growth hormone by monoclonal antibodies to the hypothalamic growth hormone-releasing factor (GRF). *Endocrinology* **111**: 2147-2148

Wood SM, Ch'ng JL, Adams EF, Webster J, Joplin GF, Mashiter K, Bloom SR (1983) Abnormalities of growth hormone release in response to human pancreatic growth hormone-releasing factor (GRF1-44) in acromegaly and hypopituitarism. *Br Med J* **286**: 1687-1691

Advances in Growth Hormone and Growth Factor Research,
edited by E.E. Müller, D. Cocchi and V. Locatelli
Pythagora Press, Roma-Milano and Springer Verlag, Berlin-Heidelberg © 1989

The hypersecreting somatotroph: functional and morphological aspects*

A. Spada[1], M. Bassetti[2], M. Arosio[1], L. Vallar[2], F. R. Elahi[2], and G. Giannattasio[2]

[1] *Department of Endocrinology, School of Medicine; and* [2]*CNR Center of Cytopharmacology, Department of Pharmacology, University of Milan, Milan, Italy*

Patients with acromegaly often show peculiar alterations of the GH secretory pattern. In particular, alterations in the sensitivity to physiological releasing (GHRH) or inhibiting (SRIF) hormones, together with abnormal responsiveness to hypothalamic agents such as dopamine (DA) and TRH or GnRH, are frequently observed in this disease. Cosecretion of pituitary hormones other than GH, such as prolactin, glycoprotein α–subunit, and TSH, may also occur. However, the underlying biochemical events and morphological correlates have not so far been extensively investigated. This prompted us to investigate the following aspects:
— Alterations of receptor regulation
— Alterations of postreceptor "coupling" processes
— Transmembrane signaling pathology
— Colocalization of different hormones in the same cell

ALTERATIONS OF RECEPTOR REGULATION

It is well known that the great majority of GH–secreting adenomas are sensitive to the specific releasing and inhibiting hormones both *in vivo* (Arosio et al. 1985; Faglia et al. 1985; Gelato et al. 1985) and *in vitro* (Lamberts et al. 1984; Spada et al. 1984). As far as GHRH is concerned, important alterations of the GH secretory pattern in response to this peptide are present in acromegaly. In fact, contrary to

* This work was partially supported by a grant from the Italian National Research Council, Special Project Oncology.

normal subjects, in acromegalic patients the repeated administration of GHRH (50 µg iv every 2 h) is able to elicit GH release after each injection (Spada et al. 1987). As it has recently been demonstrated that GHRH induces desensitization to its own action in normal somatotrophs (Bilezikjian and Vale 1984), we investigated whether this physiological regulatory mechanism might be lost in adenomatous somatotrophs.

A 4-h pre–exposure of cultured normal somatotrophs to 10 nM GHRH induced a clear reduction of GH response to the subsequent challenge with the peptide, while in adenomatous somatotrophs this procedure did not modify GH release (Fig.

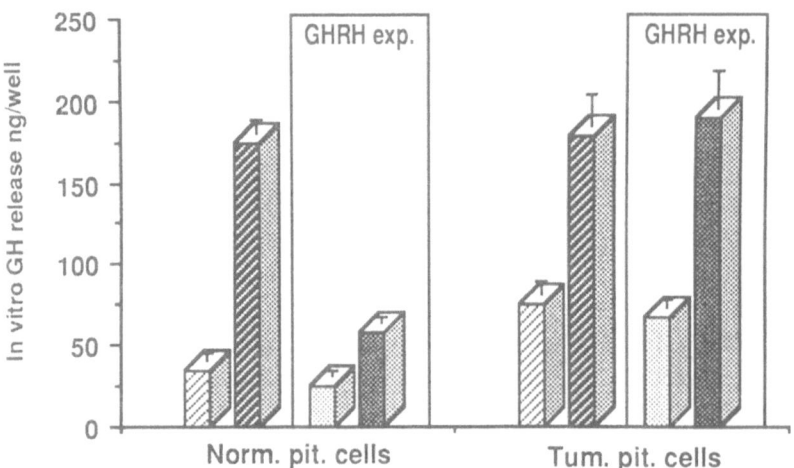

Fig. 1. Effect of 4-h exposure to 10 nM GHRH on basal GH release and subsequent challenge with GHRH (10 n*M*) from cultured normal and adenomatous somatotrophs.

1). These data demonstrate that a lack of desensitization to GHRH occurs in adenomatous somatotrophs.

Since GHRH is an important stimulator of both GH secretion and somatotroph replication (Billestrup et al. 1986), it is tempting to speculate that an impairment of this regulatory mechanism may have some importance in the development of GH–secreting adenomas. Although direct experimental demonstrations are not available, it seems likely that a similar lack of desensitization also occurs for SRIF receptors. In fact, escape phenomena are not observed in acromegalic patients during long–term somatostatin treatment (Lamberts et al. 1987).

ALTERATIONS OF POSTRECEPTOR "COUPLING" PROCESSES

It is well established that DA is able to reduce GH release in a high proportion of GH–secreting adenomas, directly acting on adenomatous somatotrophs. The pharmacological characteristics of the dopaminergic receptor on tumoral somatotrophs are those of the D_2 receptor, typically present on normal lactotrophs (Kebabian and Calne 1979). In fact, ergot derivatives are potent DA agonists and sulpiride is a potent DA antagonist (Liuzzi et al. 1974; Arosio et al. 1980; Faglia et al. 1985). Since the D_2 receptor of lactotrophs is coupled with the inhibition of adenylate cyclase (AC) activity (Giannattasio et al. 1981; Swennen and Denef 1982), we investigated whether this intracellular transduction mechanism of the DA signal was also operating in tumoral somatotrophs. As shown in Figure 2, DA at concentrations effective in inhibiting AC activity in membranes from lactotroph–enriched preparations was completely ineffective in membranes from GH–secret-

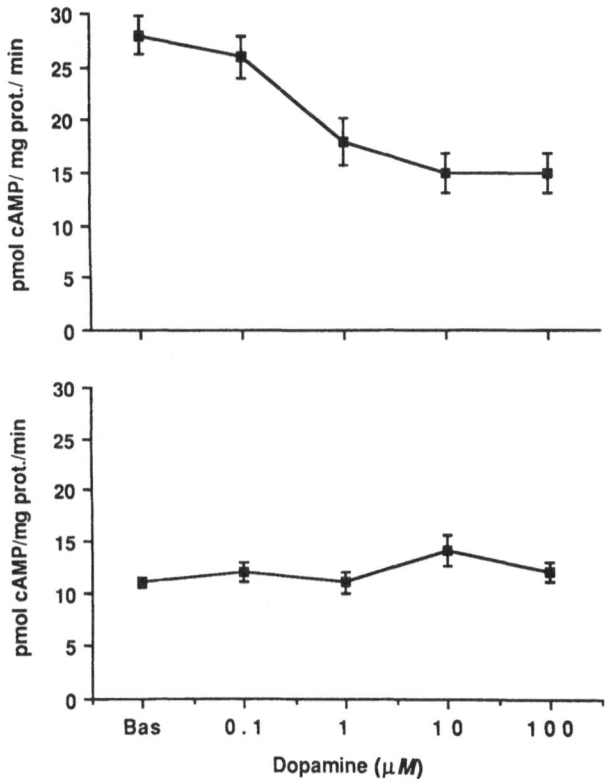

Fig. 2. Effect of different concentrations of DA on adenylate cyclase activity on membrane preparations from lactotroph–enriched populations (*upper panel*) and GH–secreting adenomas (*lower panel*).

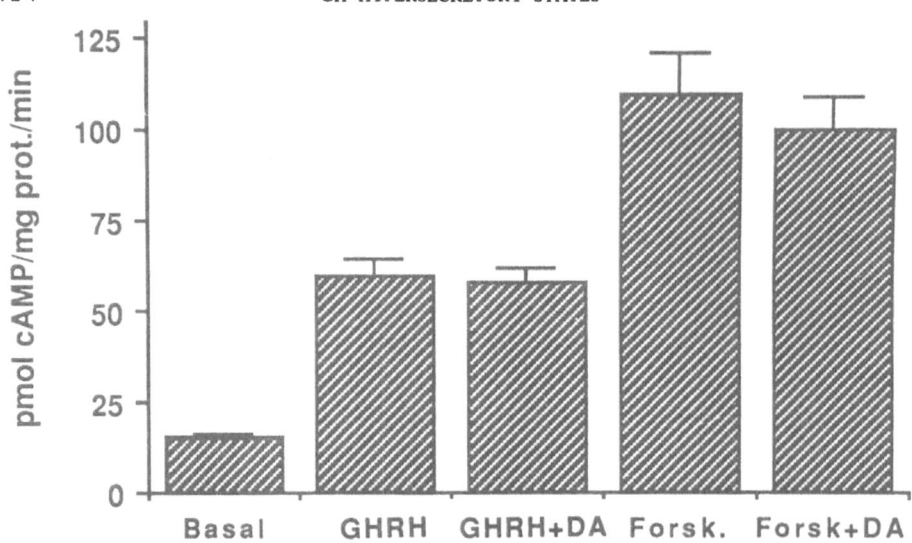

Fig. 3. Effect of GHRH (0.1 μ*M*) alone and in combination with DA (1 μ*M*), and forskolin (Forsk, 10 μ*M*) alone and in combination with DA (1 μ*M*) on adenylate cyclase activity of membrane preparations from GH–secreting adenomas.

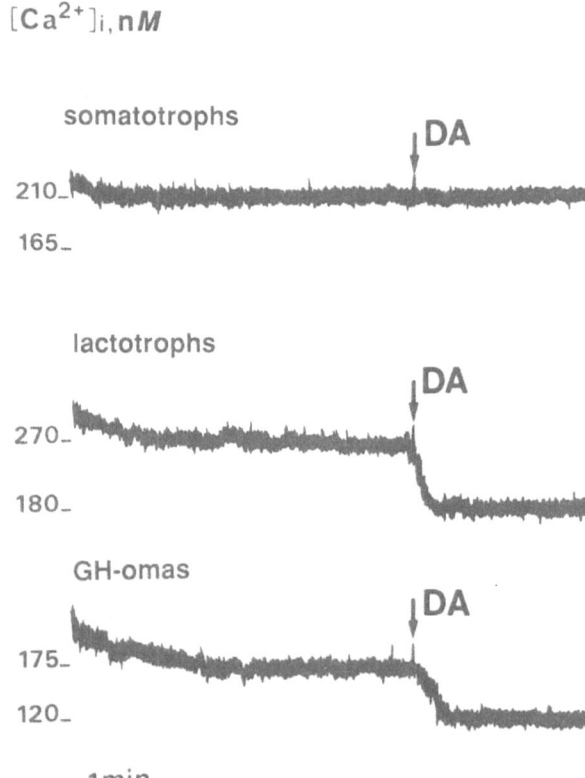

Fig. 4. Fluorescence traces of cells loaded with Fura–2. Cells were inserted in a thermostatically controlled cuvette, and when indicated, DA (1 μ*M*) was added. GH–omas, GH–secreting adenomas.

ing adenomas. Moreover, no inhibitory effect was detectable even when DA was tested on AC already stimulated by GHRH and forskolin (Fig. 3).

Since we had recently demonstrated that the spectrum of intracellular signals elicited in lactotrophs by the activation of D_2 receptors includes the inhibition of both cAMP and calcium (Malgaroli et al. 1987), the possible effect of DA on cytosolic free Ca^{2+} concentrations ($[Ca^{2+}]i$) was evaluated in adenomatous somatotrophs. For this study, cells were loaded with Fura–2, a fluorescent indicator highly selective for $[Ca^{2+}]i$, and $[Ca^{2+}]i$ calculated according to Grynkiewicz et al. (1985). As shown in Figure 4, DA did not modify resting $[Ca^{2+}]i$ levels in normal somatotrophs, in which DA did not exert any secretory effect. In contrast, DA (1 μM) significantly reduced $[Ca^{2+}]i$ levels both in rat lactotrophs and in human tumoral somatotrophs.

Taken together, these results suggest that D_2 receptors present on tumoral somatotrophs, contrary to those on lactotrophs, are not coupled with the adenylate cyclase–cAMP system. In these cells the transduction of the DA signal involves the inhibition of Ca^{2+} influx.

TRANSMEMBRANE SIGNALING PATHOLOGY

It is well known that important processes such as hormone secretion and cell replication are cAMP–dependent functions in somatotrophs (Spada et al. 1984; Billestrup et al. 1986). On this line, GHRH and SRIF regulate GH secretion by modulating the intracellular cAMP levels (Spada et al. 1984).

We studied a large number of GH–secreting tumors, and two groups of adenomas with different secretory activities and biochemical characteristics were identified (Vallar et al. 1987). Some adenomas (group 1) showed *in vitro* GH release and intracellular cAMP levels similar to those observed in normal pituitary. The values of AC activity measured in membranes from these tumors were also within the normal range. In contrast, in group–2 adenomas (which represent about 30% of the tumors studied) the *in vitro* GH release, the intracellular cAMP levels, and the AC activity were extremely high (Fig. 5). Moreover, while GHRH stimulated GH release and AC activity in group 1 tumors, the high levels of GH and AC were not further increased by the peptide in group 2 (Fig. 6). In group–2 tumors important alterations of AC regulation were present. In fact, while guanine nucleotides and fluoride stimulated AC in group–1 tumors — which occurs in all the cell systems so far reported (Spiegel and Downs 1981) — these agents exerted an inhibitory effect in group 2, likely interacting with the inhibitory regulatory protein of the cyclase (Gi). SRIF inhibited AC activity in both groups (Fig. 7). In group–2 tumors the stimulatory protein of the cyclase (Gs) was also unaffected by cholera toxin. In fact, the toxin did not ADP–ribosylate Gs and, by consequence, did not stimulate AC activity.

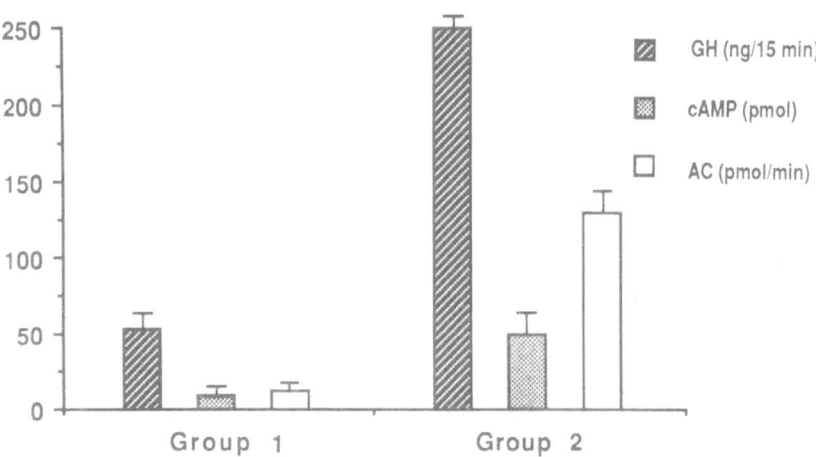

Fig. 5. In vitro GH release (*hatched bars*), intracellular cAMP levels (*dotted bars*) in cultured cells, and adenylate cyclase activity (*open bars*) in membrane preparations from group–1 and –2 GH–secreting adenomas.

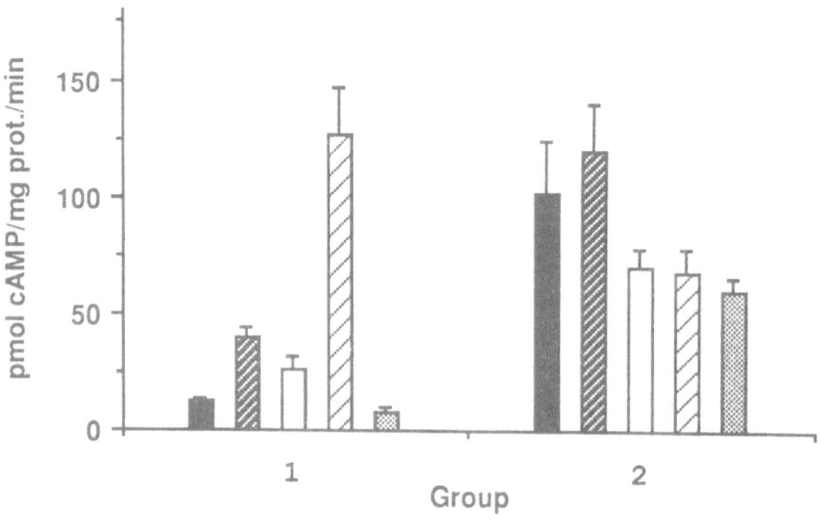

Fig. 6. Effect of different agents on adenylate cyclase activity from group–1 and –2 adenomas. *Closed bars*, basal activity; *hatched bars*, GHRH 1 µM; *open bars*, GTP 0.1 µM; *lightly hatched bars*, fluoride 10 mM; *dotted bars*, SRIF 0.1 µM.

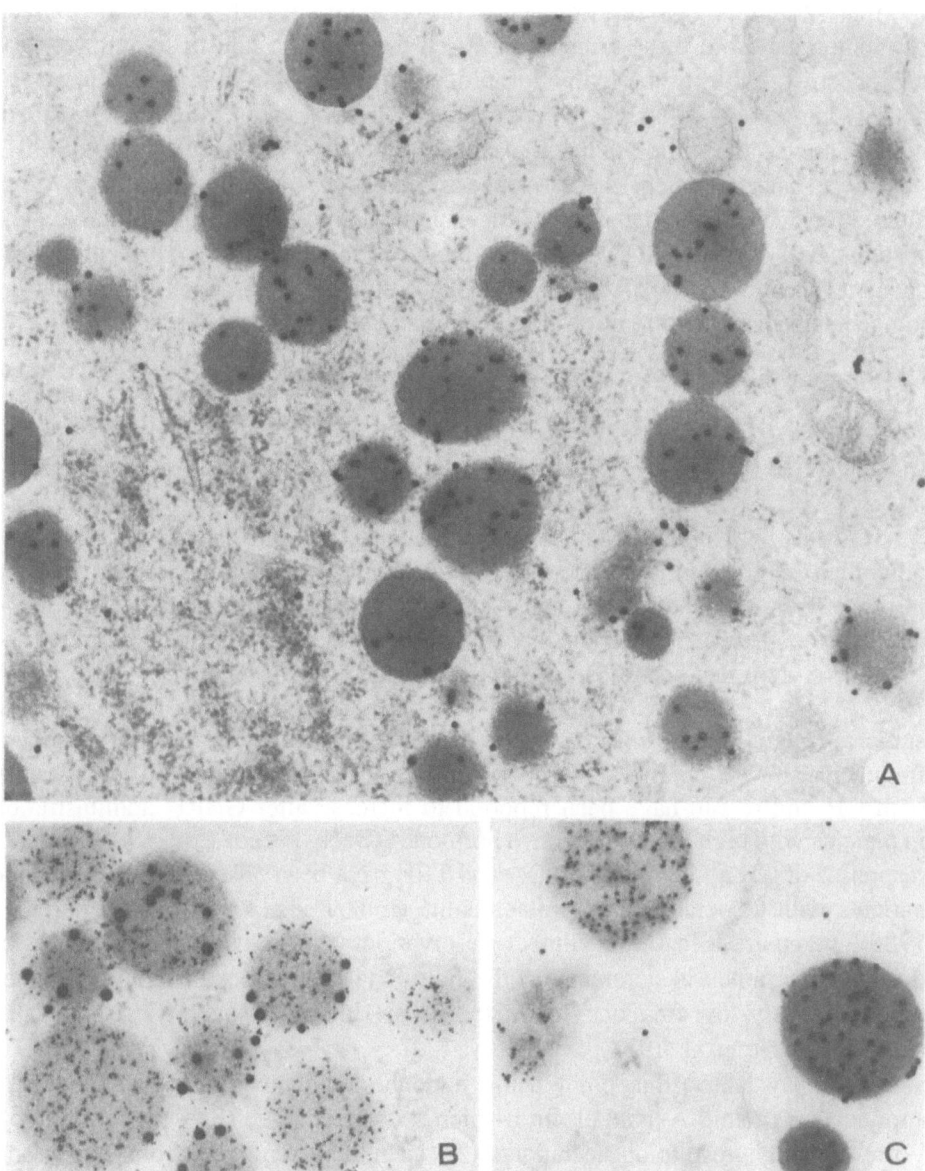

Fig. 7. A–C. Electron micrographs of pituitary adenomas from acromegalic patients. **A** cell containing secretory granules positive only for GH (*large particles*) (x 45,000); **B** cell containing granules positive only for prolactin (*small particles*) and granules positive for both prolactin and GH (x 50,000); **C** cell containing two types of granules, positive only for GH (*large particles*) and positive only for prolactin (*small particles*) (x 60,000).

The molecular alterations responsible for the derangement in the AC control in group–2 tumors have been proved to reside in Gs. In fact, 1% cholate extracts of membranes from these tumors (in which the catalytic unit of the enzyme has been destroyed during the extraction procedure) were able to reconstitute an AC activity in cyc–S49 cells (which are genetically lacking in Gs; Northup et al. 1980) which showed the typical altered response pattern present in group–2 tumors.

These data document the existence in some GH–secreting adenomas of an altered Gs, which causes marked changes in the activity and regulation of AC and sustains both elevated cAMP levels and GH secretion. Moreover, taking into account the importance of cAMP in somatotropic cell replication, it is tempting to speculate that this alteration might play a role in tumoral growth.

COLOCALIZATION OF DIFFERENT HORMONES IN THE SAME CELL

It is well known that in acromegaly the associated hypersecretion of hormones other than GH frequently occurs. In fact, high levels of prolactin have been reported in about 30%–40% of acromegalic patients (de Pablo et al. 1981; Faglia et al. 1985) while in about 10–20% associated hypersecretion of glycoprotein hormone α–subunit (α–sub) has been observed (Mac Farlane et al. 1982; Beck–Peccoz et al. 1985). High levels of TSH were reported in a small proportion of acromegalic subjects (Faglia et al. 1987). It is worth noting that the hypersecreted hormones frequently show a parallel pattern of response to secretory stimuli. For instance, both GH and α–sub have been reported to increase after GHRH administration in patients with high levels of the two hormones (Beck–Peccoz et al. 1985). Parallel responses of GH and prolactin to DA and TRH have been observed in acromegalic patients with associated hyperprolactinemia (Liuzzi et al. 1974; Lamberts et al. 1985). In contrast to these clinical observations suggestive for cosecretion of different hormones in acromegaly, the traditional immunocytochemistry studies reported a very low frequency of mixed cells (Horvath et al. 1981; Ishikawa et al. 1983; Kanie et al. 1983).

We re–evaluated this point using a double immunocytochemical labeling, applying the protein A–gold electron–microscope technique (Bassetti et al. 1986). With this high–resolution technique, 22 adenomas removed from hyper–and normoprolactinemic acromegalics were studied. Eleven adenomas were composed of cells positive only for GH, four of a large majority of cells positive for GH and a small number of cells positive for prolactin and seven adenomas had an important proportion of cells positive for prolactin (from 20% to 80%). Of the nine acromegalic patients with associated hyperprolactinemia, six had adenomas positive for GH and prolactin, while three had pure GH–secreting adenomas.

The great majority (about 80%) of cells positive for prolactin were somatomammotrophs, being also positive for GH. The ultrastructural appearance of somatomammotrophic cells was quite variable: some cells contained both pure GH granules and mixed granules positive for GH and prolactin, while in some cells granules positive only for prolactin were intermingled at random with granules positive only for GH (Fig. 7).

Utilizing anti–GH and anti–α–sub antibodies, the copresence of GH and α–sub in the same cell was demonstrated in adenomas removed from patients with associated hypersecretion of α–sub (Beck–Peccoz et al. 1985). Similar data were observed in one patient with high levels of GH and TSH (Beck–Peccoz et al. 1986).

Taken together, these results indicate that the copresence of two or more hormones in the same cell is not a rare event in human GH–secreting adenomas. This can account for the simultaneous release of all the hypersecreted hormones in response to a single releasing hormone frequently observed in acromegalic patients.

The biological events which make pituitary cells and particular somatotrophs capable of producing two hormones even markedly different in chemical composition remain to be elucidated.

CONCLUSIONS

A large number of human GH–secreting adenomas have been studied by different approaches, and important alterations, such as lack of desensitization to GHRH action, uncoupling of D_2 receptors to adenylate cyclase, molecular alterations of the regulatory protein of adenylate cyclase (Gs), and copresence of GH and other pituitary hormones (i.e., prolactin, α–subunit, TSH) in the same cell, have been demonstrated to occur in tumoral somatotrophs. These findings may contribute to a better knowledge of pathogenetic mechanisms underlying GH–secreting adenoma formation and growth.

Acknowledgment
We thank Dr. G. Faglia for critical revision of the manuscript.

REFERENCES

Arosio M, Moriondo P, Travaglini P, Ambrosi B, Beck-Peccoz P, Conti Puglisi F, Secchi F, Faglia G (1980) Modifications in serum growth hormone concentration induced by sulpiride in acromegalic patients pretreated with dopamine, bromocriptine and metergoline. *J Clin Endocrinol Metab* **51**: 454-461

Arosio M, Ambrosi B, Guglielmino L, Faglia G (1985) Human pancreatic growth hormone-releasing factor (hpGRF-44) in acromegaly before and after adenomectomy. Modifications induced by somatostatin (GHRIH) infusion. *J Endocrinol Invest* **8**: 449-453

Bassetti M, Spada A, Arosio M, Vallar L, Brina M, Giannattasio G (1986) Morphological studies on mixed growth hormone (GH)- and prolactin (PRL)-secreting human pituitary adenomas. Coexistence of GH and PRL in the same secretory granule. *J Clin Endocrinol Metab* **62**: 1093-1100

Beck-Peccoz P, Bassetti M, Spada A, Medri G, Arosio M, Giannattasio G, Faglia G (1985) Glycoprotein hormone α-subunit and GH coexistence in the same tumoral cell. *J Clin Endocrinol Metab* **61**: 541-548

Beck-Peccoz P, Piscitelli G, Amr S, Ballabio M, Bassetti M, Giannattasio G, Spada A, Nissim M, Weintraub BD, Faglia G (1986) Endocrine and morphological studies on pituitary adenomas secreting growth hormone, thyrotropin (TSH) and α-subunit; evidence for secretion of TSH with increased bioactivity. *J Clin Endocrinol Metab* **62**: 704-711

Bilezikjian LM, Vale WW (1984) Chronic exposure of cultured rat anterior pituitary cells to GHRH causes partial loss of responsiveness to GHRH. *Endocrinology* **115**: 2032-2034

Billestrup N, Swanson LW, Vale W (1986) Growth hormone-releasing factor stimulates proliferation of somatotrophs *in vitro*. *Proc Natl Acad Sci USA* **83**: 6854-6857

de Pablo F, Eastman RC, Roth J, Gorden P (1981) Plasma prolactin in acromegaly before and after treatment. *J Clin Endocrinol Metab* **53**: 344-352

Faglia G, Arosio M, Ambrosi B (1985) Recent advances in diagnosis and treatment of acromegaly. In: Imura H (ed) *The Pituitary Gland*. Raven, New York, pp. 363-404

Faglia G, Beck-Peccoz P, Piscitelli G, Medri G (1987) Inappropriate secretion of thyrotropin by the pituitary. *Horm Res* **26**: 79-99

Gelato ML, Merriam GR, Vance MC, Goldman JA, Webb C, Evans WS, Rock J, Oldfield EH, Molitch ME, Rivier J, Vale W, Reichlin S, Frohman LA, Loriaux DL, Thorner MO (1985) Effect of growth hormone-releasing factor on growth hormone secretion in acromegaly. *J Clin Endocrinol Metab* **60**: 251-259

Giannattasio G, De Ferrari ME, Spada A (1981) Dopamine-inhibited adenylate cyclase in female rat adenohypophysis. *Life Sci* **28**: 1605-1612

Grynkiewicz G, Poenie M, Tsien RY (1985) A new generation of Ca^{2+} indicators with greatly improved fluorescence properties. *J Biol Chem* **260**: 3440-3450

Horvath E, Kovacs K, Singer W, Smyth HS, Killinger DW, Ezrin C, Weiss MH (1981) Acidophilic stem cell adenoma of the human pituitary. Clinicopathological analysis of 15 cases. *Cancer* **47**: 761-771

Ishikawa H, Nogami H, Kamio M, Suzuki T (1983) Single secretory granules contain both GH and prolactin in pituitary mixed type of adenoma. *Virchows Arch* **399**: 221-226

Kanie N, Kageyama N, Kuwayama A, Nakane T, Watanabe M, Kawaoi A (1983) Pituitary adenomaas in acromegalic patients: an immunohistochemical and endocrinological study with special reference to prolactin-secreting adenomas. *J Clin Endocrinol Metab* **57**: 1093-1101

Kebabian JW, Calne DB (1979) Multiple receptors for dopamine. *Nature* **277**: 93-96

Lamberts SWJ, Verleun T, Oosterom R (1984) The interrelationship between the effects of somatostatin and human pancreatic growth hormone-releasing factor on growth hormone release by cultured pituitary tumor cells from patients with acromegaly. *J Clin Endocrinol Metab* **58**: 250-254

Lamberts SWJ, Klijn JGM, Voonhoven CCJ, Stefanko SZ (1985) Different responses of growth hormone secretion to guanfacine, bromocriptine and thyrotropin-releasing hormone in acromegalic patients with pure growth hormone (GH)-containing and mixed GH/prolactin-containing pituitary adenomas. *J Clin Endocrinol Metab* **60**: 1148-1153

Lamberts SWJ, Uitterlinden P, del Pozo E (1987) SMS 201-995 induces a continuous decline in circulating growth hormone and somatomedin-C levels during therapy of acromegalic patients for over two years. *J Clin Endocrinol Metab* **65**: 703-711

Liuzzi A, Chiodini PG, Botalla L, Silvestrini F, Müller EE (1974) Growth hormone-releasing activity of TRH and GH-lowering effect of dopaminergic drugs in acromegaly: homogeneity of the two responses. *J Clin Endocrinol Metab* **39**: 871-876

Mac Farlane IA, Beardwell CG, Shalet SM, Ainslie G, Rankin E (1982) Glycoprotein hormone α-subunit secretion in patients with pituitary adenomas: influences of TRH, LRH and bromocriptine. *Acta Endocrinol* (Copenh) **99**: 487-492

Malgaroli A, Vallar L, Reza Elahi F, Pozzan T, Spada A, Meldolesi J (1987) Dopamine inhibits cytosolic Ca^{2+} increases in rat lactotroph cells. Evidence of a dual mechanism of action. *J Biol Chem* **262**: 13920-13927

Northup JK, Sterweis PC, Smigel MD, Scheifer LS, Ross EM, Gilman AG (1980) Purification of the regulatory component of adenylate cyclase. *Proc Natl Acad Sci USA* **77**: 6516-6520

Spada A, Vallar L, Giannattasio G (1984) Presence of an adeylate cyclase dually regulated by somatostatin and human pancreatic growth hormone-releasing factor in GH-secreting cells. *Endocrinology* **115**: 1203-1209

Spada A, Reza Elahi F, Arosio M, Sartorio A, Guglielmino L, Vallar L, Faglia G (1987) Lack of desensitization of adenomatous somatotrophs to growth hormone-releasing hormone in acromegaly. *J Clin Endocrinol Metab* **64**: 585-591

Spiegel AM, Downs RWJ (1981) Guanine nucleotides: key regulators of hormone receptor-adenylate cyclase interaction. *Endocr Rev* **2**: 275-305

Swennen I, Denef C (1982) Physiological concentrations of dopamine decrease adenosine-3',5'-monophosphate: levels in cultured rat anterior pituitary cells and enriched populations of lactotrophs; evidence for a causal relationship to inhibition of prolactin release. *Endocrinology* **111**: 398-405

Vallar L, Spada A, Giannattasio G (1987) Altered Gs and adenylate cyclase activity in human GH-secreting pituitary adenomas. *Nature* **330**: 566-568

Advances in Growth Hormone and Growth Factor Research,
edited by E.E. Müller, D. Cocchi and V. Locatelli
Pythagora Press, Roma-Milano and Springer Verlag, Berlin-Heidelberg © 1989

Medical treatment of acromegaly – Dopa-minergic agonists and long-acting somatostatin*

P.G. Chiodini[1], R. Cozzi[1], D. Dallabonzana[1], G. Oppizzi[1], G. Verde[1], MM. Petroncini[1], A. Liuzzi[1], E. Boccardi[2], and I. Lancranjan[3]

[1] *Divisione di Endocrinologia;* [2] *Servizio di Neuroradiologia, Ospedale Niguarda, Milano, Italy;* [3] *Department of Neuroendocrinology, Sandoz Ltd., Basel, Switzerland.*

The medical treatment of acromegaly derives its values from the fact that neurosurgey and radiation therapy may be, in this disease, frequently unsuccessful. Indeed, the frequency with which neurosurgery cures acromegaly depends on the size of the tumor: those patients with tumors less than 10 mm in diameter (microadenomas) have an approximately 80% chance of cure; in contrast, with bigger tumors, particularly when invasive, the cure rate may be as low as 52%.

In addition, surgery can induce hypopituitarism, and the recurrence of the disease is not an uncommon event.

Conventional external irradiation is an effective treatment in patients with acromegaly who have relatively modest GH elevations. The normalization of GH levels occurs in 30%-80% of the patients within 6-24 months but may sometimes be delayed as long as 5-10 years.

Pituitary implantation of yttrium or proton-beam therapy produces a satisfactory response in 27%-50% of patients but a high proportion are rendered hypopituitary.

No successful medical treatment was found until 1972, when Liuzzi et al. observed that the dopamine agonist L-dopa could lower GH levels in about 50% of patients. The goal of a stable pharmacological suppression of GH levels in acromegaly was achieved (Liuzzi et al. 1974) by the use of the long-acting dopaminergic drug bromocriptine (Br). However, only about 30% of patients

* This study was supported by the Italian National Research Council Special Project Oncology, grant no. 86.00675.44.

achieve normalization of GH and somatomedin- C (Sm-C) levels, and multiple daily administrations of Br or other dopamine agonists may be required.

Recently, new slow-release formulations of Br (Parlodel LAR and Parlodel MR) and a somatostatin analog with long-lasting inhibitory activity on GH release (SMS 201-995) have been made available for clinical use. These agents seem to improve the effectiveness of the pharmacological treatment of this disease.

The aim of this paper is to summarize the great number of studies and our own experience on dopamine agonists (DAs) and on SMS 201-995 (SMS) in order to establish the present status of the medical treatment of acromegaly).

EFFECTS OF DOPAMINE AGONISTS ON GH AND SM-C LEVELS

Although all the DAs capable of directly stimulating dopamine receptors inhibit GH release in acromegaly, only a few of them have been used in the medical treatment of this disease due to their prolonged activity. Br is the most widely used drug; other DAs, such as lisuride (Liuzzi et al. 1978), pergolide (Kendall-Taylor et al. 1982), or CU-32085 (Besser et al. 1983), however, have been proven equally effective in reducing GH levels in acromegaly, whereas a lower GH inhibitory effect is provided by metergoline (Chiodini et al. 1976). Recently, we have had some experience with a derivative of lisuride, transdihydrolisuride. This drug is of particular interest since it possesses a GH - lowering effect slightly weaker than that of lisuride but with a remarkably lower incidence of side effects (Dallabonzana et al. 1986).

As far as the data on Br are concerned, the results of most studies in the literature (Thorner et al. 1975; Chiodini et al. 1975; Summers et al. 1975; Sachdev et al. 1975; Schwinn et al. 1976; Belforte et al. 1977; Quabbe 1981; Wass et al. 1982a) show that chronic administration of the drug can consistently reduce GH levels in acromegaly.

Some discrepancies emerge, however, when one evalutates the percentage of patients who have benefited from the treatment. This is due mainly to the lack of common criteria for assessing treatment effectiveness. In our opinion, the treatment may be considered effective when GH levels are reduced to at least 50% below the pretreatment values. Clearly, this does not imply a normalization of GH levels. In our series of 96 patients collected since 1974, a 50% reduction of GH levels has been achieved in 56% of the patients during Br treatment but a normalization of hormone levels in only 23% of the 96. This figure agrees with the data from most of the studies in the literature. In particular, Quabbe (1981) has reported that 27% of his 238 patients had GH levels normalized by Br. Lindholm et al. (1981) claimed that Br is ineffective in acromegaly, since no differences in the GH response to an oral glucose test during placebo or Br treatment were observed.

However, this test may be unaltered by Br treatment, despite lowering of basal GH levels; therefore, the conclusions of this study seem untenable.

The problem of the dose regimen has also been widely discussed. Wass et al. (1982a) reported that by increasing the doses of Br up to 60 mg/day the percentage of responsive patients may be increased. Other studies have demonstrated that the maximal dose of Br to be used is 20 mg/day (Sachdev et al. 1973). We agree with the latter conclusion, since in our series doses higher than 20 mg/day produced neither a more marked decrease of GH levels in patients responsive to lower doses nor a GH-lowering effect in unresponsive patients.

The GH-lowering effect of DAs can be maintained even for years in those patients in whom the side effects do not prevent the administration of effective drug doses. In fact, patients treated for up to 10 years with Br continue to have a suppression of GH and Sm-C levels and improvement of clinical symptoms.

Although it has been reported that after stopping the treatment GH levels may be lower than those measured in pretreatment conditions (Thorner et al. 1975), we have constantly found that, even after several years of treatment, drug withdrawal is followed (within 24-48 h) by a sharp increase of GH levels to the basal values.

In addition, the TRH-induced GH release is not impaired by the chronic administration of DAs (Schwinn et al. 1976; Belforte et al. 1977). This finding suggests that DAs impair only the release but not the synthesis of GH in acromegaly.

The dopaminergic treatment is accompanied by an improvement of clinical features and of the metabolic parameters altered by the raised GH levels, such as glucose tolerance or hyperhydroxyprolinuria. A widely discussed question is whether these changes are linked to the GH-lowering effect of DAs. Although these clinical and metabolic parameters generally improve only in the GH-responsive patients, there is evidence that they can also ameliorate independent of the lowering of GH levels.

To investigate this problem, Wass et al. (1982b) studied plasma Sm-C levels in acromegalic patients treated with Br, and classified theree groups of patients. The first group showed a clear clinical and metabolic improvement associated with a fall in circulating GH and Sm-C levels. The second group showed no clinical, metabolic, or hormonal changes. The third group was the most interesting, since GH levels did not change but there was a consistent clinical and metabolic amelioration and a clear-cut fall in plasma Sm-C levels.

The findings of this last group have been explained by the fact that Br has been shown to alter the proportion between the monomeric and the polymeric, less-active form of circulating GH in favor of the latter. Since both the GH forms, monomeric and polymeric, are measured by radioimmunoassay methods, a shift in the proportion of one form to another would not necessarily be recorded by changes in immunoreactive GH levels.

Wass et al. (1982b) suggested, therefore, that Sm-C levels correlate better with the clinical response to Br than GH levels do. However, Nortier et al. (1985) were unable to confirm this finding; in their series, six patients showed clinical improvement with hardly any reduction in the GH secretion, and Sm-C levels remained unchanged. Accordingly, our results obtained in 20 patients (Oppizzi et al. 1986) indicate a close dependency of Sm-C levels on plasma GH levels, although similar GH levels may have different somatomedin-stimulating activity.

In conclusion, it is difficult for us to explain why acromegaly may clinically improve during Br treatment in spite of apparently unchanged GH levels. On considering also the data of Moses et al. (1981), who reported a good clinical response to chronic Br treatment in six of seven acromegalic patients with considerably decreased GH levels, and only a small decrease of plasma Sm-C levels in five of them, we feel that caution should be used if only Sm-C levels are utilized to assess disease activity in acromegalic patients.

EFFECTS OF DOPAMINERGIC DRUGS ON TUMOR SIZE

The major problem to be solved in order to establish a clear-cut role of medical treatment in the therapeutic approach to GH-secreting adenomas is whether these drugs, in addition to their activity on GH secretion, can also affect the growth of the tumor, as they do for prolactin-secreting adenomas.

This problem was raised for the first time by the report of Wass et al. (1979), showing an improvement of visual field defects in acromegalic patients under Br treatment. To date, several studies have described the effects of DAs on the size of GH-secreting adenomas, but the number of acromegalics evaluated in each study is limited and the results obtained are conflicting (McGregor et al. 1979; Spark et al. 1982; Wollensen et al. 1982). It has been discussed whether the shrinkage of the tumor could be due to an effect of Br on the lactotrophic portion of the tumor. However, Besser and Wass (1984) reported that also in the presence of normal prolactin levels, about 45% of GH-secreting tumors have a significant size reduction. They claimed that the size changes are less dramatic because pure GH-secreting tumors are smaller than the ones that secrete both prolactin and GH.

In the past few years we have studied 32 patients, selected on the basis of the presence of large (greater than 10 mm in diameter) adenomas, treated with DAs for 8 - 30 months (mean 12.6); seven of these 32 patients were markedly (i.e., more than 200 ng/ml) hyperprolactinemic.

Computed tomography (CT) revealed a reduction of the tumor size in only two of the 25 normoprolactinemic but in four of the seven hyperprolactinemic patients. If the data on GH levels and tumor size are compared in the normoprolactinemic patients, it appears that no tumor shrinkage was obtained in any of the GH-nonresponsive patients, and that the two patients whose tumors shrank were GH responders. However, in 12 of 18 GH-responsive patients the tumor size was not affected by the treatment. In the hyperprolactinemic patients the shrinkage of the tumor occurred independent of their GH responsiveness.

Collectively, these data demonstrate that: (a) tumor size reduction is much more frequent in hyperprolactinemic acromegalics than in normoprolactinemic (56% vs 8%); (b) the suppression of the GH hypersecretion is a necessary but not sufficient condition to obtain a reduction of the tumor, at least in patients with normal prolactin levels; and (c) the mechanism which brought about the size reduction of the GH-secreting tumors, although widely discussed, is still unknown. On considering the greater incidence of tumor size reduction in hyperprolactinemic compared with normoprolactinemic acromegalics, we suggest that DAs exert their effects by reducing, similar to what has been shown for macroprolactinomas, the cellular volume of prolactin-secreting cells largely represented in a mixed GH- and prolactin – cell adenoma.

These tumors, indeed, which constitute 25% of all GH-secreting adenomas, are associated not only with elevated but, less frequently, also with normal prolactin levels (Melmed et al. 1983).

EFFECT OF NEW LONG-ACTING PREPARATIONS OF BROMOCRIPTINE

In an attempt to avoid the problem of multiple daily administration and to reduce the fluctuation of GH levels, new formulations of Br provided with longer bioavailability have recently been proposed. We have studied the effectiveness of two of these in acromegalic patients.

Parlodel LAR (Long-Acting Repeatable)

Fifty milligrams of the injectable form of Br were administered monthly for 1-4 months to five patients responsive to the oral form of this drug. GH levels were determined at days 0, 7, 14, and 28 after the first injection and then at 28-day intervals. A CT was performed basally and 28 days after the first injection.

A significant reduction of mean daily GH was obtained until the 28th day, with a maximal inhibition on the 4th day (Fig. 1) Superimposable results were obtained when the drug injection was repeated. CT did not reveal any reductions in tumor size.

Parlodel MR (Modified Release)

Ten Br-responsive patients were treated with increasing doses of Parlodel MR (MR) (2.5, 5, and 10 mg p.o. once daily at 9 p.m. for 1-3 months), and plasma samples for GH determination were taken at day 0 and then at 15-day intervals, 12 and 24 h after drug intake.

GH levels were found to be significantly reduced below the pretreatment value 12 h after the ingestion of MR at each of the three doses studied. Only the dose of 10 mg kept GH levels still significantly reduced after 24 h. Sm-C levels were significantly lower than baseline during treatment with 5 and 10 mg (Fig. 2).

Collectively, the chronic administration of these new agents, by obtaining a more prolonged GH suppression with minimal side effects, may improve the compliance to the treatment; however, their overall effects are not different from those obtained by the use of the currently available DA agents.

TREATMENT WITH SMS 201-995

Several studies (Plewe et al. 1984; Lamberts et al. 1985; Cozzi et al. 1986; Chiodini et al. 1987) on the therapeutic value of a somatostatin analog (SMS 201-995), provided with a powerful and long-lasting inhibitory effect on GH secretion, have already been performed and have produced evidence that this agent reduces GH levels in acromegalic patients and reverses the metabolic consequences of the hypersecretion of this hormone. Daily dosages of 100-1500 μg given in two or

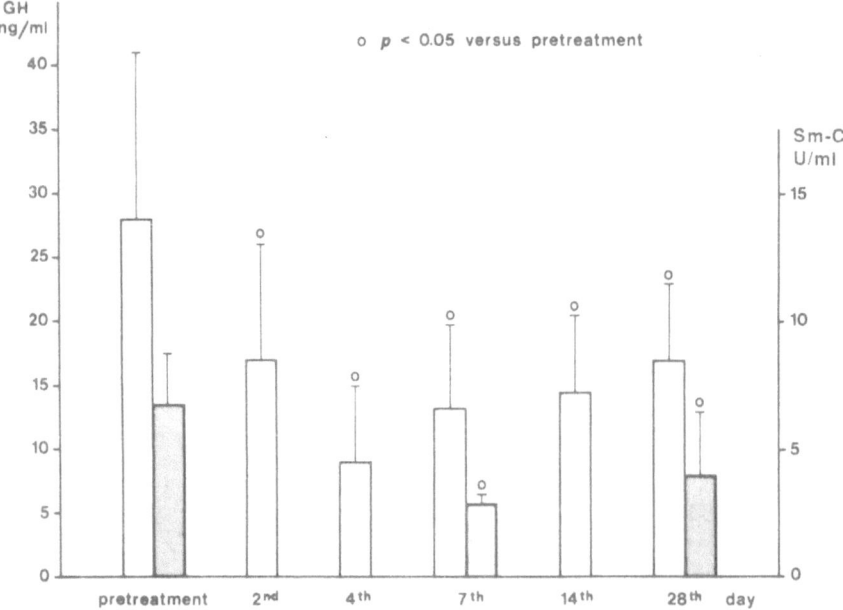

Fig. 1. Mean GH and Sm-C levels after Parlodel LAR (50 mg monthly) in five acromegalic patients.

three subcutaneous injections have been reported to reduce GH and Sm-C levels in the majority of patients studied, with a wide intersubject variability. The effect of the drug is already evident at the beginning of the treatment and remains unchanged as long as SMS is administrered.

We have studied the effects of SMS given either by multiple subcutaneous injections (MI) or by continuous infusion (CI) with a minipump on a large series of acromegalic patients during long-term treatment. In addition, in a few patients the effects of a combined treatment with SMS and Br were evaluated.

Multiple Daily Injections

We have treated 26 patients with SMS by MI at doses of 100-1500 µg daily for 3-27 months. Basal plasma GH (55.8 ± 8 ng/ml) and Sm-C levels (8.6 ± 4 units/ml) significantly decreased by the second day and fell after 15 days of treatment to 12 ± 3 ng/ml (p <0.01) and to 2.5 ± 0.6 units/ml (p <0.05) respectively. Thereafter,

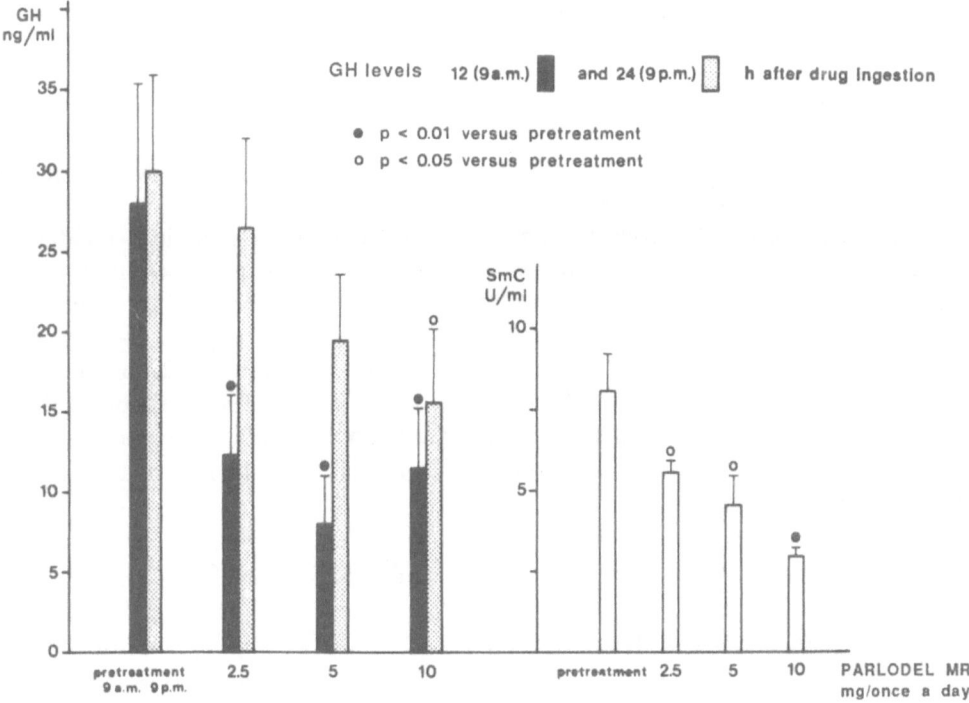

Fig 2. GH and Sm-C levels before and during treatment with Parlodel MR in ten acromegalic patients.

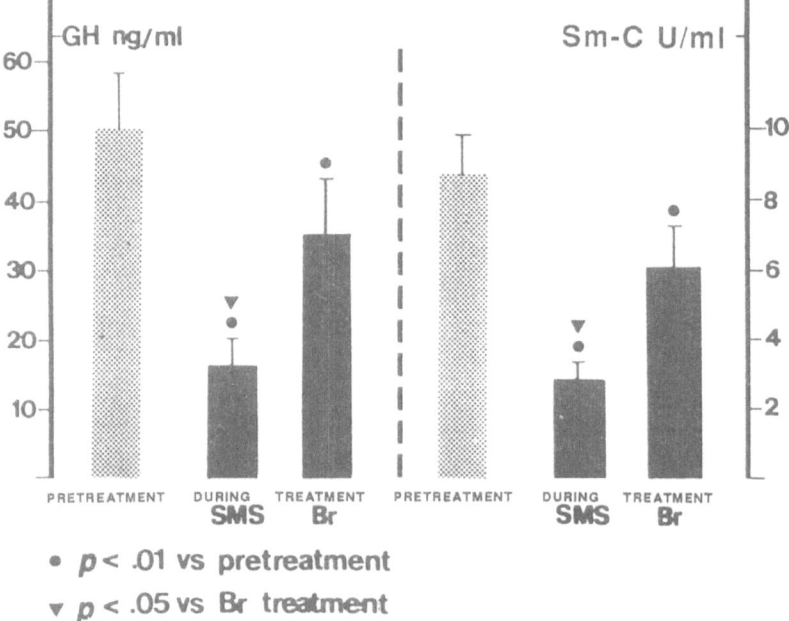

Fig 3. GH and Sm-C suppression during chronic treatment with SMS 201-995 and Br in 13 acromegalic patients.

there were no further significant changes of GH and Sm-C levels, even in the patients treated for up to 27 months.

The mean daily GH profile showed a maximal GH inhibition within 2-3 h after each SMS administration and a progressive return close to preinjection values within 6-12 h. This GH pattern was less evident, however, when 100 μg or more tid were used; indeed, GH levels were in the normal range throughout the day in three of the nine patients treated with these dosages. Collectively, 22 of the 26 patients (85%) had their GH and Sm-C levels reduced by at least 50% by SMS. In a few patients a true refractoriness to the drug activity could be demonstrated; one patient in particular failed to reduce her GH levels even when given 1500 μg/day of the drug.

Comparison Between Bromocriptine and SMS 201-995

A comparison between the long-term effects of Br (20 mg/day) and of SMS (100-300 μg/day) performed in 13 of these 26 patients revealed that SMS induces a greater GH and Sm-C inhibition than Br does (Fig. 3). Plasma GH and Sm-C levels were lower during SMS than during Br in 11 of the 13 patients; in addition, only two patients were unresponsive to SMS, whereas 50% of the patients did not respond to Br.

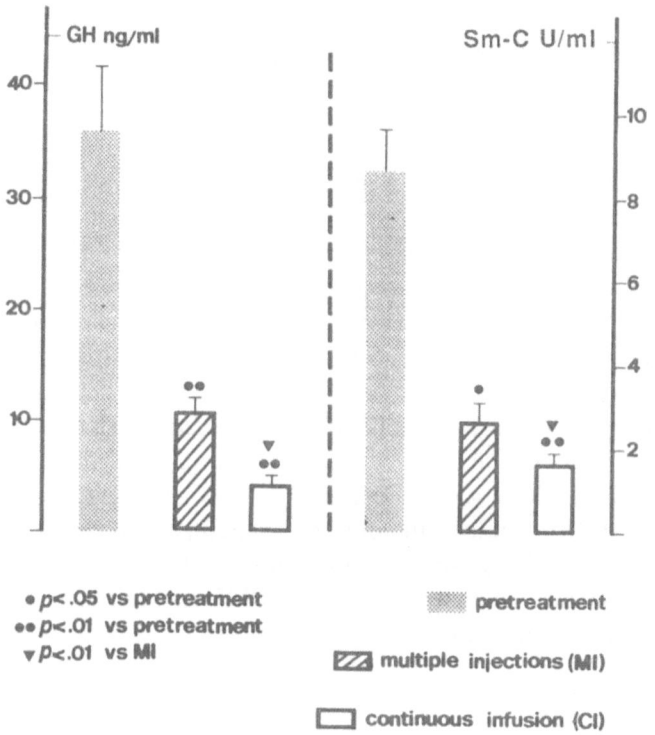

Fig. 4. GH and Sm-C levels (mean and SEM) before and after 15 days of treatment with SMS 201-995 by multiple injections or by continuous infusion in nine acromegalic patients.

Continuous Infusion of SMS 201-995

Long-term treatment was given to nine patients treated either by MI or CI with the same dosage of SMS (100-600 µg daily) for at least 2 months: GH and Sm-C levels significantly fell with either kind of treatment within the second day but, during CI, a significantly more marked decrease of GH and Sm-C was observed (Fig. 4).

The analytical evaluation of the data revealed normalization of GH levels over the day and of Sm-C in all cases during CI, but in no case during MI. The prolongation of CI treatment for 12-18 months in four patients showed a stabilization of GH and Sm-C levels within the normal range, but the discontinuation of the therapy was followed by the return of GH and Sm-C to pretreatment levels within 24-36 h just as after MI treatment withdrawal.

Effects of SMS 201-995 Treatment on the Clinical and Metabolic Parameters

The signs and symptoms of the disease ameliorated in all patients whose GH levels were decreased even if not normalized during SMS treatment, both by MI and by CI.

Improvement of the metabolic alterations related to GH hypersecretion also occurred. Serum procollagen III propeptide (PIIIP) levels fell (P<0.01) from 16.5 ± 1.8 to 8.4 ± 0.5 ng/ml. Overt diabetes improved, and in two of the three diabetics insulin therapy could be stopped. Slight and transient postprandial hyperglycemia occurred in six patients in whom glucose tolerance had previously been normal. Fasting and postprandial IRI levels did not significantly change during treatment.

CT scans in comparable coronal sections revealed a reduction in tumor size in ten of the 20 patients with evidence of a large adenoma (eight during MI and two during CI). This figure of tumor size reduction, also observed by other authors (Lamberts et al. 1985; Barnard 1986; Tolis et al. 1986) and greater than that observed during Br treatment, suggests that SMS also exerts an antitumoral effect on GH-secreting adenomas.

Combined Treatment with SMS 201-995 and Br

Since Br, unlike SMS, does not inhibit the GH-releasig effect of GHRH, different pathways mediate the GH-lowering activity of the two drugs. This finding prompted us to evaluate the therapeutic activity of combined treatment with these two agents in four patients. We have thus observed that low doses of SMS, given by MI, and of Br induced a significantly greater reduction of daily plasma GH and Sm-C levels compared with treatment with either drug at the same dosages; the levels became normal in two of the patients so treated.

CONCLUSIONS

In accordance with the results obtained in smaller series (Plewe et al. 1984; Ch'ng et al. 1985; Lamberts et al. 1985; Landolt et al. 1985), chronic treatment with SMS induced a long-term GH and Sm-C suppression in 85% of our patients. This figure is far greater than that obtained with DAs.

The degree of the GH suppression in relation to the dosage was unchanged, even during treatment prolonged for up to 27 months, indicating that a desensitization of the pituitary receptors for somatostatin does not occur during SMS treatment. The findings that plasma GH increased again 6-12 h after each SMS dose and that the withdrawal of SMS was always followed by a fast return of plasma GH and Sm-C levels near to the pretreatment levels suggest that SMS may inhibit the release rather than the synthesis of GH.

The results of the dose-response study indicate that a satisfactory control of hypersecretion during SMS by MI is dependent on both the timing of administration and the dose, since a stable normalization of GH levels could be obtained only with 100 μg tid.

A stable GH suppression to normal levels throughout the day may not be necessary however, for the control of the disease, since plasma Sm-C fell into the normal range as well in seven patients whose GH levels had not been normalized. In addition, an impressive improvement of clinical symptoms and of metabolic alterations due to GH hypersecretion were observed in most patients during the treatment, even when plasma GH and Sm-C were not normalized.

We did not find any consistent long-term changes of glucose tolerance in the patients with previously normal blood glucose, in agreement with the data reported by Lamberts et al. (1985), and IRI levels were not suppressed by SMS treatment as might have been expected. On the contrary, diabetes improved, probably as a consequence of the GH suppression induced by SMS.

The comparative evaluation of the results obtained with SMS and Br is of clinical and pathophysiological interest: 100-300 μg/day of SMS were actually more effective in reducing GH and Sm-C levels than 20 mg/day of Br. Since, in our experience, higher Br dosages do not obtain any further suppression of GH levels, we can conclude that the chronic GH-lowering effect of SMS is more pronounced than that of Br. However, in some patients partially responsive either to Br or to SMS the combined treatment seems to have additional effects.

In previous studies (Lamberts et al. 1985; Landolt et al. 1985; Barnard et al. 1986; Tolis et al. 1986) a slight reduction of the tumor size was observed during chronic SMS administration. In the present series the tumor shrank in ten of 20 patients so far examined by CT. These findings indicate that SMS may exert an antitumoral effect on the GH-secreting adenomas, but this problem requires more extensive studies.

In conclusion, SMS is a more effective agent than Br in the medical treatment of acromegaly and may be considered, especially when given by continuous infusion, the drug of first choice. In fact, the percentage of patients whose GH and Sm-C are normalized or at least markedly reduced is far higher than that well established for the dopamine agonists. In addition, tumor shrinkage is, in our experience, more frequent with SMS than with these compounds (Table 1).

However, the long-acting dopaminergic compounds can represent an effective and more practical alternative to SMS treatment for selected patients.

Since SMS and Br act by different mechanism to lower GH levels they may have complementary roles in the medical treatment of acromegaly.

Although the ideal form of medical treatment of acromegaly is still to be established, the appropriate use of different pharmacological tools may allow a good control of the disease.

Therefore, it is likely that the prophecy of Cushing is valid: in 1932 he wrote: "The time will come, ere long perhaps, when the biochemists will show us how to cure most of the common functional adenomas of this gland."

Table 1. Overall evaluation of bromocriptine and SMS 201-995 effects in acromegaly

| | GH levels | | Tumor reduction | Clinical and metabolic improvement |
	Reduction	Normalization		
Bromocriptine (7.5–20 mg/day)	55/98 (56%)	23/98 (23%)	6/32 (18%)	58/98 (59%)
SMS 201–995 (100–1500 µg/day)	22/26 (85%)	9/26 (35%)	10/20 (50%)	22/26 (85%)

REFERENCES

Barnard L (1986) Reduction of pituitary tumor size in acromegaly treated with a somatostatin analogue (SMS 201-995). 68th Meeting of Endocrine Society, June 25-27, Anaheim (abstract 983)

Belforte L, Camanni F, Chiodini PG, Liuzzi A, Massara F, Molinatti GM, Muller EE, Silvestrini F (1977) Long-term treatment with 2Br-alpha ergocryptine in acromegaly. *Acta Endocrinol* (Copenh) **85**: 235-241

Besser GM, Wass JAH (1984) The medical treatment of acromegaly. In: Black PL, Zervas NT, Ridgway EC, Martin JB (eds) *Secretory Tumors of the Pituitary Gland.* Raven, New York, pp 155-168

Besser GM, Wass JAH, Grossman A, Moult PJA, Boulox P (1983) Hormonal and clinical aspects of dopamine agonists. In: Calne DB, Horowski H, McDonald JM, Wuttke W (eds) *Lisuride and Other Dopamine Agonists.* Raven, New York, pp 239-254

Ch'ng LJC, Sandler LM, Kraenzli ME, Burrin JM, Joplin JF, Bloom SR (1985) Long-term treatment of acromegaly with a long-acting analog of somatostatin. *Br Med J* **290**: 284-287

Chiodini PG, Liuzzi A, Botalla L, Oppizzi G, Müller EE, Silvestrini F (1975) Stable reduction of plasma growth hormone (hGH) levels during chronic administration of 2Br-alpha-ergocryptine (CB 154) in acromegalic patients. *J Clin Endocrinol Metab* **40**: 705-708

Chiodini PG, Liuzzi A, Müller EE, Botalla L, Cremascoli G, Oppizzi G, Verde G, Silvestrini F (1976) Inhibitory effect of an ergoline derivative, methergoline, on growth hormone and prolactin levels in acromegaly. *J Clin Endocrinol Metab* **43**: 956-961

Chiodini PG, Cozzi R, Dallabonzaa D, Oppizzi G, Verde G, Petroncini M, Liuzzi A, del Pozo E (1987) Medical treatment of acromegaly with SMS 201-995, a somatostatin analog: a comparison with bromocriptine. *J Clin Endocrinol Metab* 64: 447-450

Cozzi R, Chiodini PG, Dallabonzana D, Petroncini M, Verde G, Oppizzi G, Liuzzi A (1986) Effect of SMS 201-995 administered by minipump infusion in acromegaly. International Conference on Somatostatin. Washington, May 6-8, 1986, p 64

Cushing A (1932) The basophil adenomas of the pituitary body and their clinical manifestations (pituary basophilism). *Bull Jonhs Hopkins Hosp* 50: 137

Dallabonzana D, Liuzzi A, Oppizzi G, Cozzi R, Verde G, Chiodini PG, Rainer E, Dorow R, Horowski H (1986) Chronic treatment of pathological hyperprolactinemia and acromegaly with the new ergot derivative terguride. *J Clin Endocrinol Metab* 63: 1002-1007

Dunn P, Donald R, Espiner A (1977) Bromocriptine suppression of plasma growth hormone in acromegaly. *Clin Endocrinol* (Oxf) 7: 273-276

Kendall-Taylor P, Upstill-Goddard G (1982) The effect of pergolide in acromegaly. 64th Meeting of Endocrine Society, San Francisco, June 16-18 (abstract 98)

Lamberts SWJ, Uitterlinden P, Verschoor L, van Dongen KJ, del Pozo E (1985) Long-term treatment of acromegaly with the somatostatin analogue SMS 201-995. *N Engl J Med* 313: 1576-1578

Landolt AM, Osterwalder V, Jantzer R, Stuckmann G (1985) The medical treatment of acromegaly with lisuride. *Neuroendocr Lett* 2: 93-96

Lindohlm J, Riishede J, Vestergaard S, Hummer L, Faber O, Hagen C (1981) No effect of bromocriptine in acromegaly. *N Engl J Med* 304: 1450-1453

Liuzzi A, Chiodini PG, Botalla L, Cremascoli G, Silvestrini F (1972) Inhibitory effect of L-dopa on GH release in acromegalic patients. *J Clin Endocrinol Metab* 35: 941-943

Liuzzi A, Chiodini PG, Botalla L, Cremascoli G, Müller EE, Silvestrini F (1974) Growth hormone (GH)-releasing activity of TRH and GH-lowering effect of dopaminergic drugs in acromegaly: homogeneity in the two responses. *J Clin Endocrinol Metab* 38: 910-912

Liuzzi A, Chiodini PG, Oppizzi G, Botalla L, Verde G, De Stefano G, Colussi G, Graf KJ (1978) Lisuride hydrogen maleate: evidence for a long-lasting dopaminergic activity in humans. *J Clin Endocrinol Metab* 46: 196-200

McGregor AM, Scanlon MF, Hall R, Hall K (1979) Effects of bromocriptine on pituitary tumour size. *Br Med J* 2: 700-703

Melmed SL, Braunstein GD, Horvath E, Ezrin C, Kovacs K (1983) The pathophysiology of acromegaly. *Endocr Rev* 4: 271-297

Moses AC, Molitch ME, Sawin CT, Jackson IMD, Biller BJ, Furlanetto R, Reichlin S (1981) Bromocriptine therapy in acromegaly: use on patients resistant to conventional therapy and effect on serum levels of somatomedin-C. *J Clin Endocrinol Metab* 53: 752-756

Nortier JWR, Croughs RJM, Thijssen JHH, Schwarz F (1985) Bromocriptine therapy in acromegaly: effects on plasma GH levels, somatomedin-C levels and clinical activity. *Clin Endocrinol* (Oxf) 22: 209-217

Oppizzi G, Petroncini MM, Dallabonzana D, Cozzi R, Verde G, Chiodini PG, Liuzzi A (1986) Relationship between somatomedin-C and growth hormone levels in acromegaly: basal and dynamic evaluation. *J Clin Endocrinol Metab* 63: 1348-1353

Plewe G, Beyer J, Krause U, Neufeld M, del Pozo E (1984) Long-acting and selective suppression of growth hormone secretion by somatostatin analogue SMS 201-995 in acromegaly. *Lancet* ii: 782-784

Quabbe HJ (1981) Treatment results in 235 patients with acromegaly. A report of the Acromegaly Study Group. *Acta Endocrinol* (Copenh) [Suppl] **240**: 66

Sachdev J, Tunbridge WMG, Weightman DR, Gomez-Pan A, Hall R (1975) Bromocriptine therapy in acromegaly. *Lancet* ii: 1164-1167

Schwinn G, Schwark H, McIntosh C, Milstray HR, Willms B, Köbberling J (1976) Effect of the dopamine receptor-blocking agent pimozide on the GH response to arginine and exercise and on the spontaneous growth hormone fluctuations. *J Clin Endocrinol Metab* **43**: 1183-1187

Spark RF, Baker R, Bienford C, Borgland R (1982) Bromocriptine reduces tumor size and hypersecretion. Requiem for surgery? *JAMA* **247**: 311-316

Summers VK, Hipkin, Diver Mj, Davis JC (1975) Treatment of acromegaly with bromocriptine. *J Clin Endocrinol Metab* **40**: 494-497

Thorner MO, Chait A, Aitken M, Benker G, Bloom SR, Mortimer CH, Sanders P, Stuart Mason A, Besser GM (1975) Bromocriptine treatment of acromegaly. *Br Med J* **1**: 299-303

Tolis G, Yotis A, Malachtari S, Rigas G, Hortoglou A, del Pozo E, Andreu I, Papavasileiu C (1986) Follow-up of acromegalic patients treated with an octapeptide somatostatin analogue for over a year. 68th Meeting of Endocrine Society, June 25-27 Anaheim (abstract 746)

Wass JAH, Moult PJ, Thorner MO, Dacie JE, Charlesworth M, Jones ME, Besser GM (1979) Size reduction of pituitary tumours in patients with prolactinomas and acromegaly treated with and without radiotherapy. *Lancet* ii: 66-70

Wass JAH, Williams J, Charlesworth M, Kindsley DPE, Halliday AM, Doniach I, Rees LN, MacDonald VI, Besser GM (1982a) Bromocriptine in management of large pituitary tumours. *Br Med J* **284**: 1908-1912

Wass JAH, Clemmons DR, Underwood LE, Barrow I, Besser GM, Van Vyk JJ (1982b) Changes in circulating somatomedin-C levels in bromocriptine-treated acromegaly. *Clin Endocrinol* (Oxf) **17**: 369-373

Wollensen F, Andersen I, Karle K (1982) Size reduction of extrasellar pituitary tumors during bromocriptine treatment. *Ann Intern Med* **96**: 281-286

Advances in Growth Hormone and Growth Factor Research,
edited by E.E. Müller, D. Cocchi and V. Locatelli
Pythagora Press, Roma-Milano and Springer Verlag, Berlin-Heidelberg © 1989

Effect of Sandostatin on glucose metabolism and counterregulatory mechanisms: studies in normal volunteers

E. del Pozo[1], C. Sieber[1], M. Neufeld[1], H. Berthold[1], and P.H. Althoff[2]

[1] *Experimental Therapeutics Department, Clinical Research, Sandoz Ltd., Basel, Switzerland; and* [2] *University Medical Clinic, Frankfurt am Main, Federal Republic of Germany*

The stepwise modification of the active core of natural somatostatin (SRIF) by Bauer et al. (1982) led to the synthesis of a potent octapeptide, SMS 201-995 (Sandostatin). Studies conducted in experimental animals have confirmed its GH-inhibitory effect, and in human trials Sandostatin was found to exhibit an action profile similar to that of the native peptide (del Pozo et al. 1986), but the elimination half-life following subcutaneous administration was 113 min (del Pozo et al. 1986), far beyond the 2-3 min reported for SRIF (Sheppard et al. 1979).

Because of its profound action on pancreatic endocrine and exocrine function, the effect of Sandostatin on glucose homeostasis has been investigated in normal subjects, in patients with insulin-dependent diabetes mellitus, and in acromegalics (Spinas et al. 1985; del Pozo et al. 1986, 1987; del Pozo and Kutz 1987; Plewe et al. 1987). The present report reviews the results of previous investigations and evaluates the effect of Sandostatin on insulin counterregulation in normal individuals under various experimental conditions.

EFFECT OF SANDOSTATIN ON BASAL AND STIMULATED B-CELL FUNCTION

Studies conducted in the mid 1970s have shown that the infusion of SRIF to normal fasting subjects leads to a biphasic blood glucose profile. Thus, an initial decline is followed by transient increase to values clearly above control levels (Gottesman et al. 1982), which can be explained by a dual mechanism consisting of early glucagon suppression as the cause of the moderate fall in circulating glucose, followed by inhibition of insulin release and decrease in peripheral glucose utilization, although this latter mechanism has recently been contested (Meneilly et al. 1987).

In patients on exogenous insulin replacement, metabolic changes induced by SRIF are partially compensated. However, in subjects with normal pancreatic endocrine function or in acromegaly, repeated exposure to this peptide can give rise to marked hyperglycemic fluctuations through inhibition of insulin secretion (del Pozo et al. 1987; Lamberts et al. 1987). In the following sections, the effects of Sandostatin on postprandial glucose control and on the modulation of counter-regulatory mechanisms are analyzed.

Figure 1 presents the glucose and insulin profiles recorded during infusions of 5, 10, and 20 µg/h Sandostatin to normal volunteers. Hypoinsulinemia is accentuated with increasing doses to produce reduction up to 78% of basal (del Pozo and Kutz 1987). Blood glucose presents little variation, although a tendency to adopt a sinusoidal profile can be clearly recognized with the higher doses. This indicates that untreated subjects require little insulin to maintain basal glucose within the normal range. In turn, young subjects exhibiting constantly elevated basal blood glucose are suspect of having lost most of their pancreatic B-cell functional capacity, a typical feature of insulin-dependent diabetes.

The situation is quite different when the ability to release insulin by the B cells is challenged by a meal containing carbohydrates during insulin blockade with somatostatin. In order to assess the effect of subcutaneous administration of Sandostatin on the hormonal changes induced by a standard mixed meal, a three-step study was conducted in normal volunteers. Hormonal parameters measured were plasma insulin and glucagon, and blood glucose. The composition of food consisted of 30 g protein, 75 g carbohydrates, and 50 g fat. In a first trial, single subcutaneous administration of 100 g Sandostatin to six male subjects just before food ingestion elicited a reduction in circulating insulin followed by a marked increase in blood glucose (Fig. 2, upper panel). As expected, the postprandial glucose peak was delayed due to retardation of intestinal absorption of sugar (del Pozo and Kutz 1987). Reactive hyperglycemia was considerably attenuated in the second step, when food was ingested 180 min after injection (Fig. 2, middle panel). Indeed, mean values of integrated blood glucose profiles (Fig. 2, right upper and middle panels) were reduced by 60% in comparison with concentrations recorded when Sandostatin was injected immediately before the meal. In addition, parallel GH and insulin measurements have previously shown the inhibitory effect of Sandostatin on the latter to be short-lived (del Pozo et al. 1986). Thus, appropriate timing between administration of the peptide and food ingestion may assist in controlling the effect of transient insulinopenia.

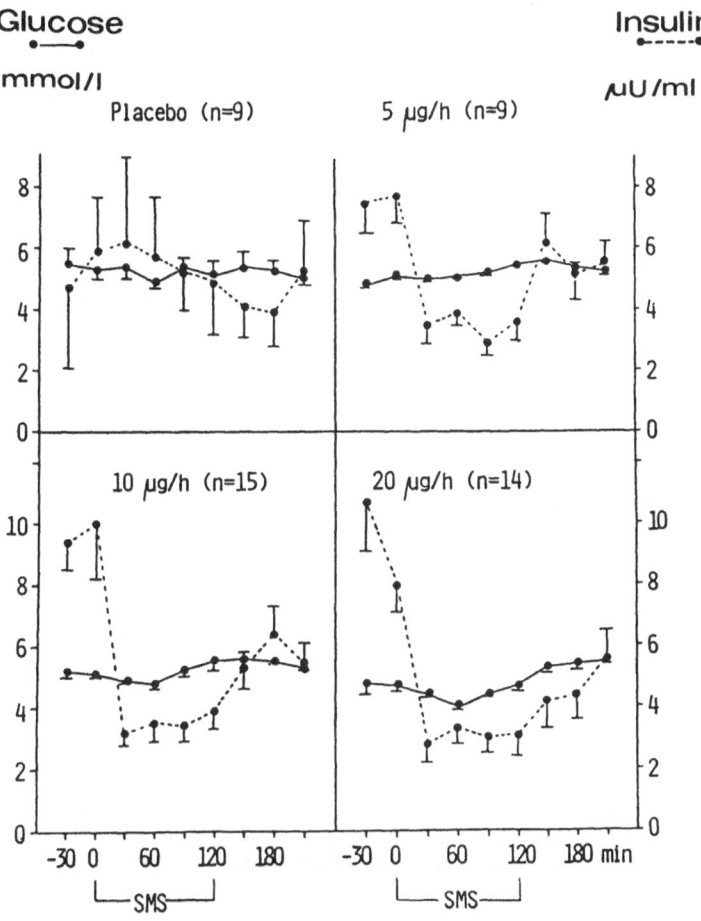

Fig. 1. Glucose and insulin profiles recorded during infusion of varying doses of Sandostatin to normal volunteers (explanation in text).

In a third step, the glucagon response to a single subcutaneous injection of 100 μg Sandostatin was studied in ten male volunteers following the ingestion of 300 g of lean beef. During the control period there was little change in blood glucose, accompanied by a modest elevation in circulating insulin (Fig. 2, lower panel). The injection of Sandostatin prevented the insulin discharge, and the integrated blood glucose values increased accordingly. This increment, however, represented only a small fraction when compared with the glucose reaction recorded after the mixed meal.

As expected, hyperglucagonemia induced by a protein meal was completely abolished by Sandostatin (Fig. 3), whereas the ingestion of carbohydrates, protein, and fat had disclosed only a discrete elevation. This latter effect, not shown in the

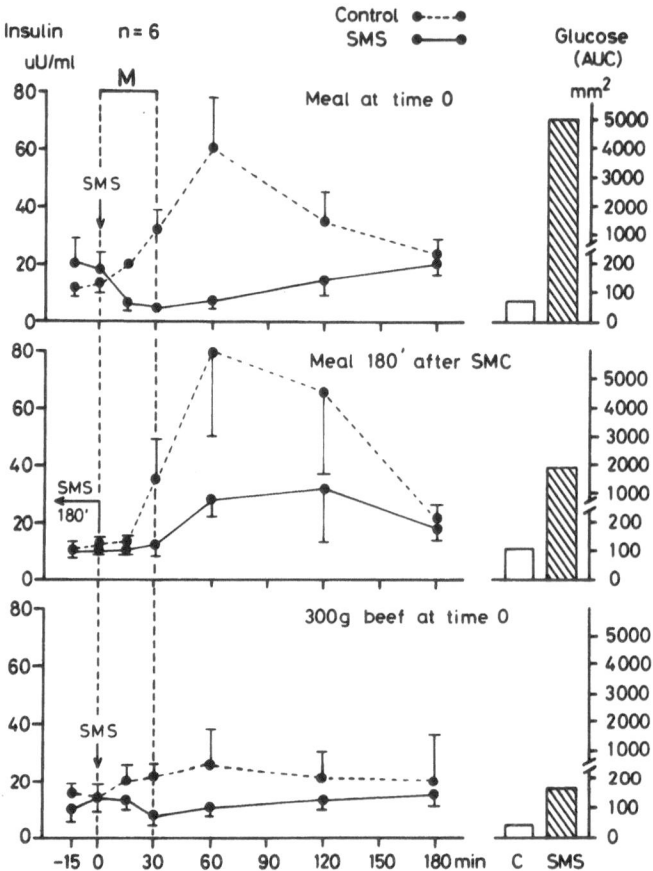

Fig. 2. Plasma insulin and glucose measured in a three-step trial using normal volunteers. *Upper panel,* Sandostatin (SMS) administered just before food ingestion; *middle panel,* Sandostatin administered 180 min before food ingestion; *lower panel,* Sandostatin administered following ingestion of lean beef.

figure, was also antagonized by Sandostatin. The degree of inhibition and the significance level were higher when glucagon release was induced by protein, since carbohydrates are known to dampen glucagon release.

It can be concluded that timely administration of Sandostatin and a reduction in the carbohydrate content of diet will markedly attenuate alterations in glucose homeostasis subsequent to the inhibitory effect of this peptide on B-cell function. Since natural somatostatin and the recently synthesized derivatives are potent antagonists of somatotropin and glucagon, two hormones known to oppose insulin action, the following section deals with the effect of Sandostatin on insulin counterregulatory mechanisms.

Glucagon

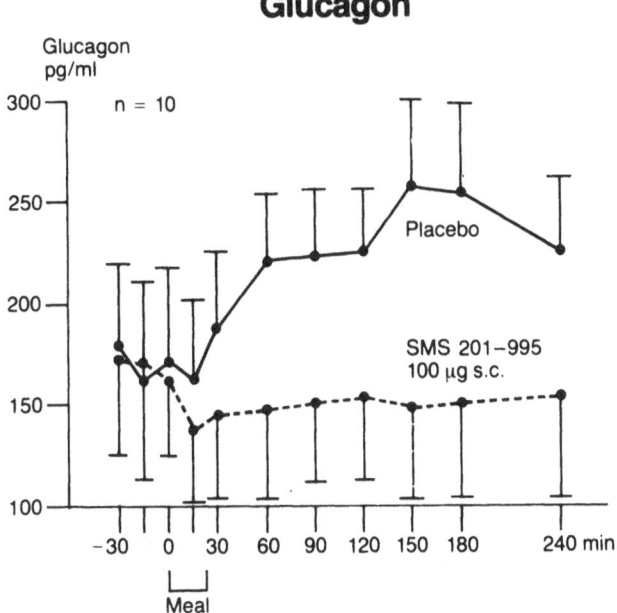

Fig. 3. Inhibitory effect of Sandostatin (SMS 201-995) on hyperglucagonemia induced by a protein meal.

STUDY OF COUNTERREGULATORY MECHANISMS FOLLOWING INSULIN HYPOGLYCEMIA AND SANDOSTATIN ADMINISTRATION

Maintenance of glucose homeostasis is the result of a complex mechanism, involving also short-loop connections between pancreatic A, B, and D cells (Gottesman et al. 1982). In human beings recovery from insulin-induced hypoglycemia is triggered by counterregulatory hormones, especially glucagon (Cryer 1981) and epinephrine, whereas norepinephrine seems to play a minor role (Welle et al. 1980; de Feo et al. 1983). Other antagonists such as growth hormone and corticosteroids are characterized by their relatively slow onset of action (Amiel et al. 1987; Schwartz et al. 1987). In patients with insulin-dependent diabetes (IDDM) the glucagon response to hypolycemia is impaired or even absent, probably through a lack of local adrenergic nerve activation within the pancreas (Hisatomi et al. 1985). Recovery from hypoglycemia is therefore dependent on adequate epinephrine and growth hormone secretion for insulin counteraction.

The availability of somatostatin derivatives with prolonged action has motivated a series of studies on the control of glucose metabolism. This was based on the antiketogenic effect mediated through glucagon suppression and a better stabilization of sugar balance through blockade of GH release. However, the effect of somatostatin on cortisol and catecholamine release as an important aspect of insulin counterregulation has not been thoroughly investigated.

In an experimental design eight normal male subjects were rendered hypoglycemic with an intravenous bolus injection of 0.075 units/kg of regular insulin (Rosak et al. 1986). The study was conducted under basal conditions and following an intravenous infusion of 50 μg Sandostatin. Results are depicted in Figure 4. Hypoglycemia was accentuated following injection of the peptide, probably as an expression of the known potentiating action of somatostatin on insulin (Gerich et al. 1981). Cortisol and catecholamine responses to the fall in blood glucose were enhanced by Sandostatin, reflecting integrity of counterregulatory mechanisms and adequate reserve capacity. Particularly epinephrine, instrumental factor in compensating insulin hypoglycemia (Cryer 1981), exhibited a significant increase above control values.

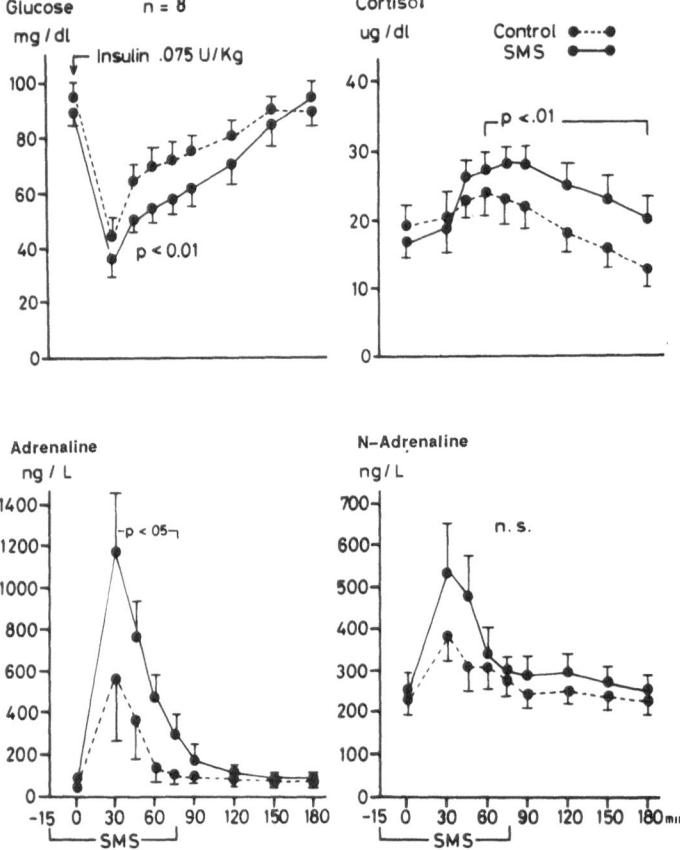

Fig. 4. Effects of Sandostatin on counterregulatory mechanisms following insulin hypoglycemia in eight normal male subjects (explanation in text).

Even small fluctuations of blood glucose levels within the physiological range are sufficient to provoke counterregulatory responses of catecholamines, which nevertheless can lie within the normal or near normal range in healthy (Santiago et al. 1980) and diabetic subjects (de Fronzo et al. 1980). This increase in catecholamine secretion seems to be produced by neurogenic mechanisms (Khalil et al. 1986). Therefore, caution is advised in the use of somatostatins in IDDM patients in whom epinephrine synthesis may be impaired by concomitant neuropathy (Kleinbaum and Shamoon 1983) or secondary to β-adrenergic blocking agents (Popp et al. 1982), since these subjects can experience more prolonged and severe insulin-induced hypoglycemic episodes (Rosak et al. 1986).

REFERENCES

Amiel SA, Simonson DC, Tamborlane WV, de Fronzo RA, Sherwin RS (1987) Rate of glucose fall does not affect counterregulatory hormone responses to hypoglycemia in normal and diabetic humans. *Diabetes* **36**: 518-522

Bauer W, Briner U, Doepfner W, Haller R, Huguenin R, Marbach P, Petcher TJ, Pless J (1982) SMS 201-995: a very potent and selective octapeptide analogue of somatostatin with prolonged action. *Life Sci* **31**: 1133-1140

Cryer PE (1981) Glucose counterregulation in man. *Diabetes* **30**: 261-264

de Feo P, Bolli G, Perriello G, de Cosmo S, Compagnucci P, Angeletti G, Santeusanio F, Gerich J, Motolese M, 'Brunetti P (1983) The adrenergic contribution to glucose counterregulation in type-1 diabetes mellitus. *Diabetes* **32**: 887-893

de Fronzo RA, Hendler R, Christensen N (1980) Stimulation of counterregulatory hormonal responses in diabetic man by a fall in glucose concentration. *Diabetes* **29**: 125-131

del Pozo E, Neufeld M, Schlüter K, Tortosa F, Clarenbach P, Bieder E, Wendel L, Nüesch E, Marbach P, Cramer H, Kerp L (1986) Endocrine profile of a long-acting somatostatin derivative, SMS 201-995. Study in normal volunteers following subcutaneous administration. *Acta Endocrinol* (Copenh) **111**: 433-439

del Pozo E, Lamberts SWJ, Sieber C, Gomez-Pan A (1987) Effect of a long-acting somatostatin analogue (SMS 201-995) on glucose homeostasis in type-1 diabetes and in acromegaly. In: Reichlin S (ed) *Somatostatin, Basic and Clinical Status*. Plenum, New York, pp 293-302

del Pozo E, Kutz K (1987) Pharmacological properties and effect on glucose homeostasis of a somatostatin derivative (SMS 201-995): studies in humans. *Prog Endocr Res Ther* **3**: 207-214

Gerich J, Haymond M, Rizza R, Verdonk C, Miles J (1981) Hormonal and substrate determinants of hepatic glucose production in man. In: Veneziale CM (ed) *The Regulation of Carbohydrate Formation and Utilization in Mammals*. University Park Press, Baltimore, pp 419-457

Gottesman IS, Mandarino LJ, Gerich JE (1982) Somatostatin: its role in health and disease. *Spec Top Endocrinol Metab* **4**: 177-243

Hisatomi A, Maruyama H, Orci L, Vasko M, Unger RH (1985) Adrenergically mediated intrapancreatic control of the glucagon response to glucopenia in the isolated rat pancreas. *J Clin Invest* **85**: 420-426

Khalil F, Marley PD, Livett BG (1986) Elevation in plasma catecholamines in response to insulin stress is under both neuronal and non-neuronal control. *Endocrinology* **119**: 159-167

Kleinbaum J, Shamoon H (1983) Impaired counterregulation of hypoglycemia in insulin-dependent diabetes mellitus. *Diabetes* **32**: 493-498

Lamberts SWJ, Uitterlinden P, del Pozo E (1987) SMS 201-995 induces a continuous decline in circulating growth hormone and somatomedin-C levels during therapy of acromegalic patients for over two years. *J Clin Endocrinol Metab* **65**: 703-710

Meneilly GS, Dariush E, Minaker KL, Rowe JW (1987) Somatostatin does not alter insulin-mediated glucose disposal. *J Clin Endocrinol Metab* **65**: 364-367

Plewe G, Noelken G, Krause U, Beyer J, del Pozo E (1987) Suppression of growth hormone and somatomedin-C by long-acting somatostatin analogue SMS 201-995 in type-1 diabetes mellitus. *Horm Res* **27**: 7-12

Popp DA, Shah SD, Cryer PE (1982) Role of epinephrine-mediated beta-adrenergic mechanisms in hypoglycemic glucose counterregulation and posthypoglycemic hyperglycemia in insulin-dependent diabetes mellitus. *J Clin Invest* **69**: 315-326

Rosak C, Althoff PH, Fassbinder W, del Pozo E, Boehm BO (1986) Veränderungen der hormonellen Gegenregulation nach insulininduzierter Hypoglykämie bei gleichzeitiger Gabe des Somatostatinanalogons SMS 201-995. *Med Klin* **24**: 773-778

Santiago JV, Clarke WL, Shah SD, Cryer PE (1980) Epinephrine, norepinephrine, glucagon, and growth hormone release in association with physiological decrements in the plasma glucose concentration in normal and diabetic man. *J Clin Endocrinol Metab* **51**: 877-883

Schwartz NS, Clutter WE, Shah SD, Cryer PE (1987) Glycemic thresholds for activation of glucose counterregulatory systems are higher than the threshold for symptoms. *J Clin Invest* **79**: 777-781

Serrano-Rios M, Navascues I, Saban J, Ordoñez A, Sevilla F, del Pozo E (1986) Somatostatin analogue SMS 201-995 and insulin needs in insulin-dependent diabetic patients studied by means of an artificial pancreas. *J Clin Endocrinol Metab* **63**: 1071-1074

Sheppard M, Shapiro B, Berelowitz M, Pimstone B (1979) Metabolic clearance and plasma half-disappearance time of exogenous somatostatin in man. *J Clin Endocrinol Metab* **48**: 50-53

Spinas GA, Bock A, Keller U (1985) Reduced postprandial hyperglycemia after subcutaneous injection of a somatostatin analogue (SMS 201-995) in insulin-dependent diabetes mellitus. *Diabetes Care* **8**: 429-435

Welle S, Lilavathana U, Campbell RG (1980) Increased plasma norepinephrine concentration and metabolic rates following glucose ingestion in man. *Metabolism* **29**: 806-809

Advances in Growth Hormone and Growth Factor Research,
edited by E.E. Müller, D. Cocchi and V. Locatelli
Pythagora Press, Roma-Milano and Springer Verlag, Berlin-Heidelberg © 1989

Neuroregulatory abnormalities of growth hormone secretion

S. REICHLIN

Endocrine Division, Department of Medicine, New England Medical Center, Tufts University School of Medicine, Boston, Massachusetts, U.S.A.

Although GH-secreting tumors of the pituitary and the rarer ectopic GHRH-secreting tumors are the most obvious and serious disorders of GH hypersecretion in man, there are a number of pathological states in which excess secretion of GH occurs, due mainly to abnormalities in the normal neuroendocrine basis of GH regulation (Table 1; see Molitch 1986 for review). These GH secretory states differ in their pathobiological implications; the most common and important is diabetes, in which GH hypersecretion aggravates metabolic abnormalities and possibly contributes to the adverse course of retinal microangiopathy. Other metabolic disorders in which GH hypersecretion occurs are starvation (Hintz et al. 1978; Clemons et al. 1981) including anorexia nervosa (Garfinkel 1987), hepatic cirrhosis (Shankar et al. 1986), and renal failure (Bessarione et al. 1987). GH hypersecretion is a cardinal feature of Laron-type dwarfism (Laron et al. 1968), is commonly seen in the diencephalic syndrome (Drop et al. 1980), and occurs in stress (Glick et al. 1965; Reichlin 1968). One rare cause of GH hypersecretion is the syndrome of polyostotic fibrous dysplasia, in which an intrinsic growth-regulatory disorder, expressed in multiple tissues including the pituitary, can lead to the syndrome of acromegaly (Kovacs et al. 1984; Rodman et al. 1988). In this paper I have focused on GH dysregulation in diabetes, because more is known about this disorder than about the other conditions and because it may be the most common form of GH hypersecretion.

The understanding of the mechanism underlying the various forms of GH dysregulation should be put in the context of the currently accepted schema for hypothalamic pituitary GH control (Fig. 1). According to current textbook teaching (Martin and Reichlin 1987; Müller 1987), GH secretion rate is recognized to be the result of the interaction of the stimulating effects of GHRH with the inhibitory

Table 1. Regulatory abnormalities of growth hormone hypersecretion

---

Diabetes mellitus
Malnutrition
Hepatic Cirrhosis
Renal failure
Stress
Laron-type dwarfism
Polyostotic fibrous dysplasia

---

effects of somatostatin. Much new data, including observations reported at this meeting, indicate strongly that GH, acting either directly or by way of somatomedin-C, exerts feedback control of GHRH and somatostatin secretion; somatomedin-C inhibits release of GH at the level of the pituitary. Hypophysectomy is followed by a decrease in hypothalamic somatostatin content (Patel 1979) and a decline in hypothalamic somatostatin mRNA concentration (Rogers et al. 1987). GH administration restores hypothalamic somatostatin content (Patel 1979). The effects of GH and somatomedin-C in stimulating hypothalamic somatostatin secretion have previously been demonstrated *in vitro* (Berelowitz et al. 1981). With respect to feedback of GH (and/or somatomedin-C) on hypothalamic GHRH, data presented by Frohman

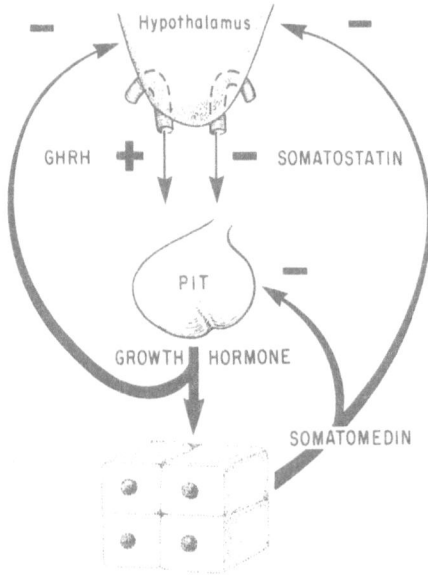

Fig. 1. Schema of GH secretion control, emphasizing the interaction between central nervous factors (stimulated by GHRH, inhibited by somatostatin) with peripherally acting somatomedin-C (IGF-I), which acts directly on the pituitary to inhibit basal and stimulated secretion. Somatomedin-C also acts on the hypothalamus to stimulate somatostatin release; the effects on GHRH are unknown. (From Martin and Reichlin 1987, with permission).

(Chomczynski et al. 1987), Mayo (Mayo, this volume), and de Gennaro (de Gennaro et al. 1987) and their respective collaborators indicate that hypophysectomy leads to an increase in GHRH mRNA, an effect returned toward normal by doses of GH that restore growth to normal. The function of somatomedin-C as a GH-secretion regulator has important implications for the understanding of GH dysregulation in metabolic disorders. Previously accumulated data, together with new information gleaned at this meeting, indicate that somatomedin-C acts directly at the level of the pituitary to suppress GH secretion and synthesis, both under basal states and in response to GHRH. (Wehrenberg et al. 1982; Yamashita and Melmed 1986). Zapf and colleagues (Zapf et al., this volume) have reported that 1 week's administration of recombinant IGF-1 to two men led in one case to the suppression of GHRH-induced GH release, the first experimental proof in man of the hypothesis (drawn from the studies of the elevated GH levels of the somatomedin-C-deficient Laron-type dwarf) that somatomedin-C exerts negative feedback control on GH secretion (Daughaday et al. 1969; Hintz et al. 1978).

However, the inferences to be drawn from studies of somatomedin-C as a GH regulator in metabolic disorders must take into account the fact that most of the somatomedin-C circulating in the blood is complexed to at least two binding proteins which may modify access to either pituitary or hypothalamic regulatory sites (Hintz 1984; Brismar et al. 1987). Moreover, data presented at this meeting by Blum et al. indicated that, under certain conditions, somatomedin-C is a more potent mitogen when complexed to its binding protein (Blum et al. 1987).Whether this applies to neuroregulation is unknown — an important issue when trying to explain GH dysregulation in diabetes.

GH HYPERSECRETION IN DIABETES

As first pointed out by Hansen and colleagues (Hansen 1970; Hansen and Johansen 1970; Johansen and Hansen 1971), poorly controlled type-1 diabetics demonstrate markedly abnormal patterns of GH secretion, including spontaneous hypersecretion at night, excessive release during exercise (Hansen 1970), resistance to suppression by hyperglycemia (Yde 1969; Press et al. 1984a,b), and increased responsiveness to the α_2 -agonist clonidine (Topper et al. 1985). Many of the patients display GH levels that are in the acromegalic range but do not show signs of acromegaly. The abnormalities of GH secretion clearly contribute to insulin resistance, as shown by the effects of GH infusion (Press et al. 1984a), and probably play a role in the pathogenesis of the dawn phenomenon, the early morning worsening of carbohydrate tolerance (Gerich 1984; Campbell et al. 1985; Koivisto et al. 1986).

Since Poulsen's description (Poulsen 1953) of a woman who underwent a complete remission of severe diabetic retinopathy following spontaneous hemorrhagic infarction of the pituitary, it has also been believed that GH is involved in the course of this severe complication of the disease (Lundbaek et al. 1970a,b; Merimee et al. 1978, 1983; Rabin et al. 1984). The severity of retinopathy is correlated with GH levels, and ablation or disease-induced destruction of the pituitary leads to a decrease in severity of the disorder.

Among the most convincing arguments for the importance of GH in diabetic retinopathy is the demonstration in ateliotic dwarfs (many of whom have diabetes in addition to GH deficiency) of relative resistance to this complication (Merimee et al. 1978, 1983), although «background» retinopathy (in contrast to active proliferative retinopathy) has been observed in at least one pituitary dwarf with diabetes (Rabin et al. 1984). Hence, the elucidation of this form of dysregulation is not a trivial pursuit, because it may be the most important form of GH hypersecretion encountered in clinical medicine.

The basis of the GH regulatory abnormality in diabetes is unknown. One aspect, however, is abundantly clear. The abnormality is secondary to the state of poor control and is completely normalized when blood sugar has been properly regulated (Tamborlane et al. 1979; Simonson et al. 1985; Tamborlane and Sherwin 1986; Hermansen et al. 1987). Hence, the defect is due to either sustained hyperglycemia or relative insulinopenia, or both. Theoretical possibilities to explain this abnormality are that (a) GHRH is secreted in excess, (b) somatostatin secretion is deficient, (c) somatomedin-C is deficient or its action is opposed by circulating factors. None of these hypotheses have been definitively established, and there are arguments in favor of each.

Several observations point to a secretory abnormality of GHRH. These include the demonstration of an increased number and amplitude of GH secretory spikes in poorly controlled diabetes (Simonson et al. 1985; Hermansen et al. 1987) and the relative resistance of GH secretion to administration of somatostatin analogues. In our own studies, carried out in collaboration with Drs. Tom Segerson and Lloyd Wilcox, two young women with rapidly progressing diabetic retinopathy were treated with increasing doses of SMS 201-995 (Fig. 2, Table 2). Diabetic regulation in these patients was relatively poor because of the brittleness of their glucoregulation and poor dietary compliance. Despite the administration of increasing amounts of SMS 201-995 up to 1200 µg per day (in divided doses), neither case

Fig. 2. Poorly controlled diabetics often show marked hypersecretion of GH, as illustrated by this 24-h sampling of plasma GH in a 23-year-old woman with severe, progressive retinopathy. Her long history of poor metabolic control is reflected in the elevated hemoglobin A_1C values (see Table 2). Efforts to treat her with increasing doses of the long-acting somatostatin analog SMS 201-995 required the relati-vely large dose of 1200 µg per day to normalize her mean 24-h GH secretory profile, but even this large dose did not reduce her GH levels to below normal.

Table 2. Effect of SMS 201-995 on GH, Sm-C, and Hgb A₁C in two diabetic women

Interval (mos)	Dose of SMS (μg)	Total	Mean GH (ng/ml)	Sm-C (U/ml)	Hgb A₁C (%)
			Patient 1		
	0	0	4.2 ± 3.7	1.0	17.1
	50 b.i.d.	100	4.0 ± 4.5	0.73	15.6
2	100 b.i.d.	200	3.0 ± 2.6	1.60	–
4	150 4 i.d.	600	4.3 ± 5.3	1.70	13.0
9	300 4 i.d.	1200	1.6 ± 1.0	1.60	11.0
			Patient 2		
	0	0	2.4 ± 1.5	0.83	–
	50 b.i.d.	100	2.5 ± 2.0	0.76	12.6
2	100 b.i.d.	200	3.5 ± 4.6	0.80	9.9
4	150 4 i.d.	600	2.5 ± 2.6	0.91	11.5
9	300 4 i.d.	1200	1.5 ± 1.0	–	–

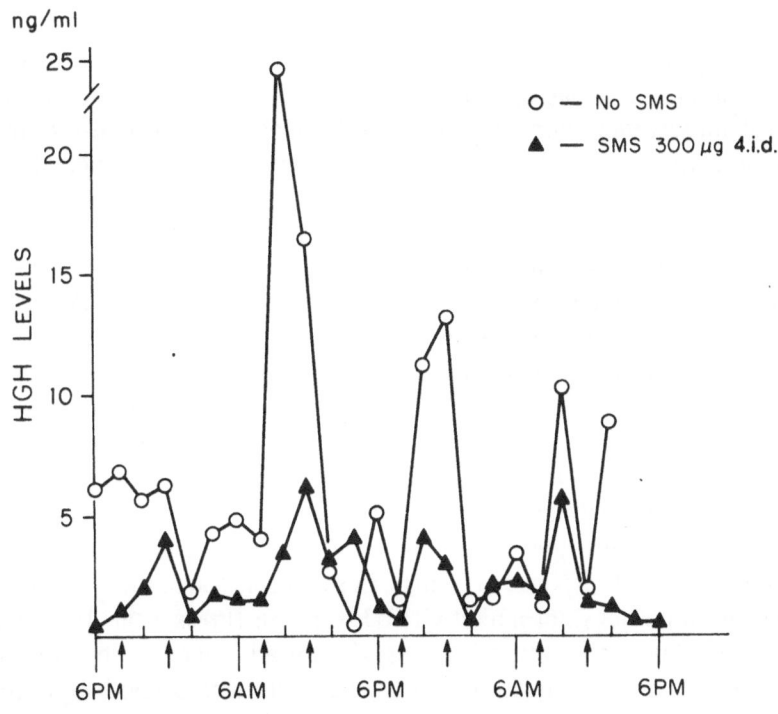

showed suppression below the normal level. Much lower doses commonly normalize GH in the majority of acromegalics. On the other hand, in several reports, SMS 201-995 administration significantly lowered GH levels in diabetics using relatively small doses, but these were all well-controlled patients. For example, in the studies of Spinas et al. in well-controlled diabetics (as judged by almost normal hemoglobin A_1C values (Spinas et al. 1985), 100 µg of SMS 201-995 lowered GH levels significantly below normal, and in the work of Davies et al. 50 µg of SMS 201-995 reduced peak nocturnal GH from 26.1 m units/l to 3.7 m units/l (Davies et al. 1986). It remains to be seen whether complete GH deficiency can be produced with any tolerable dose of somatostatin in poorly controlled diabetics.

A second possible explanation for the GH hypersecretory abnormality in poorly controlled diabetes is a relative deficiency in endogenous somatostatin secretion. Again, data pertaining to this hypothesis are quite indirect and depend mainly on the interpretation of the mechanism of cholinergic involvement in GH regulation. Administration of pirenzepine, a cholinergic antagonist, brings about virtually complete suppression of the excessive nocturnal secretion of GH in both normal and diabetic individuals (Mendelson et al. 1978; Davis and Davis 1986; Peters et al. 1986; Page et al. 1987). Muscarinic cholinergic antagonists like pirenzipine and atropine block all types of induced GH secretory responses in man (Mendelson et al. 1978; Delitala et al. 1983; Casanueva et al 1984; Taylor et al. 1985; Peters et al. 1986), with the possible exception of the response to hypoglycemia (Evans et al. 1985), including the response to administration of GHRH (Massara et al. 1984; Jordan et al. 1986). Since these agents do not influence GH secretion from isolated pituitaries, it has been inferred that the effect is mediated by increased somatostatin secretion. Further support for this idea is the work of Richardson and colleagues, who showed that the addition of acetylcholine to hypothalamic slices *in vitro* inhibited the release of somatostatin (Richardson et al. 1980). It could be argued from this finding that muscarinic antagonists disinhibit the normal suppressive effects of somatostatin on pituitary GH release. On the other hand, in studies carried out using diencephalic cultures, our group (Fig. 3), and Peterfreund and Vale (1983) found that acetylcholine stimulates somatostatin secretion. The basis for this discrepancy is unknown.

If somatostatin secretion is excessive in poorly controlled diabetics, it is logical that reponses to GHRH would be suppressed. In fact, response to a standard dose of GHRH in either type-1 or type-2 diabetics, even those in poor control, is normal (Press et al. 1984b; Pietschmann et al. 1987). However, there is a crucial difference, in that normals whose glucose levels are elevated by glucose infusion show a marked inhibition of response to GHRH, in sharp contrast to the diabetic, in whom glucose elevation does not inhibit GHRH response (Press et al. 1984b). It could be argued that hyperglycemia stimulates somatostatin secretion in normals but not in the diabetic, and that this effect is mediated through a cholinergic link.

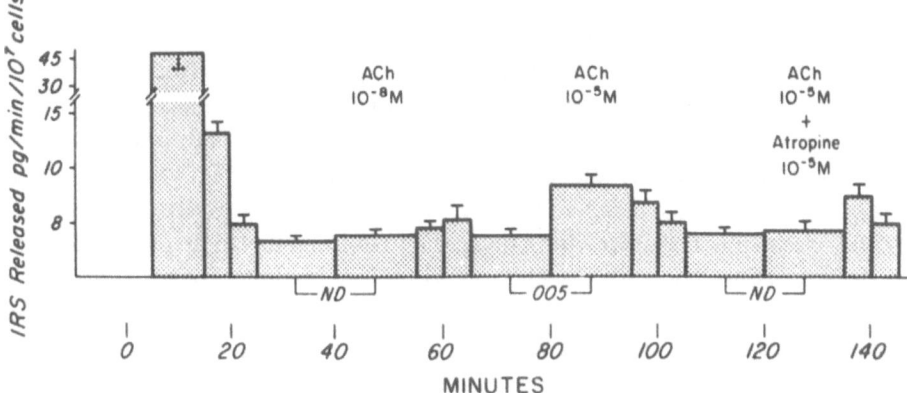

Fig. 3. Effect of acetylcholine on release of somatostatin from dispersed cerebral cortical cells in culture.

The role of central glucoreceptors in regulation of GH secretion has been well established from the earliest days of work on GH regulation, using the radioimmunoassay. Insulin-induced hypoglycemia and the administration of 2-deoxyglucose systemically (Glick et al. 1965) or directly into the lateral hypothalamus stimulate GH release and Blanco and I showed that the central administration of glucose in the squirrel monkey blocked reflex GH release induced by insulin-induced hypoglycemia (Blanco et al. 1966). In the rat, hypoglycemia stimulates the release of somatostatin (Painson et al. 1984) (accounting for most of the GH suppression seen in this species). In hypothalamic fragments (Robbins 1983) and in a diencephalic cell culture system (Lengyel et al. 1984) the switch from high to low glucose levels stimulates somatostatin release.However, an increase in glucose in the media surrounding the hypothalamic cells does not increase somatostatin secretion. Hence, due both to presumed species differences in responsiveness to hypoglycemia, and to the inevitable limitations of *in vitro* work, no clear conclusion can be drawn as to the role of central glucose levels in the regulation of somatostatin secretion in man.

Although it would seem reasonable to postulate that in diabetes there is an abnormality of GH regulation due to chronic alteration in hypothalamic responsivity to hyperglycemia, one cannot exclude the possibility that these findings are due to a relative or an absolute central deficiency of insulin. Insulin itself (Baskin et al. 1983; Morley 1987) and insulin receptors (Hendricks et al. 1984; Lowe and LeRoith 1986) have been identified in the brain; insulin has been shown to enter the cerebral compartment after systemic injection (Wallum et al. 1987); intraventricular insulin injection has been shown to produce anorexia (Woods et al. 1979), and centrally administered antisera to insulin increase food intake (Strubbe and Mein 1977). All of the techniques used to normalize glucose levels (to demonstrate restoration of GH secretory dynamics) have utilized insulin injections.

GH hypersecretion could also come about from a deficiency of somatomedin-C. The somatomedin deficiency hypothesis is supported by the frequent occurrence of growth retardation in young children with diabetes, despite seemingly adequate blood levels of GH (Winter et al. 1979; Salardi et al. 1986, 1987). In poorly controlled older juvenile diabetics, serum immunoassayable somatomedin-C is variously reported as being reduced or normal and, if low, to normalize after aggressive insulin therapy (Tamborlane et al. 1981). In the adult, even poor control of diabetes is not apparently accompanied by low immunoreactive somatomedin-C levels. Guastamacchia et al. reported that somatomedin-C levels were strikingly low in both type-1 and type-2 diabetes, but that the severity of vascular changes were not correlated with the severity of somatomedin-C depression (Guastamacchia et al. 1987). In addition, the principal somatomedin-C binding protein is reported to be markedly elevated in poorly controlled diabetes, and the changes do correlate with the severity and duration of the disease (Brismar et al. 1987). The extent to which these interactions might modulate central and peripheral GH regulation is unknown.

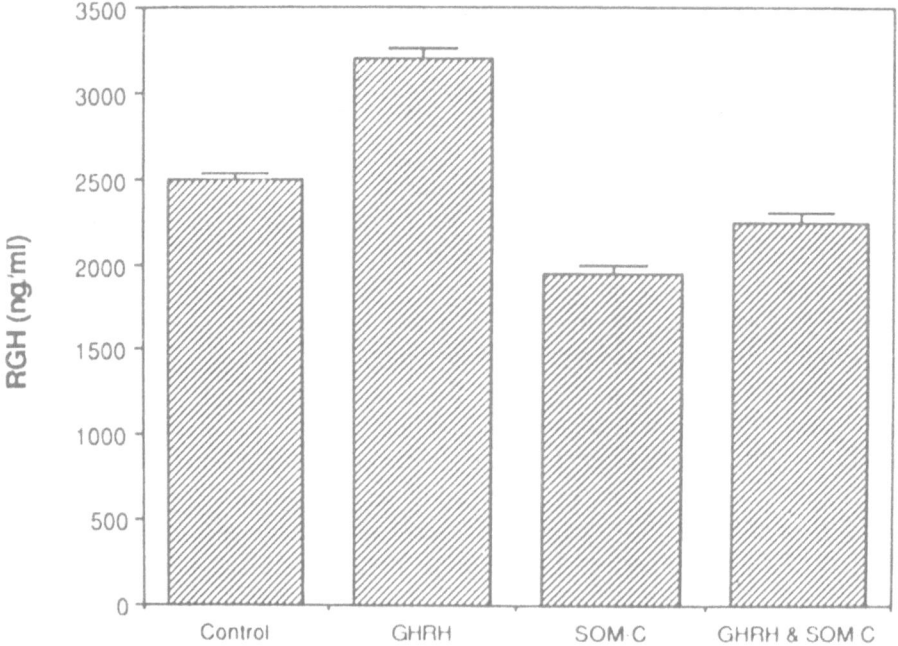

Fig. 4. Inhibitory effect of somatomedin-C (SOM-C) on GH response to GHRH.

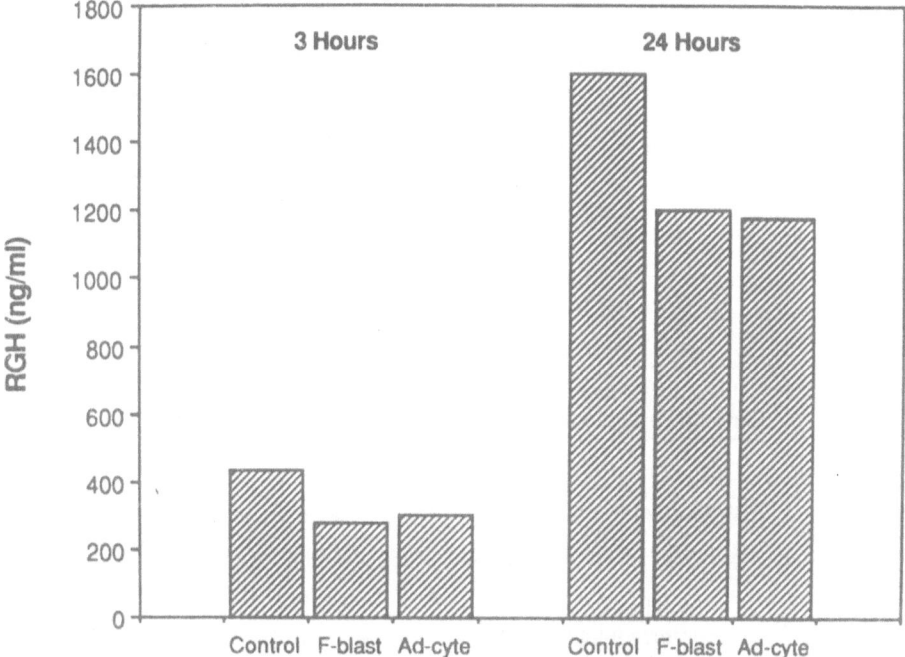

Fig. 5. Incubation of anterior pituitary cells in medium conditioned by adipocytes or by fibroblasts leads to inhibition of responses to GHRH. Times designated are periods of pre-incubation with conditioned media.

A further complexity in the somatomedin-C control system has emerged from studies of human and experimental diabetes. Phillips and colleagues have dem-onstrated a somatomedin-C inhibitor in the blood of poorly controlled diabetics, both rat and human (Phillips and Scholz 1982; Phillips 1986). This substance, apparently a protein, is reduced when diabetic control is restored to normal. In Phillip's work, the bioassay for somatomedin-C has been the cartilage sulfation assay, but effects on the pituitary have not been evaluated as yet. Somatomedin-C (IGF-1), and IGF-2 are present in the brain as endogenous regulators (Noguchi et al. 1987).

We have been studying the interaction of somatomedin-C and GHRH on GH secretion in isolated rat pituitary cells and, in our laboratory, have confirmed that somatomedin-C is inhibitory to GH release (Fig. 4). We have also found that medium conditioned by previous exposure to either fibroblasts or 3 T3 adipocytes is inhibitory to GH release (Fig. 5). The identity of this factor it unknown, but it may be somatomedin-C since this substance is known to be secreted by fibroblasts.

In an attempt to place these disparate elements in a relevant context, I propose a model, shown in Figure 6, to explain how peripheral and central GH regulatory functions may be operative in poorly controlled diabetes.

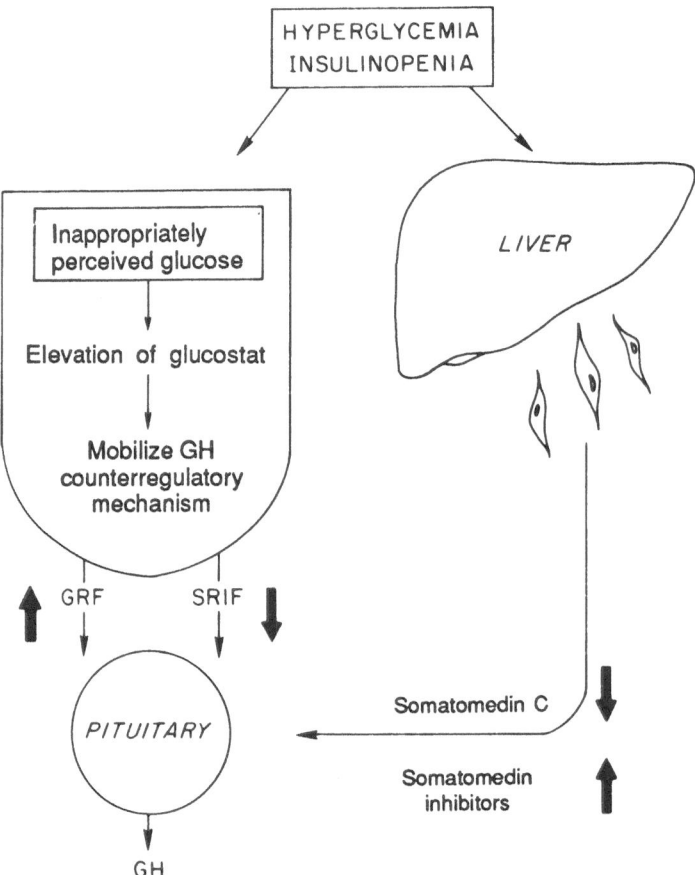

Fig. 6. Proposed mechanisms of GH dysregulation in poorly controlled diabetes. It is postulated that inappropriately perceived blood glucose concentration due to central insulin deficiency and/or persistent hyperglycemia resets (elevates) the central "Glucostat". The hypothalamus therefore interprets its information to mean that there is a central glucopenia, and then mobilizes counterregulatory mechanisms. These include increased GHRH secretion and decreased somatostatin secretion. Studies with muscarinic antagonists indicate that the effect may be mediated by altering cholinergic tone. In addition, the peripheral metabolic abnormalities in diabetes lead to a decrease in the concentration of somatomedin-C, alterations in the concentration of somatomedin-C binding proteins, and the increase of somatomedin-C inhibitory proteins, all of which tend to remove a normally tonic suppressive influence on the pituitary and a factor that normally stimulates somatostatin from the hypothalamus.

GH DYSREGULATION IN OBESITY

Paradoxically, the hyposecretion of GH in obesity may shed light on the mechanism of GH hypersecretion in poorly controlled diabetes. Beginning with the original studies of Roth, Glick, Berson and Yalow (Glick et al. 1965) and confirmed by every group who have subsequently studied the question, patients with obesity have beeen shown to have blunted or absent GH secretory responses to stimuli such as hypoglycemia, sleep, or exercise (Bell et al. 1970; El-Khodary et al. 1971; Fingerhut and Krieger 1974; Garlaschi et al. 1975; Laurian et al. 1975). In our own studies, the response to exercise in elderly women (Dawson-Hughes et al. 1986) or in men receiving benzodiazepines (Fig. 7a) has been shown to be strictly a function of body weight (Fig.7b). In the latter studies, carried out in collaboration with Drs. Timothy Stryker and David Greenblatt, individuals who were 15% or more above ideal body weight were found to have blunted response.

The mechanism underlying this response is not known, but recent work indicateds that the pituitary itself becomes relatively unresponsive to injections of GHRH (Fig. 8). Williams and colleagues (Williams et al. 1984) have shown that morbidly obese individuals are refractory to GHRH, and that this reponse is normalized when weight is restored to normal. More recently, Vance et al. have shown that spontaneous GH secretion and responsiveness to GHRH in the obese rapidly returns to normal during brief periods of starvation (Vance et al. 1987). I would propose (Fig. 9) that the state of calorie satiation and hyperinsulinemia in the obese person leads to a resetting of the brain glucostat, resulting in the inhibition of glucose-mobilizing neuroendocrine responses. In contrast to the poorly controlled diabetic, somatostatin is secreted in excess while GHRH secretion is reduced.

The operation of peripheral circulating factors in this reaction must be considered. In addition to the possible changes in somatomedin-C levels, which are normal or elevated in obesity, there may be as yet unrecognized changes in somatomedin-C inhibitory compounds and changes in circulating products of fat cells. For example, we have found that medium conditioned by fat cells is inhibitory to GH release (Fig. 5), and Flier and Spiegelman have recently shown that fat cells synthesize a secretory protein (adipsin) which circulates in the blood, and which is reduced in several forms of obesity (Flier et al. 1987). The effects on GH secretion of this substance (or others secreted by fat cells) in starvation have not as yet been evaluated. These potential metabolic regulators of the pituitary may be involved in the etiopathogenesis of the disorders in which GH secretion is abnormal. The peripheral metabolic signal responsible for the flattened GH response in obesity appears to act through a central element as well. Mazza et al. have reported that cholinergic enhancement with pyridostigmine stimulated GH release in obese adults and children (Mazza et al. 1987).

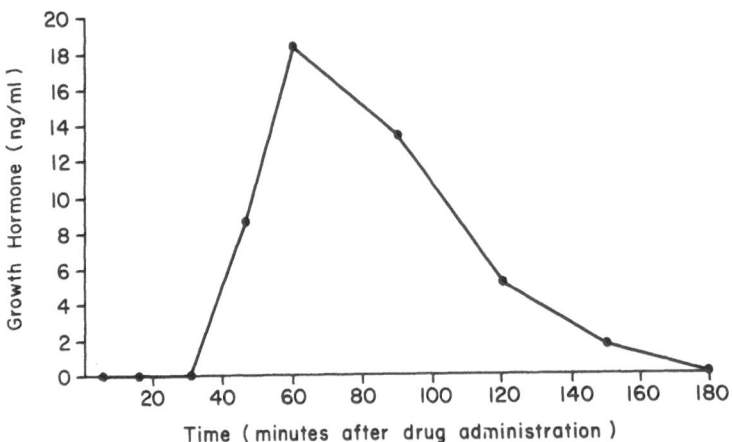

Fig. 7a. Benzodiazepines stimulate the release of GH in man, as illustrated in this figure.

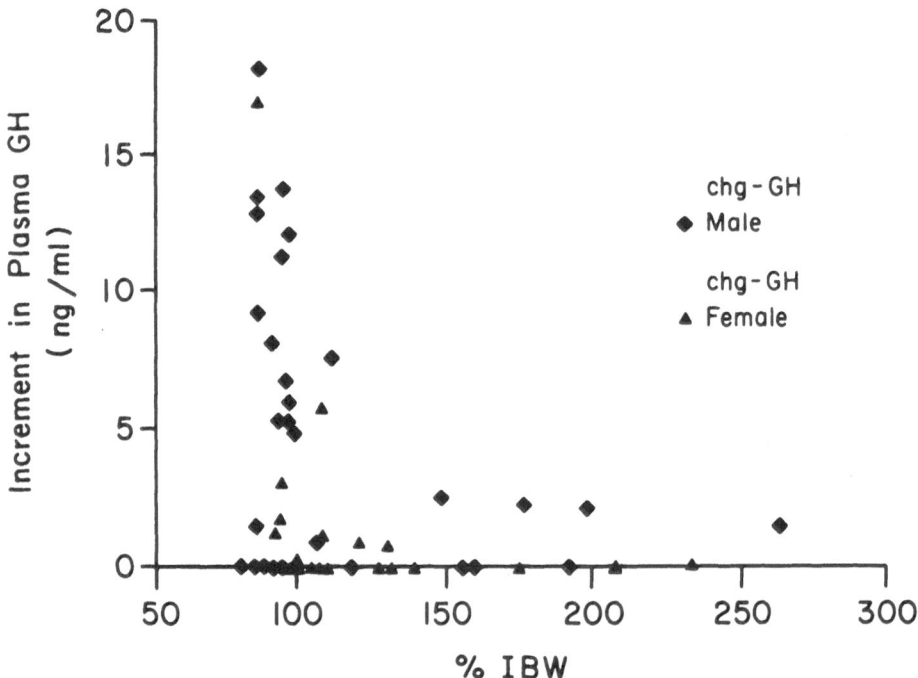

Fig. 7b. The response is determined strictly by body weight. Chg-GH, change in growth hormone; IBW, ideal body weight.

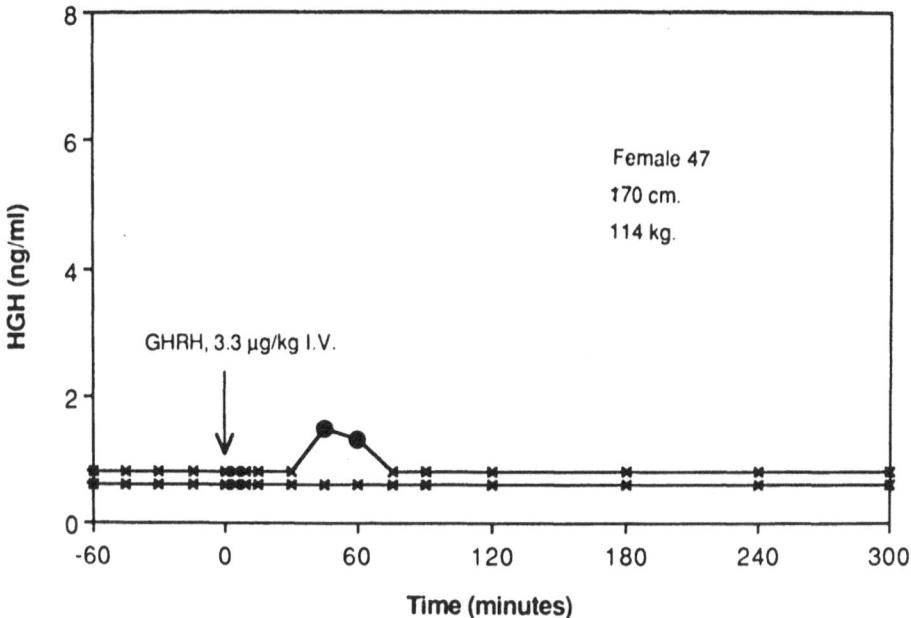

Fig. 8. Suppression of GH response to injected GHRH in an obese woman.

SUMMARY

In clinical medicine, states of GH hypersecretion are most commonly associated with metabolic disturbances, including starvation, anorexia nervosa, hepatic cirrhosis, uremia, and poorly controlled diabetes. It is proposed that these abnormalities are due primarily to alterations in the feedback effects of somatomedin-C (IGF-I, interacting with circulating IGF-I binding proteins and inhibitors) on the somatotroph and on the GHRH and somatostatinergic systems. Because the abnormalities in diabetes are corrected with cholinergic antagonists, it is likely that the principle target of the metabolic regulators is on the central neural control mechanisms. It is postulated that the changes in central regulation are due to an altered «set point» of the central glucostat mechanism. It is also proposed that the GH hyposecretion of the obese is due to the operation of the same central mechanisms, but in the zone of central calorie-insulin satiation.

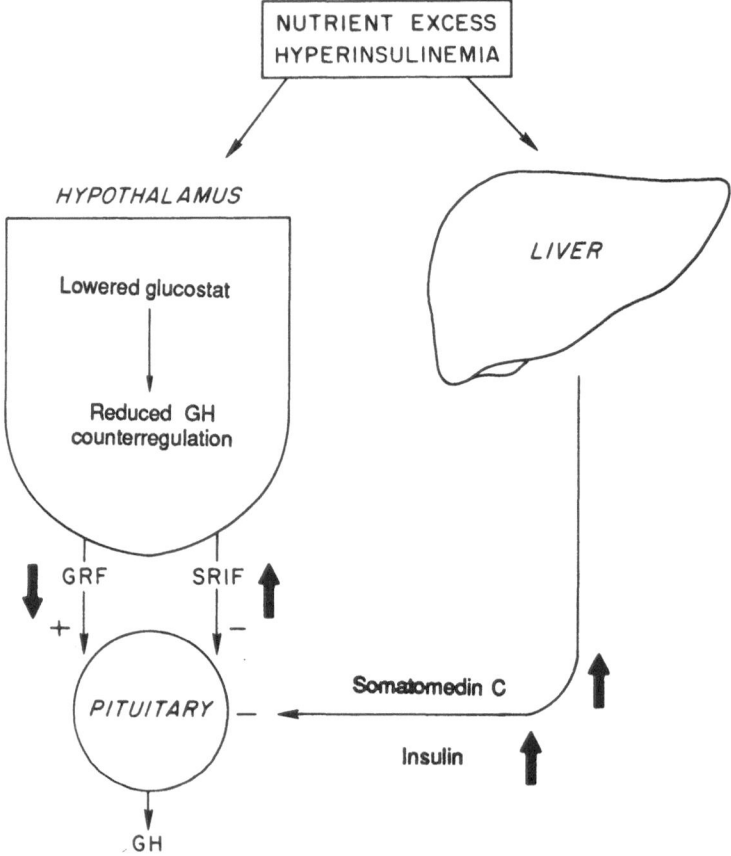

Fig. 9. GH secretory abnormalities observed in obesity are generally reciprocal to those seen in poorly controlled diabetes and therefore may shed light on the pathophysiology of GH regulation in diabetes.

REFERENCES

Baskin DG, Woods SC, West DB, van Houton M, Posner BI, Dorsa DM, Porte D Jr (1983) Immunocytochemical detection of insulin in rat hypothalamus and its possible uptake from cerebrospinal fluid. *Endocrinology* **113**: 1818-1825

Bell JP, Donald RA, Espiner EA (1970) Pituitary response to insulin-induced hypoglycemia in obese subjects before and after fasting. *J Clin Endocrinol Metab* **31**: 546-551

Berelowitz M, Szabo M, Frohman LA, Firestone S, Chu L (1981) Somatomedin-C mediates growth hormone negative feedback by effects on both the hypothalamus and the pituitary. *Science* **212**: 1279-1281

Bessarione D, Perfumo F, Giusti M, Ginevri F, Mazzocchi G, Gusmano R, Giordano G

(1987) Growth hormone response to growth hormone-releasing hormone in normal and uraemic children. Comparison with hypoglycaemia following insulin administration. *Acta Endocrinol* (Copenh) **114**: 5-11

Blanco S, Schalch DS, Reichlin S (1966) Control of growth hormone secretion by gluco-receptors in the hypothalamic-pituitary unit (abstr). *Fed Proc* **251**: 191

Blum WF, Jenne E, Kietzmann K, Bierich JR, Ranke MB (1987) Insulin-like growth factor I is a better mitogen if complexed with its binding protein. *J Endocrinol Invest* **10** [Suppl 2]: 25

Brismar K, Gutniak M, Werner S, Hall K (1987) Somatomedin binding protein in diabetes mellitus. *J Endocrinol Invest* **10** [Suppl 2]: 28

Campbell PJ, Bolli GB, Cryer PE, Gerich JE (1985) Pathogenesis of the dawn phenomenon in patients with insulin-dependent diabetes mellitus. *N Engl J Med* **312**: 1473-1479

Casanueva FF, Villanueva L, Cabranes JA, Cabezas-Cerrato J, Fernandez-Cruz A (1984) Cholinergic mediation of GH secretion elicited by arginine, clonidine and physical exercise in man. *J Clin Endocrinol Metab* **59**: 526-530

Chomczynski P, Downs TR, Frohman LA (1987) Feedback regulation of growth hormone (GH)-releasing hormone (GRH) gene expression by growth hormone in rat hypothalamus. 69th Annual Meeting Endocrine Society (abstract 617)

Clemmons DR, Klibanski A, Underwood LE, McArthur SW, Ridgway EC, Beitins IZ, Van Wyk JJ (1981) Reduction of plasma immunoreactive somatomedin-C during fasting in humans. *J Clin Endocrinol Metab* **53**: 1247-1250

Daughaday WH, Laron Z, Pertzelan A, Heins JN (1969) Defective sulfation factor generation: a possible etiological link in dwarfism. *Trans Assoc Am Physicians* **82**: 129-140

Davies RR, Turner SJ, Orskov H, Alberti KGMM, Johnston DG (1986) The response of patients with type-1 (insulin-dependent) diabetes to a single night-time injection of SMS 201-995 (abstr). *Diabetologia* **29**: 118A

Davis B, Davis K (1986) Effect of propantheline on GH and blood glucose levels. *Lancet* i: 1382

Dawson-Hughes B, Stern D, Goldman J, Reichlin S (1986) Regulation of growth hormone and somatomedin-C secretion in postmenopausal women: effect of physiological estrogen replacement. *J Clin Endocrinol Metab* **63**: 424-432

De Gennaro V, Cattaneo E, Müller EE, Maggi A (1987) Growth hormone regulation of growth hormone-releasing hormone gene expression. *J Endocrinol Invest* **10** [Suppl 2]: 82

Delitala G, Maioli M, Pacifico A, Brianda SA, Palermo M, Mannelli M (1983) Cholinergic receptor control mechanisms for L-dopa-,apomorphine. and clonidine-induced growth hormone secretion in man. *J Clin Endocrinol Metab* **57**: 1145-1149

Drop SL, Guyda HJ, Colle E (1980) Inappropriate growth hormone release in the diencephalic syndrome of childhood: case report and 4-year endocrinological follow-up. *Clin Endocrinol* (Oxf) **13**: 181-187

El-Khodary AZ, Ball MF, Stein B, Canary JJ (1971) Effect of weight loss on the growth hormone response to arginine infusion in obesity. *J Clin Endocrinol Metab* **32**: 42-51

Evans PJ, Dieguez C, Foord S, Peters JR, Hall R, Scanlon MF (1985) The effect of cholinergic blockade on the growth hormone and prolactin response to insulin hypoglycemia. *Clin Endocrinol* (Oxf) **22**: 733-735

Fingerhut M, Krieger DT (1974) Plasma growth hormone response to L-dopa in obese subjects. *Metabolism* **23**: 267-271

Flier JS, Cook KS, Usher P, Spiegelman BM (1987) Severely impaired adipsin expression in genetic and acquired obesity. *Science* **237**: 405-408

Garfinkel PE. Anorexia nervosa: an overview of hypothalamic-pituitary function (1984) In: Brown GM, Koslow SH, Reichlin S (eds) *Neuroendocrinology and Psychiatric Disorder*. Raven, New York, pp 301-314

Garlaschi C, DiNatale B, Del Guercio MJ, Caccamo A, Gargantini L, Chiumello G (1975) Effect of physical exercise on secretion of growth hormone, glucagon and cortisol in obese and diabetic children. *Diabetes* **24**: 758-761

Gerich JE (1984) Role of growth hormone in diabetes mellitus. *N Engl J Med* **310**: 848-850

Glick SM, Roth J, Yalow RD, Berson SA (1985) The regulation of growth hormone secretion. *Recent Prog Horm Res* **21**: 241-283

Guastamacchia E, Nardelli GM, Di Paolo S, DePergola G, Ciapolillo A, Lacasella R, Santoro G (1987) Somatomedin-C (Sm-C) in type-1 and type-2 diabetics: correlation with existence of retinopathy. *J Endocrinol Invest* **10** [Suppl 2]: 29

Hansen AP (1970) Abnormal serum growth hormone response to exercise in juvenile diabetics. *J Clin Invest* **49**: 1467-1478

Hansen AP, Johansen K (1970) Diurnal patterns of blood glucose, serum free fatty acids, insulin, glucagon and growth hormone in normal juvenile diabetics. *Diabetologia* **6**: 27-33

Hendricks SA, Agardh CD, Taylor SI, Roth J (1984) Unique features of the insulin receptor in rat brain. *J Neurochem* **43**: 1302-1309

Hermansen K, Møller A, Christensen CK, Christiansen JS, Schmitz O, Ørskov H, Alberti KGMM, Mogensen CE (1987) Diurnal plasma profiles of metabolite and hormone concentration in insulin-dependent diabetic patients during conventional insulin treatment and continuous subcutaneous insulin infusion. *Acta Endocrinol* (Copenh) **114**: 433-439

Himsworth RL, Carmel RW, Frantz AG (1972) The location of the chemotransmitter controlling growth hormone secretion during hypoglycemia in primates. *Endocrinology* **91**: 217-226

Hintz RL (1984) Plasma forms of somatomedin and the binding protein phenomenon. *Clin Endocrinol Metab* **13**: 31-42

Hintz RL, Suskind R, Amatayakul K, Thanangkul O, Olson R (1978) Plasma somatomedin and growth hormone values in children with protein-calorie malnutrition. *J Pediatr* **92**: 153-156

Johansen K, Hansen AP (1971) Diurnal serum growth hormone levels in poorly and well-controlled juvenile diabetics. *Diabetes* **20**: 239-245

Jordan V, Dieguez C, Lafaffian I, Rodriguez-Arnao MD, Gomez-Pan A, Hall R, Scanlon MF (1986) Influence of dopaminergic, adrenergic and cholinergic blockade and TRH administration on GH responses to GRF 1-29. *Clin Endocrinol* (Oxf) **24**: 291-298

Kohner EM, Hamilton AM, Joplin GF, Fraser TR (1976) Florid diabetic retinopathy and its response to treatment by photocoagulation or pituitary ablation. *Diabetes* **25**: 104-110

Koivisto VA, Yki-Jarvinen H, Helve E, Karonen S, Pelkonen R (1986) Pathogenesis and prevention of the dawn phenomenon in diabetic patients treated with CSII. *Diabetes* **35**: 78-82

Kovacs K, Horvath E, Thorner MO, Rogol AD (1984) Mammosomatotroph hyperplasia associated with acromegaly and hyperprolactinemia in a patient with the McCune-

Albright syndrome. A histologic, immunocytologic and ultrastructural study of the surgically removed adenohypophysis. *Virchows Arch* **403**: 77-86

Laron Z, Pertzelan A, Karp M (1968) Pituitary dwarfism with high serum levels of growth hormone. *Isr J Med Sci* **4**: 883-894

Laurian L, Oberman A, Ayalon D (1975) Under-responsiveness of growth hormone secretion after L-dopa and deep sleep stimulation in obese subjects. *Isr J Med Sci* **11**: 482-487

Lengyel AJ, Grossman A, Nieuwenhuysen-Kruseman AC, Ackland J, Rees LH, Besser M (1984) Glucose modulation of somatostatin and LH-RH release from rat hypothalamic fragments *in vitro. Neuroendocrinology* **39**: 31-38

Lowe W, LeRoith D (1986) Insulin receptors from guinea pig liver and brain: structural and functional studies. *Endocrinology* **118**: 1669-1677

Lundbaek K, Jensen VA, Steen Olsen T, Ørskov H, Juel Christensen N, Johansen K, Prage Hansen A, Osterey R (1970) Diabetes, diabetic angiopathy and growth hormone. *Lancet* **ii**: 131-133

Lundbaek K, Malmros R, Andersen H, et al. (1970) Hypophysectomy for diabetic angiopathy: a controlled clinical trial. In: Goldberg M, Fine S (eds) *Treatment of Diabetes Retinopathy*. (PHS) publication No. 890 Governement Printing Office Washington DC, pp 291-312

Martin JB, Reichlin S. (1987) *Clinical Neuroendocrinology*, 2nd edn. Davis, Philadelphia

Massara F, Ghigo E, Goffi S, Molinatti GM, Müller EE, Camanni F, (1984) Blockade of hp GRF 40-induced growth hormone release in normal men by a cholinergic muscarinic antagonist. *J Clin Endocrinol Metab* **58**: 1025-1026

Mazza E, Ghigo E, Imperiale E, Molinatti P, Corrias A, DeSanctis C, Camanni F (1987) Effect of cholinergic enhancement on GH secretion in obese adults and children. *J Endocrinol Invest* **10** [Suppl 2]: 58

Mendelson WB, Sitaram N, Wyatt RJ, Gillin JC (1978) Methscopolamine inhibition of sleep-related growth hormone secretion. Evidence for a cholinergic secretory mechanism. *J Clin Invest* **61**: 1683-1690

Merimee TJ (1978) A follow-up study of vascular disease in growth hormone- deficient dwarfs with diabetes. *N Engl J Med* **298**: 1217-1222

Merimee TJ, Zapf J, Froesch ER (1983) Insulin-like growth factors, studies in diabetics with and without retinopathy. *N Engl J Med* **309**: 527-530

Molitch ME (1986) Growth hormone hypersecretory states. In: Raiti S, Tolman RA (eds) *Human Growth Hormone*. Plenum, New York, pp 29-50

Morley JE (1987) Neuropeptide regulation of appetite and weight. *Endocr Rev* **8**: 256-287

Müller EE (1987) Neural control of somatotropic function. *Physiol Rev* **67**: 962-1053

Noguchi T, Kurata LM, Sgisaki T (1987) Presence of a somatomedin-C-immunoreactive substance in the central nervous system: immunohistochemical mapping studies. *Neuroendocrinology* **46**: 277-282

Page MD, Koppeschaar HPF, Dieguez C, Gibbs JT, Hall R, Peters JR, Scanlon MF (1987) Cholinergic muscarinic receptor blockade with pirenzepine abolishes slow-wave sleep-related growth hormone release in young patients with insulin-dependent diabetes mellitus. *Clin Endocrinol* (Oxf) **26**: 355-359

Painson JC, Ling N, Tannenbaum GS (1984) Effects of intracellular glucopenia on pulsatile growth hormone (GH) secretion: mediation by GH-releasing factor (GRF) and somatostatin (SRIF). Proc of the 7th International Congress of Endocrinology, Quebec City, July 1984, A1496 (abstr)

Patel YC (1979) Growth hormone stimulates hypothalamic somatostatin. *Life Sci* **24**: 1589-1593

Peterfreund R, Vale W (1983) Muscarinic cholinergic stimulation of somatostatin secretion from cultured brain cells. *Endocrinology* **112**: 526-534

Peters JR, Evans PJ, Page MD, Hall R, Gibbs JT, Dieguez C, Scanlon MF (1986) Cholinergic muscarinic receptor blockade with pirenzepine abolishes slow-wave sleep-related growth hormone release in normal adult males. *Clin Endocrinol* (Oxf) **25**: 213-217

Phillips LS (1986) Nutrition, somatomedins, and the brain. *Metabolism* **35**: 78-87

Phillips LS, Scholz TD (1982) Nutrition and somatomedin. IX. Blunting of insulin-like activity by inhibitor in diabetic rat serum. *Diabetes* **31**: 97-104

Pietschmann P, Schernthaner G, Prskavec F, Gisinger C, Freyler H (1987) No evidence for increased growth hormone responses to growth hormone-releasing hormone in patients with diabetic retinopathy. *Diabetes* **36**: 159-162

Poulsen JE (1953) Recovery from retinopathy in a case of diabetes with Simmond's disease. *Diabetes* **2**: 7-12

Press M, Tamborlane WV, Sherwin RS (1984a) Importance of raised growth hormone levels in the metabolic derangements of diabetes. *N Engl J Med* **310**: 810-815

Press M, Tamborlane WV, Thorner MO, Vale W, Rivier J, Gertner JM, Sherwin RS (1984b) Pituitary response to growth hormone-releasing factor in diabetes, failure of glucose-mediated suppression. *Diabetes* **33**: 804-806

Rabin D, Bloomgarden ZT, Feman SS, Davis TQ (1984) Development of diabetic complications despite the absence of growth hormone in a patient with post-pancreatectomy diabetes. *N Engl J Med* **310**: 837-839

Reichlin S (1968) Hypothalamic control of growth hormone secretion and the response to stress. In: *Endocrinology and Human Behaviour*. Oxford University Press, London, pp 256-283

Richardson SB, Hollander CS, D'Eletto R, Greenleaf PW, Thaw C (1980) Acetylcholine inhibits the release of somatostatin from rat hypothalamus *in vitro. Endocrinology* **107**: 122-129

Robbins RJ (1983) Influence of glucose on somatostatin synthesis and secretion in isolated cerebral cortical cells. *J Neurochem* **40**: 1430-1434

Rodman EF, Shucart WA, Dayal Y, Adelman LS, Reichlin S (1988) Micronodular pituitary hyperplasia in acromegaly with polyostotic fibrous dysplasia; in preparation

Rogers KV, Vician L, Steiner RA, Clifton KD (1987) Reduced preprosomatostatin messenger ribonucleic acid in the periventricular nucleus of hypophysectomized rats determined by quantitative in situ hybridization. *Endocrinology* **121**: 90-93

Salardi S, Cacciari E, Ballardini D, Righetti F, Capelli M, Cicognani A, Zucchini S, Natali G, Tassinari D (1986) Relationships between growth factors (somatomedin-C and growth hormone) and body development, metabolic control, and retinal changes in children and adolescents with IDDM. *Diabetes* **35**: 832-836

Salardi S, Tonioli S, Tassoni P, Tellarini M, Mazzanti L, Cacciari E (1987) Growth and growth factors in diabetes mellitus. *Arch Dis Child* **62**: 57-62

Shankar TP, Fredi JL, Himmelstein S, Solomon SS, Duckworth WC (1986) Elevated growth hormone levels and insulin resistance in patients with cirrhosis of the liver. *Am J Med Sci* **291**: 248-254

Simonson DC, Tamborlane WV, DeFronzo RA, Sherwin RS (1985) Intensive insulin therapy reduces counterregulatory hormone responses to hypoglycemia in patients with type-I diabetes. *Ann Intern Med* **103**: 184-190

Spinas GA, Bock A, Keller U (1985) Reduced postprandial hyperglycemia after subcutaneous injection of a somatostatin-analogue (SMS 201-995) in insulin-dependent diabetes mellitus. *Diabetes Care* **8**: 429-435

Strubbe JH, Mein CG (1977) Increased feeding in response to bilateral injections of insulin antibodies in the VMH. *Physiol Behav* **19**: 309

Tamborlane WV, Sherwin RS (1986) Effect of insulin on growth hormone-induced metabolic derangements in diabetes. *Metabolism* **35**: 956-959

Tamborlane WV, Sherwin RS, Koivisto V, Hendler R, Genel M, Felig P (1979) Normalization of the growth hormone and catecholamine response to exercise in juvenile-onset diabetic subjects treated with a portable insulin-infusion pump. *Diabetes* **28**: 785-788

Tamborlane WV, Hintz RL, Bergam M, Genel M, Fehg P, Sherwin RS (1981) Insulin-infusion pump treatment of diabetes: influences of improved metabolic control on plasma somatomedin levels. *N Engl J Med* **305**: 303-307

Taylor BJ, Smith PJ, Brook CGD (1985) Inhibition of physiological growth hormone secretion by atropine. *Clin Endocrinol* (Oxf) **22**: 497-501

Topper E, Gertner J, Amiel S, Press M, Genel M, Tamborlane WV (1985) Deranged alpha-adrenergic regulation of growth hormone secretion in poorly controlled diabetes: reversal of the exaggerated response to clonidine after continuous subcutaneous insulin infusion. *Pediatr Res* **19**: 534-536

Vance ML, Faria ACS, Thorner MO (1987) Growth hormone during fasting: enhancement of endogenous secretion, pulsatile release and rhythms. 69th Annual Meeting Endocrine Society, p 78 (abstr)

Wallum BJ, Taborsky GJ, Porte D Jr, Figlewicz DP Jr, Jacobson L, Beard JC, Ward WK, Dorsa D (1987) Cerebrospinal fluid insulin levels increase during intravenous insulin infusions in man. *J Clin Endocrinol Metab* **64**: 190-194

Wehrenberg WB, Ling N, Böhlen P, Esch F, Brazeau P, Guillemin R (1983) Physiological roles of somatocrinin and somatostatin in the regulation of growth hormone secretion. *Biochem Biophys Res Commun* **109**: 562-567

Williams T, Berelowitz M, Joffe SN, Thorner MO, Rivier J, Vale W, Frohman LA (1984) Impaired growth hormone response to growth hormone-releasing factor in obesity: a pituitary defect reversible with weight reduction. *N Engl J Med* **311**: 1403-1407

Winter RJ, Phillips LS, Klein MN, Traisman HS, Green OC (1979) Somatomedin activity and diabetic control in children with insulin-dependent diabetes. *Diabetes* **28**: 952-954

Woods SC Jr, Lotter EC, McKay LD, Porte D Jr (1979) Chronic intracerebroventricular infusion of insulin reduces food intake and body weight of baboons. *Nature* **282**: 503-505

Yamashita S, Melmed S (1986) Insulin-like growth factor-I action on rat anterior pituitary cells: suppression of growth hormone secretion and messenger ribonucleic acid levels. *Endocrinology* **118**: 176-182

Yde H (1969) Abnormal growth hormone response to ingestion of glucose in juvenile diabetics. *Acta Med Scand* **186**: 449-504

Advances in Growth Hormone and Growth Factor Research,
edited by E.E. Müller, D. Cocchi and V. Locatelli
Pythagora Press, Roma-Milano and Springer Verlag, Berlin-Heidlberg © 1989

Perspectives on growth hormone and growth factor research- Post-congress reflections

H.G. Friesen

Department of Physiology University of Manitoba, Winnipeg, Canada

The International Congress on Advances in Growth Hormone and Growth Factor Research was held in the venerable Main Building of the State University of Milan on Via Festa del Perdono, September 28-30, 1987. This Congress was the fifth conference on the subject of growth hormone in a series inaugurated in Milan in the fall of 1967. Although I have not attended all the intervening conferences, it was my privilege to attend the first in 1967, as well as the one held 20 years later. In responding to Professor Müller's invitation to write a personal perspective on the 1987 conference, I have chosen to look backwards, to recall what was known about growth hormone in 1967 and to compare and contrast that with the reports on the exciting new observations which were presented at this international forum.

The first Congress was especially memorable because it was my first visit to Europe. The excitement generated by the introduction to another people, a new culture, heightened my interest, and the meeting left an indelible impression.It is worth recalling for the record some of the outstanding endocrinologists who presented their research results in 1967 and to note how much progress truly has been achieved in a relatively short time. This perspective is helpful, because it then becomes clear how rudimentary and incomplete our understanding was 20 years ago of the control of growth hormone secretion and the mechanism of action of growth hormone. Implicit in this comparison with current knowledge must be the conclusion that while recognizing great progress we must also acknowledge how much remains to be clarified. The challenge to discovery remains as strong as ever, although the questions to be addressed and the tools and techniques which will be employed have changed substantially.

In 1967, one individual who played a leading role, both in the organization and in the scientific presentation, was Professor C.H. Li. Although he was a member of the Scientific Organizing Committee in 1987, he was unable to attend the

Congress, and it was with great regret that we learned of his untimely death in December 1987. Professor Li represented the vital link between the earlier «giants» of endocrinology in Berkeley, epitomized by Herbert Evans, and the modern era of endocrinology,where hormones could be defined in precise chemical terms. The Hormone Research Laboratory at the University of California at San Francisco, which Professor Li directed for many years, was recognized worldwide for its seminal discoveries. This included, among other things, the purification and chemical characterization of virtually all pituitary hormones including growth hormone. Along with Raben, Li was the first to purify human growth hormone and to establish its efficacy in the treatment of growth hormone-deficient children.It was this discovery in 1958 that gave impetus to growth hormone research and led to human pituitary collection programs in several countries. Professor Li went on to establish the amino acid sequence of growth hormone and, several years later, the chemical synthesis of human growth hormone, a tour de force at the time, tarnished slightly by the belated recognition that one of the tryptic peptides had been misassigned in the proffered sequence. Nevertheless, his accomplishments and contributions to the field of growth hormone research stand as one of the triumphs of modern endocrinology.

Just a few years prior to 1967, the first specific and sensitive methods for measuring circulating levels of growth hormone had been developed by Berson and Yalow. Prior to that time, any understanding of the physiological and pharmacological factors regulating growth hormone secretion, or for that matter almost all hormones was based on inference. The diagnosis of growth hormone deficiency and hypersecretory states was made largely on clinical grounds. Thus, it might be understood why there was so much excitement, why so many presentations in 1967 featured reports on measuring growth hormone concentrations clinically and experimentally, and why Dr. Berson's views were so widely sought. Now we take for granted the routine measurement in the circulation of virtually any hormone.

The subject of neuroendocrine regulation of growth hormone secretion in 1967 was in its infancy — after all, the essential tool for monitoring growth hormone secretion, the RIA, had just been established. In the intervening years, a remarkable series of discoveries led to the identification of the two major hypothalamic regulators, somatostatin and somatocrinin, or growth hormone – releasing hormone (GHRH), and the contribution each makes to the minute-to-minute regulation of growth hormone secretion. As well, a variety of pharmacological agents which influence growth hormone secretion have been studied to define which neurotransmitters modulate and regulate growth hormone secretion, often by increasing or decreasing GHRH or somatostatin. New analogues of somatostatin have been developed. For the past several years these have been used to inhibit growth hormone secretion as well as to shrink tumors in patients with acromegaly and other disorders. One analogue, Sandostatin, is also being examined in other clinical settings (e.g., diabetes) to see if any therapeutic benefit is evident upon selective inhibition of growth hormone secretion. Similarly, the use of GHRH as a diagnostic tool for distinguishing pituitary and hypothalamic causes of growth hormone deficiency has been explored. Also, initial results on the role and usefulness of this agent as an alternative therapy to growth hormone in the treatment of growth hormone deficiency are being reported.

In the quest to understand the mechanism of action of growth hormone many findings of significance have been reported, two of which are seminal. The primary observation in 1957 by Salmon and Daughaday that growth hormone administration triggers a circulating mediator, sulfation factor or somatomedin, stimulated hundreds of subsequent studies. Independently, the Swiss group led by Froesch and Zapf pursued studies of nonsuppressible insulin-like growth factor I and II. They established that IGF-I was identical with somatomedin-C, that this peptide was growth hormone dependent, and that it caused growth in hypophysectomized rats, confirming an essential element of the somatomedin hypothesis, namely, that the growth-promoting effect of growth hormone was mediated by a circulating peptide, somatomedin. Subsequent studies established that many tissues in addition to the liver produced IGF-I in response to growth hormone administration, and that it is the local tissue production of IGF-I which is of greatest importance in mediating the action of growth hormone. As well, the measurement of IGF-I levels in patients with acromegaly or growth hormone deficiency has been helpful in confirming the diagnosis and assessing the response to therapy.

Although the primary finding that growth hormone stimulated production of sulfation factor was reported in 1957, progress at the time of the 1967 Congress was modest. Today, the structure of IGF is known, the gene has been cloned and expressed, and sensitive methods for measuring IGFs are widely available.

The revolution in biology that has occurred in the 20-year period between the first and the present Congress has also made its impact felt on growth hormone research. Indeed, the results of the application of molecular biology tools to so many areas of growth hormone research were the major and dominant revelation of this Congress.

The genes for human growth hormone and its variants, as well as the placental lactogen (chorionic somatomammotropin), have been isolated and characterized and assigned to a 58-kb region on chromosome 17. Great interest is presently focused, and will be in the future on defining the regulatory regions on the 5' flanking region of human growth hormone which controls the transcription of the GH gene. It is evident from these studies that several tissue-specific factors determine the selective expression of human growth hormone in the somatotropes and that additional regulatory regions mediate the effects of GHRH and other secretagogues. I would anticipate that by the time of the next Congress the "trans" acting factors that act at the regulatory regions of the growth hormone gene either to enhance or silence gene expression will have been isolated and cloned.

The genes for IGF-I and IGF-II have been cloned and characterized. One presumes, as is the case for the growth hormone gene, that there are regulatory regions on the IGF genes which respond to the growth hormone receptor-activation signaling mechanism with enhanced transcription and/or greater IGF-mRNA stability to promote IGF synthesis and secretion, but few details of this process are as yet available. The heterogeneity of the multiple forms and variants of the IGFs and their binding proteins are being clarified.

For growth hormone to initiate an effect it must interact with its receptor. Although it was rumored that the growth hormone receptor had been cloned, no reports on this subject were presented. However, the recent publications outlining the successful cloning and characterization of the receptor have once more transformed a theoretical concept into precise chemical detail (Leung et al. 1987; Wallis 1987). We await with interest the outcome of a host of experiments that have now become feasible.

Science is technology driven, and the conference provided ample evidence for the validity of this dictum. Southern, Northern, and Western blot analysis, *in situ* hybridization and immunohistochemistry, cDNA libraries, expression vectors, etc. – these were everywhere; yet 20 years ago they were not even part of the vocabulary of science.

Similarly, a whole new generation of growth factors has been uncovered by the new biology and linked to another generation of growth factors. I refer to the exciting work on oncogenes. It was a great privilege to hear and perceive the exuberance of the grand lady of science and Nobel laureate, Professor Rita Levi Montalcini, as she shared the results of her recent research on nerve growth factor (NGF). Especially striking was her finding that postnatal growth was profoundly inhibited when antibodies to NGF were administered to fetuses at specific stages of development. What is the mechanism of this effect, and do we have to incorporate fetal NGF into the scheme of factors controlling body growth?

Three other members of the growth factor family were discussed, namely epidermal growth factor (EGF), platelet-derived growth factor (PDGF), and transforming growth factors α and β, the latter two forms having only a common first name but quite distinctive structures and receptors and, one presumes, unique function. The relationship of several of these growth factors and their receptors to oncogenes is now widely recognized, as is the overexpression of a number of oncogenes in a variety of tumors and cancer cell lines. What is less clear and remains a challenge is what role each plays in normal growth and cell function. Because of this deficiency in knowledge, we tend as endocrinologists to revert back into thinking of endocrinology of growth in the simpler, more familiar terms, namely, a repertoire consisting of growth hormone, IGFs, sex steroids, and thyroid hormones. The basis for incorporating the plethora of cellular growth factors and oncogenes acting often in an autocrine and paracrine fashion into a more accurate and comprehensive picture of growth remains an elusive but essential goal for all students of the science of growth.

It was also the new biology which furnished recombinant-DNA-produced human growth hormone, providing for the first time a sufficient amount of human growth hormone for treatment of all growth hormone-deficient children. This new reality has also been responsible for stimulating a large number of studies through-out the world to reexamine and redefine growth hormone-deficiency states as well as to define which children may benefit from growth hormone therapy and what is the optimal route and frequency of administration. Definitive answers to these important questions are still not available and are likely to be vigorously debated for some time to come.

Regrettably, there were no presentations on the use of recombinant growth hormone to stimulate milk production in cows and the mechanism responsible for the increase. The economic impact of this is clearly great, as is evident from the ban imposed by the European Community on the use of growth hormone for this purpose, given the political sensitivity of farm surpluses and subsidies.

Genetic engineering techniques have also made available a supply of IGF-I for experimental studies. Thus, the first clinical studies are being reported on the administration of IGF-I for experimental studies which are beginning to define more precisely the biological and metabolic effects of IGF-I. It is evident from these studies that IGF-I has about 7% the potency of insulin in causing hypogly-cemia. Moreover, it is also clear that while IGF-I can increase linear growth in hypophysectomized animals it is less potent than growth hormone in doing so, and that, in addition, the effects of IGF-I and growth hormone on some target tissues such as muscle differ strikingly.

In my analysis I have not given as much prominence to the neuroregulation of growth hormone secretion and to the clinical studies of growth hormone as they properly received at the Congres. Indeed, for this analysis I commend the reader to Dr. S. Reichlin's masterful overview. My decision was to focus on the molecular biology of growth hormone research as the highlight of the fifth Congress. It would have been imprudent to have attempted to forecast the direction and developments of growth hormone research in the next two decades immediately after the first Congress, and I will not attempt such a rash initiative now. I have every confidence that the next two decades of research on growth and the presentations at the twelfth Congress in the twenty-first century will be every bit as exciting and remarkable as were the discoveries featured at the fifth Congress. Although many of the people and the methodologies on growth factors will be new at that time, when the event is held once more in that charming setting of the State University of Milan, the rich tradition of outstanding science and wonderful hospitality are assured.

REFERENCES

Leung DW, Spencer SA, Cachianes G, Hammonds RG, Collins C, Henzel WJ, Barnard R, Waters MJ, Wood WI (1987) Growth hormone receptor and serum binding protein purification, cloning and expression. *Nature* **330**: 537-543

Wallis M (1987) Growth hormone receptor cloned. *Nature* **330**: 521-522

Author Index

Subject Index